Life Span Nutrition
Conception through Life

Life Span Nutrition
Conception through Life

Sharon Rady Rolfes   Linda Kelly DeBruyne
Eleanor Noss Whitney, editor

West Publishing Company

ST. PAUL   NEW YORK   LOS ANGELES   SAN FRANCISCO

Composition: Carlisle Communications
Copyediting: Mary Berry, Naples Editing Service
Cover Image: Photo by Cliff Feulner/The Image Bank
Interior and Original Cover Design: Roslyn Stendahl, Dapper Design
Illustrations: Ted Bollmann and Rolin Graphics
Index: Jo-Anne Naples, Naples Editing Service

50 W. Kellogg Boulevard
P.O. Box 64526
St. Paul, MN 55164-1003

Printed in the United States of America
97 96 95 94 93 92 91 90 8 7 6 5 4 3 2 1 0
Library of Congress Cataloging-in-Publication Data

Rolfes, Sharon Rady.
Life span nutrition : conception through life / Sharon Rady Rolfes, Linda Kelly DeBruyne; Eleanor Noss Whitney, editor.
p. cm.
ISBN 0-314-66811-X
1. Nutrition. 2. DeBruyne, Linda K. 3. Whitney, Eleanor Noss. I. Title.
[DNLM: 1. Nutrition. QU 145 R747L]
QP141.R64 1990
613.2—dc20
DNLM/DLC
for Library of Congress

89-70730
CIP

PHOTO CREDITS
21 Elizabeth Crews; **102** Courtesy of Dr. James Hanson/ University of Iowa Hospitals and Clinics; **115** (top left, top right, bottom left) Petit Format/Nestle/Photo Researchers, Inc. (bottom right) Gabor Demjen/Stock, Boston; **116** Peter Menzel/Stock, Boston; **133** Hazel Hankin/Stock, Boston; **174** Elizabeth Crews; **176–177** From "The Breast-feeding Guide for Working Mothers" courtesy of Mead Johnson Nutritionals; **184** Elizabeth Crews; **217** From C. Conn, The Specialist in General Practice, 2nd ed. (Philadelphia: Saunders, 1957); **259** Anthony M. Vannelli; **494** Courtesy of U.S. Department of Agriculture

CREDITS FOR OPENING SCULPTURES
Chapter 1 *Family Group* by Henry Moore. 1948–49. Bronze. Photo by Lee Boltin. Reproduced with permission from the Hakone Open-Air Museum, Kanagawa-ken, Japan.
Chapter 2 *The Kiss* by Auguste Rodin. Giraudon/Art Resource, N.Y.
Chapter 3 *Pregnant Woman* by Pablo Picasso. 1950. Bronze. 41¼" by 7⅝" by 6¼". Gift of Mrs. Bertram Smith. Collection, The Museum of Modern Art, New York.
Chapter 4 *First Born* by Hugo Robus. Private collection care of Forum Gallery.
Chapter 5 Detail from *Fountain of the Seasons* by Christian Petersen. 1940. Limestone. Courtesy of Iowa State University.
Chapter 6 *Father and Son* by Paul T. Granlund, sculptor-in-residence at Gustavus Adolphus College in St. Peter, Minnesota. 1968. 30 inches.
Chapter 7 *Children* by Paul T. Granlund. 1963. 22 inches.
Chapter 8 *Girl on a Swing* by Richard Fleischner. 1966–67. Polyester resin, fiberglass, and polypropylene rope. 69½ × 15 × 30¼ inches. Courtesy of the National Trust for Historic Preservation, Nelson A. Rockefeller Collection. Photo by Charles Uht.
Chapter 9 *Singing Man* by Ernst Barlach. Reproduced with permission from the Fine Arts Collection of the Museum Ludwig, Cologne, Germany and Ernst and Hans Barlach GBR, Lizenzverwaltung Ratzeburg. Photo from the Rheinisches Bildarchiv, Cologne.
Chapter 10 *Married Love* by Oscar Nemon. Photo courtesy of J.C. Nichols Company. Sculpture located in the Country Club Plaza, Kansas City, Missouri.

This book focuses on the nourishment of individuals and families, but all of its concerns fall within the larger context of the whole human family—of over five billion individuals and the earth that nourishes us all. We therefore dedicate this book to the Earth herself in the hope that all who read it will remember to take care of her as she cares for us.

Nourishment means more than providing food; it also means giving love. Anyone joining my family at one of our gatherings would see an abundance of both. I dedicate this book to all the members of my family—including those who have departed and those who have yet to arrive.

Sharon

To the memory of my grandmother, Helen Garfunkel, whose commitment to God and family was steadfast, and whose firm but loving ways taught me to respect and appreciate life.

Linda

Contents in Brief

Contents

▶ Preface

This book brings together research from over a thousand scientific articles to contribute to the understanding of how nutrition influences people throughout their lives. In relaying this wealth of knowledge, we have tried to maintain a writing style that is both educational and enjoyable to read. Because our readers will be interested in life span nutrition for both professional and personal reasons, we have conveyed information on two levels. Those who plan to serve others as dietitians, nutritionists, nutrition educators, and caretakers of special age groups should find within these pages the information they need to design programs that will meet their clients' needs. Those who want simply to learn how to provide the best possible nutrition for themselves and for their own families should find that information translated into practical pointers that will assist them in this task.

The chapters introduce and present the stages of the life span chronologically. First, Chapter 1 provides an overview of nutrition principles applicable to all stages. Chapter 2 then begins the life span with a look at the needs of the prospective mother and the factors that surround reproduction and conception. Chapters 3 and 4 describe the crucial time of pregnancy—first, showing the tremendous impact nutrition has at this time for good or ill; then, offering the nutrition information necessary to support a normal pregnancy. Chapter 5 compares and contrasts breastfeeding and formula feeding as means of delivering the infant's single most important food and presents nutrition for the lactating woman as well. Chapter 6 examines the infant's other nutrition needs and the ways they change during the all-important first year of life; Chapter 7 then follows the course of childhood to puberty. Chapter 8 continues the story into adolescence and traces the effects of new activities and influences on teenagers' lives. Chapter 9 examines adulthood from the point of view of health promotion and the prevention of disease, and Chapter 10 describes the nutrition needs of the later years. Each chapter is organized in roughly the same way so that the reader can know what to expect. For every stage of life, the reader can learn the characteristics of normal growth and development, the basics of nutrition assessment, the most common nutritional deficiencies seen, nutrient needs and practical means of delivering nutrition, and nutrition implications of lifestyle factors typical of that stage.

Interspersed with the chapters are Focal Points devoted to special topics relevant to phases of the life span. No Focal Point follows the introductory chapter, so the numbering begins with Focal Point 2, which describes how diet therapy can make a difference, even prior to conception, in the case of inborn errors of metabolism. Focal Point 3 describes the devastating effects that alcohol intake during pregnancy can cause. Focal Point 4 provides research findings on the effects of vegetarianism practiced at vulnerable times of life.

Focal Point 5 examines the details of tooth development and research on the cariogenicity of foods. Focal Point 6 describes the nutrition impacts of common disorders that most children experience at times—infections and fever, diarrhea, and constipation. Focal Point 7 explores obesity—its development, effects, and treatment—with a special focus on children and prevention. Focal Point 8 examines the roles drugs, alcohol, and tobacco play in the lives of teenagers and in relation to nutrition. Finally, Focal Point 9 offers provocative research findings on the effects of nutrition on the brain as it ages.

The book provides additional special features. Definitions of terms appear adjacent to places where they are first used in the text. Practical Points offer applications of nutrition information to people's daily lives. The appendixes provide tables, charts, and forms used in nutrition assessment; the RDA, the U.S. RDA, and the RNI; the Four Food Group Plan and its Canadian and vegetarian versions; the composition of infant formulas; the composition of commonly used vitamin-mineral supplements for various age groups; and nutrition resources.

A few notes on style: we have used the term *parent* to refer to many kinds of caretakers, knowing that in many situations it is not literally a biologic parent who may be responsible for a child's nutrition. We have addressed portions of the text directly to this person, even though not all readers are presently in that role in life. In our view, these elements of style make for pleasanter reading than the generic passive voice, and we hope readers will not take them amiss.

We have learned and benefitted from writing this book, and have found that the information in it serves us well in our lives, both as professionals and as parents and family members. We hope it serves you equally well.

Sharon Rady Rolfes
Linda Kelly DeBruyne
Eleanor Noss Whitney
January 1990

▶ Acknowledgments

We must begin our thank you list with Gary Woodruff for his encouragement to write this book in the first place and Ellie Whitney for her many contributions to its content and style. Our editors, Peter Marshall, Becky Tollerson, and Laura Mezner Nelson deserve a round of applause for the care they gave this book in coordinating its review and production. We are grateful to Linda Patton for her efficient library research work; Ted Bollmann for his skilled artistry; Elisa Malo for her patient and competent word processing; Ledean Joyner for her assistance with a variety of production tasks; Mary Berry for her copyediting talents; Jo-Anne Naples for her careful attention to the index; and Emily Autumn for her assistance in art research.

We give thanks daily for our children, Lyle, Zak, Tyler, and Marni, who have given us many experiences that brought the information in this book to life. We appreciate the love and enthusiastic support of "the Toms", and of our associates Fran and Ellie in helping us to complete this project.

Finally, we thank our reviewers for enhancing the quality and accuracy of this text:

Lindsey Allen
University of Connecticut

Kathryn Anderson
Florida State University

Dee Baxter
Georgia State University

Patsy Brannon
University of Arizona

Sarah Burroughs
California Polytechnic State University-San Luis Obispo

Christine Condit
Appalachian State University

Conrad Demsky
Andrews University

Gail Disney
University of Tennessee

Carolyn Dunn
University of North Carolina-Greensboro

Karen Hauersperger
Presbyterian Hospital, School of Nursing, Charlotte, NC

H. W. Hwang
Louisiana State University

Michael Jenkins
Kent State University

Bob Keith
Auburn University

Hanif Khan
Central Michigan University

Bernice Kopel
Oklahoma State University

Edith Lerner
Case Western Reserve University

Nina Mercer
University of Guelph

Jackie O'Palka
Montana State University

Jean Peters
Oregon State University

Jackie Runyan
Iowa State University

Nancy Sheard
University of Massachusetts-Amherst

Anne Smith
The Ohio State University

Elvita Smith
San Jose State University

Sam Smith
University of New Hampshire

Virginia Utermohlen
Cornell University

Nutrition Overview

1

Family Group by Henry Moore.

Each person enters this world with a unique genetic map that determines the primary ways in which physical and mental characteristics will develop throughout life. A person must accept many of these characteristics without option for change, but can change others within genetically defined limits. One of several ways a person can ensure the optimal growth, maintenance, and general health of the body is through proper nutrition. Ideally, nutrients supplied by the diet will more than adequately cover losses and the requirements incurred by the physiological demands of growth, reproduction, lactation, disease, and aging.

All people—pregnant and lactating women, infants, children, adolescents, and adults—need the same nutrients. The same general principles govern the selection and delivery of those nutrients for all people, yet at the same time, nutrient needs differ in some particular ways for each time of life. This chapter both reviews the ways in which nutrition is the same for everyone and introduces the factors that make nutrition different in each phase of life. It also provides the basics of assessment of nutrition status, in preparation for the later sections on assessment that give special consideration to different times of life.

Nutrition Basics

No matter what stage of the life span, the nutrients people need are, in general, the same. Standards for nutrient intakes are set in the same ways, and diets are planned according to the same general principles. This section provides a brief review of these subjects.

The Nutrients

Both foods and the human body are composed of nutrients and other materials. The six classes of nutrients are water, carbohydrate, fat, protein, vitamins, and minerals. Four of the six classes of nutrients (carbohydrate, fat, protein, and vitamins) are organic, while the other two (minerals and water) are inorganic. During metabolism, three of the four organic nutrients (carbohydrate, fat, and protein) provide energy the body can use. In contrast, vitamins, minerals, and water do not yield energy in the human body.

energy-yielding nutrients: the nutrients that break down to yield energy the body can use: carbohydrate, fat, and protein.

The energy-yielding nutrients, carbohydrate, fat, and protein, are vital, for life demands continual replenishment of the energy spent daily. When metabolized in the body, the energy-yielding nutrients break down; that is, their carbon and hydrogen atoms (and others) are split apart and combined with oxygen, yielding carbon dioxide and water, materials that must be excreted. As the nutrients break down, they release energy. Some of this energy is released as heat, some is transferred into other compounds (including fat) that compose the structures of the body cells, and some is used as fuel for activities. The amount of energy the energy-yielding nutrients release is measured in kcalories.

kcalorie: a unit by which energy is measured; the amount of heat necessary to raise the temperature of 1 kg of water 1° C.

1 g carbohydrate = 4 kcal.
1 g protein = 4 kcal.
1 g fat = 9 kcal.

A gram of carbohydrate provides 4 kcalories; a gram of protein also provides 4; and a gram of fat provides 9 kcalories.* The energy content of a

*In light of current research, the 9 kcal/g for fat may be revised upwards; K. Donato and D. M. Hegsted, Efficiency of utilization of various sources of energy for growth, *Proceedings of the National Academy of Sciences* 82 (1985): 4866–4870.

food thus depends on how much carbohydrate, fat, and protein it contains. If the body does not use these nutrients to fuel metabolic and physical activities, it rearranges them (and the energy they contain) into storage compounds such as glycogen and fat, and stores them for use between meals. If more energy is taken in than is expended—no matter from which of the three energy-yielding nutrients—the result is weight gain. Figure 1–1 shows the ways the body uses carbohydrate, fat, and protein when they are supplied in excess of energy needs, as well as when an energy deficit is occurring.

Alcohol also is metabolized in the body to yield energy (7 kcalories per gram). When taken in excess of energy need, alcohol, too, is converted to body fat and stored, but when alcohol contributes a substantial portion of the energy in a person's diet, it is harmful. (Focal Point 8 is devoted partly to the effects of alcohol on nutrition.)

1 g alcohol = 7 kcal.

The vitamins are also organic, and the body may break them down, but it does not extract usable energy from them. Their role is rather to help the body use the energy-yielding nutrients and conduct its metabolic activities. The vitamins are divided into two classes: some are soluble in water (the B vitamins and vitamin C) and others, in fat (vitamins A, D, E, and K). Each of the 13 different vitamins plays its own special roles in metabolism.

vitamin: an essential organic nutrient required in small amounts.

Minerals are inorganic elements, and some are essential in human nutrition. Some minerals are still being studied to determine whether they are essential. Still other minerals are known not to be nutrients, but are important to nutrition because they displace the mineral nutrients in the body, causing deranged body functions—a problem exemplified by the discussion of lead toxicity in Chapter 7.

mineral: a small, inorganic atom or molecule; some minerals are essential nutrients required in small amounts.

The roles of the vitamins and minerals are diverse and numerous enough to fill whole books on nutrition. Table 1–1 summarizes some basic information about them.

Water constitutes about 60 percent of an adult's body weight and a higher percentage of a child's. In addition to the obvious dietary source, water itself, nearly all foods contain water. Water is also generated from the energy-yielding nutrients in foods during metabolism (recall that the carbon and hydrogen in these nutrients combine with oxygen to yield carbon dioxide and water). Water from these three sources normally balances perfectly with daily water losses.

Recommended Nutrient and Energy Intakes

The set of standards used for nutrient and energy intakes in the United States is the Recommended Dietary Allowances (RDA); those selected for use on food labels are the U.S. RDA. The sections that follow cover each of these in a general way, and the chapters to come discuss specific nutrient recommendations for each age group. The Canadian standards are the Recommended Nutrient Intakes (RNI), presented in Appendix B.

The RDA The RDA are standards for nutrient intakes created by a committee of scientists appointed to the task by the National Academy of Sciences.[1] The RDA are revised periodically, and the latest revision is presented on the inside front cover of this book and in Appendix B. They are the levels of intakes of essential nutrients that are judged to be adequate to meet the known nutrient needs of practically all healthy people in the United States under usual

RDA: Recommended Dietary Allowances, set by the Food and Nutrition Board of the National Academy of Sciences as standards for nutrient intake. They cover the needs of most healthy individuals under usual environmental conditions.

Figure 1–1 Feasting and Fasting

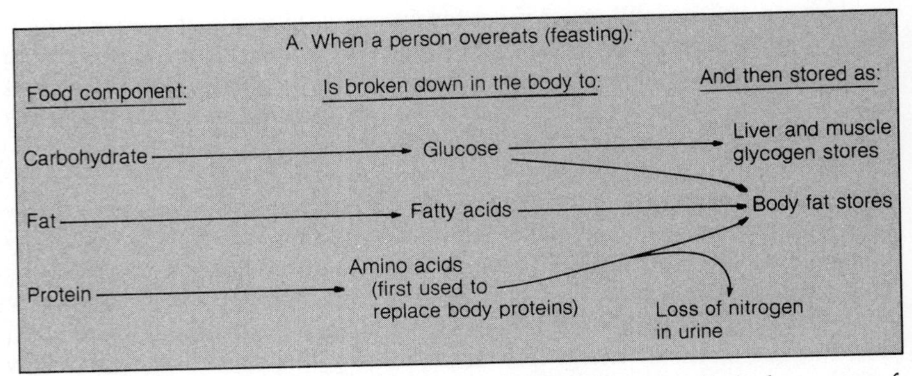

A. When a person eats in excess of energy needs, the body stores limited amounts of glycogen and larger quantities of fat.

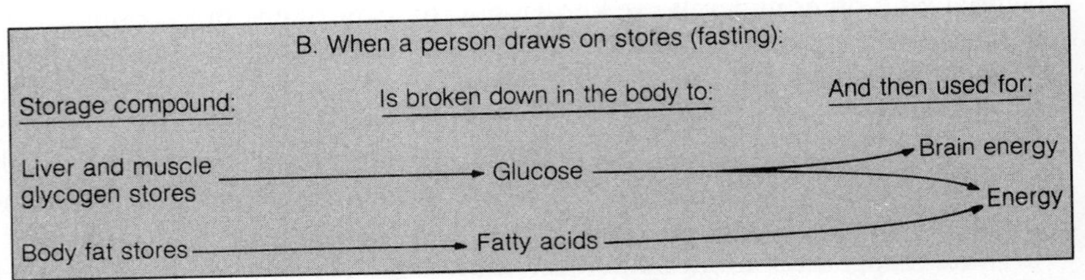

B. When food is unavailable to provide energy, the body draws on its glycogen and fat stores for energy.

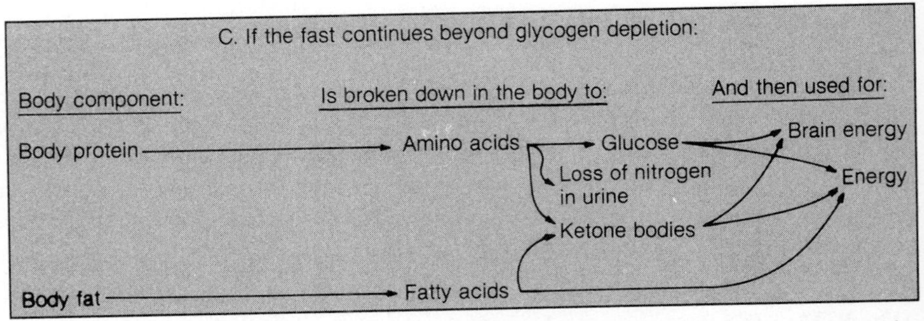

C. When glycogen stores are depleted, the body begins to break down its protein (muscle and lean tissue) to amino acids for glucose synthesis needed for brain and nervous system energy. In addition, the liver converts fats to ketone bodies, which serve as an alternative energy source for the brain, thus slowing the breakdown of body protein.

Table 1–1 The Vitamins and Minerals

Vitamins	Chief Functions in the Body	Significant Sources
Vitamin A (retinol, retinal, retinoic acid); precursor is provitamin A carotenoids such as beta-carotene	Vision; maintenance of cornea, epithelial cells, mucous membranes, skin; bone and tooth growth; reproduction; hormone synthesis and regulation; immunity; cancer protection	Retinol: fortified milk, cheese, cream, butter, fortified margarine, eggs, liver Beta-carotene: spinach and other dark leafy greens, broccoli, deep orange fruits (apricots, cantaloupe) and vegetables (squash, carrots, sweet potatoes, pumpkin)
Vitamin D (calciferol, cholecalciferol, 1,25-dihydroxy vitamin D); precursor is the body's own cholesterol	Mineralization of bones (raises calcium and phosphorus blood levels by increasing absorption from digestive tract, withdrawing calcium from bones, stimulating retention by kidneys)	Self-synthesis with sunlight; fortified milk, fortified margarine, eggs, liver, fish
Vitamin E (alpha-tocopherol, tocopherol, tocotrienol)	Antioxidant (detoxification of strong oxidants), stabilization of cell membranes, regulation of oxidation reactions, protection of polyunsaturated fatty acids (PUFA) and vitamin A	Plant oils (margarine, salad dressings, shortenings), green and leafy vegetables, wheat germ, whole-grain products, liver, egg yolk, nuts, seeds
Vitamin K (phylloquinone, naphthoquinone)	Synthesis of blood-clotting proteins and a blood protein that regulates blood calcium	Bacterial synthesis[a] in the digestive tract; liver, green leafy vegetables, cabbage-type vegetables, milk
Thiamin (vitamin B_1)	Part of TPP (thiamin pyrophosphate), a coenzyme used in energy metabolism, supports normal appetite and nervous system function	Occurs in all nutritious foods in moderate amounts; pork, ham bacon, liver, whole grains, legumes, nuts
Riboflavin (vitamin B_2)	Part of FMN (flavin mononucleotide) and FAD (flavin adenine dinucleotide), coenzymes used in energy metabolism, supports normal vision and skin health	Milk, yogurt, cottage cheese, meat, leafy green vegetables, whole-grain or enriched breads and cereals
Niacin (nicotinic acid, nicotinamide, niacinamide, vitamin B_3); precursor is dietary tryptophan	Part of NAD (nicotinamide adenine dinucleotide) and NADP (its phosphate form), coenzymes used in energy metabolism; supports health of skin, nervous system, and digestive system	Milk, eggs, meat, poultry, fish, whole-grain and enriched breads and cereals, nuts, and all protein-containing foods
Vitamin B_6 (pyridoxine, pyridoxal, pyridoxamine)	Part of PLP (pyridoxal phosphate) and PMP (pyridoxamine phosphate), coenzymes used in amino acid and fatty acid metabolism, helps to convert tryptophan to niacin, helps to make red blood cells	Green and leafy vegetables, meats, fish, poultry, shellfish, legumes, fruits, whole grains
Folate (folacin, folic acid, pteroylglutamic acid)	Part of THF (tetrahydrofolate) and DHF (dihydrofolate), coenzymes used in new cell synthesis	Leafy green vegetables, legumes, seeds, liver
Vitamin B_{12} (cobalamin and related forms)	Part of methylcobalamin and deoxyadenocobalamin, coenzymes used in new cell synthesis, helps to maintain nerve cells	Animal products (meat, fish, poultry, shellfish, milk, cheese, eggs)

[a]Vitamin K needs cannot be met from bacterial synthesis alone, however it is a potentially important source in the jejunum and ileum where absorption efficiency ranges from 40% to 70%.

Continued

Table 1–1 *Continued*

Vitamins	Chief Functions in the Body	Significant Sources
Pantothenic acid	Part of Coenzyme A, which is used in energy metabolism	Widespread in foods
Biotin	Part of a coenyzme used in energy metabolism, fat synthesis, amino acid metabolism, and glycogen synthesis	Widepread in foods
Vitamin C (ascorbic acid)	Collagen synthesis (strengthens blood vessel walls, forms scar tissue, matrix for bone growth), antioxidant, thyroxin synthesis, amino acid metabolism, strengthens resistance to infection, helps in absorption of iron	Citrus fruits, cabbage-type vegetables, dark green vegetables, cantaloupe, strawberries, peppers, lettuce, tomatoes, potatoes, papayas, mangos
Major Minerals		
Calcium	The principal mineral of bones and teeth; also involved in normal muscle (including heart muscle) contraction and relaxation, proper nerve functioning, blood clotting, blood pressure, and immune defenses	Milk and milk products, small fish (with bones), tofu (bean curd), greens (broccoli, chard), legumes
Phosphorus	A principal mineral of bones and teeth; part of every cell, important in the genetic material, as part of phospholipids, in energy transfer, and as buffering systems that maintain acid-base balance	All animal tissues
Magnesium	Involved in bone mineralization, the building of protein, enzyme action, normal muscular contraction, transmission of nerve impulses, and maintenance of teeth	Nuts, legumes, whole grains, dark green vegetables, seafood, chocolate, cocoa
Sodium	An electrolyte that maintains normal fluid balance and acid-base balance; assists in nerve impulse transmission	Salt, soy sauce; moderate quantities in whole, unprocessed foods; large amounts in processed foods
Chloride	An electrolyte that maintains normal fluid balance and acid-base balance; part of the hydrochloric acid found in the stomach, necessary for proper digestion	Salt, soy sauce; moderate quantities in whole, unprocessed foods; large amounts in processed foods
Potassium	An electrolyte that maintains normal fluid balance and acid-base balance; facilitates many reactions, including the making of protein; supports cell integrity; assists in the transmission of nerve impulses and the contraction of muscles, including the heart	All whole foods: meats, milk, fruits, vegetables, grains, legumes
Sulfur	As part of proteins, stabilizes their shape by forming sulfur-sulfur bridges; part of the vitamins biotin and thiamin and the hormone insulin; involved in the body's detoxification processes (combines with toxic substances to form harmless compounds)	All protein-containing foods

Continued

Table 1–1 *Continued*

Trace Minerals	Chief Functions in the Body	Significant Sources
Iodine	A component of the thyroid hormone thyroxin, which helps to regulate growth, development, and metabolic rate	Iodized salt, seafood, plants grown in most parts of the country and animals fed those plantsTrace Minerals
Iron	Part of the protein hemoglobin, which carries oxygen from place to place in the body; part of the protein myoglobin in muscles, which makes oxygen available for muscle contraction; necessary for the utilization of energy as part of the cells' metabolic machinery	Red meats, fish, poultry, shellfish, eggs, legumes, dried fruits
Zinc	Part of many enzymes; associated with the hormone insulin; involved in making genetic material and proteins, immune reactions, transport of vitamin A, taste perception, wound healing, the making of sperm, and the normal development of the fetus	Protein-containing foods: meats, fish, poultry, grains, vegetables
Copper	Necessary for the absorption and use of iron in the formation of hemoglobin, part of several enzymes, a factor that helps to form the protective covering of nerves	Meats, drinking water
Fluoride	An element involved in the formation of bones and teeth, helps to make teeth resistant to decay	Drinking water (if naturally fluoride containing or fluoridated), tea, seafood
Selenium	Part of an enzyme that works with vitamin E to protect body compounds from oxidation	Seafood, meat, grains
Chromium	Associated with insulin and required for the release of energy from glucose	Meats, unrefined foods, fats, vegetable oils
Cobalt	Part of vitamin B_{12} and therefore involved in nerve cell function and in the process of blood formation	Vitamin B_{12}–containing foods (meats, milk and milk products)
Molybdenum	Facilitator, with enzymes, of many cell processes	Legumes, cereals, organ meats
Manganese	Facilitator, with enzymes, of many cell processes	Widespread in foods

environmental stresses.[2] Diet planners use them primarily as guidelines to aid in the evaluation and planning of diets for groups of people, although they can also use the RDA to estimate the probable risk of deficiency for an individual when intakes are averaged over a sufficient length of time. Different RDA apply to different age groups, and from age 11 onwards, the RDA differ for the two sexes as well. They are based on available scientific evidence to the greatest extent possible, and they are recommendations, not requirements. With the exception of the energy allowances, the RDA include a substantial margin of safety. Healthy individuals whose needs are higher than the average are covered by the RDA, but people with medical problems may have nutrient needs that exceed the RDA and require individual evaluation.

With careful planning, the RDA for every nutrient can be met from a variety of available foods. This is a difficult goal to achieve on a daily basis, and even though the RDA are expressed in terms of *per day*, they should be interpreted as *average* intakes over time. The length of time varies for each nutrient, depending on the body's use and storage of the nutrient. For most nutrients (such as thiamin), the RDA covers average intakes over at least three days; for others (such as vitamin A and vitamin B_{12}), the average might be several months. The RDA are generous, and although they do not necessarily cover every individual for every nutrient, they should not be exceeded by much. Some nutrients can be toxic at intakes only slightly above the RDA, and people's tolerances for high doses of nutrients vary.

The RDA for energy, in contrast to those for nutrients, are not generous, but are set at the mean of food energy needs of the populations for which they are intended. They cover the energy expended at rest, in light to moderate physical activity, and as a result of thermogenesis. The latest version of the RDA energy table provides the mean for each age-sex group, based on median heights and weights.

Protein is the only energy-yielding nutrient for which an RDA has been established. The authorities setting the RDA assume that a person will receive a certain minimum amount of energy from the protein that is required, and then will consume sufficient carbohydrate and fat to meet the energy RDA.

Applications of the RDA are expanding. Their widespread use requires an awareness of their appropriate applications and limitations. These three comments seem most worthy of mention:

- The RDA are intended to be met through diets composed of a variety of foods. In this way, all other nutrients for which RDA have not been established will also be covered.
- The RDA are not minimal requirements, nor are they optimal levels of intake. They are safe and adequate recommendations that include a generous margin of safety.
- The RDA are most appropriately used for populations, but can be used to estimate probable risk of deficiency for individuals if comparison is made over a sufficient length of time.[3]

U.S. RDA: United States Recommended Daily Allowances. The U.S. RDA are used on labels, and in most instances, they are the highest RDA for an age/sex group for each nutrient.

The U.S. RDA To avoid confusion, a person using the RDA must understand how they differ from the U.S. RDA. The U.S. RDA are a set of figures chosen by the Food and Drug Administration (the FDA) to use as standards on food labels, so that nutrient contents of foods can be expressed on labels as percentages of those standards. The intent was to assist consumers in evaluating the nutrient contents of foods for themselves and at the same time to spare them the burden of learning the different units in which nutrient amounts are expressed. Thus all nutrient amounts in a food, whether originally measured in micrograms, milligrams, grams, or RE, can be expressed as "percent of U.S. RDA."

Four sets of U.S. RDA were developed for different groups of people—infants, children, adults, and pregnant and lactating women. The most commonly used of these is the U.S. RDA for adults. The one for infants is used for formulas. Supplements designed for children and for pregnant and lactating women may refer to the U.S. RDA for these groups on their labels. The U.S. RDA for adults are shown in Table 1–2; the complete U.S. RDA are in Appendix B.

Diet-Planning Guides

Knowing the individual nutrients and their recommended intakes enables planners to juggle available foods to create diets that supply all the needed nutrients in the appropriate amounts for good health. Planners think in terms of adequacy (obtaining all nutrients in sufficient quantities), balance (ensuring that all food types are represented in reasonable proportions), variety (seeking nutrients from different food sources to obtain all possible trace elements and

Table 1–2 The U.S. RDA for Adults

Nutrient	U.S. RDA
Nutrients That *Must* Appear on the Label[a]	
Protein (g), PER ≥ casein[b]	45
Protein (g), PER < casein	65
Vitamin A (RE)	1,000
Vitamin C (ascorbic acid) (mg)	60
Thiamin (vitamin B_1) (mg)	1.5
Riboflavin (vitamin B_2) (mg)	1.7
Niacin (mg)	20
Calcium (g)	1
Iron (mg)	18
Nutrients That *May* Appear on the Label	
Vitamin D (IU)	400
Vitamin E (IU)	30
Vitamin B_6 (mg)	2
Folate (folic acid, folacin) (mg)	0.4
Vitamin B_{12} (μg)	6
Phosphorus (g)	1
Iodine (μg)	150
Magnesium (mg)	400
Zinc (mg)	15
Copper (mg)	2
Biotin (mg)	0.3
Pantothenic acid (mg)	10

[a]Whenever nutrition labeling is required.
[b]PER is an index of protein quality.

Source: Adapted from *Nutrition Labels and U.S. RDA,* (an FDA Consumer Memo), DHHS (FDA) publication no. 81-2146 (Washington, D.C.: Government Printing Office, 1981).

nutrient density: a characteristic of a food. A nutrient-dense food provides a high quantity (relative to need) of one or (preferably) several essential nutrients, with a small quantity (relative to need) of kcalories.

intakes to appropriate amounts). The single principle that helps most to accomplish all of these objectives is to select a variety of foods of high nutrient density.

Several tools are useful in diet planning. Most commonly used are food group plans, which specify a certain number of portions of foods from each of several groups each day; and exchange patterns, which identify the foods in each group and provide standards for portion sizes. The food group plan in widest use, the Four Food Group Plan, is presented in Appendix C together with its vegetarian and Canadian versions; special food group plans for special age groups are presented in tables in the chapters to come. The most commonly used exchange system is that designed by the American Dietetic Association and the American Diabetes Association, which make it available as a booklet.* All of the food portions on each of its lists provide approximately the same number of kcalories and the same amounts of energy-yielding nutrients (protein, fat, and carbohydrate).

Dietary Goals and Guidelines

A nutritious diet fulfills two criteria. It provides enough of the essentials, and it avoids excess. The planner evaluating a diet thus has two questions to ask. One is: How does the diet compare with the RDA (or the RNI), which makes specific recommendations for food energy, protein, vitamins, and minerals? The second question is: How well does the diet control intakes of total energy, fat, cholesterol, salt, sugar, and alcohol? The RDA do not address this second set of concerns; those concerns are, instead, the subject of a variety of other recommendations collectively called dietary guidelines.

Several sets of dietary guidelines have originated from the awareness that overnutrition contributes to the illnesses many people suffer from today: heart disease, cancer, diabetes, liver disease, and others. Among the sets of recommendations published in the United States have been the *Dietary Goals for the United States* (1977); the *Dietary Guidelines for Americans* (1980); *The Surgeon General's Report on Nutrition and Health* (1988); and the National Research Council (NRC) report, *Diet and Health: Implications for Reducing Chronic Disease Risk* (1989). Table 1–3 presents a summary of the recommendations published by the U.S. government since 1977. These sets of guidelines differ somewhat from one another, but reflect more agreement than disagreement. All of them emphasize prevention of overnutrition and disease.

The two newest sets of recommendations include points of key importance to the life span, the subject of this book. For instance, as Table 1–3 shows, the Surgeon General's report recommends consumption of iron-rich foods––advice that is especially significant for women and children, whose iron intakes often fall short of recommendations. The NRC report includes a recommendation to maintain adequate calcium intake, asserting that adequate intakes of this nutrient may help minimize adult bone loss, especially in women. The NRC report also recommends maintaining an optimal intake of fluoride, particularly

*The booklet *Exchange Lists for Meal Planning* can be obtained by sending $1.25, plus $3.00 shipping and handling fee, to the American Dietetic Association, P.O. Box 97215, Chicago, IL 60678–7215.

Table 1–3 Dietary Recommendations for the General Public

Year	Agency[b]	Publication	Recommendations[a]							
			Variety	*Maintain Ideal Body Weight*	*Include Starch and Fiber*	*Limit Sugar*[c]	*Limit Fat*[c]	*Limit Cholesterol*	*Limit Salt*	*Limit Alcohol*
1977	U.S. Senate	*Dietary Goals for the U.S.*		+	+	+	+	+	+	
1979	DHEW	*Healthy People: The Surgeon General's Report on Health Promotion and Disease Prevention*	+	+	+	+	+	+	+	+
1979	DHEW/NCI	*Statement on Diet, Nutrition, and Cancer—Prudent Interim Principles*	+	+	+		+			+
1980	USDA/DHHS	*Dietary Guidelines for Americans*	+	+	+	+	+	+	+	+
1980	DHHS	*National 1990 Nutrition Objectives*	+	+	+	+	+	+	+	+
1984	DHHS/NHLBI	*Recommendations for Control of High Blood Pressure*		+			+		+	+
1985	USDA/DHHS	*Dietary Guidelines for Amercians, 2nd edition*	+	+	+	+	+	+	+	+
1986	DHHS/NCI	*Cancer Control Nutrition Objectives for the Nation: 1985–2000*		+	+		+			+
1987	DHHS/NHLBI	*National Cholesterol Education Program Guidelines*	+	+	+		+	+		+
1988	DHHS/NCI	*Dietary Guidelines for Cancer Prevention*	+	+	+		+		+	+
1988	DHEW	*The Surgeon General's Report on Nutrition and Health*	+	+	+	+	+	+	+	+
1989	NRC	*Diet and Health: Implications for Reducing Chronic Disease Risk*	+	+	+	+	+	+	+	+

[a]Other recommendations include the following: increase consumption of foods containing vitamins and minerals (DHHS/NCI, 1986); increase physical activity (USDA/DHHS, 1980, 1985; DHHS, 1980); reduce intake of salt-cured or smoked foods (DHHS/NCI, 1988); use appropriate sources of fluoride, limit consumption and frequency of use of foods high in sugar, increase consumption of foods high in calcium, and consume foods that are good sources of iron (DHEW, 1988); maintain adequate calcium intake, avoid supplements in excess of RDA, and maintain optimal fluoride intake (NRC, 1989).

[b]U.S. Senate = U.S. Senate Select Committee on Nutrition and Human Needs; DHEW = Department of Health, Education, and Welfare; NCI = National Cancer Institute; USDA = United States Department of Agriculture; DHHS = Department of Health and Human Services; NHLBI = National Heart, Lung, and Blood Institute; NRC = National Research Council.

[c]Recommendations prior to 1977 were for *inclusion* in the daily diet, as opposed to subsequent recommendations to *limit* intake.

Source: Adapted from *The Surgeon General's Report on Nutrition and Health: Summary and Recommendations*, DHHS (PHS) publication no. 88-50211 (Washington, D.C.: Government Printing Office 1988), Appendix C.

during the growing years, underscoring the importance of this nutrient in the development of caries-resistant teeth in children. Details of the NRC report are presented in Table 1–4.

Nutrition scientists are of divergent opinions about providing guidelines such as these to advise the public on diet. Some feel justified in offering concrete advice; others feel that advice can reasonably be given not to the public at large but only to individuals addressed singly. The two points of view might be characterized as the preventive approach and the medical approach. The first of these positions emphasizes everyone's health, and if it errs, it does so on the side of urging more people to make more changes than may be necessary. The object is to help prevent anyone's developing disease from any preventable diet-related cause. Proponents believe that since it cannot be predicted just which individuals may be susceptible to dietary factors related to diseases, the same advice should be offered to all. The second position represents the more conservative, traditional approach. It emphasizes the cure of already-existing disease rather than the prevention of potential disease. The guidelines the government has offered to the public represent the preventive

Table 1–4 Dietary Recommendations of the NRC Report

- Reduce total *fat* intake to 30% or less of kcalories. Reduce saturated fatty acid intake to less than 10% of kcalories, and the intake of cholesterol to less than 300 mg daily.[a]
- Increase intake of starches and other *complex carbohydrates*.[b]
- Maintain *protein* intake at moderate levels.[c]
- Balance food intake and physical activity to maintain appropriate *body weight*.
- For those who drink *alcoholic beverages*, limit consumption to the equivalent of less than 1 oz of pure alcohol in a single day.[d] Pregnant women should avoid alcoholic beverages.
- Limit total daily intake of *salt* (sodium chloride) to 6 g or less.[e]
- Maintain adequate *calcium* intake.
- Avoid taking dietary *supplements* in excess of the RDA in any one day.
- Maintain an optimal intake of *fluoride*, particularly during the years of primary and secondary tooth formation and growth.

[a]The intake of fat and cholesterol can be reduced by substituting fish, poultry without skin, lean meats, and low-fat or nonfat dairy products for fatty meats and whole-milk dairy products; by choosing more vegetables, fruits, cereals, and legumes; and by limiting oils, fats, egg yolks, and fried and other fatty foods.
[b]Every day eat five or more servings of a combination of vegetables and fruits, especially green and yellow vegetables and citrus fruits, and six or more daily servings of a combination of breads, cereals, and legumes.
[c]Meet at least the RDA for protein; do not exceed twice the RDA.
[d]The committee does not recommend alcohol consumption. One ounce of pure alcohol is the equivalent of two cans of beer, two small glasses of wine, or two average cocktails.
[e]Limit the use of salt in cooking, and avoid adding it to food at the table. Salty, highly processed salty, salt-preserved, and salt-pickled foods should be consumed sparingly.

Source: Adapted from the National Academy of Sciences report, *Diet and Health: Implications for Reducing Chronic Disease Risk*, which was produced by the Committee on Diet and Health of the Food and Nutrition Board of the National Research Council and partially reprinted verbatim in *Nutrition Reviews* 47 (1989): 142–149.

approach, and if the guidelines err, it is in advising more people than necessary to follow their diet advice. Much evidence supports their recommendations as healthy, though. While some people may maintain good health without following such advice, most people would be well served to heed it.

Special Concerns for Specific Stages of Life

As mentioned previously, people need the same nutrients throughout their lives, but the nutrient amounts and the ways in which they demand special attention vary. Personal factors affecting nutrient needs include age, sex, genetic history, and lifestyle habits such as exercise.

This text focuses on the nutrient needs of people in each stage of life—prior to conception, pregnany, lactation, infancy, childhood, adolescence, and adulthood—and on the special nutrition attention each stage requires. A brief look at some nutrition concerns of specific times of life will set the stage for the chapters to come.

Take body weight, for example. It is important in every stage, but for different reasons at different times. Women who enter pregnancy at the appropriate weight and who gain at a certain rate have the best chance of bearing full-term, healthy infants of normal weight—infants whose health and survival outlooks are optimal. An infant's weight at birth is a predictor of the infant's future health status. A child's weight is of interest as a measure of growth, and underweight is therefore a matter of concern. So is overweight: childhood obesity urgently needs preventive attention if the adult years are to find this health hazard manageable. Weight in adults ties significantly to many different disease risks.

Take iron status for another example of the different emphases in different stages of life. During the growing years, iron is critical for the building of new blood cells. During pregnancy, a woman has to consume enough iron to provide fetal stores to last for a half year after birth. After that half year, added dietary iron is crucial for the continued maintenance of iron status; anemia is common in these years. At menstruation, iron needs increase substantially for girls, and at menopause, they are cut again.

Calcium needs also receive focus at every stage and are important for different reasons from one time to the next. Throughout early life, calcium contributes to the growth of bone. Later, in adulthood, adequate calcium intakes may help defend against hypertension. In later life, continued calcium ingestion may help to defend against the health impact of inevitable adult bone loss.

Just as nutrient needs are important for different reasons at different times of life, so too is the avoidance of excesses. For example, fat and cholesterol intakes are important at all times, but in different ways. For adults, limiting fat and cholesterol intakes is important to reduce the risk of cardiovascular disease, especially for men. However, it may be important for infants and children not to have too little fat in their diets. Normal growth and development may be impaired when dietary restrictions of fat and cholesterol are imposed on children.[4] It also may be important not to have too much; Chapter 7 discusses the controversy over children's fat intakes. The chapters to come deal with the key nutrients and with the dietary recommendations that apply to each stage of life.

Nutrition Assessment Basics

To give all the details of nutrition assessment procedures would entail writing another textbook. However, any student of nutrition should know the basics of a proper nutrition assessment procedure, for two reasons.

First, anyone seeking to deliver competent health care must be able to evaluate nutrition status. Health care facilities must make nutrition assessment a routine part of the initial workup on every client—infants, children, pregnant women, and all others—to ensure their normal growth and healthy development. Medical care must also employ nutrition assessment techniques, or refer all clients to specialists skilled in them, so that nutrition handicaps will not hinder responses to medical treatment and recovery from illnesses.

Second, anyone seeking to obtain competent health care needs to know what to expect in the way of nutrition care. Nutrition is such a popular subject that many unqualified people claim it as their province. Fraudulent practices in nutrition are even more abundant than they have been in the past, and they have always been rampant. Thousands of people with fake nutrition degrees, more than in any other field, are offering useless advice.[5] The knowledgeable consumer needs to know enough about what to expect in a nutrition assessment so as to be able to avoid incompetent care.

The same basic nutrition assessment principles apply to all stages of the life span, but the details of emphasis differ from one stage to the next. The basics are presented here; the particulars applicable to different stages of life are in the chapters to come. To avoid clutter, many of the tables and charts routinely used have been collected into Appendix A at the end of this book; the letter *A* in references to "Table A–1," "Figure A–1," "Form A–1," etc., should alert the reader to turn to that appendix, if desired.

Nutrition assessment evaluates many factors that influence or reflect nutrition status, using many sources. The assessor, usually a registered dietitian (R.D.) or a physician trained in clinical nutrition, uses:

- ▸ Historical data.
- ▸ Anthropometric measurements.
- ▸ Physical examinations.
- ▸ Biochemical analyses.

Each of these methods involves collecting data in a variety of ways, and the assessor interprets each finding in relation to the others in order to create a total picture.

The accurate gathering of this information and its careful interpretation are the basis for a meaningful evaluation. Gathering information is a time-consuming process, though, and time is often a rare commodity in the health care setting. A strategic compromise is to screen clients by collecting preliminary data. Data such as height, weight, and hematocrit are easy to obtain and can alert health care workers to potential problems. Nutrition screening identifies clients who will require additional, detailed nutrition assessment (see Figure 1–2).

nutrition screening: the use of preliminary nutrition assessment techniques to identify people who are malnourished or who are at risk for malnutrition.

Figure 1–2 Nutrition Screening
A preliminary evaluation screens for possible nutrition disorders. If results of initial tests are abnormal, follow-up tests are conducted to further evaluate the disorder. If results of these tests indicate no nutrition disorder, a reevaluation is conducted at a later time. If results indicate abnormalities, steps are taken to correct them.

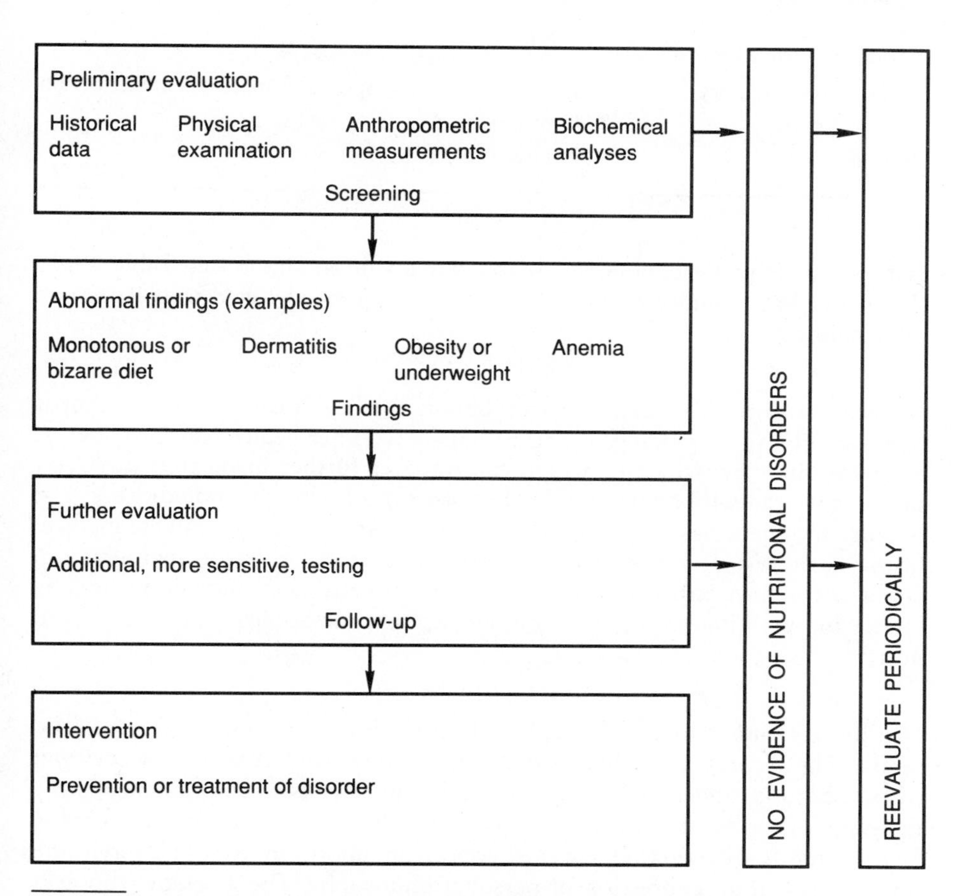

Source: Adapted from S. J. Fomon and coauthors, *Nutritional Disorders of Children: Prevention, Screening, and Followup,* DHEW (HSA) publication no. 76-5612 (Washington, D.C.: Government Printing Office, 1976), inside front cover.

Historical Data

Clues about present nutrition status become evident with a review of a person's historical data (see Table 1–5). Even when data are subjective, they reveal important facts about a person. A thorough history provides a sense of the whole person. An adept history taker uses the interview not only to gather facts, but also to establish rapport, exploring several aspects of a person's history: medical history, diet, socioeconomic circumstances, and drug use. A complete history can identify many different kinds of risk

Table 1–5 Nutrition Assessment: Historical Data Routinely Used

History	Identifies
Medical history	Medical conditions that affect nutrition status
Diet history 24-hour recall Food diary Food frequency record	Nutrient intake excesses or deficiencies
Socioeconomic history	Personal, financial, and environmental influences on food intake
Drug history	Medications that affect nutrition status

factors that may alert the assessor to poor nutrition status (see Table 1–6). (Form A-1 in Appendix A shows the kinds of questions asked to collect such information.)

Medical history The assessor can obtain medical histories from records completed by the attending physician, nurse, or other health care provider. A personal interview with the person can uncover further medical information that might otherwise be overlooked because no one else has thought to ask or because the client has failed to report it earlier. An accurate, complete medical history can reveal conditions that place a client at risk for malnutrition. Diseases can have either immediate or long-term effects on nutrition status by interfering with ingestion, digestion, absorption, metabolism, or excretion of nutrients.

Diet history and analysis A diet history provides a record of a person's food intake. The accurate recording of such data requires skill. A trained interviewer often uses food models or photos and measuring devices to help clients identify the types of foods and quantities consumed. The assessor also needs to know how the foods are prepared. Food choices are an important part of lifestyle and often represent an expression of personal philosophy. The assessor who asks nonjudgmental questions about food intake encourages trust and increases the likelihood of obtaining accurate information.

Methods of obtaining food intake data include the 24-hour recall, the food frequency checklist, and the food diary. (Typical forms used for these three types of diet histories are Form A–2, A–3, and A–4.) The 24-hour recall provides data on one day only and is commonly used in nutrition surveys to obtain estimates of the typical food intakes of large numbers of people in given populations. Its usefulness is limited in that it does not provide enough accurate information to allow generalizations about an individual's usual food intake. This limitation is partially overcome when 24-hour recalls are collected on several nonconsecutive days.[6]

An advantage of the 24-hour recall is that it is easy to obtain. It is also more likely to provide accurate data, at least about the past 24 hours, than a person's estimates of average intakes over long times. However, the previous day's intake may not be typical; the person may be unable to report portion

Table 1–6 Nutrition Assessment: Risk Factors for Poor Nutrition Status

Medical History	Diet History	Socioeconomic History of Family	Drug History
Alcoholism	Anorexia nervosa	Eating alone	Antibiotics
Anorexia nervosa	Bulimia	Inadequate food budget	Anticancer agents
Cancer	Frequent eating out	Inadequate food preparation facilities	Anticonvulsants
Chewing or swallowing difficulties (including poorly fitted dentures, dental caries, and missing teeth)	Inadequate food intake	Inadequate food storage facilities	Antihypertensive agents
Circulatory problems	Intravenous fluids (other than total parenteral nutrition) for 10 or more days	Poor education	Catabolic steroids
Constipation	No intake for 10 or more days	Poor self-concept	Oral contraceptives
Diabetes	Poor appetite	Transportation unavailable	Vitamin and other nutrient preparations
Diarrhea	Restricted or fad diets		
Diseases of the GI tract			
Drug addiction			
Fever			
Heart disease			
Hormonal imbalance			
Hyperlipidemia			
Hypertension			
Infection			
Kidney disease			
Liver disease			
Lung disease			
Mental retardation or deterioration			
Multiple pregnancies			
Nausea			
Neurologic disorders			
Overweight			
Pancreatic insufficiency			
Paralysis			
Physical disability			
Pregnancy			
Radiation therapy			
Recent major illness			
Recent major surgery			
Recent weight loss or gain			
Smoking of cigarettes			
Surgery of the GI tract			
Trauma			
Ulcers			
Underweight			
Vomiting			

sizes accurately; and the person may even conceal facts about foods eaten. As a result, sometimes the information gathered in a 24-hour recall does not truly reflect a person's typical intake.

Another approach is to use a food frequency checklist. The purpose of this record is to ascertain how often an individual eats a specific type of food per day, week, or month. This information is especially useful in helping to pinpoint food groups that may be missing from the diet. That a person ate no vegetables yesterday may not seem particularly significant, but that the person never eats vegetables should ring a bell warning of poor nutrition status.

Used together with the 24-hour recall, the food frequency record is especially useful. Each permits the assessor to double-check against the other the accuracy of the information obtained.

Still another alternative is the food diary, which includes not only food eaten but also time of day, place eaten, others present, and mood. A food diary can help both the assessor and client to determine factors associated with eating that may affect dietary balance and adequacy. Food diaries work well with cooperative people but require considerable time and effort on their part.

A prime advantage of the food diary is that the diary keeper assumes an active role and may for the first time become aware of personal food habits and begin to assume responsibility for them. It also provides the assessor with an accurate picture of the diary keeper's lifestyle and factors that affect food intake. For these reasons, a food diary can be particularly useful in outpatient counseling for such nutrition problems as overweight, underweight, or food allergy. The major disadvantages stem from poor compliance in recording the data and conscious or unconscious changes in eating habits that may occur while the person is keeping the diary.

After collecting food intake data, the assessor must estimate nutrient intakes, either informally or by using food composition tables. An informal estimate is possible only if the assessor has enough prior experience with formal calculations to "see" nutrient amounts in reported food intakes without calculations. Even then, such an informal analysis is best followed by a spot check for key nutrients by actual calculation.

The formal calculation can be performed either manually (by looking up each food in a table of food composition, recording its nutrients, and adding them up) or by using a computer diet analysis program. The assessor then compares the intakes with standards such as the RDA.

Used with skill, diet histories can be superbly informative about a client's nutrient intakes. The skillful assessor uses them with their limitations in mind. For example, a computer diet analysis tends to imply an accuracy greater than is possible to obtain from data as uncertain as those that provide the starting information. Nutrient contents of foods listed in tables of food composition or stored in computer data bases are averages, and for some nutrients, complete data are not available. In addition, the available data on nutrient contents of foods are simply that; they do not state the amounts of nutrients a person actually absorbs. Iron is a case in point: its availability from a given meal may vary from as high as 50 percent to less than 2 percent, depending on the relative amounts of heme iron, nonheme iron, vitamin C, meat, fish, and poultry eaten at the meal and on the presence of inhibitors of iron absorption such as tea, coffee, and nuts.[7]

Furthermore, there is uncertainty about portion sizes. The person who reports eating "a serving" of greens may not distinguish between a quarter cup and two whole cups; only trained individuals can accurately report serving sizes. Children tend to remember the serving sizes of foods they like as being larger than serving sizes of foods they dislike.[8]

Thus there are many sources of error in the comparison of reported nutrient intakes with nutrient needs. Most history takers learn to use shortcut systems to obtain rough estimates of nutrient intakes and then use calculation methods to pinpoint suspected nutrient deficiencies or imbalances.

An estimate of nutrient intakes from a diet history, when combined with other sources of information, allows the assessor to confirm or eliminate the possibility of suspected nutrition problems. Consider the concerns during special stages of life mentioned earlier. Body weight was the first one. When a person is overweight or underweight, the assessor can use the diet history to detect an excessive or deficient food energy intake. Other areas of concern were calcium and iron status. A food frequency checklist may be especially valuable for assessing dietary calcium and iron, since the primary food source of each of these nutrients is a single food group: milk and milk products for calcium, and meat and meat alternates for iron. The absence or scarcity of these foods on a food frequency checklist will alert a skilled assessor to obtain more information about calcium and iron intakes; to look at other factors that influence calcium status, as described in Chapter 2; and to proceed with tests of iron status, as described in Appendix A. If high fat and cholesterol intakes are apparent from the diet history, this alerts the assessor to evaluate other indicators of cardiovascular disease risk and provide appropriate diet advice.

The assessor must constantly remember that each person digests, absorbs, metabolizes, and excretes nutrients in a unique way; individual needs vary. Intakes of nutrients identified by diet histories are only pieces of a puzzle that must be put together with other indicators of nutrition status in order to extract meaning.

Socioeconomic history Socioeconomic factors profoundly affect nutrition status. The people a client lives with, the client's education, the ethnic group, and many other such factors all influence the person's food choices. Also, in general, the quality of the diet declines as income falls. At some point, the ability to purchase the foods required to meet nutrient needs is lost; an inadequate income puts an adequate diet out of reach. Agencies use poverty indexes to identify people at risk for poor nutrition and to qualify people for government food assistance programs.

Low income affects not only the power to purchase foods but also the ability to shop for, store, and cook them. A skilled assessor will note whether a person has transportation to a grocery store that sells a sufficient variety of low-cost foods, and whether the person has access to a refrigerator and stove.

Drug history The important interactions of foods and drugs require that special attention be paid to any drugs taken routinely. Hundreds of drugs interact with nutrients, making imbalances or deficiencies likely. Table 1–7 describes mechanisms of food and drug interactions.

Table 1–7 Mechanisms of Food and Drug Interactions

Drugs Can Change Food Intake By	
Altering the appetite. Interfering with taste or smell. Inducing nausea or vomiting. Causing sores or inflammation of the mouth.	
Drugs Can Change Nutrient Absorption By	*Foods Can Change Drug Absorption By*
Changing the acidity of the digestive tract. Altering digestive juices. Altering motility of the digestive tract. Inactivating enzyme systems. Damaging mucosal cells.	Changing the acidity of the digestive tract. Stimulating secretion of digestive juices. Altering motility of the digestive tract. Binding to drugs.
Drugs Can Change Nutrient Metabolism By	*Foods Can Change Drug Metabolism By*
Acting as structural analogues. Interfering with metabolic enzyme systems. Binding to nutrients.	Interfering with a drug's action. Contributing pharmacologically active substances (example: tyramine).
Drugs Can Change Nutrient Excretion By	*Foods Can Change Drug Excretion By*
Altering reabsorption in the kidneys. Displacing nutrients from their plasma protein carriers.	Changing the acidity of the urine.

The chance that an adverse drug-nutrient interaction will result in nutrient deficiencies increases if the drug is taken for a long time, if the person is taking several drugs simultaneously, or if the person is in poor nutrition status when drug treatment begins. (Form A–5 is used to elicit the necessary information regarding drugs, and Table A–1 summarizes the effects on nutrition that are most likely to occur in reaction to the drugs most commonly taken.)

In a formal assessment, if a person is taking any drug, the assessor records on the drug history form the name of the drug; the dose, frequency, and duration of intake; the reason for taking the drug; and signs of any adverse effects. This information becomes one of the puzzle pieces that must be fitted together to obtain a picture of the person's nutrition status.

anthropometric: relating to measurement of the physical characteristics of the body, such as height and weight.
anthropos = human
metric = measuring

Anthropometric Measurements

Anthropometrics are physical measurements that provide an indirect assessment of body composition and development (see Table 1–8). They are useful

Table 1–8 Nutrition Assessment: Anthropometric Measurements Routinely Used

Measurement	Reflects
Height-weight	Overnutrition and undernutrition Growth in children
Recent weight change	Overnutrition and undernutrition
Fatfold	Subcutaneous fat and total body fat
Head circumference	Brain growth and development in children

primarily for two purposes: first, to evaluate the progress of growth in pregnant women, infants, children, and adolescents; and second, to detect undernutrition and obesity in all age groups.

In using anthropometry, health care providers compare measurements taken on an individual with standards specific for gender and age or with previous measures of the individual. The standards (given in Appendix A) derive from measurements taken on populations; measurements taken periodically and compared with previous measurements reveal changes in an individual's status.

Anthropometric measurements are easy to take and require minimal equipment. However, the skills of the measurer limit the accuracy and value of these measurements. Mastering the correct techniques requires proper instruction and practice to ensure reliability. Furthermore, significant changes in measurements are slow to occur in adults. When changes do occur, they represent prolonged dietary practices.

Height and weight Height and weight are the most commonly used anthropometric measurements. Length measurements for infants and children up to age three and height measurements for children over three are particularly valuable in assessing growth, and therefore nutrition status. Chapters 6 and 7 describe preferred methods of measuring height in infants and children. For adults, height measurements help to estimate desirable weight and to interpret other assessment data.

Unfortunately, many health care providers merely ask clients how tall they are, rather than measuring their height. Self-reported height is often inaccurate and should be used only as a last resort when measurement is impractical (in the case of an uncooperative client, an emergency hospital admission, or the like).

For measuring weight, the beam balance is most accurate (see Figure 1–3). Bathroom scales are inaccurate and inappropriate in a professional setting. To make repeated measures useful, standardized conditions are necessary. Each weighing should take place at the same time of day (preferably before breakfast), in the same amount of clothing (without shoes), after the person has voided, and on the same scale. As with all measurements, the assessor records observed weight immediately; accuracy to the nearest 0.1 kilogram or ¼ pound is especially important for infants and small children, since they weigh much less than adults.

Assessment of weight for height requires comparison with standards. The standards used for infants and children, the growth charts, are described in their respective chapters. For adults, the health care provider typically com-

Figure 1–3 Weight Measurement of an Older Child or Adult
Whenever possible, children and adults are measured on beam balance scales to ensure accuracy.

pares weights with weight-for-height tables (such as Table A–2), which are specific for height, gender, and frame size. To use the height-weight tables, the assessor refers to a table of frame sizes, such as the one based on elbow breadth (Table A–3) or the one that compares wrist circumference with height (Figure A–1 and Table A–4). The height and weight tables suggest an appropriate weight range, rather than pinpointing one ideal weight. This is a good reminder that there is no one perfect weight for anyone.

Most assessors use either the 1959 or the 1983 height-weight tables available from the Metropolitan Life Insurance Company. Height-weight tables available prior to these reported the average weights of people at the time they purchased life insurance, and revealed that older people were heavier. This indicated that as people age, they gain weight. Then, as the risks of excessive body weight became apparent, health professionals challenged the use of average weights as standards, on the basis that averages used as standards reflect what is, and not necessarily what is best. Research showed that below-average weights correlated with the greatest longevity. Later tables, including the 1959 tables of "desirable" weight, indicated weight ranges based on average weights *and* on weights correlating with the greatest longevity.[9]

The 1983 edition of the tables raised the standards. The standard weights were higher than those on the 1959 tables, but still fell below the average weights of the general population. They reflected not average weights but the weight ranges that correlated with the lowest mortality rates of people who had purchased life insurance. However, they excluded people with major diseases, such as heart disease and cancer.[10] A major criticism of the 1983 tables is that by including only healthy obese people they underrepresent mortality risks due to the diseases associated with obesity. Another criticism is that data based solely on mortality rates ignore possible nonfatal consequences of obesity, such as chronic illnesses and accidents.

Researchers note limitations of both the 1959 and the 1983 tables.[11] These limitations include the following:

- Frame size measurements were not taken on the populations used to generate the standards. No standards exist for frame sizes, and no reliable means of measuring them exist.
- Measurements of men and women 60 years of age or older were not included.
- The population selected was not a random sample.

Despite these limitations, the height-weight tables are useful for helping to identify both underweight and obesity.

A standard derived from height and weight, which is especially useful for estimating the risk to health associated with obesity, is the body mass index (BMI). Figure A–2 presents a pair of charts for determining the BMI and helps users to see that any individual's weight falls at some point on a continuum. The farther toward either extreme the weight falls, the more urgently it requires correction, but there is a broad zone between the extremes within which many weights are acceptable.

body mass index (BMI): an index of a person's weight in relation to height, determined by dividing the weight, in kilograms, by the square of the height, in meters:

$$BMI = \frac{\text{weight (kg)}}{\text{height}^2\text{ (m)}}.$$

Fatfold measurements Measures of body fatness are more meaningful than measures of mere weight, especially for helping determine whether weight is appropriate, and also for purposes of assessing disease risks. A practical method for estimating body fatness is to use a fatfold caliper—a device that measures the thickness of a fold of fat on the back of the arm, below the shoulder blade, on the side of the waist, or elsewhere. Approximately half of the fat in the body is directly beneath the skin, and its thickness reflects total body fat. The fatfold test is a practical diagnostic procedure in the hands of a trained person. Body fat determined by fatfold measures correlates well with other, more sophisticated measures of total body fat, such as underwater weight, radioactive potassium count, and total body water. Figure A–3 shows how to measure the triceps fatfold, and triceps fatfold percentiles are given in Table A–5.

A major limitation of the fatfold test is that fat may be thicker under the skin in one area than in another. A pinch at the side of the waistline may not yield the same measurement as a pinch on the back of the arm. This limitation can be overcome by taking fatfold measures at several (often three) different places on the body.

The procedure for taking three fatfold measures is as follows. Take all fatfold measurements on the right side of the body while the person is standing. The fatfold sites for men are the chest (a diagonal fold midway between the shoulder crease and nipple), abdomen (a vertical fold just to the side of the umbilicus), and thigh (a vertical fold on the front of the thigh midway between the hip and the knee). For women the sites are the triceps (a vertical fold on the back of the upper arm, midway between the shoulder and the elbow), suprailium (just above the hip bone in line with the middle of the armpit), and thigh (a vertical fold on the front of the thigh midway between the hip and the knee).

Measure each fatfold site three times and take the average of the two closest measures as the final value. Derive percent body fat by summing the three fatfold values and looking them up on Tables A–6 for women and Table A–7 for men.

Successive fatfold measures taken on the same person become meaningful only when long times are involved, for fat stores increase and decrease slowly. Short-term changes in subcutaneous fat are undetectable. Thus single measures are generally useful only for comparison with standards; successive measures are needed for following long-term changes in body composition.

Physical Examinations

Besides taking anthropometric measurements, the nutrition assessor can also use a physical examination to search for signs of nutrient deficiency or toxicity. Such an examination requires knowledge and skill. Many physical signs are nonspecific; they can reflect any of several nutrient deficiencies, as well as conditions not related to nutrition. For example, cracked lips may be caused by sunburn, windburn, dehydration, or any of several B vitamin deficiencies. For this reason, physical findings are especially unreliable, by themselves, for diagnosis of a nutrition problem. Instead, their value is in revealing possible problems for other assessment techniques to confirm, or in confirming conclusions drawn from other assessment measures.

With this limitation understood, physical symptoms can be most informative; they communicate much information about nutritional health.

Many tissues and organs can reflect signs of malnutrition. Physical signs of malnutrition appear most rapidly in parts of the body where cell replacement occurs at the highest rates, such as in the hair, skin, and digestive tract (including the mouth and tongue). Table 1–9 lists general physical signs of malnutrition. (Table A–8 lists the signs of specific vitamin and mineral imbalances.)

Biochemical Analyses

All of the approaches to nutrition assessment discussed so far are external approaches. Biochemical tests help to determine what is happening to the body internally. Most are based on analysis of blood and urine samples, which

The **serum** is the watery portion of the blood that remains after removal of the cells and clot-forming material; **plasma** is the fluid that remains when unclotted blood is centrifuged. In most cases, serum and plasma concentrations are similar. Lab technicians usually prefer serum samples, because plasma samples occasionally clog mechanical blood analyzers.

Table 1–9 Nutrition Assessment: General Physical Signs of Malnutrition

Body System	Normal	Malnutrition
Hair	Shiny, firm in the scalp	Dull, brittle, dry, loose; falls out
Eyes	Bright, clear pink membranes that adjust easily to light	Pale membranes; spots; redness; adjust slowly to darkness
Teeth and gums	No pain or caries, gums firm, teeth bright	Missing, discolored, decayed teeth; gums bleed easily and are swollen and spongy
Face	Good complexion	Off-color, scaly, flaky, cracked skin
Glands	No lumps	Swollen at front of neck, cheeks
Tongue	Red, bumpy, rough	Sore, smooth, purplish, swollen
Skin	Smooth, firm, good color	Dry, rough, spotty; "sandpaper" feel or sores; lack of fat under skin
Nails	Firm, pink	Spoon shaped, brittle, ridged
Internal systems	Normal heart rate, heart rhythm, and blood pressure; normal digestive function; normal reflexes and psychological development	Abnormal heart rate, heart rhythm, or blood pressure; enlarged liver, spleen; abnormal digestion; burning, tingling of hands, feet; loss of balance, coordination
Muscles and bones	Good muscle tone, good posture, long bones straight	"Wasted" appearance of muscles; swollen bumps on skull or ends of bones; small bumps on ribs; bowed legs or knock-knees

contain nutrients, enzymes, and metabolites (end products) that reflect nutrition status.

Interpretation of biochemical data requires skill. Long metabolic sequences lead to the production of the end products and metabolites seen in blood and urine. No single test can reveal nutrition status; many factors influence laboratory tests. The low blood concentration of a nutrient may reflect a primary deficiency of that nutrient, but it may also be secondary to the deficiency of one or several other nutrients or to a disease. However, taken together with other assessment data, laboratory test results help to make a total picture that becomes clear with careful interpretation. They are especially useful in helping to detect subclinical malnutrition by uncovering early signs of malnutrition before the clinical signs of a classic deficiency disease appear.

subclinical deficiency: a nutrient deficiency in the early stages, before the outward signs have appeared.

Laboratory tests most commonly performed to detect nutrition-related health status include measures of protein-energy status, blood lipid measures, blood glucose determinations, and measures of vitamin and mineral status. Protein-energy malnutrition (PEM) is uncommon in the United States, but is the world's number-one malnutrition problem. Laboratory tests used to detect its presence and evaluate its severity include measures of serum total protein, serum albumin, serum transferrin, total lymphocyte count, and urinary creatinine excretion. A complete discussion of the assessment of PEM appears in Appendix A; standards for evaluating PEM in children are in Chapter 7.

The determination of blood cholesterol concentration provides a way of identifying people who are at risk for cardiovascular disease. Table 7–5 in Chapter 7 shows percentile classifications of blood cholesterol concentrations for children, and Table 9–4 shows initial classifications for adults and recommended follow-up.

Tests of blood glucose can identify people who are at risk for diabetes mellitus. A glucose concentration of greater than 140 milligrams per 100 milliliters blood on more than one occasion establishes a positive diagnosis for adults. Another tool for diagnosing diabetes is the glucose tolerance test (see Table 1–10). Diabetes during pregnancy (discussed in Chapter 4) poses special concerns for both the mother and infant. Norms for pregnancy are presented in Chapter 4.

Laboratory tests are also used to assess vitamin and mineral status. Table A–9 summarizes the biochemical tests used for this purpose. These tests are particularly useful in assessing vitamin and mineral status when combined with dietary histories and physical findings. Vitamin and mineral levels present in the blood and urine sometimes reflect recent intakes rather than long-term intakes. This makes detecting subclinical deficiencies difficult. Furthermore, many nutrients interact; the amounts of other nutrients in the body can affect a lab value for a particular nutrient. It is also important to remember that nonnutrient conditions influence both physical signs and biochemical measures.

The assessment of calcium status is difficult, because no evidence of a developing calcium deficiency can be found in a blood sample; blood calcium remains normal no matter what the bone content may be. Even an X ray of the bones reveals a calcium deficit only when it is so advanced as to be virtually irreversible. Assessment of calcium status therefore depends on two kinds of histories. Diet history information can provide clues to low calcium intakes, and personal medical history information can reveal risk factors for excessive adult bone loss.

Table 1–10 Nutrition Assessment: Standards for Glucose Tolerance Test Results[a]

Time (minutes)	Upper Limits of Normal (mg/100 ml)	Values Suggestive of Diabetes (mg/100 ml)
0	115	140
60	200	>200
120	140	200

[a]This is one of many variations of the glucose tolerance test.
Note: The person follows a high-carbohydrate diet for at least three days before the test and cannot take medication, smoke cigarettes, or exercise during the test. The test may not be valid if the person is emotionally or psychologically stressed. Blood is drawn after an overnight fast to obtain the fasting glucose concentration. The person then receives a measured amount of flavored glucose to drink. Blood is taken again at regular intervals of 30 to 60 min (various clinics use different intervals). This table lists the upper limits considered normal and the values suggestive of diabetes for each time interval. At least two of the glucose concentrations should exceed the latter values in order for diabetes to be diagnosed. A person whose glucose concentrations fall between the upper limits of normal and the values suggestive of diabetes is diagnosed as having impaired glucose tolerance.

Because iron deficiency is an extremely common nutrient deficiency disease and affects people at all stages of life, its assessment is particularly important. Indicators of iron status change from stage to stage of the life span, so laboratory diagnosis of iron deficiency for infants and children requires the use of age-specific standards, while for pregnant women, standards specific to pregnancy are used.

For practical reasons, clinicians define iron status in terms of hemoglobin concentration. (Hemoglobin is the iron-containing pigment of the red blood cells, whose function is to carry oxygen.) Hemoglobin is relatively easy to measure. Unfortunately, limited hemoglobin production is the final stage of iron deficiency development (see Table 1–11). If a dietary insufficiency of iron limits hemoglobin production so much as to become detectable from a hemoglobin measure, then the assessor can conclude that iron must clearly have become insufficient for the production of other iron-containing

Table 1–11 Nutrition Assessment: Biochemical Tests Used to Detect Iron Deficiency

Stage	Detected By	Reflects
Stores depleted	↓ Serum ferritin	Liver and bone marrow stores
Transport iron diminished	↓ Transferrin saturation	Iron-binding capacity, serum iron
Hemoglobin production limited	↑ Erythrocyte protoporphyrin	Detectable anemia, microcytosis

Source: Adapted from P. R. Dallman, Diagnostic criteria for iron deficiency, in *Iron Nutrition Revisited—Infancy, Childhood, Adolescence,* report of the 82nd Ross Laboratories Conference on Pediatric Research (Columbus, Ohio: Ross, 1981).

molecules. A clear picture of iron status, especially of early deficiency, requires several biochemical measures (Tables A–13 through A–18), considered within the framework of the three stages of iron deficiency. Appendix A provides further discussion of the assessment of iron status.

Considerable knowledge has been gained from research into the nutrient needs and from the nutrition assessment of people in various stages of life. This text describes that knowledge, applies it to help improve people's nutrition status and health, and explores questions yet to be answered.

Chapter 1 Notes

1. H. Kamin, Status of the 10th edition of the Recommended Dietary Allowances—Prospects for the future, *American Journal of Clinical Nutrition* 41 (1985): 165–170.
2. Food and Nutrition Board, *Recommended Dietary Allowances*, 10th ed. (Washington, D. C.: National Academy of Sciences, 1989).
3. Food and Nutrition Board, 1989.
4. M. T. Pugliese and coauthors, Parental health beliefs as a cause of non-organic failure to thrive, *Pediatrics* 80 (1987): 175–182.
5. S. Barrett, Why licensing of "nutritionists" is needed, *Nutrition Forum*, May 1985, p. 40.
6. K. J. Morgan and coauthors, Collection of food intake data: An evaluation of methods, *Journal of the American Dietetic Association* 87 (1987): 888–896.
7. E. R. Monsen and coauthors, Estimation of available dietary iron, *American Journal of Clinical Nutrition* 31 (1978): 134–141.
8. J. T. Dwyer, E. A. Krall, K. A. Coleman, The problem of memory in nutritional epidemiology research, *Journal of the American Dietetic Association* 87 (1987): 1509–1512.
9. E. S. Weigley, Average? Ideal? Desirable? A brief overview of height-weight tables in the United States, *Journal of the American Dietetic Association* 84 (1984): 417–423.
10. *Build Study 1979* (Chicago: Society of Actuaries and Association of Life Insurance Medical Directors of America, 1980), as cited in Weigley, 1984.
11. N. Robinett-Weiss and coauthors, The Metropolitan Height-Weight Tables: Perspectives for use, *Journal of the American Dietetic Association* 84 (1984): 1480–1481.

Nutrition Prior to Conception

2

The Kiss by Auguste Rodin.

A woman's body accomplishes remarkable feats. For approximately the first decade of life, her body's primary focus is on its own growth and maintenance. Then her body begins to offer ova for reproduction, and after fertilization, it provides support for the development of a whole new human being. Even after the infant's birth, the woman's body offers nourishment for several more months. The health of the woman's body prior to conception influences her fertility, the health of the infants she may later conceive and bear, and her own health later in life.

conception: the union of the male spermatozoon and the female ovum; fertilization.

fertility: the capacity of a woman to produce a normal ovum periodically and of a man to produce normal spermatozoa; the ability to reproduce.

ovum: the female reproductive cell, capable of developing into a new organism upon fertilization, commonly referred to as an egg; *ova* is the plural.
ovum = egg

spermatozoon (sper-mat-oh-ZOH-on): a male reproductive cell, capable of fertilizing an ovum; *spermatozoa* is the plural.
spermatos = seed
zoon = life

zygote: the term for the product of the union of ovum and spermatozoon for the first two weeks after fertilization.

The union of an ovum and spermatozoon in the creation of a zygote begins the development of a human being, and a multitude of new experiences for the parents. The time prior to this moment provides a unique opportunity for a woman to prepare herself physically, mentally, and emotionally for the many changes to come. To prepare the mother's body as the most suitable environment in which a fetus will develop requires establishing healthful habits.

A discussion on preconception nutrition must, by its nature, focus primarily on women—prior to and between pregnancies. (Chapters 9 and 10 focus on the nutritional needs of both women and men throughout the adult years.) Prepregnant women's needs are different from those of men, from those of pregnant and lactating women, and from those of women after the childbearing years. A man's nutrition may affect his fertility and possibly the genetic contributions that he makes to his children, and it is discussed later in this chapter; but a woman's body is the environment in which new human beings are grown, and it is through women that nutrition prior to conception exerts the greater influence. The better a woman takes care of herself nutritionally before and between pregnancies, the more successful her pregnancies are likely to be.

This chapter begins with women's nutrition in general as a background for special topics related to preconception nutrition's effects on pregnancy. Where necessary, it looks ahead to explain a few details about pregnancy itself, in order to illustrate why nutrition *prior* to pregnancy is so important. The accompanying miniglossary defines some terms used to demarcate periods in the lives of women and infants. Figure 2–1 places some of these terms on a time line.

Figure 2–1 Terms for Stages Surrounding Pregnancy and Birth

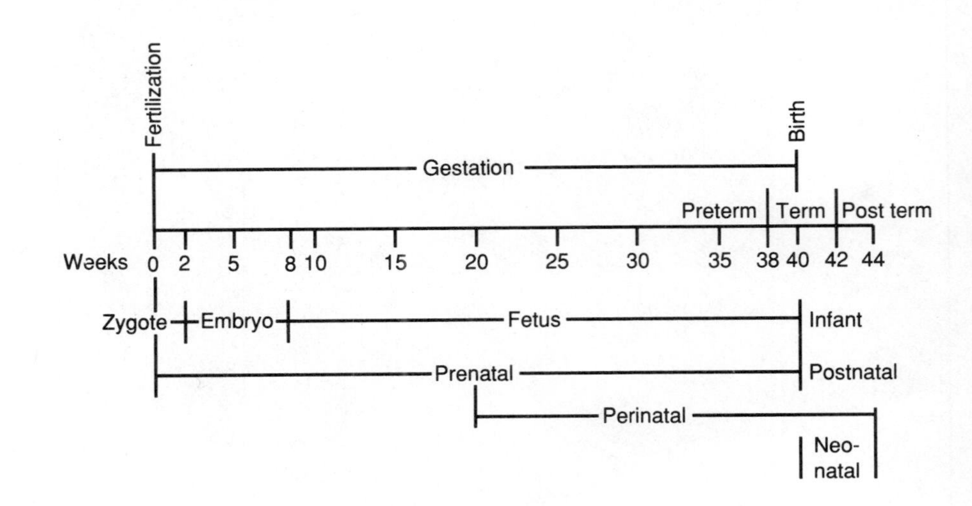

Miniglossary of Pregnancy and Birth Terms

These terms describe a woman's pregnancy status before, during, and after:

pregravid (pre-GRAV-id): before pregnancy.
pre = before
gravid = pregnant

gravid (GRAV-id): pregnant. A **gravida** (GRAV-ih-da) is a pregnant woman; **gravidity** is pregnancy. A woman during her first pregnancy is a **primigravida** (PRY-mee-gravida). A woman who has been pregnant two or more times is a **multigravida.**

postpartum: after childbirth.
post = after
partus = birth

These terms describe the number of a woman's pregnancies from the smallest to the largest:

nullipara (nul-LIP-ah-ra): a woman who has borne no children; the adjective is **nulliparous** (nul-LIP-ah-rus).
null = none
parere = to bear

primipara (pry-MIP-ah-ra): a woman who has borne, or is giving birth to her first child; the adjective is **primiparous** (pry-MIP-ah-rus).

multipara (mul-TIP-ah-ra): a woman who has borne more than one infant, regardless of infant survival; the adjective is **multiparous** (mul-TIP-ah-rus).

These terms describe the time surrounding birth:

prenatal: concerning the time before birth.
natal = birth

perinatal: concerning the time preceding, during, or after birth.
peri = around

neonatal: concerning the first four weeks after birth; a newborn infant is a **neonate.**
neos = new

postnatal: occurring after birth.

These terms describe the stages of intrauterine development.

zygote: (already defined earlier) the product of conception from zero to two weeks.

embryo: the offspring from two weeks to eight weeks.

fetus: the offspring from eight weeks to term (nine months).

These terms describe an infant's gestational age at birth:

The term **gestation** refers to the period from conception to birth; for human beings, normal length of gestation is from 38 to 42 weeks.

preterm: an infant born prior to the 38th week of gestation; also referred to as a **premature** infant.

term: an infant born between the 38th and 42nd weeks of gestation.

post term: an infant born after the 42nd week of gestation.

Nutrition and Women's Health

A major aim of many dietary recommendations today is to prevent disease for the general population. They include, but do not focus specifically on, prepregnant women. Much attention focuses on preventing cardiovascular disease, the number one killer of all adults, that is more often fatal, earlier, in men than in women. To this end, current guidelines recommend avoiding too much fat and sodium and advocate a diet that limits high-fat foods such as red meats and dairy products (see Chapters 1 and 9).

The recommendations to limit fat and sodium might benefit everyone threatened with cardiovascular disease, but women have other specific needs in their reproductive years, and their hormones help to protect them to some extent from this particular disease. If they need to limit fat and sodium, they need to do so in a different way from men. Men may benefit from limiting red meats and dairy products, but women following such advice may suffer iron and calcium deficiencies. In short, men and women face different nutrition- and diet-related problems. A task force for the ADA has therefore developed a set of recommendations to promote health and prevent disease specifically for women (see Practical Point: Nutrition Recommendations for Women).[1] Any woman wanting to do all that she can to make a future pregnancy healthy should review these recommendations with an eye to improving her diet wherever necessary. Appendix C describes the Four Food Group Plan for diet planning.

Nutrition Recommendations for Women: a set of dietary guidelines designed by the ADA specifically to meet the needs of women.

Appropriate Body Weight

The ADA recommendation to maintain healthy body weight is old advice that includes a valuable new concept—that of body weight's being described as "healthy" rather than as "ideal" or "desirable." Prior to pregnancy, a woman determining her appropriate body weight needs to consider her optimal health, not cultural norms, which may be unrealistic.

▸▸PRACTICAL POINT

Nutrition Recommendations for Women

1. Eat a variety of foods daily from all major food groups:
   - 3 to 4 servings of low-fat dairy foods.
   - 2 servings of low-fat meat/meat alternates.
   - 4 servings of vegetables/fruits.
   - 4 servings of whole-grain breads/cereals.
2. Maintain a healthy body weight.
   - For adults who need to lose weight, do so safely and effectively by not going below 10 kcal/lb of present weight, not skipping meals, and increasing physical activity (exercising).

▸▸ PRACTICAL POINT *continued*

- ▸ Gain weight, if necessary, by increasing kcaloric intake and exercising in moderation.

3. Exercise regularly (three days per week).[a]
4. Limit total fat consumption to no more than one-third of daily kcalories:
   - ▸ Select a variety of fat sources: saturated, polyunsaturated, and monounsaturated.[b]
   - ▸ Limit nonfood-group foods, such as margarine, butter, cooking oils, salad dressings, cookies, cakes, and cream.
   - ▸ Choose low-fat selections of meat and milk food groups.
5. Eat at least one-half of daily kcalories from carbohydrates:
   - ▸ Select complex-carbohydrate foods, such as beans, peas, pasta, vegetables, nuts, and seeds.
6. Eat a variety of fiber-rich foods:
   - ▸ Make daily selections from fresh fruits, vegetables, legumes (navy, pinto, and kidney beans), and whole grains (brown rice, oatmeal, and oat and wheat bran).
   - ▸ Increase intake of fiber gradually.
   - ▸ Avoid excess fiber intake,[c] especially from one source.
7. Include 3 to 4 daily servings of calcium-rich foods:
   - ▸ Consume low-fat milk, yogurt, and cheese.
   - ▸ Use low-fat milk, yogurt, and cheese in cooking.
   - ▸ Eat broccoli, sardines with bones, canned salmon with bones, and collard greens.
8. Include plenty of iron-rich foods:
   - ▸ Make daily selections from lean meat, liver, prunes, pinto and kidney beans, spinach, leafy green vegetables, and enriched and whole-grain breads/cereals.
9. Limit intake of salt and salt-containing foods:[d]
   - ▸ Limit the addition of salt in food preparation.
   - ▸ Limit use of the saltshaker.
10. Rely on foods for necessary nutrients, using vitamin and mineral supplements only under specific circumstances.
11. If you drink alcoholic beverages, limit alcohol intake to one to two drinks daily.[e]
12. Avoid smoking.
13. Adjust diet, exercise, and other health promotion behaviors to correspond with your own identified risk factors:
    - ▸ Remember that diet is only one risk factor; heredity, lifestyle, and environment are other factors.
    - ▸ Consult a physician with questions about risk factors.

▸▸PRACTICAL POINT *continued*

14. If you have questions about the adequacy of your diet, consult a registered dietitian.

---

[a]Chapter 8 expands on the benefits of regular exercise.
[b]The American Heart Association recommends a total fat intake of 30% or less of total kcalories, with 10% or less coming from saturated fat.
[c]The ADA recommendation is 20 to 35 g of fiber per day; Position of the American Dietetic Association: Health implications of dietary fiber, *Journal of the American Dietetic Association* 88 (1988): 216.
[d]Omit foods that are high in salt, such as pickles, luncheon meats, sardines, potato chips, and soy sauce. Originally, sodium restriction was advised, but this recommendation is omitted here in light of the finding that salt specifically, and not sodium-containing products, raises blood pressure; T. W. Kurtz, H. A. Al-Bander, and C. Morris, "Salt-sensitive" essential hypertension in men: Is the sodium ion alone important? *New England Journal of Medicine* 317 (1987): 1043–1048.
[e]Focal Point 8 discusses alcohol metabolism and the health risks associated with drinking alcoholic beverages. Focal Point 3 highlights the dangers associated with alcohol use during pregnancy.
Source: Adapted from The American Dietetic Association's Nutrition Recommendations for Women, *Journal of the American Dietetic Association* 86 (1986): 1663–1664. Reprinted with permission.

Three suboptimal states can adversely affect a woman's prospects of enjoying an optimal pregnancy: obesity, undernutrition, and unbalanced nutrition. Obesity can adversely affect fetal development and render childbirth difficult. Undernutrition or unbalanced nutrition can make it impossible for a woman even to conceive, and if she does, can impair the course and outcome of her pregnancy. Later sections will describe these problems; the point here is that any woman who hopes to become pregnant in the future is well advised to prepare by achieving and maintaining a healthy weight prior to pregnancy.

The sections on "Undernutrition" and "Obesity" on pp. 60–62 describe the impact of prepregnancy weight on pregnancy.

To lose weight, the ADA advises a woman to follow a specific rule of thumb—the "10-kcalorie rule" (see Practical Point: 10-kCalorie Rule). The ADA's purpose in offering a minimum number of kcalories for weight loss is to ensure that the woman can make a selection of foods that will deliver all nutrients in adequate amounts. A woman consuming 10 kcalories per pound (of her present body weight) can gradually lose weight and, with the appropriate food selections, will not compromise her health.

10-kcalorie rule: a rule presented in the ADA Nutrition Recommendations for Women that establishes the number of kcalories a woman should consume to achieve gradual weight loss and dietary adequacy—10 kcal/lb of present body weight.

▸▸PRACTICAL POINT

10-kCalorie Rule

If a woman wants to find out how low to set her energy intake in order to lose weight, she has two options. One is to estimate her total energy expenditure and then subtract a number of kcalories per day from that to arrive at a desired rate of weight loss. (An example is to subtract 500 kcal/day to lose a pound a week.) The other alternative is to use a simpler, often safer, rule of thumb suggested by the ADA—the 10-kcalorie rule.[a]

Conversion Factors
1 kg = 2.2 lb.
1 lb body fat = 3500 kcal.

▸▸ PRACTICAL POINT *continued*

Consider the first alternative, using a sedentary woman who weighs 135 lb (61 kg). To estimate her energy expenditure, she would have to combine estimates for her basal metabolism and voluntary muscular activity. To estimate her basal metabolic energy output, she would use the factor 0.9 kcal/kg body weight/hour:

$$61 \text{ kg} \times 0.9 \text{ kcal/kg/hour} \times 24 \text{ hour/day} = 1318 \text{ kcal/day.}$$

To approximate energy output for physical activity for a sedentary person, she would add 50% of the estimate for basal metabolism:

$$1318 \text{ kcal/day} \times 50\% = 659 \text{ kcal/day.}$$

Her total estimated energy output, then, is:

$$1318 \text{ kcal/day} + 659 \text{ kcal/day} = 1977 \text{ kcal/day.}$$

Next, she would pick a rate of weight loss. For each pound of weight loss per week, she would subtract 500 kcal/day. The woman in this example wants to lose 2 lb/week, so she would calculate:

$$1977 \text{ kcal/day} - 1000 \text{ kcal/day} = 977 \text{ kcal/day.}$$

This method provides no guidelines to ensure a safe rate of weight loss or minimum number of kcalories, but guidelines are often offered along with it: "Do not lose more than 1 to 2 lb per week"; or "Eat no less than 1200 kcal/day." The woman therefore chooses to eat 1200 kcal/day.

Now compare the 10-kcalorie rule for weight loss for another sedentary, 135-pound woman:

$$10 \text{ kcal/day} \times 135 \text{ lb} = 1350 \text{ kcal/day.}$$

On this intake, the woman will lose between 1 and 2 lb/week. Her energy intake and rate of weight loss will not be much different from those of the first woman, but she arrived at them differently. The first woman picked a rate of weight loss and adjusted energy intake accordingly; the second picked an energy intake, and the rate of weight loss adjusted accordingly.

The 10-kcalorie rule limits weight loss to the maximum safe rate. It requires smaller women to lose weight less rapidly, for safety's sake, and it permits more-rapid weight loss at higher body weights. For example, a 110-lb woman must have no less than 1100 kcal/day, and may lose 1 lb/week. A 220-lb woman must have 2200 kcal/day and may lose 2 lb/week. She should adjust her energy intake downwards at intervals as she loses weight, and her weight-loss rate will decrease accordingly. Each weight-loss rate is safe for the person concerned, and each energy intake is high enough to be realistically achievable and to provide dietary adequacy.

---

[a]The American Dietetic Association's Nutrition Recommendations for Women, *Journal of the American Dietetic Association* 86 (1986): 1663–1664.

To gain weight, an underweight woman is advised to add nutrient-dense foods to an already adequate diet. One way she can do this is to adopt the diet recommended for a woman who is already pregnant (see Chapter 4).

As for unbalanced nutrition, this, too, needs to be corrected. Dietary habits are easy to observe, and the woman can take the opportunity to evaluate hers prior to pregnancy. A dietary history taken by a dietitian or a dietary intake recorded by the woman and assessed by a dietitian can determine where improvements are possible. By learning to eat nutrient-dense foods prior to pregnancy, a woman can establish, in advance, eating habits that will support a healthy pregnancy later without excessive weight gain.

Problem Nutrients

The diets women typically eat provide some nutrients in ample quantities. Vitamin C and protein are examples. Other nutrients seem harder to get—notably, calcium, iron, vitamin B_6, folate, magnesium, and zinc. The ADA recommendations specifically address two of these—calcium and iron.

Calcium Ninety-nine percent of the body's calcium resides in the skeletal system. Bone may appear to be static, but like other body tissues, it is always changing—being broken down and re-formed. Bone continuously exchanges calcium with its surrounding fluids. During times of inadequate calcium intake, the bones serve as a source of calcium for the tissues' needs. With repeated withdrawals, the calcium stores become depleted, threatening the integrity of the bones.

Figure 2–2 shows the three phases of bone development throughout life. From birth to approximately age 20, the bones are actively growing by modifying their length, width, and shape. This rapid growing phase overlaps with the next period of peak bone mass development that occurs between the ages of 12 and 40.[2] During this next period, skeletal mass increases. Bones grow both thicker and denser by remodeling, a maintenance and repair process

Figure 2–2 Phases of Bone Development throughout Life
The bones actively grow from birth to approximately age 20. The active growth phase overlaps with the next phase of peak bone mass development that occurs between the ages of 12 and 40. The final phase, when bone resorption exceeds formation, begins between ages 30 and 40 and continues throughout life.

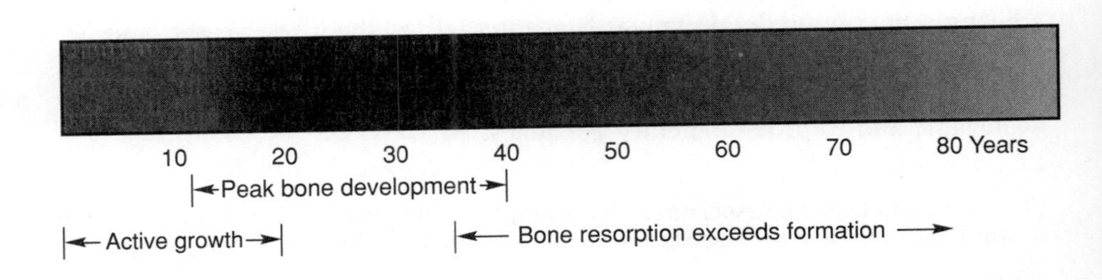

involving the loss of existing bone and the deposition of new bone. The final phase, which begins between 30 and 40 years of age and continues throughout life, finds bone loss exceeding new bone formation.

Nearly all people experience some bone loss as they grow older—especially women after menopause or surgical removal of the ovaries (which often accompanies hysterectomy). This suggests that hormonal changes are responsible. Because women's bones are less dense than men's and their hormonal changes accelerate losses, the extensive adult bone loss of osteoporosis is eight times more common in women than in men. Osteoporosis leads to crippling deformities and irreparable fractures that often hasten death.

osteoporosis: reduced density of the bones.
osteo = bones
porosis = porous

Bone loss is a natural process that women cannot completely prevent, but they may be able to forestall its impact by achieving maximal bone mass before and during the childbearing years. Many factors determine a woman's bone mass; probably most important among them are her heredity and the extent to which she puts demands on her bones by exercising and by maintaining a healthy body mass.[3] Surprisingly, dietary calcium may be less important than has been thought in the past, for people are able to adapt to wide ranges of calcium intake.[4] However, adaptation to a changed calcium intake occurs slowly over a period of months or years, so for those who ingest large quantities of calcium early in life, it may be desirable to maintain these intakes throughout life. Since milk and milk products are the primary sources of calcium in our culture, and because drinking milk is a habit best learned early, Chapter 7, "Nutrition during Childhood," discusses solutions to such problems as milk dislikes, milk allergies, and lactose intolerance.

The average calcium intakes of women of childbearing age are consistently less than the recommended intakes, as Figure 2–3 illustrates. The calcium intakes of many individuals are even lower than the average. It thus appears that many people must be incurring calcium losses from their bones throughout adult life, but this may not be the case. The RDA for calcium, as for all nutrients, is set high enough to cover the small percentage of people whose needs are considerably higher than the average. Many individuals may be able to maintain balance on substantially lower intakes than this. Calcium balance is difficult to measure, but some research on both animals and human beings indicates that, given enough time, people can adapt to intakes of 600, 400, or even fewer milligrams a day without incurring losses of minerals from their bones.[5] Still, below an unknown minimum which may differ from one person to the next, low calcium intakes must be a risk factor for osteoporosis. Other risk factors include:

- Being female.
- Being Caucasian or Asian.
- Having a family history of excessive adult bone loss.
- Being underweight.
- Having a small frame size.
- Leading a sedentary life.
- Consuming more than one alcoholic beverage per day.
- Smoking.

Figure 2–3 Daily Calcium Intakes of Females Compared with their RDA
Women's average calcium intakes tend to fall lower throughout life. In hopes of preventing this decline, and in the hope that increased calcium intakes *may* help forestall the adult bone loss that leads to osteoporosis, some authorities favor a recommendation higher than the RDA.[a] The dotted line indicates that the calcium needs of pregnancy and lactation are higher than those of nonpregnant women.

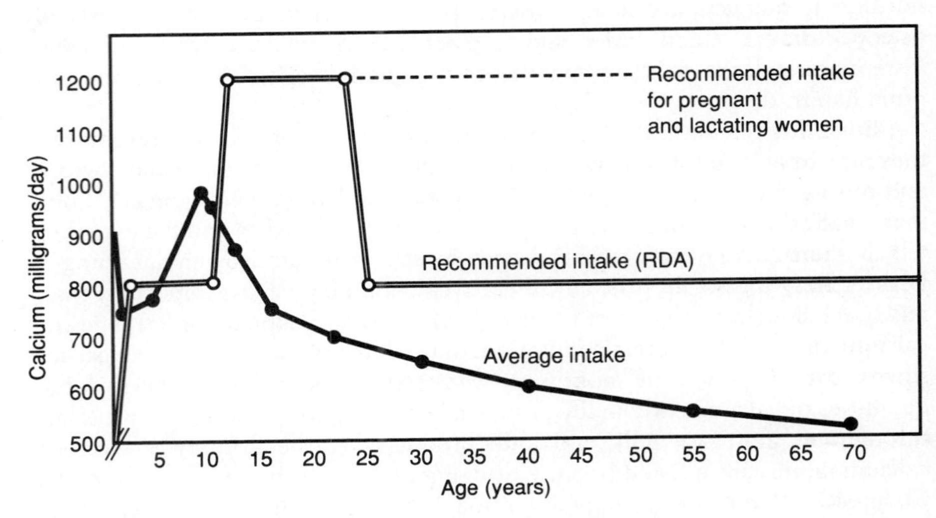

[a]Osteoporosis: National Institutes of Health Consensus Development Conference Statement, April 1984 (Washington, D.C.: Government Printing Office, 1984).

Source: Adapted from National Center for Health Statistics, *Dietary Intake Source Data: United States, 1976–1980,* DHHS publication no. (PHS) 83–1681 (Washington, D.C.: Government Printing Office, March 1983).

- Consuming caffeine equivalent to that in more than two cups of coffee or tea per day.
- Leading a stressful life.
- Being postmenopausal.

Chapter 4 discusses calcium in pregnancy: the mother's needs, her absorption, and the transfer from mother to fetus.

For women facing future pregnancies, calcium needs will intensify. Women require calcium above their customary intakes throughout pregnancy, even though most of it does not transfer to the fetus until the last trimester.

To help maximize calcium stores before the start of pregnancy, the ADA recommendation suggests that a woman eat 3 to 4 servings of calcium-rich foods daily. To consume enough calcium from dairy products without exceeding her kcalorie allowance, a woman must select, with care, mostly fat-free and low-fat items. The upcoming section on supplements describes calcium supplement usage, absorption, and risks.

Iron Iron functions as a component of hemoglobin, myoglobin, and many other proteins. The body's total iron content also includes iron stores. With inadequate iron intake, the stores serve as a source of iron to meet the body's needs. Only after the depletion of iron stores do hemoglobin concentrations begin to fall. Therefore, iron depletion progresses quite far before this standard blood test can diagnose it (details are in Appendix A).

During the procreative years, women are in a precarious state with respect to iron sufficiency. Their normal iron losses exceed those of men because of repeated menstrual blood losses. In addition, the average iron intakes of women are consistently lower than the recommended intake, as Figure 2–4 illustrates. The combination of inadequate iron intake and the natural iron loss that occurs in the procreative years makes it a challenge to obtain sufficient iron. All too often, women enter pregnancy with depleted iron stores.

During pregnancy, the increase in blood volume and the demands of the fetus will drain the iron stores further. Fetal withdrawals of iron will be made from maternal iron stores regardless of the mother's iron status. Thus the hemoglobin concentration in a newborn may be normal when the mother's stores are low, and even when she is overtly anemic. The consequences of maternal iron deficiency are evident as increased risks of poor pregnancy outcomes—fetal deaths, preterm births, low birthweights, and medical abnormalities.[6]

A woman planning to become pregnant can benefit from a nutrition assessment to determine her iron status. If deficient, she will want to correct the problem before she becomes pregnant. Otherwise, the stress of iron-deficiency anemia can seriously impair both her physical health and her emotional adjustment.

Figure 2–4 Daily Iron Intakes of Females Compared with their RDA
Women's average iron intakes fall below recommendations, throughout early and middle life.

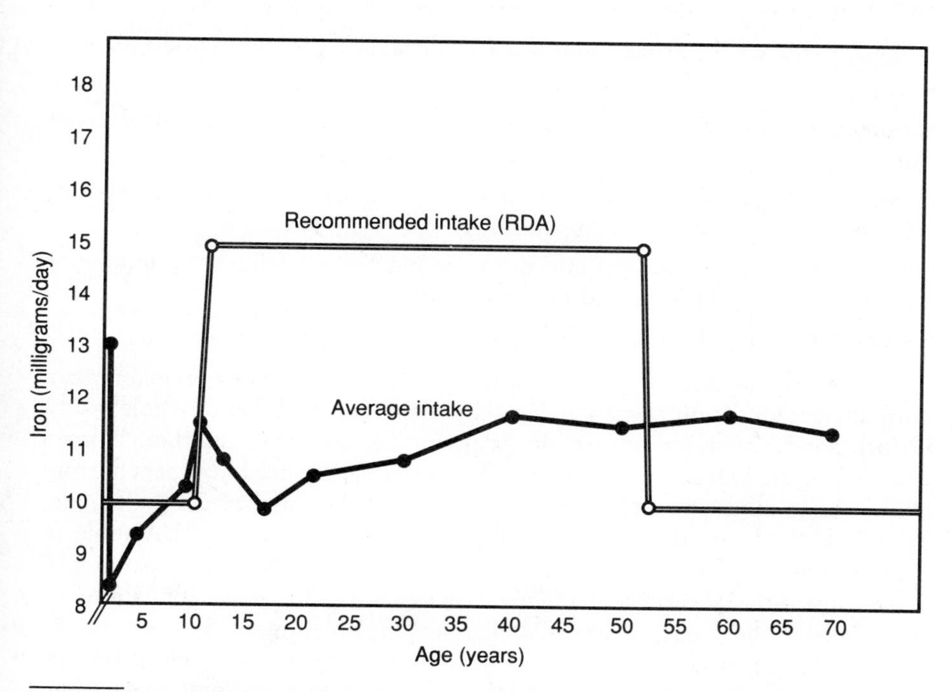

Source: Adapted from National Center for Health Statistics, *Dietary Intake Source Data: United States, 1976–1980*, DHHS publication no. (PHS) 83–1681 (Washington, D.C.: Government Printing Office, March 1983).

The best way to *prevent* iron deficiency is to eat iron-rich foods regularly, in the context of meals that enhance iron absorption. The best way to *treat* iron deficiency may be to combine iron-rich foods and supplements. (A discussion of iron supplements follows this section.)

Other nutrients The ADA recommendations include special mention of calcium and iron, but other nutrients may also be lacking in the diets of many women. For example, the average zinc intakes of women between the ages of 19 and 55 are about 75 percent of the RDA.[7] Wise selections of foods can offer substantial quantities of more than one nutrient to rectify several problems with one solution. However, no one food or food group can correct for all inadequacies. Eating a variety of foods from each of the food groups emerges as the best plan.

Supplements for Women

The ADA recommendations urge women to rely on foods for necessary nutrients, using vitamin and mineral supplements only under specific circumstances. To determine whether an individual is at risk for a nutrient deficiency requires a complete nutrition assessment. When such an assessment so indicates, a health care provider can recommend an appropriate supplement. The following list acknowledges that these specific conditions may justify the taking of supplements by some prepregnant women, including:[8]

- Women with low food energy intakes, such as habitual dieters.
- Women who eat bizarre or monotonous diets, such as some food faddists.
- Women with illnesses that take away the appetite.
- Women with illnesses that impair absorption of nutrients—including diseases of the liver, gallbladder, pancreas, and digestive system.
- Women taking medications that interfere with the body's use of specific nutrients (see Table A-1 in Appendix A).
- Women who have diseases, infections, or injuries, or who have undergone surgery resulting in increased metabolic needs.
- Women who eat all-plant diets (vegans).

Except for women in these circumstances, women prior to pregnancy can normally get all the nutrients they need by eating a varied diet of whole foods. Unfortunately, many do not eat this way, for one reason or another. Women who do not eat well enough to receive the nutrients they need may benefit from multivitamin-mineral supplements. A rule of thumb when selecting a supplement is to find one that provides all of the nutrients of the RDA table in amounts smaller than, equal to, or very close to the RDA for that person.

As mentioned earlier, inadequate calcium intake may contribute to age-related bone loss. However, calcium supplementation has yet to be validated by research as effective in preventing osteoporosis.[9] Nevertheless, supplementation may be a way—and, in fact, the only way—some women can meet their recommended intakes of calcium. In the absence of final answers on this question, it may, at least for the present, be prudent to add to the previous list of justifications for supplement use women whose calcium intakes are too low to forestall extensive bone loss.

Most healthy people absorb calcium as well from calcium carbonate, calcium acetate, calcium lactate, calcium gluconate, and calcium citrate as they do from whole milk.[10] A consumer could refine the choice among these by comparing other variables such as cost and number of tablets to be ingested. Calcium carbonate contains the highest percentage of calcium (40 percent) and therefore can meet daily needs with the smallest number of tablets. Consumers should avoid calcium-rich preparations of bonemeal or dolomite (limestone), because they may contain unsafe levels of contaminants such as lead, arsenic, cadmium, or mercury. Consumers using calcium supplements combined with vitamin D must limit the quantity of vitamin D they are taking to the RDA or below in order to avoid toxic doses.

Iron is like many other nutrients in that foods meet the body's requirements better than supplements do. Still, many women are unable to meet their iron requirements with foods alone, especially if they are limiting their food energy intakes, and might benefit from supplements providing the recommended 15 milligrams.[11] The body absorbs the ferrous form of an iron supplement efficiently, and taking the supplement with meals enhances absorption.[12] An addition to the previous list of circumstances that may justify the taking of a supplement follows:

- Women who bleed excessively during menstruation.

The taking of individual mineral supplements requires caution. Minerals compete for binding sites and thus interfere with one another's bioavailability. For example, a calcium phosphate dibasic supplement inhibits magnesium absorption. If it is fortified with magnesium, then it interferes with both calcium and iron absorption.[13] Calcium carbonate and calcium hydroxyapatite supplements also interfere with iron absorption when taken with meals.[14] Likewise, iron supplements may impair folate, zinc, copper, and selenium status, although research findings conflict.[15] In light of these and other interactions, and in view of the failure of supplements to enhance nutrition status as foods do, it seems appropriate to repeat that foods are preferred to supplements as nutrient sources.

In addition to adversely affecting the bioavailability of other nutrients, supplement use can cause other problems with varying degrees of severity. For example, iron supplements cause heartburn and constipation in some people. These side effects diminish if the iron is taken in small doses with foods, and if the total daily intake is less than 120 milligrams.

In concluding this section on supplements, it seems appropriate to repeat that foods are preferable. Certainly, for the best possible health prior to pregnancy, women need to pay attention to their nutrient intakes. Hand in hand with nutrition goes another lifestyle factor that enhances health—fitness.

Fitness

The ADA task force on women's recommendations recognizes the health advantages of fitness, and recommends regular exercise. Exercise burns kcalories and, if a balanced program is chosen, improves muscular strength, endurance, and flexibility. In addition, aerobic exercises such as jogging, swimming, cycling, and dancing provide the added benefits of improving cardiovascular and respiratory endurance and resistance to disease. Regular exercise provides psychological benefits as well. These include a positive self-image, a sense of well-being, and a positive attitude in general—all important to a person whose body is about to be temporarily transformed by a pregnancy.

Exercise also improves a woman's nutrition status. Bones do not passively accumulate calcium from foods; physical activity helps bones to store calcium and become dense, strong, and able to carry more weight. Physical activity also develops the lean body tissue that serves as a storage site for iron and other nutrients. In pregnancy, a healthy circulatory system developed in response to exercise will be efficient at transporting nutrients through the maternal body and to the fetus.

A healthy woman conditioned to physical activity prior to pregnancy can normally continue exercising throughout her pregnancy. However, conception is not a good time to begin strenuous workouts. A woman who wants to be physically active when she becomes pregnant needs to become physically active *beforehand*. She must establish a baseline of exercise intensity, duration, and frequency that will represent the maximum for the duration of any future pregnancy. Chapter 8 presents a section on "Fitness for Teens and Adults," because it is in the teen years that the habit of regular exercise, voluntarily undertaken, should become firmly established.

Nutrition and Women's Cycles

The physiological changes and hormonal shifts women experience create special nutrient needs. The following section focuses on the nonpregnant, reproductive time in a woman's life—a time of menstrual cycles.

The Menstrual Cycle

The average woman experiences 500 menstrual cycles in her lifetime, losing more than 17 liters of blood and 6500 milligrams of iron.[16] No doubt, the physical losses incurred by menstrual cycles impose on such a woman tremendous needs, not only of iron, but of all nutrients. Furthermore, the constantly changing hormonal and metabolic activities of the cycle make women's food energy needs change in concert. An understanding of the interrelationships between the hormones and organs involved in a typical menstrual cycle will lay the foundation for an understanding of the nutrition needs of women during this phase of life.

During puberty, a woman's monthly cycles begin—an event known as menarche. Hormones synchronize and coordinate the events of the monthly menstrual cycle. These hormones affect one another's activities and elicit responses from the sex organs, and from the body's other organs and tissues. Figure 2–5 illustrates how the hormones' fluctuating concentrations account for the events of the month.

menarche (men-ARK): the onset of menstruation; the first menstrual period, usually occurring between the ages of 11 and 14.
men = month
arche = beginning

Figure 2–5 The Menstrual Cycle
Changes in hormone concentrations elicit ovarian follicle development and changes in the uterine lining during the menstrual cycle.

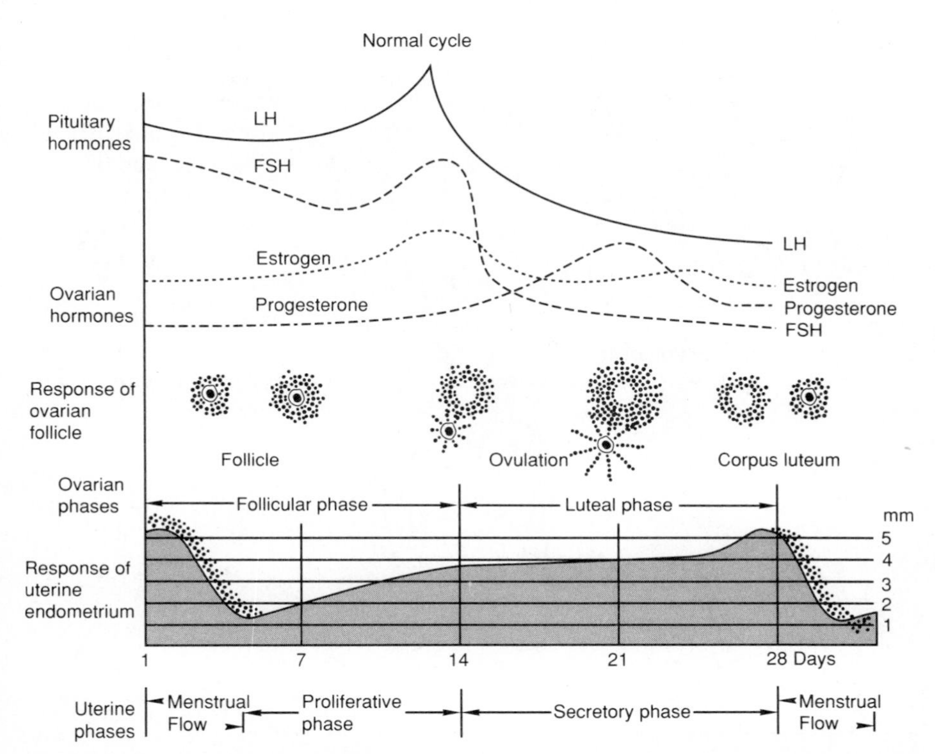

The most dramatic physical changes occur in the ovaries and the uterus (see Figure 2–6). Female infants are born with about 1 million ova in each ovary. Of these ova, relatively few will reach maturity and participate in a menstrual cycle. Fewer still will undergo fertilization and develop into new individuals. Most ova will degenerate over the years, leaving none at the time of menopause.

In the ovary, an ovum, surrounded by a layer of follicular cells, enlarges and develops into a mature follicle. The hormone FSH (follicle-stimulating hormone), which dominates the beginning of the cycle, encourages the growth of the follicle. When the follicle reaches maturity, a sharp rise in the hormone LH (luteinizing hormone) triggers its rupture, thus releasing the ovum. This event, ovulation, occurs approximately 14 days prior to the next menstrual flow. The process of follicle development takes approximately two weeks and is called the follicular phase of the menstrual cycle (see Miniglossary of Menstrual Cycle Phases).

In response to the high levels of LH, the ruptured follicle transforms into a structure known as the corpus luteum. Without ovum fertilization, the corpus luteum gradually develops to maturity and then rapidly degenerates. (If pregnancy does occur, the corpus luteum remains active.) The process of corpus luteum development and degeneration takes roughly 10 to 14 days and is called the luteal phase of the menstrual cycle.

ovaries: the two glands that produce ova and female hormones.

uterus: the muscular organ within which the embryo and fetus develop from the time of implantation to birth.

menopause: the time in a woman's life when menstrual activity ceases, usually between the ages of 35 and 55.
pause = cessation

follicle: a small, saclike structure in the ovary, consisting of an ovum surrounded by epithelial cells that secrete the hormone estrogen.

FSH, or **follicle-stimulating hormone:** a hormone from the pituitary gland that stimulates the growth of follicles.

LH, or **luteinizing** (LOO-tin-eye-zing) **hormone:** a hormone from the pituitary gland that stimulates development of the ruptured follicle into the corpus luteum and signals ovulation.

ovulation: the ripening and rupturing of the mature follicle and subsequent release of the ovum.

corpus luteum (CORE-pus LOO-tee-um): a mass of glandular tissue that develops from a ruptured follicle and secretes hormones.
corpus = body
luteum = yellow

Figure 2–6 The Ovaries, Oviducts, and Uterus
During ovulation, an ovary releases an ovum into an oviduct that connects with the uterus.

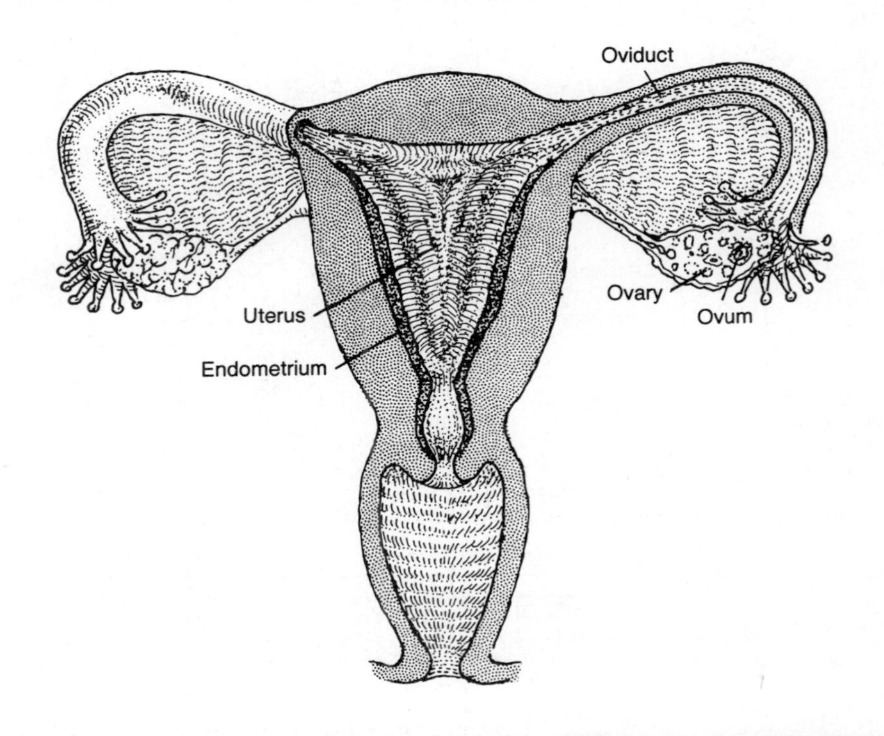

Miniglossary of Menstrual Cycle Phases

Several names apply to the phases of the menstrual cycle. One set of names reflects the events occurring within the ovaries; another, those occurring in the uterus. Figure 2–5 reveals that the ovary's follicular, or preovulatory, phase coincides with menstrual flow and with the uterus's proliferative phase. Similarly (see Figure 2–5), the ovary's luteal, or postovulatory, phase coincides with the uterus's secretory phase. Events within the ovaries:

follicular (preovulatory) phase: ovarian events of follicle development prior to ovulation; coincides with uterine menstrual flow and the uterine proliferative phase.

luteal (postovulatory) phase: ovarian events of corpus luteum development from ovulation to menstrual flow; coincides with uterine secretory phase.

Events within the uterus:

menstrual flow: the periodic discharge of blood, disintegrated endometrial cells, and gland secretions from the uterus.

proliferative phase: uterine events of endometrium development prior to ovulation; coincides with ovarian follicular phase.

secretory phase: uterine events of endometrium disintegration that lead to menstrual flow; coincides with ovarian luteal phase.

Both the follicle and the corpus luteum secrete estrogen. The times of maximum activity of these two structures are evident in the two rises in estrogen concentration during the cycle (review Figure 2–5). During the last half of the follicular phase, the rapidly rising levels of estrogen signal the pituitary to reduce FSH production (negative feedback) and to increase LH production (positive feedback). During the luteal phase, the rise in estrogen does not stimulate the surge of LH, as it did in the follicular phase. This is because the corpus luteum is secreting progesterone, as well as estrogen. The progesterone-estrogen combination inhibits FSH and LH release, thus preventing follicle development. With corpus luteum degeneration, progesterone and estrogen levels diminish, FSH and LH levels begin to rise, and the cycle repeats itself.

estrogen (ESS-tro-jen): one of the female sex hormones produced in the ovaries, responsible for sexual development and for regulation of the menstrual cycle.

progesterone (pro-JESS-teh-rone): a female hormone produced in the ovaries, responsible for changes in the uterine endometrium in the luteal phase of the menstrual cycle.

In the uterus, profound changes occur in response to progesterone and estrogen. The rising levels of estrogen during the first half of the menstrual cycle stimulate the growth of the uterine lining, the endometrium. The endometrium thickens and develops an extensive vascular system in preparation for the pregnancy, should fertilization occur. The proliferative phase of endometrium development lasts about ten days, ceasing with ovulation. Without fertilization, the endometrium begins to disintegrate—the secretory phase. Finally, the endometrium lining begins to be shed. Menstrual "blood" is the endometrium lining, which leaves the body within four or five days. At the end of menstruation, the endometrium has returned to its minimal state.

endometrium (en-doe-MEE-tree-um): the membrane lining the inner surface of the uterus.

The average menstrual cycle lasts 28 days, give or take a few days. The days are usually counted from the first day of bleeding, because it is easy to detect.

Many physiological activities support the menstrual cycle's many events and have nutrition implications. Basal body temperature and basal metabolic rate fluctuate with the changing events of the menstrual cycle. Basal body temperature drops during the follicular phase and rises during the luteal phase.[17] The temperature begins its rise with the surge in LH. It continues to climb after the LH peak, coinciding with the increase of serum progesterone concentrations, which raise body temperature. Secretion of progesterone during the menstrual cycle also coincides with changes in basal metabolic rate.[18]

These changes in basal body temperature and basal metabolic rate may affect food intake and body weight changes. Women tend to eat more food per day for the half-month after ovulation (prior to menstruation) than during the half-month before ovulation (after menstruation). One study noted that mean energy intake prior to menstruation was 500 kcalories per day greater than after menstruation; another reported a smaller, but still significant, difference of 200 to 300 kcalories per day.[19] A follow-up study revealed that the source of the additional kcalories was carbohydrate-rich foods.[20] Another study found that basal metabolic rates followed the same pattern as the energy intakes reported in the first study, but that the kcaloric difference was only 350 kcalories.[21] The 150-kcalorie difference between the studies may arise from differences in their designs. The first study examined the food intake patterns of women in normal life situations, whereas the second study examined energy expenditures in a metabolic unit.

These studies suggest that eating more food prior to menstruation may be a woman's most appropriate response to the rise in basal metabolic rate that occurs at that time. If any harm results from eating extra food, the harm comes from failing to reduce intakes to go with the postmenstrual decline in metabolic rate. A woman can best respond to her body's signals by carefully selecting a well-balanced diet that provides extra food energy prior to menstruation and returns to a lower food energy intake following menstruation.

Premenstrual Syndrome

premenstrual syndrome (PMS): a cluster of physical, emotional, and psychological symptoms that occur prior to menstruation and diminish during or after menstruation.

During the 1980s, interest in women's health issues heightened, and the problem of premenstrual syndrome (PMS) became a popular news item—even though its existence had preceded its name by centuries. As is often the case when health problems make a news splash, companies and clinics were quick to offer "solutions," including nutrient supplements, before scientific research could provide answers to questions about the possible causes and treatments of PMS.

PMS is a cluster of physical, emotional, and psychological symptoms that some women experience prior to their menstrual periods. A woman suffering from PMS may complain of any or all of the following symptoms: cramps or aches in the abdomen; back pain; headaches; acne; swelling of the face and limbs associated with water retention; weight gain; food cravings, especially for sweets; constipation; breast swelling and tenderness; diarrhea; fatigue; and mood changes, including nervousness, anxiety, irritability, and depression. Researchers are attempting to define clusters of these symptoms, in hopes of assigning different causes to them.

Specific PMS symptoms and combinations of symptoms vary from woman to woman. The distinguishing feature of PMS is the timing of the symptoms with a woman's menstrual cycle. Most women begin to experience symptoms seven to ten days prior to menstruation, during the luteal phase of their cycles. Symptoms disappear, or at least diminish significantly, with menstruation.

Exactly how many women have PMS is unknown, primarily because the diagnosis depends solely on a woman's description of her symptoms. By keeping a symptom diary, a woman can see how her physical and emotional well-being fluctuates in relation to her menstrual cycle. Physicians may conduct diagnostic tests to eliminate the possibility that a woman's symptoms derive from another disorder, but not to confirm the diagnosis of PMS. For some women, an underlying physical or emotional disorder (such as migraine headaches or depression) surfaces premenstrually and blurs the distinction between PMS and other problems. Some women recognize a few symptoms prior to their menstrual periods as a signal, not a problem. Others become incapacitated. The intensity of the symptoms and the woman's perception of them differ for each woman.

The cause or causes of PMS remain undefined, although researchers generally agree that the hormonal changes of the menstrual cycle must be responsible.[22] Following the trail of these hormones requires masterful detective work. The hormones involved in the menstrual cycle influence the activities of a number of other hormones as well as neurotransmitters, with a multitude of effects on physical, emotional, and psychological health. Even successfully tracing the path of changes does not solve the mystery, for not all women experience PMS.

Without a full understanding of the ways PMS arises, treatment efforts flounder. In some instances, physicians prescribe remedies for specific symptoms, such as tranquilizers for anxiety. Others attempt to treat the syndrome as a whole. One obstacle researchers encounter in their efforts to find a treatment is that many women with PMS respond favorably to placebo treatments.

placebo: an inert drug or treatment used for its psychological effect.

Unproven treatments for PMS include hormones (most often, progesterone), drugs that alter hormone actions, and various nutrient supplements. In addition, over-the-counter medicines provide temporary relief of some symptoms: diuretics for fluid retention; caffeine for fatigue; and aspirin, acetaminophen, and ibuprofen for headaches. A major risk of assigning a treatment to PMS without a proper diagnosis is that physical pain associated with the menstrual period can have a wide variety of causes, some of which demand medical attention. Inflammation or infection of the lining of the uterus can cause symptoms like those ascribed to PMS, but is a life-threatening condition that demands diagnosis and treatment.

No evidence supports nutrient deficiencies as a cause of PMS, and in fact, one study using biochemical tests of nutrition status found no significant difference between women with and without PMS.[23] In no case is there evidence to support the use of supplements to alleviate PMS symptoms. Women wanting to take vitamin and mineral supplements for PMS need to be forewarned that benefits are unlikely and that risks of toxicity are possible.

Some women find that improvements in their general health benefit their PMS—for example, if they lose weight, eat a nutritious diet, exercise regularly, abstain from using tobacco and alcohol, and limit caffeine intakes. In addition, relaxation techniques may help some women to cope with the stresses of PMS. No question that these changes are advantageous to a woman's health; perhaps PMS, whatever its cause, becomes more tolerable when the body is healthier and the woman is psychologically more in control of her life.

Nutrition and Contraception

Each menstrual cycle presents a woman and her partner with the responsibility for making a choice of whether to start a pregnancy. An examination of the nutrition implications of some contraceptive methods may help a couple in their decision. The two selected here are those that have a known nutrition impact—oral contraceptives and intrauterine devices.

Oral Contraceptives

Millions of women use oral contraceptives, popularly known as the Pill, to prevent pregnancy. In its 30-year history, the Pill has become the most studied drug in the United States. The identification of a number of risk factors has prompted changes in the dosages and formulations of the Pill to produce an effective contraceptive with a wide margin of safety. As of the 1980s, pills contain one-fifth the estrogen and one-tenth the progesterone originally in the pills of the 1960s, making them as risk-free as possible, yet still effective.[24] The hormone concentrations are low, and may vary throughout the month to roughly simulate the changes that normally occur during a menstrual cycle.

Oral contraceptive use by lactating women is discussed in Chapter 5.

Such pills are known as multiphasic pills and may avert some minor side effects, such as breakthrough bleeding or, in some women, blood lipid changes.[25] Some manufacturers claim that multiphasic pills are more natural or physiologically superior to fixed-dose pills, but this claim is fallacious.[26] After all, the purpose of oral contraceptives is to alter the natural events of the menstrual cycle.

There are two types of oral contraceptives—the combination pill and the minipill. Of the women using oral contraceptives in the United States, 99 percent take the combination pill, which contains synthetic versions of the hormones estrogen and progesterone.[27] Recall from the discussion on the menstrual cycle that the ovaries naturally produce estrogen after ovulation, to suppress ovulation until after the next menstruation. The estrogen in the combination pills prevents pregnancy the same way, by suppressing ovulation.

The other type of oral contraceptive, the minipill, contains only synthetic progesterone, known as progestin. The progestin in the minipill makes the mucus surrounding the uterine opening (the cervix) less penetrable than normal to spermatozoa and may deactivate them. It also interrupts the normal preparation of the uterine endometrium and so prevents zygote implantation.

Oral contraceptives not only prevent pregnancy but also benefit a woman's reproductive system physiologically. (Table 2–1 lists the positive, as well as the negative, side effects of oral contraceptives.) Apparently, the suppression of ovulation reduces the likelihood of ovarian cancer, ovarian cysts, painful menstrual periods, and premenstrual syndrome.[28] The reduced risk of ovarian cancer is seen in women who use the Pill for as little as three months, and continues for 15 years after use ends.[29] The progesterone in the Pill offers

Table 2–1 Side Effects of Oral Contraceptives

Reduced Risk Of:	Increased Risk Of:
Benign breast disease[a]	Coronary artery disease
Endometrial cancer	Gallbladder disease
Ovarian cancer	Glucose intolerance
Ovarian cysts	High blood pressure
Painful menstrual periods	Raised cholesterol and triglycerides
Premenstrual syndrome	Reduced high-density lipoproteins (HDL)
Pregnancy	

[a]The effect of oral contraceptives on breast cancer is inconclusive; it appears that oral contraceptive use does not increase the risk for breast cancer.

Source: Adapted from B. D. Shephard, Oral contraceptives—An overview, *Journal of the Flordia Medical Association* 73 (1986): 763–767; E. L. Marut, Oral contraceptives—Who, which, when, and why? *Postgraduate Medicine* 82 (1987): 66–70; G. R. Huggins and P. K. Zucker, Oral contraceptives and neoplasia: 1987 update, *Fertility and Sterlity* 47 (1987): 733–761; D. R. Miller and coauthors, Breast cancer risk in relation to early oral contraceptive use, *Obstetrics and Gynecology 68* (1986): 863–868; R. Russell-Briefel and coauthors, Cardiovascular risk status and oral contraceptive use: United States, 1976–1980, *Preventive Medicine* 15 (1986): 352–362; P. B. Moser and coauthors, Carbohydrate tolerance and serum lipid responses to type of dietary carbohydrate and oral contraceptive use in young women, *Journal of the American College of Nutrition 5* (1986): 45–53; R. Russell-Briefel and coauthors, Impaired glucose tolerance in women using oral contraceptives: United States, 1976–1980, *Journal of Chronic Diseases* 40 (1987): 3–11; J. K. Williams, Oral contraceptives—The long-term perspective, *Journal of the Florida Medical Association* 73 (1986): 769–771; B. L. Strom, Oral contraceptives and other risk factors for gallbladder disease, *Clinical Pharmacology and Therapeutics 39* (1986): 335–341.

protection against endometrial cancer and benign breast disease.[30] The protective effect against endometrial cancer is seen in women who use the Pill for at least 12 months, and persists for 15 years after cessation of use.[31]

Research results on the relationship, if any, between oral contraceptives and breast cancer are inconclusive, primarily because breast cancer may correlate with an event, such as menarche, that occurred three to four decades earlier, and the Pill has been available for less than three decades. For now, it appears that even long-term oral contraceptive use in a young, prepregnant woman does not present a risk of breast cancer.[32]

The Pill presents negative side effects as well, most in response to its progesterone content. Progesterone is thought to be the culprit in altering blood lipids and raising the risk of coronary artery disease for Pill users. Most oral contraceptives elevate total cholesterol and triglyceride concentrations and lower high-density lipoprotein (HDL) concentrations, tipping the balance toward cardiovascular disease.[33] The risk is greatest for women over age 35 who smoke and for all women over age 45.[34] These women should find alternative contraceptive methods. For young, healthy women who do not smoke, the association between oral contraceptives and coronary artery disease disappears.[35] The progesterone in the Pill is also responsible for high blood pressure in some users, especially women who are older, multiparous, and obese. This high blood pressure reverts to normal when users stop taking the Pill.[36] Dietary measures to reduce the risk of cardiovascular disease, both that caused by high blood lipids and that caused by high blood pressure, are discussed in Chapter 9.

The progesterone in oral contraceptives alters carbohydrate metabolism.[37] Glucose tolerance diminishes, and blood glucose and insulin rise, especially in women predisposed to diabetes.[38] For most women, the changes in carbohydrate metabolism are minimal and return to normal with cessation of Pill use.[39] Because these metabolic effects are transient and reversible, they present a small risk of diabetes compared with other nonreversible factors, such as genetics.[40] Dietary measures to prevent diabetes are discussed in Chapter 9.

One risk, that of gallbladder disease, has a dose-response relationship with the estrogen content of oral contraceptives. This risk appears to be greater for young women than for women over age 40.[41] The risk of gallstone formation in oral contraceptive users stems from the progesterone-caused changes in cholesterol metabolism mentioned earlier.[42]

Oral contraceptives reduce menstrual blood flow and thereby conserve for the body all nutrients normally lost in menstrual blood—most notably iron. These contraceptives also alter the metabolism of many vitamins and minerals. Their specific effects vary with each nutrient and with the hormone concentrations of the pills. For some nutrients, oral contraceptives affect the absorption or excretion rate. For others, the rate of metabolic conversion changes the nutrient status. Whatever the mechanism, the effects on vitamins and minerals are reflected in higher or lower blood concentrations. Table 2–2 lists the effects of oral contraceptives on various nutrients.

In some cases, the effects of oral contraceptives on vitamin and mineral status may appear more positive than they are. For example, oral contraceptives raise plasma retinol concentrations but lower liver retinol concentrations. This suggests a possible redistribution of retinol in the body, and not necessarily the improved status that high plasma values might seem to imply.

Table 2–2 Effects of Oral Contraceptives on Nutrients

Nutrient	Effect on Blood Concentrations and Metabolism
Energy-yielding nutrients	
Carbohydrate	Elevated fasting glucose Elevated insulin
Lipid	Elevated triglycerides Elevated low-density lipoprotein (LDL) cholesterol Lowered high-density lipoprotein (HDL) cholesterol
Protein	Elevated coagulating proteins Lowered albumin
Vitamins	
Folate	Lowered (in some studies)
Riboflavin	Lowered (in some studies) Impaired enzyme activity (in some studies)
Vitamin A	Elevated retinol (lowered liver stores) Elevated retinol-binding proteins Lowered carotene
Vitamin B_6	Lowered
Vitamin B_{12}	Lowered in serum; normal in red blood cells
Vitamin C	Lowered in leukocytes, thrombocytes, and platelets
Vitamin E	No effect
Vitamin K	Elevated clotting factors (and lowered response to anticoagulants)
Minerals	
Copper	Elevated
Iron	Elevated
Zinc	Lowered
Water	Retained temporarily

Source: D. Dimperio, Effect of oral contraceptives on nutrient status. Address presented at the Conference on Nutrition for Pregnancy, Lactation, and Infancy on 13 February 1987, in Gainesville, Florida; L. B. Tyrer, Nutrition and the Pill, *Journal of Reproductive Medicine* 29 (1984): 547–550; and others.

In other cases, concentrations of vitamins or minerals in the bodies of oral contraceptive users are low, reflecting an increased demand in metabolic pathways, impaired absorption, increased excretion, or altered tissue distribution. However, the clinical significance of these reductions in nutrient concentrations is minor in the overall picture of factors affecting women's nutrition status.

The use of supplements to correct the nutrient fluctuations incurred by oral contraceptive use is inappropriate. If nutrient deficiencies and excesses are evident in women taking oral contraceptives, the remedy is to improve the diet.[43] Experimental evidence supports this contention. Researchers compared the adverse effects of oral contraceptives on vitamin metabolism in three groups of women.[44] One group took two multivitamin supplements a day for one week each month, the second group took one multivitamin supplement every day, and the third group took a placebo each day. The data demonstrated that routine use of multivitamin supplementation by women taking oral contraceptives is unjustified.

Water retention occurs in many women around the time of their menstrual periods and may temporarily add a few pounds to their body weights. Oral contraceptives seem to produce the same fluid-retention effect. A health care provider can evaluate this problem and recommend diuretic use if warranted, or prescribe a different brand of pill.

The gain of water weight associated with use of the Pill has led some women to believe that fat gain is a side effect of oral contraceptive use. Fear of fat gain may deter women from using oral contraceptives. One study found no difference in the weight gains of oral contraceptive users as compared with those of users of other contraceptive methods.[45] If a woman gains fat during the initial months of oral contraceptive use, she should check her diet and exercise habits, and adjust them, if necessary.

The Intrauterine Device (IUD)

The intrauterine device (IUD, or coil) is a small piece of molded plastic or plastic and metal that is inserted into the uterus. One type of IUD contains progestin. The IUD's effectiveness is not completely understood. Apparently, it induces a change in the uterine lining that interferes with the implantation of a zygote.

IUDs can cause heavy menstrual bleeding. This can be inconvenient and distressing, but rarely signifies a serious disorder. It does entail a nutrition risk, however: the development of iron-deficiency anemia due to loss of blood. Therefore, the iron status of a woman using an IUD requires regular assessment.

Nutrition, Conception, and Implantation

In nourishing themselves well early in life, a young man and woman are not only engaging in self-care but are also influencing the next generation. To understand the importance of nutrition to normal conception and fetal development, it is necessary to understand the basics of three processes: inheritance, fertilization, and implantation.

Inheritance

An ovum, the gamete produced by a female, carries within it a set of 23 chromosomes containing all of the genetic information necessary to make a human being. Each chromosome bears along its length thousands of genes, and each gene consists of coded instructions for making a single working protein—an enzyme, a structural protein, a muscle protein, or some other body protein.

gamete: a mature male or female reproductive cell; the spermatozoon or ovum. *gamein* = to marry

chromosomes: the bodies within each cell that contain the genetic material (deoxyribonucleic acid, or DNA).

genes: the basic units of hereditary information, made of DNA, that are passed from parent to offspring in the chromosomes of the gametes. Each gene codes for a protein.

somatic cells: the nonreproductive cells or tissues of the body.
soma = body

germ cells: an informal term for the cells in the ovary and testis that produce gametes.
germinate = to begin life

gonads: the primary sex organs containing the germ cells; testes in the male and ovaries in the female.
gone = seed

Each protein will determine, or help to determine, traits for the new person, such as eye color, hair color and texture, maximum height, susceptibility to various diseases, potential for muscular development, and some aspects of temperament. The exact combination of these and many other characteristics makes each individual unique. These inherited characteristics define the range within which nutrition and other environmental influences can affect a person's development.

A spermatozoon, the gamete produced by a male, also contains 23 chromosomes bearing sets of instructions for the same characteristics. When the ovum and spermatozoon merge at fertilization to form a single cell, the chromosomes from each parent line up side by side within that cell. Now there are not 23 chromosomes, but 23 *pairs*—46 chromosomes in all (see Figure 2–7).

After this union, the zygote splits into two new cells, and those cells split again. All 46 chromosomes are faithfully copied at each division so that every new cell inherits the entire set of 46 chromosomes. (To be strictly accurate, each of the somatic cells—that is, every cell in the body except the germ cells— contains the full complement of 46 chromosomes.) When the individual has become a sexually mature adult, then divisions of the germ cells in the gonads produce cells with just 23 chromosomes once again. These are, of course, the gametes, the special cells through which the individual will contribute one member of each of its pairs of chromosomes to the *next* generation. Thus a parent passes along some chromosomes from its father, and some from its mother, to its child.

An offspring always, therefore, inherits two sets of genes, or instructions for making each piece of molecular machinery—one from each parent. In many cases, the two genes code for identical products. In some cases, the two genes code for two products, both are produced, and the resulting trait is a

Figure 2–7 Gametes and Chromosomes
Human body cells contain 23 pairs of chromosomes, and gametes contain one member of each pair, or 23 single chromosomes. In these cells, only two pairs are shown. These pairs separate when the gametes are formed, and each gamete receives only one member of each pair (two chromosomes, in this diagram).

When the gametes join in fertilization to form a zygote, new pairs of chromosomes are formed. Each contains one member from the male and one from the female. Thus the zygote has a full set of 46 chromosomes, like the parent cells (four are shown in this diagram).

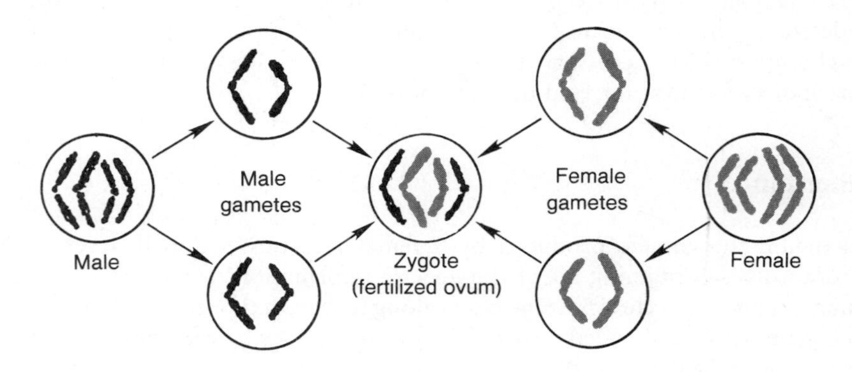

blend. In a few cases, one of the two genes may produce no product or a defective product that will not function—but the other will produce a product that will function. (The inheritance of two sets of instructions gives the new individual two chances to "get it right," in a sense.)

Occasionally, two defective genes come together in an individual, and the result is the malfunction or total absence of some gene product. One such case may present no problem if the characteristic governed by that gene product is not essential to life or health. Another such case may cause a major biochemical defect in an organism, known as an inborn error of metabolism (see Focal Point 2). Many such cases are lethal, leading to spontaneous abortions and stillbirths.

inborn error of metabolism: an inherited flaw evident as a disorder or disease present from birth.

The foregoing makes clear why some of the new individual's traits may be identical to the father's, some to the mother's, and some intermediate between the two. Some brand-new traits will also emerge: for example, traits determined by genes on chromosomes donated by the grandparents—genes that were hidden but not expressed within one or the other parent. New traits may also arise from unique combinations of gene products occurring for the first time in this new individual.

The simplest genetic arrangement to understand is the type governed by a single gene pair—the arrangement for eye color, for example. A gene on one of the chromosomes from the mother and a gene on the same location on the corresponding chromosome from the father govern this characteristic. The two genes may or may not be identical; the combination determines what the actual eye color will be. If the father's gene specifies blue-eye pigment, but the mother's codes for brown pigment, then the eyes will be brown, whether blue-eye pigment is present or not; this is because the dominant gene's product, brown pigment, literally outshadows the recessive gene's product. Thus the person who inherits one copy of each gene will have brown eyes.

dominant gene: a gene that produces a product whose effect is expressed in the appearance or functioning of the organism.

Another one of the mother's ova may not carry the message for brown eyes but may contain the recessive message for blue eyes, just as the father's spermatozoon does. (Remember, the mother also inherited two genes for eye color; she can pass along either one.) Now the products of the normally recessive genes are visible, and the person is blue eyed. Two brown-eyed people can thus have a blue-eyed child. When this happens, it is certain that while both parents possessed a dominant, brown-eye gene, both also must have carried a blue-eye gene and passed that one along to their offspring. This is one way in which "new" traits emerge in new generations.

recessive gene: a gene that produces no product, or a product whose effect is not expressed when the product from a dominant gene is present.

The way an infant's gender is determined is a variation on this theme. The father's gametes determine the sex of the infant. One of the 23 pairs of chromosomes, known as the X-Y, or sex, chromosomes, carries the sex-determining genes. People can either inherit two Xs, in which case they are female, or an X and a Y, in which case they are male. No one ever gets two Y chromosomes, because mothers, being female, have two Xs and can only donate Xs to their offspring. Males donate X chromosomes to half of their offspring, on the average, and Y chromosomes to half—thus girls and boys are born in equal numbers. In the race to fertilize the ovum, the spermatozoon that swims best or that gets through the ovum's outer covering first is the winner—and is the one that determines the infant's gender (see Figure 2–8).

Figure 2–8 Sex Inheritance
One of the 23 pairs of chromosomes, known as the X-Y, or sex, chromosomes, carries the sex-determining genes. Males contain an X and a Y; females contain two Xs. Males donate X chromosomes to half of their offspring, on the average, and Y chromosomes to half—thus boys and girls are born in equal numbers.

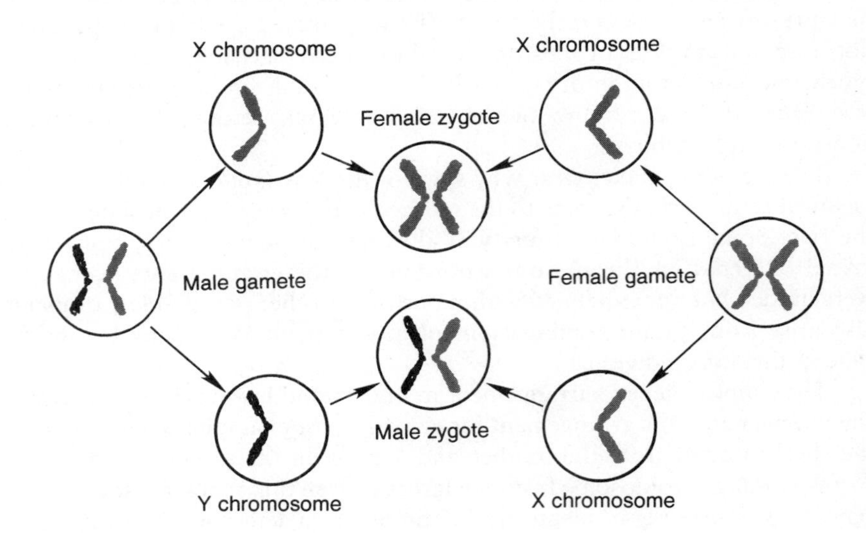

Nutrition and Inheritance

With this basic understanding, it should be possible to see how nutrition relates to conception and heredity. Genes are normally copied with amazing accuracy. This is important, for they contain instructions that are crucial to the correct making of an organism. A set of parental genes is copied each time a gamete is made. The genes are copied from the parent's germ cell into the set of 23 chromosomes to be packaged in the gamete. Then the male and female gametes unite to form a zygote with 46 chromosomes. After that, all 46 chromosomes are copied over and over again as the zygote divides and redivides to form the billions of cells of the adult organism.

mutation: a change in a cell's genetic material (DNA).

As a cell duplicates its genetic material in preparation for dividing to form two cells, every now and then it makes a mistake—a mutation. Such mistakes arise in the average gene about once in every 100 million copies. Most are immediately corrected by way of an intracellular system of comparing copies with the original, destroying defective copies, and recopying. However, a few mistakes slip by, and mutations may be transmitted to new cells. Once transmitted, (inherited), a mutation will be as faithfully copied as the original was, and will be transmitted to all future offspring of that cell. The only ways mutations are ever lost from a cell line is by changing back (an event even rarer than the mutations' genesis) or by the cells' dying out without reproducing.

mutagens: agents or events that cause genetic mutations.
mutare = to change
genesis = to produce

Mutagens, including both chemical and physical agents, cause mutations to arise in the copying of the genetic material. Among chemical mutagens are the tars in tobacco; toxins produced by bacteria; many pesticides; heavy metals; and many drugs, including both illegal and legal (prescription and over-the-counter) varieties. Among physical mutagens are several forms of radiation, including the sun's ultraviolet rays and radioactivity. (Irradiated food, however, does not contain mutagens; it has been exposed to radiation, but it does not contain any substances that, themselves, give off radiation.) Mutagens can cause not only mutations leading to inborn errors, but others leading to birth defects, cancer, possibly atherosclerosis and diabetes, and probably other diseases as well. The special class of mutagens that cause birth defects are known as teratogens.

birth defects: congenital abnormalities (present in an individual from birth), often caused by somatic mutations.

teratogens (teh-RAT-oh-gens): agents that cause somatic mutations that lead to abnormal fetal development and birth defects.
terat = monster

The earlier in development a harmful influence exerts itself, the more devastating the consequences are likely to be. Mutations in the germ cells are especially harmful, for the gametes are copied from these cells. Such mutations will be passed on to every cell of the offspring's body *and from generation to generation* through the germ cell line. They are usually recessive; typically, the gene altered by mutation produces an inactive product or no product. An individual who inherits such a defective gene will exhibit a normal trait only if the defective gene is paired with a normal gene contributed by the other parent. (Such an individual is said to be a carrier of the mutation.) However, when two carriers conceive a child, and each contributes the defective member of its pair of genes for that trait to the child, the child is unable to produce a normal product (such as an enzyme), and so has an inborn error of metabolism (see Focal Point 2). Mutations in the germ cell line may persist invisibly for many generations before their effects become apparent, but should enough of them accumulate within the human genetic material, they could condemn the human race to ever-increasing disabilities. It is for this reason, more than any other, that both women and men should, throughout their reproductive lives, avoid contact with any chemical or physical agent that causes mutations. (It is for this reason, too, that accidents involving mutagenic radiation, such as the one that took place at Chernobyl in April 1986, are so frightening.)

Teratogens cause mutations, not in the germ cell line, but in somatic cells during an individual's development. These are also harmful, but *they will affect only that individual* and will not be passed on to future generations. They can be devastating, however. If mutations occur during the first two weeks of a zygote's development, they are likely to be lethal. The reason is that each cell in a developing zygote is destined to become a large part of the body of an adult. Major abnormalities induced in whole organ systems preclude survival. Lesser abnormalities lead to birth defects. A section at the end of this chapter describes how teratogenic effects of alcohol and drugs arise. Tables 3–1 and 3–2 in Chapter 3 give examples of these effects.

Nearly all mutations are harmful. Exceptions are known, but 999 out of 1000 mutations are disadvantageous, and therefore all *mutagens* can be classed as unambiguously harmful.

Although inborn errors and many birth defects are known to be caused by mutations, it is never possible to look back in time and say when those mutations arose. They are, however, most likely to arise during times of cell division. That means that nearly all periods of early life are especially vulnerable, for nearly all involve cell division in the germ cell line. Boys and men make new spermatozoa continuously from puberty on. In young women, as soon as they are pregnant, zygotic and embryonic cell division proceed rapidly. Female fetuses develop all the ova they will release for a lifetime. Young men and women should therefore at all times avoid exposure to mutagenic radiation and contamination of food and water that might threaten the integrity of their genetic material and young offspring. To the earlier statements about obtaining a diet that is adequate, balanced, varied, and so forth should be added another descriptor: all of the foods and beverages selected should be free of contamination—in a word, safe.

Nutrition and Fertilization

Once the gametes are produced, they need to be able to meet, join, begin the production of a new individual, and settle into the uterus to develop. The meeting and joining are, of course, fertilization, or conception. Sexual intercourse from 48 hours before to 15 hours after ovulation is most likely to result in conception. The average life spans of the spermatozoon (48 hours) and the unfertilized ovum (10 to 15 hours) determine this time frame. During ovulation, the ovary discharges an ovum into the oviduct. The ovum works its way down the oviduct into the uterus over the next several days. (Figure 2- 5, presented earlier, illustrates the relationship between the ovaries, oviducts, and uterus.) The short life span of the ovum requires fertilization to occur in the oviduct, as illustrated in Figure 2–9. Poor nutrition can impair fertility and even prevent conception. Evidence supporting this assertion comes from

oviduct: one of two tubes that serve to convey the ovum from the ovary to the uterus; also called the fallopian tube.

Figure 2–9 Fertilization and the First Two Cell Divisions
The union of an ovum and spermatozoon occurs in the oviduct and creates a zygote. The original cell begins its division into many cells, changing its number but not its size.

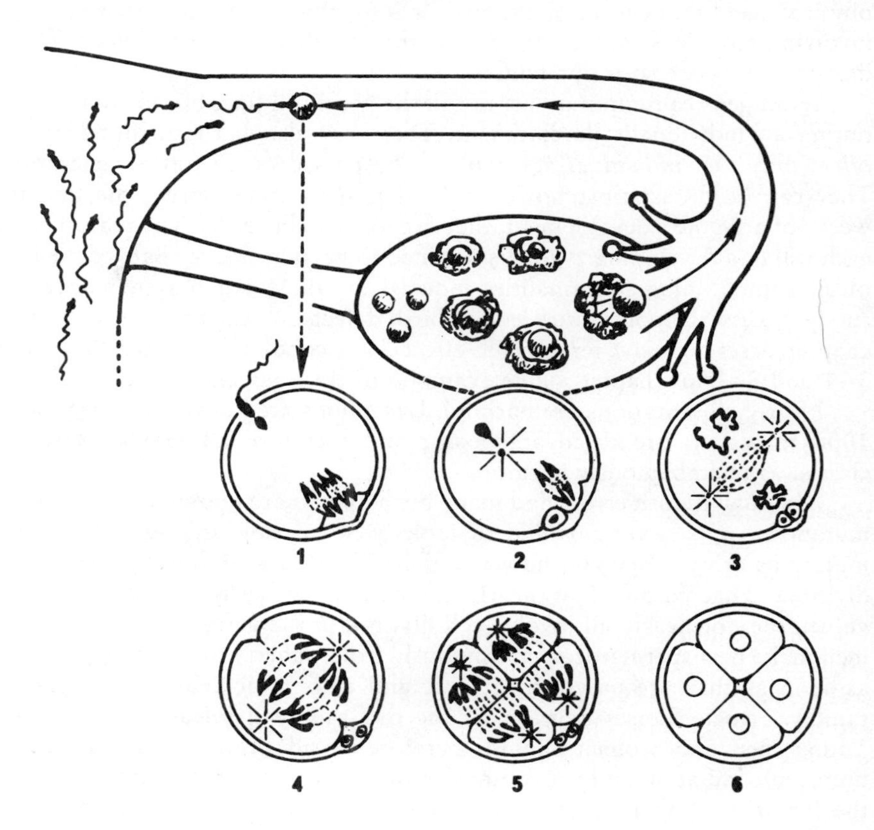

studies of famine-induced malnutrition, self-chosen underweight (from dieting), and overdoses or deficiencies of individual nutrients.

Famine-induced malnutrition and fertility Whether fertilization can take place depends both on the health of the gametes and on their environment. Malnutrition of a woman can prevent conception. The food shortages and birthrates reported in the medical records and journals of Europe before, during, and after the Second World War provide valuable records for analyzing the influence of nutrition on the fertility of women. During that war, a transportation strike restricted food supplies into Holland from September 1944 to May 1945. The resultant severe nutritional deprivation reduced fertility at a time in many people's lives when they might otherwise have conceived, as is evident from several findings.[46] The number of births declined dramatically nine months after the onset of the war-related famine to only one-third of the expected rate.[47] With the increased availability of food, birthrates began to climb, indicating that starvation-caused infertility is reversible. Meanwhile, fertility did not decline in areas with adequate food supplies, even under similar conditions of war and weather.

One precondition of fertility is a woman's ability to ovulate. During the World War II food shortages and malnutrition in Holland, approximately one-half of the female population stopped menstruating; only one-third of the women had normal menstrual cycles.[48]

Another precondition of fertility is a man's ability to produce enough viable spermatozoa. Food deprivation in men has several effects on spermatozoa, reducing their numbers, motility, and life spans.[49] Furthermore, both men and women appear to lose sexual interest during times of starvation.

These war records and other research suggest that there is a "nutritional infertility threshold."[50] Below this threshold, hormonal secretions diminish, and women develop amenorrhea, becoming infertile. Between the optimal nourishment that produces a healthy infant and the starvation that causes infertility, an intermediate zone exists that offers suboptimal prospects for pregnancy.

amenorrhea (a-MEN-oh-REE-ah): the absence, temporary or permanent, of menstrual periods; normal before puberty, after the menopause, during pregnancy, and during lactation; otherwise abnormal.
a = not
men = month
rhoia = flow

Self-induced underweight and fertility Many females who diet excessively or who engage in intense physical exercise are underweight and amenorrheic. Weight gain usually initiates menarche or restores regular menstrual cycles.

Amenorrhea is a complex phenomenon characterized by low concentrations of estrogen. Controversy surrounds the exact causative factors. Researchers question whether a minimal level of body fat is necessary for the onset and maintenance of regular menstrual cycles.[51] Amenorrhea occurs in women who lose 10 to 15 percent of their normal weight for height—approximately one-third of their body fat. Minimum weights for various heights necessary for the onset or restoration of menstruation have been identified, but it is not known whether amenorrhea depends on body weight or percent body fat. The argument for percent body fat's being the critical factor has a sound physiological basis. Fat stores affect the menstrual cycle by serving as a site for estrogen synthesis, by influencing estrogen metabolism, and by altering estrogen-binding properties of cells.[52] Dietary fat also correlates with estrogen concentrations and influences fat stores.[53]

underweight: body weight at least 10% less than standard weight for height.

Chapter 5 discusses the amenorrhea associated with lactation.

Chapter 8 discusses athletic amenorrhea, the amenorrhea of anorexia nervosa, and stress-related amenorrhea.

eumenorrhea: normal menstruation.

Whether amenorrhea occurs in response to altered metabolic signals, reduced food intake, altered body temperature, or depleted fat stores is unclear. One study found no difference in body composition or weight between women with amenorrhea and women with eumenorrhea.[54] Differences in energy intake were significant, however. The energy intake of the amenorrheic women was approximately 25 percent less than that of the eumenorrheic women, even though their energy expenditures were similar.

Amenorrhea can, of course, be due to other causes, such as psychological stress. Whatever the underlying mechanism, the amenorrhea of an underweight woman may be viewed as an adaptive response to curtail reproduction. Many animals respond to food shortages and stress with hormonal changes that suppress ovulation and thereby fertility. Thus animals do not use reproductive energies to create offspring when the environment is not conducive to their survival.

Individual nutrients and fertility In some instances, a particular nutrient deficiency or excess correlates with infertility. An example of an excess is seen in the association of elevated serum carotene concentrations (carotenemia) with amenorrhea.[55] Researchers have found that women with amenorrhea whose carotenemia reflects a diet of predominantly raw vegetables can lower their serum carotene concentrations and restore their fertility with dietary changes.[56] Other researchers, however, caution that carotene is only one of a multitude of components found in vegetables. Perhaps it is the excessive consumption of another compound commonly associated with carotene that is responsible for the effect on fertility. Women taking carotene supplements of a nonplant origin do not experience menstrual irregularities.[57]

Subclinical vitamin C deficiency may be responsible for nonspecific sperm agglutination, one cause of infertility in men. Studies of men who are unable to impregnate their wives due to sperm agglutination find consistently low serum vitamin C concentrations.[58] In one study, men with this condition took a combination supplement of vitamin C, calcium, magnesium, and manganese.[59] The results were dramatic, with all 20 men successfully impregnating their wives within the study period. It should be noted that the design of this study did not control for the possibility of a placebo effect in the men taking supplements. In an effort to isolate the specific nutrient responsible, another study provided just vitamin C supplementation of 1 gram per day.[60] Within four days, the vitamin C supplementation raised both serum and semen vitamin C concentrations and lowered the percentage of agglutinated sperm to below the level that distinguishes fertility from infertility.

Diet and sex determination It is not likely that the diet's composition strongly influences the determination of an infant's sex at conception. Researchers have, however, proposed that the woman's diet may somewhat affect the odds in favor of conceiving a boy or a girl.[61] The fluid in the vagina, through which spermatozoa swim to the waiting ovum, normally favors equally those carrying X and those carrying Y chromosomes. However, if the fluid changes, its composition may favor one over the other in the race to fertilize the ovum. The woman's diet, in turn, may influence the composition of the vaginal fluid.

Researchers attempting to find out what dietary characteristics might influence sex determination claim to have managed, in one study, to predict the sex of about 80 percent of babies correctly just by analyzing the mothers' diets.[62] In another study, mothers' diets were adjusted in an effort to influence the sex of their babies. Diets high in the minerals sodium and potassium were 80 percent successful in producing boy babies, while those high in the minerals calcium and magnesium produced girls equally successfully.[63] Foods, not supplements, were used in the studies, so it is possible that some other factor present in the foods along with those minerals was responsible for the effect, or even that something that was *missing* brought about the difference. Therefore, these studies do not suggest that taking mineral supplements is an appropriate strategy for a person who wishes to conceive a baby of a particular sex. In fact, some minerals can be toxic to developing gametes.

The studies mentioned here were preliminary, but they were seized upon by a public eager to choose the sex of their babies. Some irresponsible publications suggested tampering with women's nutrition with no regard for their nutrient needs and no proof that megadoses of any kind would not be harmful to their developing gametes. This was, of course, sensationalism—the premature drawing of conclusions from scientific findings before their implications are fully known.

Nutrition and Implantation

The zygote undergoes several cell divisions while passing through the oviduct. After reaching the uterus, it floats freely for several days, receiving nourishment from the intrauterine fluid. Cell division continues.

Simultaneously, the endometrium is preparing to receive the zygote. Conditions in the uterus at the time of conception determine whether the zygote will successfully implant and begin normal development. Failure of implantation causes the loss of the zygote, possibly even before the woman knows she is pregnant.

The uterine environment can be either hospitable or hostile. With an appropriate balance of hormones and nutrients, a zygote implants and continues its development. When drugs or other toxins are present, the endometrium may reject the zygote, or the zygote may be unable to attach to the lining. Even with successful implantation, the zygote may suffer damage that will become evident during later development. The pregnant woman cannot control all factors influencing her body's chemistry during the time surrounding implantation, but ideally, her nutrition will be optimal and her exposure to contaminants minimal at this time.

During the two weeks in which implantation takes place, the zygote divides into many cells, and these cells sort themselves into layers. The outermost layer will form the placenta and amniotic sac; the innermost cells become the embryo. These cells, in turn, become arranged in three layers—the ectoderm, mesoderm, and endoderm. Little growth in size occurs at this time, but because of the rapid cell division taking place, this is, as mentioned, a vulnerable time. Authorities agree that to provide the most nearly ideal environment possible for implantation, a woman should have been well

When a single fertilized ovum completely divides into two independent organisms, the result is monozygotic (identical) twins. When two fertilized ova develop into two independent organisms, the result is dizygotic (fraternal) twins.

implantation: the stage of development in which the zygote embeds itself in the wall of the uterus and begins to develop, during the first two weeks after conception.

ectoderm: the outermost layer of a developing embryo, which evolves into the nervous system and skin.
ecto = outside
derm = skin

mesoderm: the middle layer of a developing embryo, between the ectoderm and endoderm, which evolves into the muscular, skeletal, circulatory, and internal organ systems.
meso = middle

endoderm: the innermost layer of a developing embryo, which evolves into the glands and linings of the digestive, respiratory, and excretory systems.
endo = inside

nourished for a long time prior, should not be taking drugs of any kind (even aspirin), and should be protected from all harmful environmental exposures.

Nutrition and Future Pregnancies

Nutrition early in a woman's life not only affects her health, her ability to conceive, and the likelihood of implantation, it also affects the subsequent development of the fetus if she does manage to conceive. Like conception and implantation, early embryonic development depends both on the genetic material and on the environment. Discussion of the environment in which the embryo develops might properly be postponed to the chapters on pregnancy (and it is discussed in more detail there), but a woman's lifestyle habits and exposures *prior* to pregnancy continue to influence her body's chemistry when she is pregnant. Thus the choices a prepregnant woman makes can influence the course of a pregnancy she is not even planning at the time.

The environment can affect development in two ways—by altering the genes themselves (causing somatic mutation) or by altering their expression. Genes provide potential—that is, the capacity to reach a certain developmental level. That potential can fail to be fully realized due to poor nutrition, chemical damage, radiation, lack of oxygen, or other conditions, which can all interfere with correct gene copying or prevent the full or normal expression of the genes. Congenital abnormalities, then, can result from adverse physical or chemical conditions during development of the gametes or the fetus. Among the major nutrition-related adverse conditions are undernutrition, obesity, and alcohol and other drug use. These subjects will all be mentioned again in the chapters on pregnancy, but their prepregnancy effects, as best they can be sorted out, are discussed here.

Undernutrition

Undernutrition prior to pregnancy can have major detrimental impacts on early embryonic development, leading to low birthweight, morbidity, mortality, and retarded physical and mental development. The evidence for this comes from two sources—records taken during famines, and data on women who are underweight for other reasons (such as ill-advised dieting).

morbidity: disease.

mortality: death.

Famine at conception Famines in Europe during the Second World War not only reduced conception rates, as already mentioned, but also impaired the courses and outcomes of pregnancies when women did manage to conceive. These effects were particularly dramatic in Holland, where daily food rations for pregnant women reached a low point of 1145 kcalories and 34 grams of protein.[64] Rations for other adults were even more limited. The records showed that stillbirths and neonatal deaths were significantly higher for infants *conceived* during the famine than for those conceived prior to or following it. Apparently, many women received enough nourishment to be fertile, but not enough to have healthy infants—that is, they were in the intermediate zone

stillbirth: birth of a dead fetus.

referred to earlier. Conception of infants during times of limited food supplies consistently correlates with congenital malformations of the central nervous system.[65]

One of the tasks a woman's body performs at the start of a pregnancy is to develop a placenta, an organ of great importance. (The characteristics of the placenta are described in full in Chapter 4.) Research has demonstrated that when malnutrition occurs around the time of conception, placental cell number is reduced.[66] If the placenta fails to attain its full growth potential, it will be unable to deliver optimum nourishment to the fetus, and the infant will be born small.

If this small infant is a female, she in turn has an elevated risk of having a poor pregnancy outcome. She is more likely to have a miscarriage or a stillbirth, or to give birth to an infant with congenital malformations, with respiratory distress, or in need of neonatal intensive care.[67]

miscarriage: known medically as a spontaneous abortion, the separation of the developing fetus and the placenta from the inner wall of the uterus, resulting in the unintentional termination of the pregnancy, occurring before the beginning of the 20th week of gestation.

The intrauterine growth retardation of a female fetus is not the only factor affecting her future reproductive performance. Her nutrition during her infancy and childhood is also of crucial importance. For example, rickets or protein-energy malnutrition in childhood may lead to contracted pelvis, and in turn to problems in pregnancy and childbirth.

contracted pelvis: a pelvis whose diameter is reduced to a degree that impedes childbirth.

Thus the poor development of infants may reflect adverse conditions of their mothers' own early development. A nutritional insult to a woman during or even *before her pregnancy* can adversely affect not only her child but her *grandchild*.[68]

Underweight at conception The famines that tragically affect human health normally arise during wars or natural events such as droughts. The effects of poor nutrition can also be seen in women during times of peace and plenty when they intentionally starve themselves.

The health of, and outlook for, an infant at birth can be predicted by several indicators, as Chapter 4 describes in detail. One of the most reliable of these indicators is the infant's birthweight, and that is the one used in the studies discussed here. Several characteristics of the mother's pregnancy predict infant birthweight. Maternal weight gain during pregnancy and duration of the pregnancy are two of the most reliable of these (see Chapter 4), but a third, equally important one is maternal weight for height *prior to conception*, a subject appropriate to this chapter.

An underweight woman is at high risk for giving birth not only to a low-birthweight infant but also to a preterm infant, and perinatal mortality rates are highest for these women.[69] An underweight woman maximizes her chances of having a healthy infant by gaining weight to the standard prior to conception, or by gaining extra pounds during pregnancy.

The underweight woman who attempts to increase her weight before conceiving needs to know that weight gain is best achieved by eating a high-kcalorie diet and pursuing physical conditioning. A high-kcalorie diet alone results in fat gain; exercise ensures that the weight gained will be at least partly lean tissue. Strength training (such as sit-ups, push-ups, and repetitions with heavy weights) is the most efficient way to support weight gain efforts.

Chapters 3, 4, and 6 discuss the risks associated with preterm births and infants.

Energy intake must be high enough to support both exercise and weight gain, perhaps 700 to 1000 kcalories per day above customary intakes. If a woman eats only enough to support exercise, muscles develop at the expense of body fat. To raise food energy intake, a woman who is underweight prior to pregnancy would do well to follow the diet plans and nutrient recommendations for women who are already pregnant.

Chapter 4 provides dietary recommendations for pregnancy.

Obesity

obese: body weight of at least 20% above standard weight for height.

morbidly obese: twice or more the average standard weight for height or 100 lbs over the standard weight.

Studies examining the effects of obesity must state the criteria used to define it. Most often, researchers use 20 percent above standard body weight to define obesity's lower limit. Researchers studying morbidly obese people may use a higher cut-off point. The complications are similar, but their rate and severity intensify with increasing weight.

Approximately one of every four women between the ages of 20 and 44 is obese.[70] Like the underweight woman, the obese woman faces her own set of problems related to pregnancy and childbirth. Women who are obese at conception risk the obstetrical complications associated with hypertension, gestational diabetes, and postpartum infections.[71] (Chapter 4 discusses these obstetrical complications.) They also face the necessity of having to have labor induced by oxytocin and infants delivered by cesarean section more often than nonobese women.

oxytocin-induced labor: use of the pituitary hormone oxytocin to stimulate the uterus to contract, thus beginning the childbirth process. cesarean section:

cesarean section: removal of the fetus by an incision into the uterus, usually by way of the abdominal wall.

Maternal obesity affects not only mothers but their infants as well. These infants are likely to be born post term and to weigh more than 4000 grams (8.9 pounds).[72] Obese women are less likely to have premature infants, but if they do, their infants are likely to be large for gestational age. This may result in a misclassification of the infants as term infants, with resulting failure to recognize problems associated with prematurity.

To prevent these problems, women are advised to achieve a healthy body weight before becoming pregnant. If choosing to diet, they should postpone pregnancy; if already pregnant, they should postpone weight loss dieting. Weight loss regimens are appropriate for pregnant women only if qualified medical judgment declares their obesity an even greater risk than weight loss. (These statements are not a license for an obese pregnant woman to overeat, of course. She should eat the diet recommended for a pregnant woman.)

gastric bypass surgery: a surgical procedure intended to limit food intake; a small segment of the upper stomach is closed off from the lower portion and attached to the small intestine, thus bypassing a major portion of the stomach.

neural tube defects: any of a number of defects in the orderly formation of the neural tube during early gestation, resulting in various central nervous system disorders, such as spina bifida.

For a few morbidly obese people, gastric bypass surgery is a treatment option, but pregnancies following gastric surgery for the treatment of obesity involve risks of their own. Nutrient deficiencies are common following gastric bypass surgery, and research studies have linked the resulting dietary inadequacies with neural tube defects in infants born subsequently. Reports of these studies suggest that careful medical and dietary management during pregnancy is imperative for women who have earlier had gastric bypasses.[73] With proper dietary and medical attention, pregnancy after gastric bypass can be successful.[74] In fact, vitamin supplementation prior to and throughout pregnancy reduces the occurrence of neural tube defects.[75] The incidence of hypertension and large newborns is lower in women who have had gastric bypass surgery with adequate aftercare than in women who have remained obese.

Alcohol and Other Drugs

Focal Point 3 discusses alcohol abuse during pregnancy, but of particular interest to this chapter is alcohol abuse *prior to conception*. Even if a woman stops drinking at conception, her prior alcohol abuse will affect the development of her fetus. One study showed this clearly—infants born to women who had a history of alcohol abuse *but abstained during pregnancy* weighed less than infants born to women who did not consume alcohol.[76] Perhaps the earlier alcohol abuse led to maternal liver dysfunction and nutrient deficits that impaired fetal growth and development. This is not to say there is no point in abstaining during pregnancy. The other half of the story is that infants born to those alcohol abusers who abstained from drinking during pregnancy weighed more than infants born to women who had a similar history of alcohol abuse and continued drinking during pregnancy.

The first month of pregnancy is a critical period of fetal development. Because pregnancy confirmation usually requires five to six weeks, a woman may not even realize she is pregnant during that critical first month. Therefore, it is advisable for women who are trying to conceive or who suspect they might be pregnant to curtail their alcohol intakes to ensure a healthy start.

Other drugs, both medical and illicit, can be harmful in similar ways. Megadoses of vitamin supplements also have druglike effects. Most medications have warnings on their labels that state, "As with any drug, if you are pregnant or nursing an infant, seek the advice of a health professional before using this product." Because this warning does not appear on all harmful substances, women must warn themselves of the potential dangers of their actions. Women considering future pregnancies should include themselves along with pregnant and lactating women and heed the warning.

An example of a drug whose effects are particularly well known is Accutane. Generically known as isotretinoin, Accutane is effective in treating severe cystic acne that has been unresponsive to other treatments. Accutane bears a warning label that advises women to use an effective form of contraception beginning at least one month before the inception and continuing until one month after the termination of Accutane's use. Red stickers warn that it can cause birth defects, and pharmacists add labels that caution, "Do not become pregnant while using Accutane."

Many infants born to women taking Accutane have exhibited major birth defects.[77] That is why dermatologists must ascertain whether women consulting them for acne are sexually active, and if so, whether they are using effective contraception methods. A pregnancy test two weeks prior to the onset of treatment is imperative. Some dermatologists require monthly pregnancy tests while Accutane is taken.

A list of specific drugs to discontinue prior to pregnancy is not available. However, this one example illustrates the potent effect that a drug taken near the time of conception, even one approved by the Food and Drug Administration and prescribed by a physician, can have on the outcome of a pregnancy. The effectiveness and clearance rates of different drugs vary. Even a short-term, low dose of a dangerous drug can expose a fetus to irreparable harm. Other factors influencing the nature and severity of the harm to the fetus include the time and duration of administration. Obviously, the situation requires more caution when a woman is sexually active and conception is a possibility than when she is sexually abstinent or practicing contraception.

Among the drugs to discontinue if a woman wants to become pregnant are, of course, the oral contraceptives. Once a woman stops taking oral contraceptives, she can expect a delay of one to three months before regular ovulation and menstruation resume. If she waits for her menstrual cycle to resume before becoming pregnant, she will be able to calculate her due date. After the initial delay, conception rates become the same for these women as for others not using contraceptives. However, oral contraceptives cause no fetal damage, even if taken after conception.[78]

Much research focuses on the effects of maternal drug ingestion on conception and fetal development, but until recently, few studies had examined the effects of paternal exposure to drugs. Pregnancy begins with the union of an ovum and a spermatozoon, and it stands to reason that adverse influences on the male might affect conception. Male rats given methadone prior to mating with untreated females sired pups with low birthweights and altered behavior patterns.[79] Neonatal mortality was also high. In human males, lead, anesthetic gases, cigarettes, and caffeine are associated with adverse effects on reproduction.[80] Drugs may alter spermatozoon formation, maturation, or motility, and this can affect fertility.[81] In addition, paternal drug ingestion might alter male hormone activity, damage the spermatozoa, or, if the drug or its metabolites pass in the ejaculate, affect the intrauterine environment or the newly fertilized ovum.[82]

methadone: a synthetic analgesic drug widely used in the treatment of heroin abuse.

Reproduction is an astonishing process. That human bodies package, into cells too small to see, entire libraries of all the information necessary to make human beings inspires wonderment. To produce gametes as nearly perfect as possible, and to permit their union in the safest conditions, men and women need to treat their own bodies with respect.

With the fertilization of an ovum, a woman's body becomes an environment that nurtures another human being and releases it into this world. This is life's cycle—from infancy to childhood to adolescence to pregnancy, returning to infancy. Each of these phases of life is vital to the cycle's continuation. Men and women leave the cycle only after the reproductive years are over, when they begin their journeys into later life. Until then, people who nourish and protect their bodies do so not only for their own sakes, but also for future generations.

Chapter 2 Notes

1. The American Dietetic Association's Nutrition Recommendations for Women, *Journal of the American Dietetic Association* 86 (1986): 1663–1664.
2. R. P. Heaney and coauthors, Calcium nutrition and bone health in the elderly, *American Journal of Clinical Nutrition* 36 (1982): 986–1013.
3. L. W. Turner and E. N. Whitney, Nature versus nurture: The calcium controversy, *Nutrition Clinics,* November-December, 1989.
4. Turner and Whitney, 1989.
5. Turner and Whitney, 1989.
6. S. M. Garn, M. T. Keating, and F. Falkner, Hematological status and pregnancy outcomes, *American Journal of Clinical Nutrition* 34 (1981): 115–117.
7. Data reviewed and presented in light of 1989 RDA. B. B. Peterkin, Women's diets: 1977 and 1985, *Journal of Nutrition Education* 18 (1986): 251–257; R. L. Rizek and K. S. Tippett, Women's food consumption: Diets of American women, in 1985, *Food and Nutrition News* 61 (1989): 1–4.
8. C. W. Callaway, Statement on vitamin and mineral supplements, *American Journal of Clinical Nutrition* 46 (1987): 1076; D. Herber, W. Mertz, and R. E. Schucker, Food versus pills versus fortified foods, *Dairy Council Digest*, March–April 1987; A. E. Harper, "Nutrition insurance"—A skeptical view, *Nutrition Forum,* May 1987, pp. 33–37.
9. R. P. Heaney, Premenopausal prophylactic calcium supplementation (letter), *Journal of the American Medical Association* 245 (1981): 1362.
10. M. S. Sheikh and coauthors, Gastrointestinal absorption of calcium from milk and calcium salts, *New England Journal of Medicine* 317 (1987): 532–536.
11. Restated in terms of the 1989 RDA. Position of the American Dietetic Association: Nutrition for physical fitness and athletic performance for adults, *Journal of the American Dietetic Association* 87 (1987): 933–939.
12. F. T. O'Neil, M. T. Hynak-Hankinson, and J. Gorman, Research and application of current topics in sports nutrition, *Journal of the American Dietetic Association* 86 (1986): 1007–1015.
13. J. L. Greger, Food, supplements, and fortified foods: Scientific evaluations in regard

to toxicology and nutrient bioavailability, *Journal of the American Dietetic Association* 87 (1987): 1369–1373.

14. B. Dawson-Hughes, F. H. Seligson, and V. A. Hughes, Effects of calcium carbonate and hydroxyapatite on zinc and iron retention in postmenopausal women, *American Journal of Clinical Nutrition* 44 (1986): 83–88.
15. R. Yip and coauthors, Does iron supplementation compromise zinc nutrition in healthy infants? *American Journal of Clinical Nutrition* 42 (1985): 683–687; J. Albers, E. B. Dawson, and W. J. McGanity, Effect of elevated pre-natal iron supplementation on serum copper, zinc, and selenium levels (abstract), *American Journal of Clinical Nutrition* 43 (1986): 673.
16. E. R. Monsen, Menstrual cycle. Address presented at the 70th Annual Meeting of the American Dietetic Association in Atlanta, Georgia, 20 October 1987.
17. K. S. Moghissi, F. N. Snyer, and T. N. Evans, A composite picture of the menstrual cycle, *American Journal of Obstetrics and Gynecology* 114 (1972): 405–415.
18. S. J. Solomon, M. S. Kurzer, and D. H. Calloway, Menstrual cycle and basal metabolic rate in women, *American Journal of Clinical Nutrition* 36 (1982): 611–616.
19. S. P. Dalvit, The effect of the menstrual cycle on patterns of food intake, *American Journal of Clinical Nutrition* 34 (1981): 1811–1815; E. J. Gong, D. Garrel, and D. H. Calloway, Menstrual cycle and voluntary food intake, *American Journal of Clinical Nutrition* 49 (1989): 252–258.
20. S. P. Dalvit-McPhillips, The effect of the human menstrual cycle on nutrient intake, *Physiology and Behavior* 31 (1983): 209–212.
21. Solomon, Kurzer, and Calloway, 1982.
22. *Premenstrual Syndrome,* a report by the American Council on Science and Health, July 1985.
23. M. Mira, P. M. Stewart, and S. F. Abraham, Vitamin and trace element status in premenstrual syndrome, *American Journal of Clinical Nutrition* 47 (1988): 636–641.
24. B. D. Shephard, Oral contraceptives—An overview, *Journal of the Florida Medical Association* 73 (1986): 763–767.
25. J. W. Ellis, Multiphasic oral contraceptives—Efficacy and metabolic impact, *Journal of Reproductive Medicine* 32 (1987): 28–36.
26. E. L. Marut, Oral contraceptives—Who, which, when, and why? *Postgraduate Medicine* 82 (1987): 66–70.
27. Shephard, 1986.
28. Shephard, 1986; Marut, 1987; G. R. Huggins and P. K. Zucker, Oral contraceptives and neoplasia: 1987 update, *Fertility and Sterility* 47 (1987): 733–761.
29. The Centers for Disease Control Cancer and Steroid Hormone Study: The reduction in risk of ovarian cancer associated with oral-contraceptive use, *New England Journal of Medicine* 316 (1987): 650–655.
30. Marut, 1987; Huggins and Zucker, 1987.
31. The Centers for Disease Control Cancer and Steroid Hormone Study: Combination oral contraceptive use and the risk of endometrial cancer, *Journal of the American Medical Association* 257 (1987): 796–800; Huggins and Zucker, 1987.
32. D. R. Miller and coauthors, Breast cancer risk in relation to early oral contraceptive use, *Obstetrics and Gynecology* 68 (1986): 863–868.
33. R. Russell-Briefel and coauthors, Cardiovascular risk status and oral contraceptive use: United States, 1976–1980, *Preventive Medicine* 15 (1986): 352–362; P. B. Moser and coauthors, Carbohydrate tolerance and serum lipid responses to type of dietary carbohydrate and oral contraceptive use in young women, *Journal of the American College of Nutrition* 5 (1986): 45–53.
34. Marut, 1987; Russell-Briefel and coauthors, 1986.
35. J. K. Williams, Oral contraceptives—The long-term perspective, *Journal of the Florida Medical Association* 73 (1986): 769–771.
36. Shephard, 1986; Williams, 1986.
37. R. Russell-Briefel and coauthors, Impaired glucose tolerance in women using oral contraceptives: United States, 1976–1980, *Journal of Chronic Diseases* 40 (1987): 3–11.
38. Williams, 1986; Moser and coauthors, 1986.
39. Ellis, 1987; Russell-Briefel and coauthors, 1987.
40. Russell-Briefel and coauthors, 1987.
41. B. L. Strom, Oral contraceptives and other risk factors for gallbladder disease, *Clinical Pharmacology and Therapeutics* 39 (1986): 335–341.
42. F. Kern, Jr., and G. T. Everson, Contraceptive steroids increase cholesterol in bile: Mechanisms of action, *Journal of Lipid Research* 28 (1987): 828–839.
43. L. B. Tyrer, Nutrition and the Pill, *Journal of Reproductive Medicine* 29 (1984): 547–550.
44. K. Amatayakul and coauthors, Vitamin metabolism and the effects of multivitamin supplementation in oral contraceptive users, *Contraception* 30 (1984): 179–196.
45. S. Carpenter and L. S. Neinstein, Weight gain in adolescent and young adult oral contraceptive users, *Journal of Adolescent Health Care* 7 (1986): 342–344.
46. Z. Stein and coauthors, *Famine and Human Development: The Dutch Hunger Winter of 1944/1945* (New York: Oxford University Press, 1975).
47. C. A. Smith, The effect of wartime starvation in Holland upon pregnancy and its product, *American Journal of Obstetrics and Gynecology* 53 (1947): 599–608.
48. Smith, 1947.
49. Stein and coauthors, 1975.
50. M. Wynn and A. Wynn, The influence of nutrition on the fertility of women, *Nutrition and Health* 1 (1982): 7–13; Stein and coauthors, 1975.
51. R. E. Frisch, Fatness, menarche, and female fertility, *Perspectives in Biology and Medicine* 28 (1985): 611–633; M. E. Nelson and coauthors, Diet and bone status in amenorrheic runners, *American Journal of Clinical Nutrition* 43 (1986): 910–916.
52. Frisch, 1985; B. Goldin and coauthors, The relationship between estrogen levels and diets of Caucasian American and Oriental immigrant women, *American Journal of Clinical Nutrition* 44 (1986): 945–953.
53. Goldin and coauthors, 1986.
54. M. E. Nelson and coauthors, Diet and bone status in amenorrheic runners, *American Journal of Clinical Nutrition* 43 (1986): 910–916.
55. A. M. Frumar, D. R. Meldrum, and H. L. Judd, Hypercarotenemia in hypothalamic amenorrhea, *Fertility and Sterility* 32 (1979): 261–264; E. Kemmann, S. A. Pasquale, and R. Skaf, Amenorrhea associated with carotenemia, *Journal of the American Medical Association* 249 (1983): 926–929; P. A. Deuster and coauthors, Nutritional intakes and status of highly trained amenorrheic and eumenorrheic runners, *Fertility and Sterility* 46 (1986): 636–643.
56. Kemmann, Pasquale, and Skaf, 1983.
57. M. M. Mathews-Roth, Amenorrhea associated with carotenemia (letter), *Journal of the American Medical Association* 250 (1983): 731.
58. E. R. Gonzalez, Sperm swim singly after vitamin C therapy, *Journal of the American Medical Association* 249 (1983): 2747, 2751; W. A. Harris, T. E. Harden,

and E. B. Dawson, Apparent effect of ascorbic acid medication on semen metal levels, *Fertility and Sterility* 32 (1979): 455–459.
59. Harris, Harden, and Dawson, 1979.
60. E. B. Dawson, W. A. Harris, and W. J. McGanity, Effect of ascorbic acid on sperm fertility (abstract), *Federation Proceedings* 42 (1983): 531; Gonzalez, 1983.
61. J. Lorrain and R. Gagnon, Sélection préconceptionelle du sexe, *Union Médicale du Canada* 104 (1975): 800–803; J. Stoklowski and J. Choukroun, Preconception selection of sex in man, *Israel Journal of Medical Sciences* 17 (1981): 1061–1067; and F. Papa and coauthors, Preconceptional selection of fetal sex using an ionic method: A dietary regime, *Journal de Gynécologie, Obstétrique, et Biologie de la Reproduction* 12 (1983): 415–422.
62. Lorrain and Gagnon, 1975.
63. Stoklowski and Choukroun, 1981; Papa and coauthors, 1983.
64. M. Wynn and A. Wynn, Effects of nutrition on reproductive capability, *Nutrition and Health* 1 (1983): 165–178; Stein and coauthors, 1975.
65. C. A. Smith, Effects of maternal undernutrition upon the newborn infant in Holland (1944–1945), *Journal of Pediatrics* 30 (1947): 229–243.
66. P. Rosso, Placental growth, development, and function in relation to maternal nutrition, *Federation Proceedings* 39 (1980): 250–254.
67. E. Hackman and coauthors, Maternal birth weight and subsequent pregnancy outcome, *Journal of the American Medical Association* 250 (1983): 2016–2019.
68. Hackman and coauthors, 1983.
69. R. M. Pitkin, Assessment of nutritional status of mother, fetus, and newborn, *American Journal of Clinical Nutrition* 34 (1981): 658–668; R. L. Naeye, Weight gain and the outcome of pregnancy, *American Journal of Obstetrics and Gynecology* 135 (1979): 3–9.
70. J. Willis, The gender gap at the dinner table, a pamphlet (HHS publication no. [FDA] 84–2197) available from U.S. Government Printing Office, 5600 Fishers Lane, Rockville, MD 20857.
71. S. R. Johnson and coauthors, Maternal obesity and pregnancy, *Surgery, Gynecology, and Obstetrics* 164 (1987): 431–437.
72. Johnson and coauthors, 1987.
73. J. E. Haddow and coauthors, Neural tube defects after gastric bypass, *Lancet* 1 (1986): 1330.
74. D. S. Richards, D. K. Miller, and G. N. Goodman, Pregnancy after gastric bypass for morbid obesity, *Journal of Reproductive Medicine* 32 (1987): 172–175.
75. R. W. Smithells and coauthors, Further experience of vitamin supplementation for prevention of neural tube defect recurrences, *Lancet* 1 (1983): 1027–1031; J. Mulinare and coauthors, Periconceptional use of multivitamins and the occurrence of neural tube defects, *Journal of the American Medical Association* 260 (1988): 3141–3145.
76. R. E. Little and coauthors, Decreased birth weight in infants of alcoholic women who abstained during pregnancy, *Journal of Pediatrics* 96 (1980): 974–977.
77. *Physicians' Desk Reference*, 42nd ed. (Oradell, N.J.: Medical Economics, 1988), pp. 1705–1706.
78. After conception: Dispelling rumors about later childbearing, *Population Reports* 12 (1984): J-699–J-702.
79. L. F. Soyka and J. M. Joffe, Male mediated drug effects on offspring, *Progress in Clinical and Biological Research* 36 (1980): 49–66.
80. Soyka and Joffe, 1980.
81. L. M. Hill and F. Kleinburg, Effects of drugs and chemicals on the fetus and newborn, *Mayo Clinic Proceedings* 59 (1984): 707–716.
82. Soyka and Joffe, 1980.

▶ Focal Point 2

Inborn Errors of Metabolism

The discussion in Chapter 2 on inheritance lays the foundation for a closer look at the ramifications of a genetic error in protein synthesis. Body functions, such as metabolic reactions and transports, that depend on proteins cannot proceed when proteins are made in an insufficient quantity or when they have an abnormal structure. If an enzyme is missing or malfunctioning in the metabolic pathway that converts compound A to compound B, then compound A accumulates and compound B becomes deficient. Both the excess of compound A and the lack of compound B can lead to a variety of physical disorders and, in many cases, death. Furthermore, high concentrations of compound A become available, and low concentrations of compound B are unavailable, for use in other metabolic pathways. These consequences, in turn, create excesses and deficiencies of other metabolites that present another array of problems. The diseases that result from these "inherited biochemical blocks in normal metabolic pathways" are known as inborn errors of metabolism.[1]

In some instances, the accumulated compound is not toxic and the deficient compound is not essential, and so individuals experience no problem. In all likelihood, they will never know about the error. However, in other cases, inborn errors have severe consequences, with many of them causing mental retardation. Without proper diagnosis and treatment, they can be lethal. As is true of most medical disorders, the earlier the diagnosis and treatment, the better the prognosis.

One goal of medical research is to detect genetic defects before they cause harm. Recent advances in medical technology allow clinicians to study the developing fetus and identify abnormal conditions during gestation. One technique, amniocentesis, analyzes amniotic fluid that has been removed by a needle or syringe puncturing the amniotic sac. This technique has proven most valuable in the identification of more than 70 different inborn errors of metabolism.[2] Analysis of the amniotic fluid and cells identifies specific enzymes and measures their concentrations, revealing abnormalities. A genetic technique called restriction enzyme analysis allows scientists to locate specific genes. Such a technique holds promise of actually correcting a genetic defect in the future. For now, prenatal diagnosis of genetic diseases allows nutritional intervention during gestation.[3]

The primary treatment for most inborn errors of metabolism is nutrition. With an understanding of the biochemical pathway involved, a clinician can manipulate the diet to correct for excesses and inadequacies. Dietary management of genetic diseases restricts dietary precursors that occur prior to the error in the metabolic pathway, administers pharmacological doses of vitamins, or replaces needed products. The goal of therapy is to:

- Prevent the toxic accumulation of metabolites.
- Replace essential nutrients that are deficient as a result of the defective metabolic pathway.
- Provide a diet that supports growth and development.

To meet these three requirements is a major challenge that was unattainable until earlier this century. Increased knowledge about the body's many biochemical pathways, coupled with current technology to synthesize formulas of specific nutrient compositions, has greatly enhanced the treatment of inborn errors.

Researchers are experimenting with ways to handle the toxic metabolites that accumulate in inborn errors. One solution is to administer a compound that reacts with the toxic metabolite, thus using an alternative pathway for excretion. Such alternative pathways are proving effective in the excretion of waste nitrogen in children with inborn errors of urea synthesis.[4] Another solution is to provide an antagonist that eliminates or detoxifies the accumulated metabolites. For example, the drug penicillamine is used both as a copper chelator in Wilson's disease and to increase the solubility of urinary cystine in cystinuria.[5]

This discussion focuses primarily on the most common inborn error of metabolism—phenylketonuria (PKU)—which affects approximately 200 newborns in the United States each year. The ability to detect and treat PKU has significantly improved the lives of many people and provides an example that offers hope to those suffering from other inborn errors.

PKU is only one of several inborn errors that affect amino acid metabolism. Other disorders affect not only amino acid metabolism but also carbohydrate, lipid, and vitamin metabolism. The number of possible inborn errors is limited only by the number of possible gene mutations, for genes carry the codes to make the enzymes in the body.

Consider that a small bacterial cell carries genes for at least 1000 different enzymes and that mutations are possible in all of these genes. Human cells have 1000 times as many proteins of major importance, and in theory, any of them can appear in many different mutant forms. Not only can the genes for enzymes be affected, but also those for other proteins, such as the "pumps" that move nutrients into and out of cells, the carriers that transport them in the blood, and the structural proteins of cell membranes and connective tissue. Table FP2–1 provides a glimpse into the array of possible errors by identifying some of these genetic disorders, their defective pathways, associated symptoms, and treatments. The rest of this discussion provides a look at a few inborn errors that respond to nutrition therapy.

Classic Phenylketonuria

phenylketonuria (**PKU**): an inborn error of metabolism in which phenylalanine, an essential amino acid, cannot be converted to tyrosine. Alternative metabolites of phenylalanine (phenylketones) accumulate in the tissues, causing damage, and overflow into the urine.

Classic phenylketonuria (PKU) results from a deficiency of the enzyme phenylalanine hydroxylase. This enzyme hydroxylates the essential amino acid phenylalanine, coverting it to tyrosine (see Figure FP2–1). Without phenylalanine hydroxylase, abnormally high concentrations of phenylalanine an

Figure FP2–1 The Biochemical Pathway in PKU

Normally, the amino acid phenylalanine follows two pathways, one in the liver, the other in the kidney. In the liver, the enzyme phenylalanine hydroxylase adds a hydroxyl group (OH) to produce the amino acid tyrosine. Tyrosine, in turn, produces melanin, the pigmented compound found in skin and brain cells; the neurotransmitters epinephrine and norepinephrine; and the hormone thyroxin. In the kidney, enzymes convert phenylalanine to byproducts that are excreted.

In the liver:

Phenylalanine —(Phenylalanine hydroxylase)→ Tyrosine ——→ Melanin
Epinephrine
Norepinephrine
Thyroxin

In the kidney:

Phenylalanine ——→ Phenylpyruvic acid (a ketone body) ——→ Other phenyl acids

Individuals with PKU lack the liver enzyme phenylalanine hydroxylase, impairing conversion of phenylalanine to tyrosine. Phenylalanine accumulates in the liver and blood, reaching the kidney in abnormally high concentrations. In the kidney, an aminotransferase enzyme converts phenylalanine to the ketone body phenylpyruvic acid, which spills into the urine—thus the name phenylketonuria.

In the liver:

Phenylalanine (accumulates) —(Phenylalanine hydroxylase (deficient))→ Tyrosine (deficient)

In the kidney:

Phenylalanine (accumulates) ——→ Phenylpyruvic acid (accumulates) ——→ Other phenyl acids (accumulate)

other related compounds (phenylketones) accumulate and damage the developing nervous system. Simultaneously, the body cannot make tyrosine or other compounds (such as the hormone epinephrine) that normally derive from tyrosine. Under these conditions, tyrosine becomes an essential amino acid; that is, the body cannot make it, and therefore the diet must supply it.

PKU is a hidden disease that cannot be seen at birth, yet diagnosis and treatment beginning in the first few days of life can prevent its devastating effects. For these reasons, all newborns in the United States receive a metabolic test to screen for PKU. Tests must be conducted after infants have consumed several meals containing protein. Before the 1960s, when screening became routine, children with PKU would suffer the consequences of elevated phenylalanine concentrations. At first, the only signs are a skin rash and light skin pigmentation. Between three and six months, signs of developmental delay being to appear. The infant becomes irritable, unable to sleep restfully, and frantic. By one year, irreversible brain damage is clearly evident, and the child has already lost 40 to 50 IQ points.[6]

Table FP2–1 Inborn Errors of Metabolism

Disease Name	Defect or Deficiency	Result of Defect or Deficiency	Main Effects	Therapy
Cerebrotendinous xanthomatosis (CTX)[a]	A liver enzyme	Inability to produce bile acids effectively	Accelerated hardening of the arteries Progressive neurological disorders Cataracts Deposition of cholesterol-related chemicals in the brain	Supply bile acids
Richner-Hanhart syndrome (tyrosinemia II)[b]	Tyrosine aminotransferase	High tyrosine concentrations	Eye disorders Skin disorders	Restrict tyrosine and phenylalanine
Tyrosinosis (tyrosinemia I)[c]	Fumarylacetoacetate hydrolyase	—	Failure to thrive Vomiting; diarrhea Chronic liver disease; cirrhosis Renal tubular dysfunction Vitamin D–resistant rickets Death	
—	Ornithine transcarbamylase[d]	Inability to synthesize arginine and urea High ammonia concentrations due to altered protein metabolism	Delayed physical growth and development Extreme irritability Lethargy Muscular incoordination Seizures Vomiting Mental retardation Coma Death	Restrict protein
Hawkinsinuria[e]	—	Tyrosinemia Accumulation of toxic metabolites of tyrosine	Failure to gain weight Metabolic acidosis	—

—	Glutamate dehydrogenase[f]	High glutamate concentrations Low alpha-ketoglutarate concentrations	Neuronal degeneration	—
Propionic acidemia (PA)[g]	Propionyl-CoA carboxylase (PCC)—an enzyme required in branched-chain amino acid metabolism	High concentrations of ammonia High concentrations of glycine High concentrations of propionic acid High concentrations of organic acids Ketonuria	Vomiting Lethargy Ketoacidosis Seizures Osteoporosis Developmental retardation	Restrict total protein or restrict isoleucine, methionine, threonine, and valine Megadoses of biotin (in conjunction with other therapy)
Mevalonic aciduria[h]	Mevalonic kinase	Disruption of cholesterol (and other isoprenoid compounds) biosynthesis High mevalonate concentrations	Failure to grow and mature Cataracts Hypocholesterolemia	—
Wilson's disease (hepatolenticular degeneration)[i]	—	Accumulation of copper Low ceruloplasmin concentrations	Cirrhosis Neurological symptoms	Chelating agent (D-penicillamine) Oral zinc supplementation
—	Cobalamin R binding proteins	Low cobalamin concentrations	Microcytic hypochromic anemia	—
Megaloblastic anemia	Cobalamin TC II binding protein (transports cobalamin to tissues)	Normal plasma cobalamin concentrations	Megaloblastic anemia	Pharmacological doses of cobalamin

[a]New clue for prevention and treatment of hardening of the arteries, *Journal of the American Dietetic Association* 84 (1984): 567.
[b]Tyrosinemia II (abstract), *Journal of the American Dietetic Association* 86 (1986): 127.
[c]Tyrosinemia II, 1986.
[d]Disease carriers and lowered IQ, *Science News* 117 (1980): 166; Ornithine transcarbamylase deficiency (abstract), *Journal of the American Dietetic Association* 86 (1986): 1128–1129.
[e]Hawkinsinuria: Disorder of tyrosine metabolism (abstract), *Journal of the American Dietetic Association* 80 (1982): 82–83.
[f]Glutamate metabolism in an adult-onset degenerative neurological disorder, *Journal of the American Dietetic Association* 81 (1982): 215.
[g]P. M. Queen, P. M. Fernhoff, and P. B. Acosta, Protein and essential amino acid requirements in a child with propionic acidemia, *Journal of the American Dietetic Association* 79 (1981): 562–565.
[h]An inborn error of cholesterol biosynthesis, *Nutrition Reviews* 44 (1986): 334–336.
[i]Oral zinc therapy for Wilson's disease, *Nutrition Reviews* 42 (1984): 184–186.

Continued

Table FP2–1—*Continued*

Disease Name	Defect or Deficiency	Result of Defect or Deficiency	Main Effects	Therapy
Menke's kinky hair syndrome[j]	Metallothionein synthesis or degradation (?)	Disruption of copper metabolism Reduced activity of copper metalloenzymes Low copper concentrations (blood) High copper concentrations (cells) Low ceruloplasmin concentrations	Kinky, depigmented hair Pale skin Elastin and collagen abnormalities Hypothermia Neurological disorders Skeletal demineralization Death	—
Acrodermatitis enteropathica (AE)[k]	—	Zinc malabsorption	Dermatitis Diarrhea Alopecia Depression Delayed bone maturation Growth retardation Delayed sexual development Anemia Compromised immune system	Megadoses of zinc
Homocystinuria[l]	Cystathionine synthetase	Disruption of sulfur amino acid (cysteine, cystine, methionine) metabolism; blocks transsulfuration of methionine to cysteine High methionine concentrations and related compounds High homocysteine concentrations and related compounds Low cystathionine and cystine concentrations	Dislocation of eye lenses Mental retardation Skeletal abnormalities	Restrict methionine High cystine diet Pharamacological doses of pyridoxine (cystathionine synthetase requires pyridoxal phosphate as a coenzyme)
Lysinuric protein intolerance (LPI)[m]	Disruption of dibasic amino acid (arginine,	Low arginine concentrations	Failure to thrive Vomiting	Ornithine, arginine, or citrulline supplementation

	ornithine, lysine) absorption, transport, and excretion	Low ornithine concentrations Low lysine concentrations Prevention of urea cycle substrates from being renewed Low urea synthesis (with normal urea acid cycle enzymes) High ammonia concentrations after protein intake High orotic acid concentrations	Diarrhea Growth retardation Osteoporosis Low white blood cell concentrations Low blood platelet concentrations Enlarged liver and spleen Protein intolerance	
Hartnup disease[n]	Disruption of monoamino-monocarboxylic acid (tryptophan, alanine, asparagine, glutamine, histidine, isoleucine, leucine, phenylalanine, serine, threonine, tryptophan, tyrosine, and valine) absorption, transport, and excretion	Low amino acid concentrations Excessive amino acid excretion Minimal biosynthesis of nicotinic acid from tryptophan	Muscular incoordination due to brain disease Pellagra and associated mental disturbances	Megadoses of nicotinic acid
Cobalamin-responsive methylmalonic acidemias[o]	Methylmalonyl CoA mutase N^5-methyltetrahydrofolate-homocysteine methyltransferase	Failure to convert hydroxocobalamin to its active coenzymes	Protein intolerance Failure to thrive Episodic ketoacidosis Neurological abnormalities Death (in severe, untreated cases)	Pharmacological doses of cobalamin (1 mg/day)

[j]On the pathogenesis and clinical expression of Menke's kinky hair syndrome, *Nutrition Reviews* 39 (1981): 391–393; Menke's disease: Are we closer to learning its cause? *Nutrition Reviews* 42 (1984): 309–311.
[k]Zinc therapy of depressed cellular immunity in acrodermatitis enteropathica, *Nutrition Reviews* 39 (1981): 168–170.
[l]Pyridoxine-responsive homocystinuria, *Nutrition Reviews* 39 (1981): 16–18; M. A. Wallen and S. Packman, Nutrition and inborn errors of metabolism, in *Nutrition Update*, vol. 2, eds. J. Weininger and G. M. Briggs (New York: Wiley, 1985) pp. 71–89.
[m]Lysinuric protein intolerance: A rare cause of childhood osteoporosis, *Nutrition Reviews* 44 (1986): 110–113.
[n]Treatment of Hartnup disease with nicotinic acid, *Nutrition Reviews* 42 (1984): 251–253.
[o]Wallen and Packman, 1985.

Continued

Table FP2–1—*Continued*

Disease Name	Defect or Deficiency	Result of Defect or Deficiency	Main Effects	Therapy
Galactosemia[p]	Galactose-1-phosphate uridyl transferase	Failure to convert galactose-1-phosphate to glucose-1-phosphate	Failure to thrive Jaundice; enlarged liver Renal dysfunction Cataracts Hemolytic anemia Seizures Coma; death	Eliminate galactose from the diet (feed soybean-based formulas containing sucrose or glucose, without lactose)
Phenylketonuria (PKU)[q]	Phenylalanine hydroxylase	Failure to convert phenylalanine to tyrosine	Skin rash and light pigmentation Delayed development Neurological defects Microcephaly EEG abnormalities Seizures	Restrict protein sources; supplement with a protein-containing formula with low phenylalanine content
Malignant hyperphenylalaninemia[r]	Dihydropteridine reductase	Failure to synthesize or regenerate the cofactor tetrahydrobiopterin, which is required for phenylalanine hydoxylase, tyrosine hydroxylase, and tryptophan hydroxylase Defective synthesis of neurotransmitters (dopamine, norepinephrine, serotonin)	Neurological abnormalities	Neurotransmiter replacement Tetrahydrobiopterin administration Phenylalanine restriction (not effective alone)

Von Gierke's disease (glycogen storage disease type I)[s]	Glucose-6-phosphatase	Incomplete gluconeogenesis Incomplete glycogenolysis High concentrations of serum uric acid, cholesterol, triglycerides, and lactate	Hypoglycemia Enlarged liver Growth failure (due to low insulin production) Adiposity Hemorrhagic tendency	Provide glucose between meals and during the night (continuous nocturnal intragastric infusion)
Citrullinemia[t]	Argininosuccinic acid synthetase (urea cycle enzyme that converts citrulline to argininosuccinic acid)	Elevated ammonia concentrations Elevated citrulline concentrations	Lethargy Rapid respiration Tremors and seizures Abnormal prothrombin times Death by age 7	Feed mixture of keto-acids of essential amino acids

[p] Wallen and Packman, 1985.
[q] Wallen and Packman, 1985.
[r] Wallen and Packman, 1985.
[s] Wallen and Packman, 1985.
[t] W. L. Nyhan, Nutritional treatment of children with inborn errors of metabolism, in *Textbook of Pediatric Nutrition,* ed. R. M. Suskind (New York: Raven Press, 1981), pp. 563–576.

Nutritional Therapy

The effect of nutrition intervention in PKU is remarkable. In almost every case, dietary management can prevent the devastating array of symptoms described. Essentially, the diet restricts phenylalanine intake to a point that maintains blood phenylalanine concentrations within a safe range. As most dietitians can attest, this is more easily said than done.

Because phenylalanine is an essential amino acid, the diet cannot exclude it completely. If phenylalanine intake is too low, children suffer bone, skin, and blood disorders; growth and mental retardation; and death.[7] Therefore, the diet must strike a perfect balance between providing enough phenylalanine to support normal growth and health and not too much to cause harm. It is not that children with PKU require less phenylalanine than other children, but that they cannot handle excesses without detrimental effects. To ensure that blood phenylalanine concentrations remain within a safe range, children with PKU receive blood tests periodically and alterations in their diets when necessary. With a controlled phenylalanine intake, children with PKU can lead normal, happy lives.

To control phenylalanine intake requires strict dietary management that was impossible prior to 1958, when a special low-phenylalanine formula became commercially available.[8] This type of formula is now the primary source of energy and protein for children with PKU. Their diet excludes high-protein foods such as meat, fish, poultry, cheese, eggs, milk, nuts, dried beans, or peas. Also excluded are commercial breads and pastries made from regular flour, which has a high phenylalanine content. Basically, the diet allows foods that contain some phenylalanine, such as fruits, vegetables, and cereals, and those that contain none, such as sugar, jellies, and some candies. Clearly, it is impossible to create a diet of whole, natural foods, and children who depend primarily on a formula for their nourishment risk multiple trace mineral deficiencies.[9]

Infants receive a special casein hydrolysate formula with a low phenylalanine content. It does not contain all the phenylalanine an infant requires, and so parents supplement it with measured quantities of milk, rice cereal, and baby foods as the infant develops. Other formulas and products are available that provide a synthetic mixture of amino acids without phenylalanine. This allows older children to receive their entire phenylalanine quota from foods.[10]

People with PKU must also be aware of the phenylalanine content in products containing the sweetener aspartame. Aspartame is a combination of the two amino acids aspartic acid and phenylalanine, and therefore contributes phenylalanine to the diet. Sold under the trade name NutraSweet, aspartame is an ingredient in many foods and beverages such as powdered drink mixes, instant puddings, gelatin desserts, breakfast cereals, chewing gums, and the sweetener Equal. Products sweetened with aspartame bear a warning label for people with PKU. People with PKU need to consult with their physicians or dietitians before including aspartame in their diets.

Perhaps one of the hardest parts of this diet is in the children's sense of social isolation. From birth, children with PKU are on a "special diet" and cannot eat the foods that other children are eating. Some commercially available low-protein products, such as cookies, contain very little, if any, phenylalanine and allow children to share treats with others. Teachers, friends, and family members must understand that they cannot offer food to children with PKU before receiving permission from the children's parents. Until the

Galactosemia

Galactosemia is an inborn error of carbohydrate metabolism in which the body cannot use the monosaccharide galactose. Three enzymes are required for the conversion of galactose to glucose; in galactosemia, most commonly the enzyme galactose-1-phosphate uridyl transferase is missing or defective. When infants with galactosemia are given milk (which contains a galactose unit in each molecule of lactose), they vomit and have diarrhea. The unmetabolized galactose-1-phosphate accumulates and follows an alternative pathway to form an abnormal product that causes growth failure, liver enlargement, kidney failure, and cataracts. Infants with galactosemia experience seizures and other neurological abnormalities that lead to coma and death.[25] Early introduction of a galactose-restricted diet prevents or minimizes most of these symptoms. However, it may not prevent ovarian damage, some visual and speech problems, or other neurological abnormalities.

galactosemia: an inborn error of metabolism in which galactose cannot be metabolized normally to compounds the body can handle; an alternative metabolite accumulates in the tissues, causing damage.

Dietary adjustment in galactosemia is simpler than in PKU for a couple of reasons. First, galactose is not an essential nutrient, as is phenylalanine. The PKU diet is a balancing act of providing just enough phenylalanine for normal growth and development, without too much left over to be toxic; the galactosemia diet only needs to exclude galactose. Second, since galactose occurs only in lactose (the sugar in milk), only milk and milk products need to be restricted.

Glycogen Storage Diseases

Other genetic diseases affecting carbohydrate metabolism include the glycogen storage diseases. These diseases are characterized by abnormal glycogen deposition in liver, muscle, or both.[26]

Type I, also known as von Gierke's disease, is the most common and severe of the glycogen diseases. It is caused by a deficiency of the enzyme glucose-6-phosphatase.[27] This enzyme converts glucose-6-phosphate to glucose in the gluconeogenesis pathways and glycogenolysis pathways.[28] The body's inability to handle glucose-6-phosphate results in hypoglycemia, glycogen accumulation, and their associated metabolic consequences.

Dietary management of glycogen storage disease type I involves constant provision of carbohydrate (glucose). To accomplish this, children eat frequent (at least every three hours) high-carbohydrate meals (60 to 70 percent of total kcalories) composed primarily of starch foods.[29] To maintain glucose concentrations during sleep, children receive a special glucose polymer formula by tube feeding.

Type III, or Cori's disease, is a result of a deficiency of the glycogen debranching enzyme amylo-1,6-glucosidase, which inhibits glycogen breakdown at the branches and results in short-branched glycogen. In contrast, type IV, or Andersen's disease, is due to a deficiency of the branching enzyme, which results in abnormally long glycogen chains. Type III is a fairly harmless disorder that can be managed by diet, whereas type IV leads to liver cirrhosis and death in infancy.[30]

Vitamin Disorders

Some inborn errors affect the metabolic pathways in such a way as to raise a person's vitamin requirement. The error may be in the absorption of the vitamin; the biosynthesis, transport, or accessibility of the coenzyme form of the vitamin; or the apoenzyme protein. In such a case, the person requires pharmacological doses of the particular vitamin to overcome the error and prevent its associated detrimental effects. Table FP2–1 itemizes a few of the vitamin-responsive disorders.

As scientific understanding of human genetics and biochemistry increases, more and more inborn errors are being recognized. Understanding protein structure and function makes it possible to compensate for these defects of metabolism that otherwise would destroy the quality of life. Diet cannot always be tailored to prevent the defects of inborn errors, but in many such diseases diet can make a dramatic difference in people's lives.

Focal Point 2 Notes

1. A. E. Garrod, Inborn errors of metabolism (Croonian lectures), *Lancet* 2 (1908), as cited by M. A. Wallen and S. Packman, Nutrition and inborn errors of metabolism, in *Nutrition Update,* vol. 2, eds. J. Weininger and G. M. Briggs (New York: Wiley, 1985), pp. 71–89.
2. G. H. Lowrey, The placenta and fetal development, in *Growth and Development of Children,* 8th ed. (Chicago: Year Book Medical Publishers, 1986), pp. 53–75.
3. Wallen and Packman, 1985.
4. M. L. Batshaw and coauthors, Treatment of inborn errors of urea synthesis: Activation of alternative pathways of waste nitrogen synthesis and excretion, *New England Journal of Medicine* 306 (1982): 1387–1392.
5. Wallen and Packman, 1985.
6. Wallen and Packman, 1985.
7. P. B. Acosta and coauthors, Phenylalanine intakes of 1- to 6-year-old children with phenylketonuria undergoing therapy, *American Journal of Clinical Nutrition* 38 (1983): 694–700.
8. M. M. Hunt, H. K. Berry, and P. P. White, Phenylketonuria, adolescence, and diet, *Journal of the American Dietetic Association* 85 (1985): 1328–1334.
9. S. Stepnick-Gropper and coauthors, Trace element status of PKU children ingesting an elemental diet (abstract), *American Journal of Clinical Nutrition* 43 (1986): 676.
10. N. Reyzer, Diagnosis: PKU, *American Journal of Nursing* 78 (1978): 1895–1898.
11. S. Schild, Psychological issues in genetic counseling of phenylketonuria, in *Genetic Counseling, Psychological Dimensions,* ed. S. Kessler (New York: Academic Press, 1979), pp. 138–147.
12. Hunt, Berry, and White, 1985.
13. V. E. Schuett, E. S. Brown, and K. Michals, Reinstitution of diet therapy in PKU patients from twenty-two U.S. clinics, *American Journal of Public Health* 75 (1985): 39–42.
14. N. A. Holtzman and coauthors, Effect of age at loss of dietary control on intellectual performance and behavior of children with phenylketonuria, *New England Journal of Medicine* 314 (1986): 593–598.
15. R. Koch and E. Wenz, Phenylketonuria, in *Annual Review of Nutrition,* eds. R. E. Olson, E. Beutler, and H. P. Broquist (Palo Alto, Calif.: Annual Reviews), pp. 117–135.
16. S. E. Hogan, G. W. MacDonald, and J. T. R. Clarke, Experience with adolescents with phenylketonuria returned to phenylalanine-restricted diets, *Journal of the American Dietetic Association* 86 (1986): 1203–1207; Schuett, Brown, and Michals, 1985.
17. Schuett, Brown, and Michals, 1985.
18. Schild, 1979.
19. P. B. Acosta, Maternal PKU. Address presented at the conference Nutrition for Pregnancy, Lactation, and Infancy, Gainesville, Fla., 13 February 1987.
20. R. R. Lenke and H. L. Levy, Maternal phenylketonuria—Results of dietary therapy, *American Journal of Obstetrics and Gynecology* 5 (1982): 548–553; H. L. Levy and S. E. Waisbren, Effects of untreated maternal phenylketonuria and hyperphenylalaninemia on the fetus, *New England Journal of Medicine* 309 (1983): 1269–1274.
21. Schild, 1979.
22. Lenke and Levy, 1982; Acosta, 1987.
23. E. Drogari and coauthors, Timing of strict diet in relation to fetal damage in maternal phenylketonuria, *Lancet* 2 (1987): 927–930.
24. Wallen and Packman, 1985.
25. Wallen and Packman, 1985.
26. W. C. McMurray, *Essentials of Metabolism* (New York: Harper & Row, 1977), pp. 123–165.
27. C. C. Folk and H. L. Greene, Dietary management of type I glycogen storage disease, *Journal of the American Dietetic Association* 84 (1984): 293–301; I. E. Daeschel and coauthors, Diet and growth of children with glycogen storage disease types I and III, *Journal of the American Dietetic Association* 83 (1983): 135–141.
28. McMurray, 1977.
29. Folk and Greene, 1984.
30. McMurray, 1977.

Nutrition during Pregnancy: A Pivotal Time

3

Pregnant Woman by Pablo Picasso.

A whole new life begins at conception. Events follow fast upon one another, and nutrition plays a supportive role in all of them. Nothing could be more extraordinary than the course of a pregnancy, although oddly enough, this extraordinary time is called, simply, "normal." A chapter on nutrition during pregnancy could therefore begin by describing normal development, and then proceed to describe the nutrients needed to support that development. This chapter does not do that; the next one does.

To find out how important nutrition is to normal pregnancy, researchers have studied the effects of *malnutrition*. At no other time are those effects so clearly demonstrated as when they occur during pregnancy. That malnutrition occurs is a tragedy, and to study it is a somber endeavor, but that study throws into bold relief the tremendous importance of normal nutrition. It is unexciting to say, "Eat right," but the command takes on power in light of the consequences of malnutrition. This chapter provides the contrast that illuminates the normal course of pregnancy, the subject of Chapter 4.

Chapter 2 has already demonstrated that malnutrition *prior* to pregnancy has a major impact. Malnutrition does so by affecting a woman's nutrient stores; her health, strength, and size; her fertility; and her ability to build a normal placenta. During pregnancy, malnutrition exerts its effects by way of another mechanism—by acting on critical periods in the development of the embryo and fetus. A word of introduction on the concept of critical periods is therefore in order.

As development proceeds, intracellular activity becomes intense. Cell division requires that every organelle of the cell be duplicated—the mitochondria, the ribosomes, the vast and intricate network of intracellular membranes, the cell membrane, the nuclear membrane, the chromosomes with their extensive DNA blueprints for enzymes, and numerous other intracellular bodies. To make all these new organelles requires synthesis and assembly of a multitude of different materials—DNA, RNA, protein, lipid, and more. These syntheses are carried out by enzymes that themselves have been synthesized according to instructions from the genes—a process that requires extensive activity and coordination. Enzymes transcribe the relevant portions of DNA to make messenger RNA for delivery to the ribosomes; transfer RNAs line up along the messengers with their cargo of amino acids; enzymes link those amino acids together into new proteins; and enzymes and membrane proteins make or bring in new amino acids to make more proteins. The number of molecular events taking place in a cell about to divide is astronomical, and the orderly sequence and completion of these events are crucial to normal development. Furthermore, because each cell in an early stage of development is destined to become many cells later, any impairment of early cell multiplication may ultimately have greatly amplified effects. A time when so many chemical steps are proceeding is especially vulnerable to adverse physical and chemical influences; it is such times of intense developmental activity that are termed *critical periods*. Figure 3–1 illustrates the concept of critical periods.

critical period: a finite period during development in which certain events may occur that will have irreversible, determining effects on later developmental stages. A critical period is usually a period of cell division in a body organ.

hyperplasia: an increase in cell number.

hypertrophy: an increase in cell size.

The development of all organs occurs by the processes of cell division, during which cells are increasing dramatically in *number* (hyperplasia), and cell growth, when cell *size* increases (hypertrophy). These two processes may occur simultaneously, may overlap, or may occur in tandem. Each developing organ follows its own unique schedule.

Figure 3–1 The Concept of Critical Periods
Critical periods occur early in development. An adverse influence felt early can have a much more severe and prolonged impact than one felt later on.

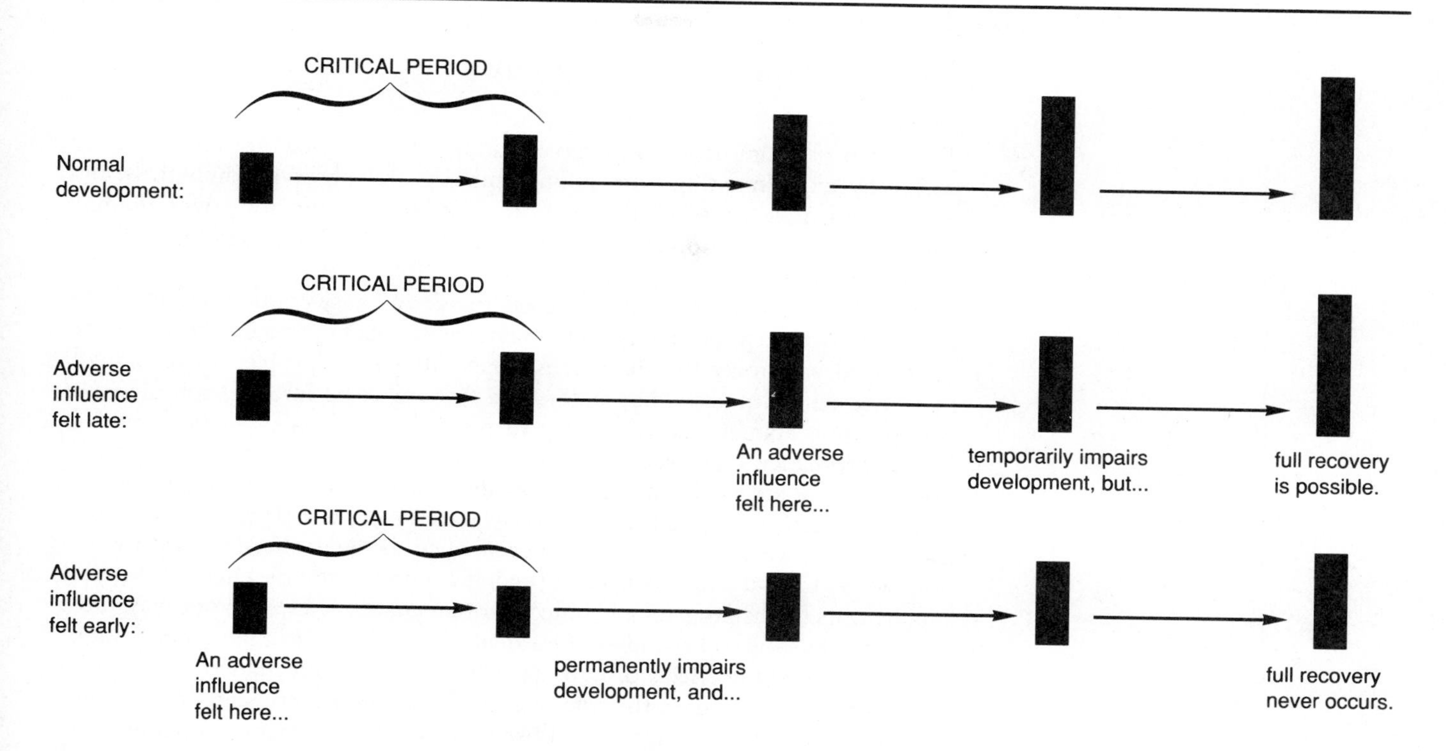

Times of increase in size are the times of most obvious growth, but critical periods of cell division and differentiation may precede them. If the organ is to reach its full potential, nutrient supplies and other environmental conditions must be optimal from the beginning. If cell division and the final cell number achieved in an organ are limited during a critical period, recovery is impossible. Early malnutrition can have irreversible effects, although they may not become fully apparent until the person reaches maturity.

Each organ and tissue is most vulnerable to nutrient deprivation or toxin interference during its own critical periods. In the fetus, for example, the heart is well developed at 16 weeks; the lungs are still immature 10 weeks later. Therefore, early malnutrition affects the heart most severely, while later malnutrition especially affects the lungs.

The phase of brain development most susceptible to malnutrition seems to be its growth spurt, the period in which brain weight increases most rapidly.[1] In human beings, the brain's growth spurt begins in midpregnancy and continues into the second year after birth.[2] Undernutrition during this period may do irreversible damage to the brain and adversely affect later behavior.

Although the concept of critical periods is valid, its simplicity may be somewhat misleading. Discrete developmental effects of single nutrient deficiencies or toxicities are seldom seen in the real world, especially in human beings. Malnutrition does not occur in discrete episodes, uncomplicated by other factors. This chapter describes the realities of malnutrition's impacts—

multifaceted, confusing, and impossible to ascribe neatly to particular nutrient deficits at particular times.

Malnutrition during Pregnancy

The effects of malnutrition during pregnancy can be studied in several different ways—by experimenting with animal subjects, by observing human beings during and after times of deprivation, and by experimenting with human beings using intervention studies. Each kind of study illuminates the problem from a different angle.

Studies of animals permit researchers to ask and answer questions about the effects of discrete episodes or types of malnutrition—but the results are, of course, guaranteed valid only for animals. Animals differ from people in size, growth rates, and length of gestation; factors like these must be carefully considered in extrapolating animal research to try to predict the effects of malnutrition on human pregnancy. Observations of human beings reveal true effects of malnutrition, but these are vastly complicated by differences among subjects, the lack of control subjects, and the effects of poverty and disease, as well as the vast human tragedy that cries out for remediation even when the most effective measures are not known. Intervention studies using people as subjects test the effects of nutrient supplements on the progress and outcome of pregnancy. Viewed in one way, these studies might seem to be studies of nutrient adequacy, but from another point of view they reveal the effects of nutrient inadequacies on pregnancy. Intervention studies do not always meet the criteria of truly rigorous experimental designs, but they do enable nutrition scientists to observe how nutrition intervention affects the health and well- being of free-living individuals.

Researchers design their studies differently, making comparisons among the studies difficult and the drawing of conclusions challenging. They can study generalized malnutrition or the effects of specific nutrient deficiencies. They can study different outcomes, such as birthweights or problem deliveries. They can review much later outcomes, such as the age of attaining sexual maturity or the ultimate height or IQ achieved. They can vary the malnutrition episodes in timing of onset, in duration, and in intensity. They can study different nutrients and groups of nutrients. Many variations and combinations of these approaches appear in the studies reported here.

This section reviews, first, studies on animals; next, observations on human beings; and third, intervention studies on human beings. The first two types of research are perhaps of greater theoretical than practical interest, but they reveal the great importance of nutrition in pregnancy and the extent of research effort dedicated to elucidating its effects. The third type of study has obvious practical applications.

Studies on Animals

Studies of nutritionally deprived animals report several significant observations. Malnourished animals bear fewer offspring per litter, the offspring are smaller, and both mother and young experience increased mortality.[3] The offspring also suffer long-term, irreversible effects, such as compromised learning ability.

Protein Studies on animals show that protein restriction during critical periods of organ development results in a reduced number of cells in the offspring at term.[4] Even after later nutrient supplementation, the liver, heart, kidney, and brain do not acquire normal numbers of cells. The effect of prenatal protein deficiency in reducing cell number in these organs is therefore irreversible.[5]

Iron Iron deficiency during critical periods of fetal brain development in animals results in a deficit of brain iron after birth—not the iron in the red blood cells passing through the brain, but the iron in the brain cells that facilitates cellular respiration and metabolism. This deficit persists despite adequate iron intake later.[6]

Zinc Biochemical changes follow rapidly on the induction of zinc deficiency. In the cases of other nutrients, a pregnant female may be able to mobilize her own stores to protect the fetus, but such is not the case with zinc. Studies on rats suggest that the mother may be unable to deliver zinc from her stores to the fetus.[7] Thus a constant dietary source of zinc is necessary in order to protect the fetus from deficiency. In rats, even short-term zinc deficiency during pregnancy produces a wide variety of congenital malformations.[8]

Other nutrients Other nutrient deficiencies have shown equally profound effects on pregnancy outcomes in animals. For example, severe deficiencies of copper and manganese correlate with fetal abnormalities and growth retardation in animals.[9]

Nutrient toxicities Abundant animal research has examined the effects of specific nutrient excesses during pregnancy. Depending on the amount and gestational stage of administration, excess vitamin A given to pregnant animals produces many types of birth defects.[10] Large doses of vitamin D may contribute to hypercalcemia in offspring, and excessive maternal intakes of vitamin C can predispose offspring to dependency on it.[11]

Animal studies enable researchers to carefully design and control experiments examining the influence of nutrient deprivation or excess on pregnancy outcome. Nutrient deficiencies and excesses in human beings seldom reach the degree of severity used in animal research. For this reason, and because studies of human beings do not afford researchers the same control as animal studies do, evidence of the effects of malnutrition during pregnancy in human beings is less consistent than that of animal studies. Despite this, studies of human beings do offer some support for the susceptibility of the human fetus to nutrient deprivation and excess.

Descriptive Studies of Human Beings

Occasionally, chance has designed conditions in the human environment that have provided information directly on human beings and have suggested directions for future studies using animals. Observations on live infants and autopsies of infants who have died in both affluent and poor societies have provided some evidence consistent with the theory that effects of nutrition deprivation in human beings are similar to some of those seen in animals. This section describes studies of immediate outcome and of later effects.

Table 3–1 summarizes some of the effects of nutrient deficiencies on pregnancy outcomes. (Table 3–2, presented later, shows the consequences of nutrient excesses.) An impaired outcome may be any of the following:

- Fetal growth retardation.
- Congenital malformations, birth defects.
- Spontaneous abortion.
- Premature birth.
- Stillbirth.
- Low birthweight.
- Reduced cell number.

low birthweight (LBW): a birthweight of 5½ lb (2500 g) or less; used as a predictor of poor health in the newborn and as a probable indicator that the mother was in poor nutrition status during and/or before pregnancy. The term **very low birthweight** describes infants that weigh 3⅓ lb (1500 g) or less.

Of these, infant birthweight is most frequently used as a predictor of future survival and health. Malnutrition, coupled with low birthweight, is the underlying or associated cause of more than half of all the deaths of children under five years of age.[12]

Table 3–1 Effects of Nutrient Deficiencies on Pregnancy Outcome

Nutrient	Possible Deficiency Effect
Energy	Low infant birthweight[a]
Folate	Miscarriage and neural tube defect[b]
Vitamin A	Congenital malformations
Vitamin D	Low infant birthweight[c]
Iron	Stillbirth, premature birth, and low infant birthweight[d]
Iodine	Cretinism (varying degrees of mental and physical retardation in the infant)[e]
Zinc	Congenital malformations[f]

[a] Z. Stein and coauthors, *Famine and Human Development: The Dutch Hunger Winter of 1944/45* (New York: Oxford University Press, 1975).
[b] R. W. Smithells and coauthors, Further experience of vitamin supplementation for prevention of neural tube defect recurrences, *Lancet* 1 (1983): 1027–1031.
[c] J. D. Maxwell and coauthors, Vitamin D supplements enhance weight gain and nutritional status in pregnant Asians, *British Journal of Obstetrics and Gynaecology* 88 (1981): 987–991.
[d] S. M. Garn, M. T. Keating, and F. Falkner, Hematological status and pregnancy outcomes, *American Journal of Clinical Nutrition* 34 (1981): 115–117.
[e] P. O. D. Pharaoh, I. H. Buttfield, and B. S. Hetzel, Neurological damage to the fetus resulting from severe iodine deficiency during pregnancy, *Lancet* 1 (1971): 308–310, as cited by L. S. Hurley, *Developmental Nutrition* (Englewood Cliffs, N.J.: Prentice-Hall, 1980), pp. 183–198.
[f] S. Jameson, Effects of zinc deficiency in human reproduction, *Acta Medica Scandinavica Supplement* 593 (1976): 3–89.

A malnourished low-birthweight infant is likely to be unable to do its job of obtaining nourishment by sucking and unable to attract its mother's attention by energetic, vigorous crying and other healthy behavior. It can therefore become an apathetic, neglected infant; this compounds the original malnutrition problem.

About 1 in every 15 infants born in the United States is a low-birthweight infant, and about one-fourth of these infants die within the first month of life.[13] Worldwide, it is estimated that one-sixth of all live infants are of low birthweight. Most of them are not preterm but are full-term infants who are small because of malnutrition.

Chapter 2 showed that malnutrition prior to pregnancy, as exemplified by the siege of Holland during World War II, reduced fertility and, for women who did get pregnant, increased the rate of stillbirths, neonatal deaths, and congenital malformations. Malnutrition during pregnancy has similar effects. An extreme condition that provided data on malnutrition during human pregnancy occurred during the German siege of Leningrad, which took place from 1941 to 1943.[14] Food during this time was inferior in quality and limited in quantity—at times severely limited. The infant mortality rate was unusually high. The more severe the food shortage, the greater the incidence of poor pregnancy outcomes. A substantial increase in the number of low-birthweight infants, premature births, and stillbirths occurred. This was a dramatic example of the effects of extreme malnutrition on human pregnancy.

In less extreme conditions of human malnutrition, it is more difficult to observe dietary effects on prenatal development. Problems in determining dietary intake accurately and in controlling all other variables that affect development complicate study design. Despite the obstacles, some relevant and interesting findings emerge.

A classic study conducted in the early 1940s found a correlation between the quality of the maternal diet and the condition of the infant at birth.[15] Of the mothers whose diets were classified as excellent or good, based on the RDA, 94 percent gave birth to infants whose condition was described as superior or good by pediatricians with no knowledge of the mothers' dietary intakes. In contrast, only 8 percent of women whose diets were classified as poor gave birth to infants in good or superior condition, a highly significant difference.

Another study examined the relationship between infant birthweight and different environmental factors, one of which was diet quality.[16] Diet quality was expressed as a nutrient adequacy ratio (NAR) index. This was determined by converting each woman's recorded intakes of 11 nutrients to percentages of the RDA and then averaging the percentages to obtain each woman's NAR index. The investigators found a significant, positive correlation between the NAR indexes of the maternal diets and the birthweights of the infants. This study provides further evidence that diet during pregnancy significantly affects the health of the infant.

The above studies focused on diet in general; others have focused on individual nutrients. Individual nutrient deficiencies can impair pregnancy's outcome. Single nutrient deficiencies rarely occur in human beings; if one nutrient is lacking, chances are great that one or more other nutrients are lacking as well. Nutrient deficiencies in human beings do not often occur in so orderly a manner as to be compatible with an experimental design. Thus, with regard to individual nutrients, a limited number of human studies is available. Important animal research already described helps to fill the gap and provides direction for studies of human beings.

Some evidence from both animals and human beings bears on iron during pregnancy. The findings on people are consistent with the hypothesis that poor maternal iron status impairs fetal and newborn iron status and, thereby, pregnancy outcome. Low maternal hemoglobin or hematocrit concentrations correlate with stillbirths, prematurity, and low birthweights.[17]

These findings are contradicted, however, by findings from other research.[18] For example, no significant difference in infant mortality and birthweight was found between iron-deficient women and women from the general obstetrical population at a large hospital. This suggests that iron transfer across the placenta to the developing fetus occurs to the extent necessary, regardless of maternal iron status. Furthermore, the way the attending health care professional clamps the umbilical cord during childbirth makes a major difference to the infant's iron status (see Chapter 4's Practical Point: Nutrition at the Time of Giving Birth).[19]

Another single nutrient deficiency that clearly affects human pregnancy outcome is folate deficiency. Folate deficiency is common in pregnant women; the incidence is as high as 50 percent in some countries.[20] The poorer the folate status at the beginning of pregnancy, the earlier deficiency symptoms appear and the more severe they become as pregnancy progresses. Folate deficiency in early pregnancy may interfere with normal placental development.[21] A severe lack of folate can lead to megaloblastic anemia, which, in turn, may cause miscarriage.[22] It is possible, but not yet confirmed, that miscarriage heralds future pregnancies with neural tube defect births. In one study of two groups of pregnant women with histories of neural tube defect births, one group received a multivitamin supplement containing folate, while the other group received no supplement. The incidence of subsequent neural tube defect births was five times greater in the unsupplemented women than in the supplemented women.[23] The researchers attributed the difference to the folate in the multivitamin supplement, because a previous double-bind randomized controlled trial of folate treatment before conception had been shown to prevent neural tube defects.[24] A link between folate deficiency and fetal growth retardation in human beings has also been demonstrated.[25]

megaloblastic anemia: the type of anemia seen in folate or vitamin B_{12} deficiency characterized by enlarged, slightly irregular red blood cells.

neural tube defect: definition on p. 62.

Zinc deficiency is teratogenic for human beings, as it is for animals. In Sweden, an association between congenital malformations in human beings and low maternal serum zinc concentrations has been found.[26] Another study examined the relationship between plasma zinc concentrations of pregnant adolescents and pregnancy outcome.[27] Mothers with plasma zinc concentrations below the sample mean gave birth to infants with undescended testes and other congenital defects.

For purposes of comparisons made in this discussion, we are assuming that serum zinc concentrations and plasma zinc concentrations are comparable measures of zinc status. An important qualification to this, however, is that as of yet, researchers have devised no single reliable method for assessing zinc status.[28] Interpretation of studies in which only one measure of zinc status is used must therefore proceed with caution. In addition, the studies described here demonstrated correlations of zinc deficiencies with defects; they did not show that the zinc deficiencies caused the defects. In light of the animal experiments showing the teratogenic effects of zinc deficiency, however, there is a real possibility that the human effects are caused by zinc malnutrition.

Malnutrition includes overnutrition as well as undernutrition, and nutrient excesses during pregnancy may also have adverse effects. In human beings, excessive maternal energy intake corresponds to fatness of the infant.[29] Many reports ascribe birth defects in infants to treatment of their mothers during pregnancy with Accutane, the high-dose vitamin A derivative used as therapy for acne, in spite of warnings about the drug.[30] In human beings and animals, large doses of vitamin C during pregnancy may condition the infant such that the risk of scurvy increases at birth when the vitamin C doses cease.[31] Evidence for the conditioning effect in human beings has been questioned, however.[32]

"conditioned" scurvy: the vitamin C–deficiency disease seen in newborn infants of mothers who ingested large doses of vitamin C during pregnancy; possibly induces an increased rate of vitamin C catabolism, which persists after birth.

Malnutrition during pregnancy not only can impair its immediate outcome, but also can induce defects that first become apparent years later, and from which developing young will never fully recover. For example, maternal fasting during pregnancy can cause later obesity in a son or daughter by depriving the brain's developing regulatory centers of glucose.[33] Another, well-known example is that mothers who are iodine deficient during pregnancy give birth to infants who later evince mental retardation. Still another is the irreversible physical and mental retardation of children born to mothers who drank alcohol during pregnancy (fetal alcohol syndrome).

Focal Point 3 discusses the effects of alcohol consumption during pregnancy on fetal development.

The irreversibility of such effects is obvious when abundant, nourishing food fed after the critical time fails to remedy the deficit. Many growth-retarded Korean orphans adopted by U.S. families after the Korean War, for example, experienced several years of catch-up growth but did not completely recover from the effects of early malnutrition.[34]

It is apparent from these studies that critical periods occur in development when either general malnutrition or deficiencies or excesses of particular nutrients cause damage to the fetus that proves irreversible even with optimal nutrition at a later time. Clearly, optimal nutrition during pregnancy enhances the likelihood of a healthy outcome.

Intervention Studies on Human Beings

Numerous intervention studies on pregnant women have been conducted worldwide. The reason is that the alarming number of low-birthweight infants demands attention. In the United States alone, the number of low-birthweight infants born in 1985 was over one-quarter of a million, or about 7 percent of total live births.[35] Worldwide, the percentage is greater.[36]

In various ways, studies of prenatal nutrition test the hypothesis that prenatal dietary supplementation of moderately malnourished women can improve pregnancy outcome. Much of the research on this subject was conducted at a time when protein was considered to be *the* nutrient lacking in the diets of poor women. For this reason, protein was the nutrient most often supplemented. The results of these studies have brought researchers slightly closer to unraveling the threads of uncertainty surrounding prenatal nutrition (for example, which food supplements at what point during gestation best promote fetal growth), but they also present new threads to untie. The results are interesting, occasionally conflicting, and often surprising.

In the following discussion, pregnant women presumed to be moderately but not overtly malnourished, based on previous pregnancy outcomes and socioeconomic status, were given daily protein-kcalorie supplements in the form of beverages or nutrient-dense foods. In the first investigation, the overall condition of the infant at birth was the outcome considered. In all the other studies, infant birthweight was the main outcome considered.

Short-term studies One study, conducted about 50 years ago, showed quite clearly the impact of prenatal nutrition on pregnancy.[37] The researchers studied the prenatal diets of 400 women. They classified the women into three groups according to diet adequacy based on specific amounts of milk, cheese, eggs, meat, vegetables, fruits, cereals, and breads eaten daily. The groups were as follows:

- *Poor diet.* Women consumed less than 60 grams of protein per day throughout pregnancy.
- *Supplemented diet.* Women with a poor diet until the fourth or fifth month were provided with additional food daily thereafter.
- *Good diet.* Women consumed between 60 and 80 grams of protein per day throughout pregnancy.

The women in the supplemented- and good-diet groups experienced better health throughout their pregnancies and had fewer complications. Miscarriages, stillbirths, and premature births occurred less frequently than in the poor-diet group.

In another study, observers noted that more low-birthweight infants than expected were being born to poor black women in New York. Because of this, researchers presumed the women to be poorly nourished and undertook a nutrient supplementation trial.[38] Based on specific criteria that identified women who were especially at risk of delivering low-birthweight infants, researchers randomly divided the women into three different treatment groups of 250 women each:

- The *supplement group* received a regular prenatal vitamin-mineral supplement plus a beverage rich in protein (40 grams protein daily plus 470 kcalories).
- The *complement group* received a regular prenatal vitamin-mineral supplement plus a beverage with much less protein and fewer kcalories (6 grams protein and 322 kcalories).
- The *unsupplemented (control) group* received a regular prenatal vitamin-mineral supplement.

Contrary to the investigators' expectations, no significant differences in birthweight were seen among the three treatment groups. In fact, the supplement group bore infants with a mean birthweight *lower* than that of those born to the control group. Even more surprising was the observation that the supplement group had more premature births and, among these births, a higher rate of neonatal deaths than the other groups. Women with histories of previous premature deliveries seemed particularly susceptible to this unexpected outcome. Thus, for some women, the supplement actually seemed to have adverse effects.

The investigators used several indexes of dietary intake to assess the women's compliance in drinking the beverage. They anticipated from the start that total compliance among the women might be an unrealistic expectation, as well as a weak point in the study. Throughout the study, the investigators used other indicators of compliance, such as 24-hour dietary recalls and urine tests for a riboflavin marker in the beverages. These indicators helped to show whether the women added the beverages to their regular food intakes as instructed; substituted the beverages for some food; or even drank the beverages at all. The researchers concluded that overall compliance was satisfactory. Despite these efforts, the study is vulnerable to criticism in that some uncertainty clouded this aspect of the design.

Believing that the women consumed essentially all beverages and ate all food as instructed, the investigators speculate that the lower birthweights of infants born to mothers in the supplement group must have been due to an adverse effect of the supplement. A possible explanation of the adverse effect is that the supplement may have contained a toxic contaminant such as lead, or too much protein.

For heavy smokers in the supplement group, the effect of supplements was favorable. Infants of the smoking women on supplements were equal in birthweight to infants of nonsmoking women on supplements, whereas smokers in the unsupplemented group had infants with reduced birthweights. The investigators concluded that for populations of women in the United States at high risk of giving birth to low-birthweight infants, supplementation of the regular maternal diet produced little or no benefit in terms of birthweight except for heavy smokers, and may have had an adverse effect overall.

In Bogotá, Colombia, a prenatal intervention study focused on pregnant women living in a poor environment whose diets were considered deficient in protein and food energy. The women were given dietary supplements of regular foods beginning in the sixth month of pregnancy.[39] Birthweights of infants of the supplemented women were higher than birthweights of controls, and the proportion of low-birthweight infants was reduced significantly. The mean increment to the women's diet (based on two 24-hour recalls) was 136 kcalories and 20 grams protein daily. A greater difference in infant weight was observed for thin women than for heavier women when compared with controls. Interestingly, the gain in birthweight for the supplemented families was seen only in male births. Unlike the New York study, this study uncovered no adverse effects of supplementation. Thus this study supports the assertion that additional food given to malnourished women in the third trimester of pregnancy can raise birthweight.

In Guatemala, a study was designed to examine the effects of protein supplements versus nonprotein supplements on infant birthweights across villages.[40] Comparison of the protein- and nonprotein-supplemented villages did not produce the expected results—that protein would be more effective than food energy in raising birthweight. The investigators then abandoned the original design in favor of comparing food energy intakes among women, regardless of the protein content of the supplements or the village in which the women lived. Ingestion by mothers of about 70 extra kcalories per day throughout pregnancy resulted in a gain of about 56 grams (2 ounces) in their infants' birthweights. This study lends support to the concept that in situations of chronic malnutrition, birthweight can be raised by sufficient kcaloric supplementation throughout pregnancy. In this study, it was food energy, not protein, that made the difference.

A study in Montreal, Canada, was conducted in the prenatal clinic of a large hospital serving low-income women.[41] Some of the women were referred to the Montreal Diet Dispensary, where they received dietary advice and food supplements. Mothers who were not referred to the diet dispensary served as controls. The diets of the women were presumed to be inadequate, but the women were not overtly malnourished. In this respect, the study was similar to the New York study.

The mean birthweight of the infants born to the women receiving food supplements was 40 grams (1.4 ounces) greater than that of controls. For women receiving food supplements, the frequency of low birthweight was 5.7 percent, compared with 6.8 percent in the controls. This difference was statistically significant. As in the Bogotá study, supplementation made a bigger difference in the birthweights of infants born to thin women (who had weighed less than 140 pounds when they conceived) than in the birthweights of infants born to heavier women. As in the New York study, some of the women who received supplements gave birth to infants with lower weights than control women. Although this difference was not statistically significant, the researchers felt it should not be discounted, and they offered the caution that for some women, nutrition supplementation might have adverse effects. They concluded that in general, though, prenatal supplementation of underweight and presumably underfed women could result in a modest rise in infant birthweight.

In a San Francisco study, women at risk for delivering low-birthweight infants were randomly assigned to one of three dietary regimens:[42]

- Normal diet plus a high-protein (80 grams) beverage supplement.
- Normal diet plus a low-protein (6 grams) beverage supplement.
- Normal diet plus a vitamin-mineral preparation.

The diets of all the women were generally adequate with respect to all nutrients. The incidence of low-birthweight infants was half the expected rate of 6 percent. The diets of the women in the high-protein group were higher in protein and food energy (100 percent of the RDA) than the diets of the women in the other two groups (80 to 90 percent of the RDA). Despite these differences, mean infant birthweights were similar for all three groups.

A study of African women reported a significant beneficial effect of prenatal food supplementation in women who would otherwise have been in negative energy balance.[43] An energy-dense food supplement in the form of biscuits, together with vitamin-fortified tea, was consumed by about 200 women as soon as pregnancy was confirmed. Mean duration of supplementation was 24 weeks. Birthweights of infants of the supplemented women were compared with birthweights of infants born during the four years preceding intervention. Supplementation was highly effective during the wet season, when food shortages and agricultural work caused negative energy balance, but was ineffective during the dry season, when women were in positive energy balance. The percentage of low-birthweight infants declined significantly, from 23.7 percent to 7.5 percent, during the wet season.

Another study evaluated the effectiveness of the Women, Infants, and Children (WIC) Supplemental Food Program on infant birthweights.[44] Several hundred women were enrolled in this study midway through pregnancy. Experimental subjects received vouchers for the purchase of protein-rich foods such as milk, eggs, and cheese, while the control subjects did not. The mean infant birthweight of the supplemented women was significantly greater than that of controls. However, when entry weights of the women were controlled for, the significance of the difference disappeared, except for WIC-supplemented women who smoked.

Another WIC evaluation study of high-risk pregnant women found the incidence of low-birthweight infants of WIC participants to be 6 percent, as expected.[45] In a group of high-risk pregnant women who qualified for WIC but did not enter the program, the incidence of low-birthweight infants was greater than 10 percent. In addition, the greater the number of WIC food vouchers received each month, the greater the mean birthweights of the infants.

The attempt to integrate the threads of information derived from these studies into the fabric of prenatal nutrition and fetal growth shows that the picture is not yet complete. It is becoming clearer, though, and to some researchers it looks like this.[46] Among women at risk of delivering low-birthweight infants, prenatal dietary supplementation can, in some instances, produce a slight rise (40 to 60 grams) in birthweights of infants. The degree of this rise seems to depend on the prior nutrition states of the women. A greater rise in birthweight is seen among thin or undernourished women, compared with heavier women. It also appears that prenatal dietary supplements can offset the fetal growth retardation caused by smoking. A protein-supplemented diet may be harmful to some women, particularly those with a previous history of bearing low-birthweight infants. In addition, protein seems no more effective than nutrient-dense food in general in raising birthweight.

The studies just discussed were of short duration. All of them examined the effect of prenatal nutrition intervention on the outcome of one pregnancy, most of them using infant birthweight as the indicator. In all cases, supplementation was initiated at some point after pregnancy was confirmed, in some cases as late as the third trimester. Most of the women studied had been at least moderately malnourished for years previously. All but one of the studies confirmed that nutrient supplementation positively affected outcomes. Infants of supplemented women weighed 1 to 3 ounces more than infants of unsupplemented women, and the incidence of low birthweights declined, or both. Earlier and longer-duration intervention produces more impressive results.

A longer-duration study Almost 200 women were involved in the Guatemalan Nutritional Supplementation study.[47] Women received either high-kcalorie (500 kcalories or more per week) or low-kcalorie (less than 500 kcalories per week) supplements containing vitamins and minerals. Some women received the high-kcalorie supplements during two consecutive pregnancies and an intervening lactation period, and their infants weighed up to 10 ounces (301 grams) more than infants of women in the low-kcalorie supplement group. Some women received high-kcalorie supplements for a shorter time—while breastfeeding a first child and during their next pregnancy—and their infants weighed up to 5 ounces (150 grams) more than infants of women in the low-kcalorie supplement group. These differences are two to three times greater than those observed in previous, shorter-term studies. The authors speculate that it is unrealistic to expect that malnutrition that develops over long periods of time can be overcome by nutrient supplementation sustained for only a few months during pregnancy, especially late in pregnancy. Earlier and longer intervention periods produce greater results.

As anticipated, the results of these studies leave many more threads to untangle. When does protein supplementation have a role in prenatal diet intervention? Why is prenatal protein supplementation harmful to some women? What does this imply for prenatal dietary intervention for these women? What effects will specific supplementation of deficient nutrients have on fetal growth? These and many other questions remain to be answered, but for the present, the need for intervention to correct inadequate diets during pregnancy is clear. The accompanying Practical Point describes present government efforts to provide assistance to pregnant women in the United States.

▸▸ PRACTICAL POINT

Maternal and Infant Assistance Programs

Pregnancy is a time of increased need. Nutrient needs increase; so do emotional and financial needs. If any of these needs are not met, the course and outcome of the pregnancy may be compromised. Frequently, low-income women may receive little, if any, prenatal care. They simply cannot afford it or are unaware of the free health services they are entitled to in their communities.

Teens, who may at first be unaware that they are pregnant, or who are trying to hide the fact that they are, may fail to seek help until late in pregnancy. Many women do not eat properly while they are pregnant because they are not aware of nutrition's importance, they do not know how to eat well, or they have too little money to purchase the food they need. Women can deal with all of these situations provided they can learn of the options available to them.

Several resources are available for anyone who may be in need of information or assistance:

- The Agriculture Extension Service provides many educational services and materials, including nutrition, food budgeting, and shopping information.
- Community hospitals and health clinics employ registered dietitians who can assist with meal planning and health care during pregnancy.
- The Women, Infants, and Children (WIC) Supplemental Food Program provides nutrition information and low-cost nutritious foods to low-income pregnant women and their children.
- The Food Stamp Program or other federal assistance programs may be available to some.

The health professional counseling such individuals can be invaluable in guiding them to the resources available.

Practices Incompatible with Pregnancy

Besides malnutrition, which presents many hazards to pregnancy, a variety of lifestyle factors can have adverse impacts. Pregnant women, given awareness of these, can make the choice to abstain from them. Table 3–2 summarizes the effects. The following sections describe the major ones.

Smoking and Smokeless Tobacco Use

Smoking never conveys a health advantage, and pregnancy dramatically magnifies the disadvantages of smoking. Smoking during pregnancy harms the mother and the development of the placenta, the embryo, the fetus, and the infant and child later in life. Smoking increases the risk of retarded development and complications at birth. Mislocation of the placenta, premature separation of the placenta, and vaginal bleeding are 92 percent more frequent among women who smoke more than one pack of cigarettes per day than among nonsmokers. Figure 3–2 shows one such effect of smoking during pregnancy. Smoking restricts the blood supply to the growing fetus and so limits the delivery of oxygen and nutrients and the removal of wastes. Tobacco smoke contains hundreds of compounds that are harmful, including nicotine and carbon monoxide. Carbon monoxide in the blood of a smoking mother may deprive the developing fetus of the oxygen necessary for optimal growth. The primary cause of most smoking-related fetal deaths is oxygen deprivation.

The risk of spontaneous abortion and neonatal death increases directly with increasing levels of maternal smoking. A positive association also exists between maternal smoking and sudden infant death syndrome (SIDS).[48] This relationship is found for frequency, quantity, and even postnatal exposure to cigarette smoke (passive smoking). The U.S. surgeon general's report concludes that "maternal smoking can be a direct cause of fetal or neonatal death in an otherwise normal infant."[49]

The U.S. surgeon general also states that of all *preventable* causes of low birthweight in the United States, smoking has the greatest impact.[50] The more the mother smokes, the greater the reduction in birthweight. On the average, infants of mothers who smoke weigh 200 grams (7.1 ounces, or almost ½ pound) less than those born to nonsmoking mothers. The deficit reflects a disproportionate lack of lean tissue—that is, body composition as well as size is abnormal. This low birthweight reflects a small-for-gestational-age profile, not preterm birth, indicating that maternal smoking directly slows fetal growth rate.[51] Fetal growth retardation of infants born to smokers is not due to other maternal factors.

Women who smoke during and after pregnancy have children with higher rates of morbidity and mortality up to five years of age.[52] Physical growth, mental development, and behavioral characteristics may be impaired in children at least up to the age of 11.[53]

The risks of smoking during pregnancy should therefore be taught from junior high school through college. Any woman who smokes and is considering pregnancy or who is already pregnant should be urged to quit or at least to cut back on the number of cigarettes smoked.

Table 3–2 Effects of Potentially Harmful Substances on the Fetus

Substance	Effect to Fetus
Cigarette smoke	Low birthweight; increased incidence of spontaneous abortion; nervous system disturbances; increased incidence of sudden infant death syndrome (SIDS); fetal death[a]
Caffeine	Central nervous system stimulant; increased incidence of spontaneous abortion[b]
Medications[c]	
Salicylates (large doses)	Pulmonary hypertension and neonatal bleeding
Acetaminophen	Renal failure
Anticonvulsants	Growth retardation and mental retardation
Oral progestogens, androgens, and estrogens	Masculinization and advanced bone age
Tetracyclines	Inhibition of bone growth; discoloration of teeth
Illicit drugs[d]	
Marijuana	Short-term irritability at birth
Heroin and methadone	Drug addiction and acute narcotic withdrawal symptoms (tremors; excessive, high-pitched crying; and disturbed sleep); low birthweight
Cocaine	Greater incidence of spontaneous abortion; uncontrolled jerking motion; paralysis; depressed interactive behavior; poor organizational response to environmental stimuli
Phencyclidine (PCP)	Facial malformations; tremors; low birthweight
Alcohol	Fetal alcohol syndrome: distorted facial features, low birthweight, impaired nervous system performance, cleft palate[e]
Nutrient excesses	
Vitamin A	Microcephaly (small head); hydrocephalus (water on the brain); spontaneous abortion[f]
Vitamin D	Hypercalcemia
Iodine	Congenital goiter
Heavy metals	
Lead	Spontaneous abortion; stillbirth; low birthweight; neurobehavioral deficits[g]
Mercury	Central nervous system damage[h]

Note: Virtually any substance may be teratogenic, depending on the dose, duration, time of exposure, and genetic makeup of the individual. Some viral infections such as rubella and herpes are teratogenic, as are irradiations of various types. This table presents but *some* of the vast array of substances known to be teratogenic in human beings.

[a]National Research Council, *Alternative Dietary Practices and Nutrition Abuses in Pregnancy, Summary Report* (Springfield, Va.: National Technical Information Service, 1982), p. 17; Pregnancy and infant health, in *Smoking and Health,* a report of the surgeon general, January 1979, available from Superintendent of Documents, U.S. Government Printing Office, Washington, DC 20202; T. A. Shepard, *Teratogenic Agents,* 3rd ed. (Baltimore: Johns Hopkins University Press, 1980), pp. 73–75.

[b]W. Srisuphan and M. B. Bracken, Caffeine consumption during pregnancy and association with late spontaneous abortion, *American Journal of Obstetrics and Gynecology* 154 (1986): 14–20.

[c]W. B. Deichmann and H. W. Gerarde, *Toxicology of Drugs and Chemicals* (New York: Academic Press, 1969), p. 683; E. M. Johnson and D. M. Kochhar, *Teratogenesis and Reproductive Toxicology* (New York: Springer-Verlag, 1983), p. 217.

[d]Johnson and Kochhar, 1983, pp. 221–235; I. J. Chasnoff and coauthors, Cocaine use in pregnancy, *New England Journal of Medicine* 313 (1985): 666–669.

[e]Johnson and Kochhar, 1983, pp. 216–217.

[f]Vitamin A and teratogenesis, *Lancet* 1 (1985): 319–320.

[g]K. N. Dietrich and coauthors, Low-level fetal lead exposure effect on neurobehavioral development in early infancy, *Pediatrics* 80 (1987): 721–730.

[h]Shepard, 1980, pp. 212–214.

Figure 3–2 Placenta Previa, an Effect of Smoking during Pregnancy
A placenta that slips down over the cervix this way may break and bleed, threatening the life of both mother and fetus.

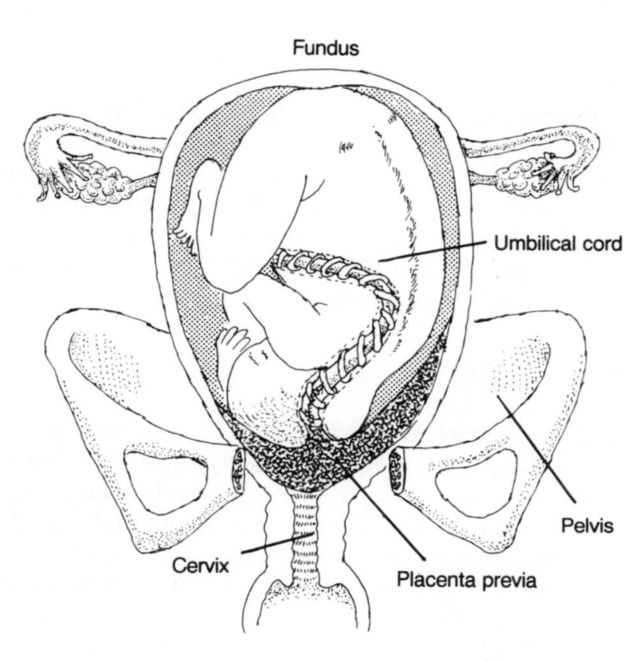

Research shows that tobacco chewing by pregnant women also affects the fetus adversely. Even in the absence of carbon monoxide inhalation, the constituents of tobacco absorbed into the bloodstream, which include lead, are damaging. Infants of mothers who chew tobacco, like those whose mothers smoke tobacco, have lower birthweights and a higher risk of fetal death than infants born to women who do not use tobacco.[54]

Caffeine

Caffeine crosses the placenta, and the developing fetus has a limited ability to metabolize it.[55] One study of pregnant women found that those who consumed more than 150 milligrams of caffeine per day were significantly more likely to experience late first- or second-trimester spontaneous abortion compared with noncaffeine users or light users (1 to 150 milligrams per day).[56] Further studies of this same design are necessary to confirm these results, but meanwhile, a prudent intake of caffeine during pregnancy would be less than 150 milligrams per day. One 6-ounce cup of coffee contains about 85 milligrams; a cup of tea or a 12-ounce cola beverage contains about 50 milligrams.[57] Caffeine is also in cocoa, some soft drinks, and some over-the-counter medicines.

The Canadian National Guidelines on Prenatal Nutrition are more liberal in their caffeine recommendation.[58] They advise moderate caffeine consumption during pregnancy and define *moderate* as less than 400 milligrams per day. For persons choosing to substitute herb teas for caffeine-containing beverages, they advise moderation in the use of herb teas during pregnancy due to potential teratogenic effects.

Medications

Drugs taken during pregnancy can cause serious congenital malformations (Accutane's effects were described in Chapter 2). Currently, more than 500,000 over-the-counter drugs are on the market. The public seems to be infatuated with pills, potions, and lotions. Many people routinely use antacids, aspirin, aspirin substitutes, and laxatives. A pregnant woman, however, should not take any pill, capsule, powder, or liquid medicine without consulting her health care provider. In chronic conditions such as epilepsy, which may require routine use of drugs, physician consultation may determine that dosages can be reduced or less powerful drugs substituted to minimize risk.

Illicit Drugs

Women should, of course, avoid the use of mind-altering drugs such as marijuana and cocaine at all times, but especially during pregnancy. In addition to effects such as interference with appetite, intellectual function, and coordination, these drugs pass easily through the placenta and are harmful to the fetus. Tetrahydrocannabinol (THC), the psychoactive ingredient in marijuana, is stored in the fatty tissue of the fetus and is potentially damaging.[59] Marijuana, like tobacco, elevates blood carbon monoxide concentrations and thus impairs oxygen delivery to the fetus.

Researchers studying more than 1000 pregnant women found that infants born to mothers who used marijuana throughout their pregnancies were 3 ounces lighter and two-tenths of an inch shorter than infants born to those who did not use marijuana.[60] More research is needed to confirm the effects of marijuana on fetal growth, but these results strengthen the assertion that marijuana adversely affects pregnancy outcome. The authors suggest that oxygen deprivation may be the mechanism by which marijuana impairs fetal growth.

A study of pregnant, cocaine-using women revealed a high incidence of spontaneous abortion compared with drug-free women.[61] In addition, infants exposed to cocaine during gestation had depressed interactive behavior and poor organizational responses to environmental stimuli. This study also suggests that infants exposed to cocaine are at risk for congenital malformations and perinatal mortality. Cocaine use during pregnancy correlates with impaired fetal growth as well.[62]

Many questions remain unanswered concerning illicit drugs and narcotics and their effects during pregnancy. What is known so far is that pregnancy and illicit drug use are a dangerous combination, to be avoided in all cases.

Alcohol

Alcohol consumption during pregnancy can cause irreversible brain damage and mental and physical retardation in the fetus. Fetal alcohol syndrome is the subject of Focal Point 3.

Nutrient Megadoses

All of the vitamins are toxic when taken in excess. The minerals are even more so, some of them at levels not far above the RDA. This is especially significant during pregnancy. The pregnant woman who is constantly being told to eat well and take care of herself may mistakenly assume that more is better regarding vitamin-mineral supplements. This is simply not true. A pregnant woman can obtain most of the vitamins and minerals she needs by eating whole foods and should take supplements only on the advice of a registered dietitian or a physician.

Whether a pregnancy is experienced by accident or by choice, the potential for many factors to exert harmful or beneficial influences on the pregnancy demonstrates what a pivotal time this is. If all goes well, and it often does, a pregnancy's full potential can be realized. Fortunately, for numerous women, control of many factors influencing pregnancy is within reach. The next chapter describes the ways in which a woman can use nutrition to weigh the odds in favor of having a healthy pregnancy and a thriving infant.

Chapter 3 Notes

1. M. Winick and P. Rosso, The effect of severe early malnutrition on cellular growth of the human brain, *Pediatric Research* 3 (1969): 181–184.
2. L. S. Hurley, *Developmental Nutrition* (Englewood Cliffs, N.J.: Prentice-Hall, 1980), pp. 94–109.
3. F. J. Zeman, Effect of protein deficiency during gestation on postnatal cellular development in the young rat, *Journal of Nutrition* 100 (1970): 530–538.
4. Zeman, 1970.
5. F. J. Zeman, R. E. Shrader, and L. H. Allen, Persistent effects of maternal protein deficiency in postnatal rats, *Nutrition Reports International* 7 (1973): 421–436.
6. P. R. Dallman, M. A. Siims, and E. C. Manies, Brain iron: Persistent deficiency following short-term iron deprivation in the young rat, *British Journal of Haematology* 31 (1975): 209.
7. L. S. Hurley and coauthors, The movement of zinc in maternal and fetal rat tissues in teratogenic zinc deficiency, *Teratology* 1 (1968): 216.
8. L. S. Hurley, J. Gowan, and H. Swenerton, Teratogenic effects of short-term and transitory zinc deficiency in rats, *Teratology* 4 (1971): 199–204.
9. L. S. Hurley, Teratogenic aspects of manganese, zinc and copper nutrition, *Physiological Reviews* 61 (1981): 249–295.
10. R. M. Pitkin, Nutritional influences during pregnancy, *Medical Clinics of North America* 61 (1977): 3–15.
11. Pitkin, 1977.
12. A. Petros-Barvazian and M. Behar, Low birthweight: What should be done to deal with this global problem? *WHO Chronicle* 32 (June 1978): 231–232; *New Trends and Approaches in the Delivery of Maternal and Child Care in Health Services,* sixth report of the WHO Expert Committee on Maternal and Child Health, as cited in *Journal of the American Dietetic Association* 71 (1977): 357.
13. National Institute of Child Health and Human Development, *Facts about Premature Birth,* HHS publication no. (NIH) 461–338-841–25324 (Washington, D.C.: Government Printing Office, 1985).
14. A. N. Antonov, Children born during the siege of Leningrad in 1942, *Journal of Pediatrics* 30 (1947): 250–259.
15. B. S. Burke and coauthors, The influence of nutrition during pregnancy upon the condition of the infant at birth, *Journal of Nutrition* 26 (1943): 569–583.
16. C. Phillipps and N. E. Johnson, The impact of quality of diet and other factors on birth weight of infants, *American Journal of Clinical Nutrition* 30 (1977): 215–225.
17. S. M. Garn, M. T. Keating, and F. Falkner, Hematological status and pregnancy outcomes, *American Journal of Clinical Nutrition* 34 (1981): 115–117.
18. J. A. Pritchard, The effects, if any, of maternal iron deficiency on pregnancy and lactation, report of the Eighty-second Ross Conference on Pediatric Research, *Iron Nutrition Revisited—Infancy, Childhood, Adolescence* (Columbus, Ohio: Ross Laboratories, 1981), pp. 89–94.
19. A. C. Yao and J. Lind, Effect of gravity on placental transfusion, *Lancet* 2 (1969): 505–508.
20. M. S. Rodriquez, A conspectus of research on folacin requirements of man, *Journal of Nutrition* 108 (1978): 1983–2103.
21. G. R. Wadsworth, Some historical aspects of knowledge about folate deficiency, *Nutrition* 27 (1973): 17–22.
22. *National Institute of Child Health and Development*, 1985; B. M. Hibbard, Folates and the fetus, *South African Medical Journal* 49 (1975): 1223–1226.
23. R. W. Smithells and coauthors, Further experience of vitamin supplementation for prevention of neural tube defect recurrences, *Lancet* 1 (1983): 1027–1031.
24. K. M. Laurence and coauthors, Double-blind randomised controlled trial of folate treatment before conception to prevent recurrence of neural tube defects, *British Medical Journal* 282 (1981): 1509–1511, as cited by M. Tolarova, Periconceptional

supplementation with vitamins and folic acid to prevent recurrence of cleft lip (letter), *Lancet* 2 (1982): 217.

25. N. Baumslag, T. Edelstein, and J. Metz, Reduction of incidence of prematurity by folic acid supplementation in pregnancy, *British Medical Journal* 1 (1970): 16–17.
26. S. Jameson, Effects of zinc deficiency in human reproduction, *Acta Medica Scandinavica Supplement* 593 (1976): 3–89.
27. F. F. Cherry and coauthors, Plasma zinc in hypertension/toxemia and other reproductive variables in adolescent pregnancy, *American Journal of Clinical Nutrition* 34 (1981): 2367–2375.
28. J. C. King, Assessment of techniques for determining human zinc requirements, *Journal of the American Dietetic Association* 86 (1986): 1523–1528.
29. J. N. Udall and coauthors, Interaction of maternal and neonatal obesity, *Pediatrics* 62 (1978): 17–21.
30. Vitamin A and teratogenesis, *Lancet* 1 (1985): 319–320.
31. E. P. Norkus and P. Rosso, Changes in ascorbic acid metabolism of the offspring following high maternal intake of this vitamin in the pregnant guinea pig, *Annals of the New York Academy of Sciences* 258 (1975): 401–409.
32. M. Levine, New concepts in the biology and biochemistry of ascorbic acid, *New England Journal of Medicine* 314 (1986): 892–902.
33. G. P. Ravelli, Z. A. Stein, and M. W. Susser, Obesity in young men after famine exposure in utero and early infancy, *New England Journal of Medicine* 295 (1976): 349–353, as cited by J. Kirtland and M. I. Gurr, Adipose tissue cellularity: A review: II. The relationship between cellularity and obesity, *International Journal of Obesity* 3 (1979): 15–55.
34. N. M. Lien, K. K. Meyer, and M. Winick, Early malnutrition and "late" adoption: A study of the effects on the development of Korean orphans adopted into American families, *American Journal of Clinical Nutrition* 30 (1977): 1734–1739.
35. National Center for Health Statistics, 3700 West Highway, Hyattsville, Md. (personal communication).
36. A. Petros-Barvazian and M. Behar, Low birthweight—What should be done to deal with this global problem? *WHO Chronicle* 32 (1978): 231–232.
37. J. H. Ebbs, F. F. Tisdall, and W. A. Scott, The influence of prenatal diet on the mother and child, *Journal of Nutrition* 22 (1941): 515–526.
38. D. Rush, Z. Stein, and M. Susser, *Diet in Pregnancy: A Randomized Controlled Trial of Prenatal Nutritional Supplementation* (New York: Alan Liss, 1979).
39. J. O. Mora and coauthors, Nutritional supplementation and the outcome of pregnancy: I. Birthweight, *American Journal of Clinical Nutrition* 32 (1979): 455–462.
40. J. P. Habicht and coauthors, Relation of maternal supplementary feeding during pregnancy to birth weight and other sociobiological factors, in Nutrition and fetal development, Proceedings of the Symposium on Nutrition and Fetal Development, 1974, *Current Concepts in Nutrition*, vol. 3, ed. M. Winick (New York: Wiley, 1974), pp. 127–146.
41. D. Rush, Nutrition services during pregnancy and birthweight: A retrospective matched-pair analysis, *Canadian Medical Association Journal* 125 (1981): 574–576.
42. S. O. Adams, G. D. Barr, and R. L. Huenemann, Effect of nutritional supplementation in pregnancy, *Journal of the American Dietetic Association* 72 (1978): 144–147.
43. A. M. Prentice and coauthors, Increased birthweight after prenatal dietary supplementation of rural African women, *American Journal of Clinical Nutrition* 46 (1987): 912–925.
44. J. Metcoff and coauthors, Effect of food supplementation (WIC) during pregnancy on birth weight, *American Journal of Clinical Nutrition* 41 (1985): 933–947.
45. E. T. Kennedy and coauthors, Evaluation of the effect of WIC supplemental feeding on birthweight, *Journal of the American Dietetic Association* 80 (1982): 220–226.
46. M. Susser, Prenatal nutrition, birthweight, and psychological development: An overview of experiments, quasi-experiments, and natural experiments in the past decade, *American Journal of Clinical Nutrition Supplement* 34 (1981): 784–803.
47. J. Villar and J. Rivera, Nutritional supplementation during two consecutive pregnancies and the interim lactation period: Effect on birth weight, *Pediatrics* 81 (1988): 51–57.
48. Pregnancy and infant health, in *Smoking and Health*, a report (January 1979), of the surgeon general available from Superintendent of Documents, U.S. Government Printing Office, Washington, D.C. 20402.
49. Pregnancy and infant health, 1979.
50. National Research Council, Food and Nutrition Board, Commission on Life Sciences, Committee on Nutrition of the Mother and Preschool Child, *Alternative Dietary Practices and Nutrition Abuses in Pregnancy, Summary Report* (Washington, D.C.: National Academy Press, 1982), p. 16.
51. Pregnancy and infant health, 1979.
52. National Research Council, 1982, p. 17.
53. Pregnancy and infant health, 1979.
54. R. C. Verma, M. Chansoriya, and K. K. Kaul, Effect of tobacco chewing by mothers on fetal outcome, *Indian Pediatrics* 20 (1983): 105–111; K. Krishina, Tobacco chewing in pregnancy, *British Journal of Obstetrics and Gynaecology* 85 (1978): 726–728.
55. National Institute of Nutrition in Canada, Caffeine: A perspective on current concerns, *Nutrition Today*, July–August 1987, pp. 36–38.
56. W. Srisuphan and M. B. Bracken, Caffeine consumption during pregnancy and association with late spontaneous abortion, *American Journal of Obstetrics and Gynecology* 154 (1986): 14–20.
57. National Institute of Nutrition in Canada, 1987.
58. Department of Health and Welfare, Federal-Provincial Subcommittee on Nutrition, Canada's National Guidelines on Prenatal Nutrition, *Nutrition Today*, July–August 1987, pp. 34–35.
59. J. C. Gampel, Marijuana and health, in *Drug Use in Society: Proceedings of the Marijuana and Health Conference* (Washington, D.C.: Council on Marijuana and Health, 1984), pp. 1–9.
60. B. Zuckerman and coauthors, Effects of maternal marijuana and cocaine use on fetal growth, *New England Journal of Medicine* 320 (1989): 762–768.
61. I. J. Chasnoff, Cocaine use in pregnancy, *New England Journal of Medicine* 313 (1985): 666–669.
62. Zuckerman and coauthors, 1989.

▶ Focal Point 3

Fetal Alcohol Syndrome

Every pregnant woman wants her child to be "normal"—to be free of mental or physical defects. When asked if she wants a boy or a girl, a pregnant woman will oftentimes reply that it does not really matter as long as the child is healthy.

Some women bear children who are not perfectly healthy, however, and often this is beyond their control. Much of an infant's development is genetically determined, and so some errors are inherited. One such disorder, PKU, is highlighted in Focal Point 2. For such a defect, prevention is impossible, but the disorder can be controlled with proper treatment. The subject of this discussion is fetal alcohol syndrome (FAS), a defect for which the reverse is true. FAS can only be prevented; no treatment is possible. In this situation, the expectant mother can take control; she can actively reject beverages that can harm her child.

fetal alcohol syndrome (FAS): the cluster of symptoms seen in an infant or child whose mother consumed excess alcohol during her pregnancy; includes mental impairment, growth retardation, and facial malformations.

Fetal Alcohol Syndrome

The many mental and physical characteristics that FAS children have in common define the syndrome. The one characteristic that all mothers of FAS infants have in common is alcohol consumption during pregnancy.

FAS is not always recognized by physicians; therefore, its incidence can only be estimated. The actual numbers may be higher than we realize. The incidence of FAS in the United States is estimated to range from 0.4 to 2.6 per 1000 live births.[1] The lower rates are seen in rural areas, where the alcoholism rate is low, and the higher rates, in metropolitan areas, where alcoholism is more prevalent.

Effects of FAS

Alcohol damages the fetus in several ways. It retards growth, impairs development of the central nervous system, and causes facial malformations. Abnormalities in these three areas are used as the minimum criteria for diagnosing FAS. Some infants suffer the consequences of alcohol consumption yet do not meet all these criteria for diagnosis. The term fetal alcohol effects (FAE) is used to describe such cases. The exact criteria for FAS are as follows:[2]

fetal alcohol effects (FAE): effects of maternal alcohol consumption on the fetus, not severe enough to be diagnosed as FAS, but still detectable.

1. *Prenatal and/or postnatal growth retardation with weight, length, and/or head circumference below the 10th percentile on growth charts.*

It is easy to see that the infant with FAS is much smaller than others, even at birth. The head circumference is smaller than would be expected, even for

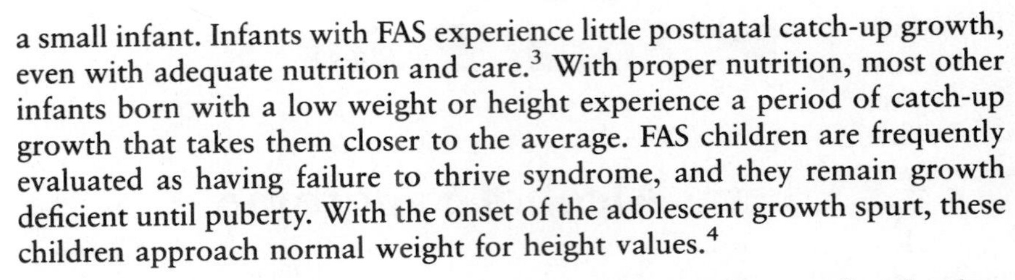

a small infant. Infants with FAS experience little postnatal catch-up growth, even with adequate nutrition and care.[3] With proper nutrition, most other infants born with a low weight or height experience a period of catch-up growth that takes them closer to the average. FAS children are frequently evaluated as having failure to thrive syndrome, and they remain growth deficient until puberty. With the onset of the adolescent growth spurt, these children approach normal weight for height values.[4]

failure to thrive syndrome: failure of a child to develop mentally and physically.

2. *Central nervous system involvement with neurologic abnormality, developmental delay, or intellectual impairment.* Signs may include:

- Low IQ.
- Poor coordination.
- Extreme nervousness.
- Irritability.
- Hyperactivity.
- Fine motor problems.
- Delayed gross motor development.

FAS is the third most common cause of mental retardation in the United States.[5] In addition to their mental disabilities, FAS children often have behavioral problems. They are irritable and fussy as infants and difficult to control as children.

3. *Characteristic facial disfigurations.*

The most obvious symptoms of FAS are the abnormal facial features, as illustrated in the photograph. These facial abnormalities reflect inadequate development of the brain, as well as of parts of the face, and they are permanent. The child does not outgrow them.

This is an important point. The FAS child's physical features are different from those of normal children, but these features alone are not a handicap. What is a handicap is that the brain's capabilities mirror the facial structure. The more severe the facial characteristics, the more severe the mental function impairment. Alcohol is responsible for both.

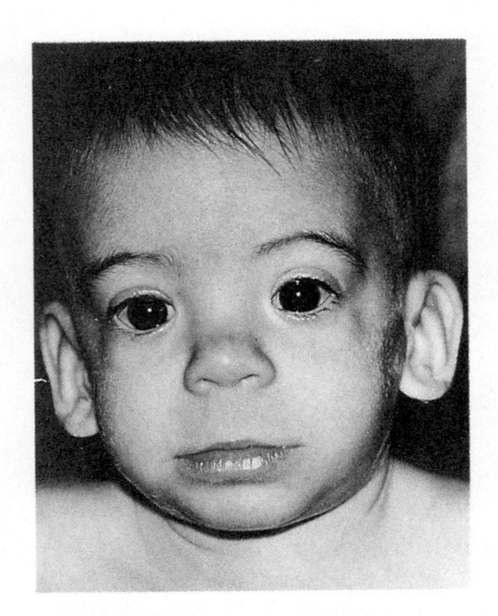

Characteristic facial abnormalities of FAS include low nasal bridge, short eyelid opening, underdeveloped groove in the center of the upper lip below the nose, thin reddish upper lip, small midface, short nose, and small head circumference.

Another effect of alcohol is that many FAS infants are too weak to suck effectively. Sucking is an infant's natural response to hunger. Sucking develops facial muscles and enables the infant to obtain nutrients from either a bottle of formula or a mother's breast. An infant who does not suck vigorously may appear apathetic to feeding. This serious feeding problem can easily lead to malnutrition, which further limits the growth of the infant, who is destined never to reach full potential.

In addition to the effects seen in the FAS child, alcohol prevents the birth of some children. Spontaneous abortions occur more frequently in women who drink more than two alcoholic beverages daily than in women who do not drink.[6]

How FAS Arises: Alcohol Metabolism

When a person drinks an alcoholic beverage, the alcohol moves rapidly from the digestive system into the blood. The alcohol-laden blood then enters the liver, which can detoxify it. There is a limit, however, to the amount of alcohol the liver can process in a given time. The potential for fetal damage arises when the liver receives more alcohol than it can handle. Alcohol that is not detoxified on the first pass through the liver enters the general circulation; once in the bloodstream, it circulates to all parts of the body.

Focal Point 8 discusses alcohol metabolism in more detail.

In a pregnant woman, alcohol freely crosses the placenta to the fetus. Fetal blood-alcohol concentrations rise until they reach an equilibrium with maternal blood-alcohol concentrations.[7] This might not seem to be a problem if the mother is functioning well (is not drunk), but while blood levels are the same for the mother and unborn child, the fetus's body is significantly smaller, and its detoxification system is less developed. Blood-alcohol concentrations tend to fall more slowly in the fetus than in the mother, and alcohol can be detected in fetal blood after it has totally disappeared from maternal blood. As Dr. C. F. Enloe, past editor of *Nutrition Today,* has observed: "After she [the mother] has become drunk she usually has a hangover. That will pass away in a few hours. For the fetus, the hangover may last a lifetime."[8]

How Much Is Too Much?

The surgeon general has issued a statement that pregnant women should drink absolutely no alcohol. Dr. Enloe agrees. So far, no "safe" level of alcohol consumption during pregnancy has been established, and there is no evidence of benefits to mother or fetus. Thus such advice is in the best interest of both mother and fetus.

Dr. Enloe creates the following hypothetical situation. Suppose a pregnant woman drinks *one* martini on an empty stomach. The alcohol will be rapidly absorbed and will quite likely be more than the liver can readily detoxify. Excess alcohol will circulate through the body to the uterus, cross the placenta, and enter the fetal bloodstream. If vital cells are being developed at that moment, the exposure to alcohol may irreversibly damage those cells.

The American Council on Science and Health (ACSH) takes a more liberal stand that many health care providers do not support. Its position statement recommends that women be cautious about alcohol use during pregnancy, and advises a daily limit of two 12-ounce beers, two 4-ounce glasses of wine, or two drinks with 1½ ounces of 80-proof liquor.[9] It maintains that the health risks associated with this level of alcohol consumption are low or nonexistent. While it is true that no risks have been observed at a low level of drinking, absence of risk cannot be conclusively established.[10]

The percentage of alcohol in distilled liquor is stated as proof: 100-proof liquor is 50% alcohol; 90 proof is 45%, and so forth. A drink is a dose of any alcoholic beverage that delivers ½ oz of pure ethanol:

- 3 to 4 oz wine.
- 10 oz wine cooler.
- 12 oz beer.
- 1 oz hard liquor (whiskey, gin, rum, vodka).

The ACSH issued this statement after a review of current scientific evidence examining the relationship between alcohol consumption and reproductive effects. While it is clear that excessive alcohol consumption is hazardous to the fetus, it is not clear at what level, if any, alcohol consumption is safe. The ACSH recommendation allows the many women who drink in this country to continue limited, occasional drinking during pregnancy without feelings of guilt.

The ACSH supports the pregnant woman who has a margarita on a Saturday night, sips a glass of champagne on her birthday, or drinks a beer at a ball game. She should not endure the nine months of her pregnancy, they say, with feelings of fear or guilt. Such stress may, indeed, be more damaging to the unborn child than an occasional alcoholic beverage.

No doubt it is important to distinguish between heavy drinking, which almost invariably affects the unborn child, and moderate drinking, which in some cases does not. The difference between the surgeon general's "no-drinks" position and the ACSH's "occasional-drinks" position is one of philosophy. There is undeniably a risk associated with *any* drinking during pregnancy; the ACSH position simply states that the risk is small enough to be negligible.

Indeed, in comparison with other risks in pregnancy, *moderate* drinking does not stand out as a *great* danger. Many other factors affect the outcome of pregnancy. The mother's socioeconomic level, age, emotional stability, diet, drug use, smoking habits, genetic makeup, and prenatal care all play at least as great a role as an alcoholic beverage or two. However, the effects of alcohol during pregnancy may be intensified by nicotine, caffeine, or other drugs. And in studies of both women and animals, the extent of fetal damage seems to correlate directly with the quantity of alcohol the mother consumes.[11]

Now it remains to define moderate drinking. A pregnant woman need not be an alcoholic in order to give birth to a baby with FAS characteristics. She only needs to drink in excess of the liver's capacity to detoxify. Reports from mothers of FAS children reveal they consumed an average of five or more drinks daily throughout their pregnancies. Does this mean that four drinks or less per day are safe? Probably not, for researchers cannot be sure they are obtaining accurate measures of alcohol consumption. Many women are reluctant to admit they drank during their pregnancies. Mothers of deformed infants often feel guilty and responsible for their children's defects. It is not surprising that they are not willing to be interviewed or to let it be known that they have understated their alcohol intakes.

Most FAS studies speak only of average intake. Little is known about the effects of periodic binge drinking. How does a drinking binge at a critical time compare with the same amount of alcohol consumed over a longer period of time? Most likely it depends on the frequency of binges, the quantity consumed, and fetal development at the time.

In conclusion, it is impossible to say whether small amounts of alcohol are safe during pregnancy. The risk associated with occasional drinking is less than that associated with heavy drinking, but why take any risk? The mother who chooses to drink during pregnancy, even moderately, is at greater risk than the mother who abstains completely.

There seems to be a gap between people's understanding of the effects of alcohol on an unborn child and their estimation of how many drinks constitutes a "moderate" level of safe drinking. One study reported that 90 percent of those surveyed were aware of the risks of drinking during pregnancy, but thought that an average of three drinks daily was safe.[12] We believe it was wise for the surgeon general to have taken such a conservative position after all.

When Is the Damage Done?

There is no time when drinking excessively is safe. The type and extent of abnormality observed in an FAS infant seems to depend on the developmental events that are occurring at the time the fetus is exposed to alcohol.

In experiments on laboratory animals, the effects of alcohol on fetal development are most marked when the female takes alcohol during the earliest period—that of organ formation. Effects also appear when a dose of alcohol elevates the female's blood-alcohol level immediately prior to conception; the amount of alcohol in the female's blood is critical. Male alcohol ingestion may also affect fertility and fetal development. Animal studies have found decreased litter size, birthweight, survival, and learning ability in the offspring of males consuming alcohol prior to conception.[13] One human study found an association between paternal alcohol intake one month prior to conception (defined as an average of two or more drinks daily or at least five drinks on one occasion) and decreased infant birthweight.[14] This relationship was independent of parents' smoking or maternal use of alcohol, caffeine, or other drugs.

Different dangers are associated with each trimester of pregnancy.[15] In the first trimester, developing organs such as the brain, heart, and kidneys may be malformed. During the second trimester, the risk of spontaneous abortion increases. During the third trimester, when the fetus is fully formed and rapidly growing, body and brain growth may be retarded.

Maternal Malnutrition

One of the health hazards associated with alcohol abuse is malnutrition. Alcohol depresses appetite because it produces euphoria, so that heavy drinkers usually eat poorly, if at all. Alcohol also attacks the digestive tract lining, causing pain and malabsorption of nutrients.

Many women who abuse alcohol can easily drink 100 grams of ethanol a day. This amount is roughly equivalent to eight beers, a pint of whiskey, or a bottle of wine. Alcohol provides 7.1 kcalories per gram, so the daily kcaloric intake from alcohol alone can be over 700 kcalories. The woman who derives so many kcalories from a source that has no nutritive value finds it difficult to obtain, from the additional kcalories she can consume, the many essential nutrients she needs to maintain her health and support the fetus's growth.

Ethanol is the alcohol in beer, wine, and liquor. For purposes of comparison, 100 g ethanol is more than 3 oz (1 oz is equivalent to 28 g). An ounce of 100-proof whiskey (1 "drink") is 50% alcohol, so it requires 6 or 7 drinks to imbibe 3 oz of pure alcohol.

Not only do alcohol abusers suffer malnutrition from lack of food, but even if they eat well, the direct effects of alcohol take their toll. Alcohol hinders the absorption, alters the metabolism, and increases the excretion of many nutrients, so that malnutrition can occur even in a well-fed drinker. Many nutrient deficiencies are associated with alcohol abuse. For example, folate deficiency is common in people with a history of alcoholism. During pregnancy, the need for this vitamin more than doubles. Women with poor nutrition status and such high nutrient needs put themselves in a precarious situation when they drink alcohol.

Thus an alcohol-abusing pregnant woman harms her unborn child not only by consuming alcohol but also by not consuming food or not getting the necessary nutrients from the food she does consume. This combination enhances the likelihood of malnutrition and a poorly developed infant. However, it is important to realize that malnutrition is not the cause of FAS. It is true that mothers of FAS children often have unbalanced diets and nutrient deficiencies. It is also true that malnutrition may augment the clinical signs seen in these children, but it is the *alcohol* that is the determining factor. An adequate diet will not prevent FAS if alcohol is abused during the pregnancy; without alcohol, the pattern of FAS is not seen.

Counseling and Prevention

The health care provider who realizes a pregnant woman drinks heavily and has no intention of quitting faces a difficult challenge. Drinking is more important to this woman than the theoretical possibility of harming an infant she has never touched and does not yet love. The task is to provide her with the information that alcohol causes birth defects, and make it clear that the connection is well established. She should understand that the only way to be sure of excluding the possibility of this kind of retardation is to abstain from drinking altogether throughout the pregnancy. The decision to abstain or not is hers, however, not the health care provider's. The responsibility of the health care provider ends when the woman has been given the information. Pushing the woman to make the decision will alienate her. She will thereafter be less inclined to confide in those trying to help her. This will also limit opportunities to guide her in other nutrition choices.

If she does decide to quit drinking, at least during the pregnancy, she deserves a pat on the back. Let her know she has made a wise decision and that the sooner she abstains, the less risk for her unborn child.

At the other extreme is the anxious woman who is usually quite careful of her health but who had a cocktail with friends during the first month of her pregnancy, before she even knew she was pregnant. Now she fears her infant will be born with mental and physical defects. In such a situation, the chances are very small that harm will have been done. The woman should be reassured that such small quantities of alcohol often do not harm the developing infant. Many drinkers, even heavy drinkers, bear normal babies. Furthermore, and even more important, the episode is in the past, and nothing can be done to change it. The choices she makes now affect the present and the future, not the past.

A pregnancy can be a long, suspenseful experience, especially if parents fear an imperfect outcome. The health care provider should minimize the things that cannot be changed and emphasize what can be done now and in the future to ensure the best possible outcome.

The one happy note in this FAS story is that of the leading causes of mental retardation, it is the only one that is totally preventable. Every female, indeed every person, should know the potential dangers of alcohol use during pregnancy. Proper health care instructions and education should begin *before* conception. FAS information should be included in all classes that discuss birth control, sexual intercourse, or pregnancy.* The message should be clear: FAS can be prevented by maternal avoidance of alcohol.

---

*An excellent ten-minute film entitled *Born Drunk: Fetal Alcohol Syndrome* is available from ABC Wide World of Learning, Inc., 1330 Avenue of the Americas, New York, NY 10019, (212) 887–5000.

Focal Point 3 Notes

1. F. L. Iber, Fetal alcohol syndrome, *Nutrition Today,* September–October 1980, pp. 4–11.
2. H. L. Rosett and L. Weiner, Alcohol and pregnancy: A clinical perspective, *Annual Review of Medicine* 36 (1985): 73–80.
3. M. Lee and J. Leichter, Alcohol and the fetus, in *Adverse Effects of Foods,* ed. E. F. P. Jelliffe and D. B. Jelliffe (New York: Plenum Press, 1982), pp. 245–251.

4. A. P. Streissguth, S. K. Clarren, and K. L. Jones, Natural history of the fetal alcohol syndrome: A 10-year follow-up of eleven patients, *Lancet* (1985): 85–91.
5. A report by the American Council on Science and Health, *Alcohol Use during Pregnancy,* 1981.
6. Iber, 1980.
7. C. F. Enloe, How alcohol affects the developing fetus, *Nutrition Today,* September–October 1980, pp. 12–15.
8. Enloe, 1980.
9. American Council on Science and Health, 1981.
10. H. L. Rosett and L. Weiner, Alcohol and pregnancy: A clinical perspective, *Annual Review of Medicine* 36 (1985): 73–80.
11. W. S. Beagle, Fetal alcohol syndrome: A review, *Journal of the American Dietetic Association* 79 (1981): 274–276.
12. R. E. Little and coauthors, Public awareness and knowledge about the risks of drinking during pregnancy in Multnomah County, Oregon, *American Journal of Public Health* 71 (1981): 312–314.
13. L. F. Soyka and J. M. Joffe, Male mediated drug effects on offspring, *Progress in Clinical and Biological Research* 36 (1980): 49–66.
14. R. E. Little and C. F. Sing, Father's drinking and infant birth weight, *Teratology* 36 (1987): 59–65.
15. Rosett and Weiner, 1985.

Nutrition during Pregnancy: The Normal Course

4

First Born by Hugo Robus.

Chapter 3 revealed the importance of nutrition to pregnancy by describing the effects of malnutrition. This chapter emphasizes the positive role adequate nutrition plays in pregnancy, and offers instruction for its management.

Between the moment of conception and the moment of birth, innumerable events determine the course and outcome of fetal development and, ultimately, the health of the newborn infant. Many of these events are beyond control, but a woman's nutrition throughout the teen and adult years, including pregnancy, is within her control, provided she has the knowledge and the means to attend to it. She needs to be motivated, though, and she will be, if she understands nutrition's importance during pregnancy. Nutrition is critical to maternal health, prenatal development, and the development of the child long after birth.

Growth and Development

Physiological events during pregnancy bring major changes to the mother's body, creating a whole new organ within it—the placenta—and producing a whole new human being. These events are reviewed here as a background for the consideration of nutrition in pregnancy.

Changes in Maternal Physiology

A woman's body changes dramatically during pregnancy. Her uterus and its supporting muscles increase in size, her breasts grow and prepare to produce milk, and her blood volume increases by half to accommodate the added load of materials to be carried. The normal weight gain of mother and fetus during pregnancy amounts to about 22 to 27 pounds.

Placental Development and Function

In the early days of pregnancy, because the zygote is so small, the nutrient-rich cells of the uterine lining provide adequate fuel and raw materials by diffusion alone. Soon, though, a specialized system is required to continue nourishing the developing embryo and fetus. The placenta develops as an interweaving of fetal and maternal blood vessels embedded in the uterine wall. A healthy placenta is indispensable to fetal development, and nutrition, in turn, is crucial to placental efficiency.

placenta: the bed of tissue that forms in the uterine wall to provide an interface between maternal and fetal circulatory systems; it is composed of villi formed from the outer membranes that surround the embryo and within which embryonic blood vessels lie; and these villi are surrounded by pools of maternal blood within the uterine wall.

The fetal portion of the placenta develops when the membranes that surround the embryo invade the uterine wall and form villi, within which embryonic blood vessels lie. The uterus responds by allowing pools of blood to form around these villi so that maternal and fetal blood will be in close proximity (see Figure 4–1).

Maternal blood never leaves the mother's circulatory system. It enters the placental tissue through small artery branches of the uterus, circulates around the fetus's villi, and leaves via the uterine vein. Similarly, fetal blood never leaves the fetal vessels. It circulates through the umbilical artery and vein and their branches within the villi. Within the spongelike endometrium, exchange of materials takes place, but no actual mingling of fetal and maternal blood occurs.

The placenta is a versatile, metabolically active organ. It transfers oxygen and nutrients to the fetus and returns waste products to the mother. Nutrients pass from maternal blood through the placenta into fetal blood; waste products move from fetal blood through the placenta into maternal blood. By exchanging oxygen, nutrients, and waste products, the placenta provides the respiratory, absorptive, and excretory functions that the fetus's lungs, gastrointestinal tract, and kidneys will provide after birth.

Far from being passive in its transport of molecules, the placenta is a highly metabolic organ with some 60 enzymes and several hormones of its own. Much like muscles or other body tissues, it uses energy fuels to support its work. The placenta metabolizes glucose for its own energy needs, as well as actively pumping it into the fetal bloodstream. It synthesizes protein and fatty acids from small precursors. It synthesizes and secretes hormones and gathers up hormones from elsewhere in the body.

Of the placental hormones, most notable are human chorionic gonadotropin (HCG) and placental lactogen.[1] Early placental secretion of these hormones leads to later placental production of progesterone and estrogen. Placental synthesis of the major estrogen of pregnancy, estriol, requires a precursor from the fetal adrenal gland. Together, the placenta and fetal adrenals produce the estriol required to maintain pregnancy.

HCG increases rapidly during early pregnancy, reaching its maximum concentration in the first eight to ten weeks, and then diminishing to a low but constant concentration at the end of the third month. Most pregnancy tests are based on the detection of this hormone in urine or plasma.

The secretion of placental lactogen starts at low levels and increases progressively. Late in pregnancy, high placental lactogen concentrations stimulate growth of the mammary glands. Placental lactogen also inhibits insulin's action on adipose and muscle tissue. Normally, insulin induces these tissues to take up amino acids, glucose, and fat, but lactogen keeps these fuels in circulation to feed the fetus and mammary glands.

Blood flow to the placenta increases dramatically during pregnancy, accounting for as much as 25 percent of the cardiac output by term.[2] The rates of maternal and fetal blood flow influence the exchange of substances across the placenta. For example, as the flow of maternal blood increases, the exchange of oxygen improves.[3] Water, oxygen, carbon dioxide, and electrolytes cross the placenta by passive diffusion, while glucose, calcium, phosphorus, magnesium, and other nutrients cross it via active transport. Fetal accumulation of these nutrients increases throughout pregnancy, becoming greatest during the third trimester.

Glucose is the main energy source for fetal growth, and as mentioned, the placenta actively transfers it to the fetus. The placenta also stores glycogen, reflecting the vital importance of maintaining the glucose supply. Placental glycogen stores increase early in pregnancy, but decline during later pregnancy as they give up their glucose to the rapidly growing fetus. By the third trimester, the fetus is producing its own insulin. This insulin promotes fetal storage of glycogen as well as of protein and fat for use after delivery.[4]

human chorionic gonadotropin (HCG): a hormone produced by the chorionic villi of the placenta that stimulates the secretion of progesterone and estrogen.

placental lactogen: a hormone produced by the placenta that stimulates the metabolism of glucose to fat.

estriol (ESS-tree-ol): the major estrogen of pregnancy, produced by the placenta.

cardiac output: the amount of blood discharged per minute from the heart.

passive diffusion: the tendency of gas, liquids, or solids to move from a region of high concentration to one of lower concentration.

active transport: the energy-requiring process of concentrating a substance inside or outside a membrane by pumping it across, usually dependent on ATP (adenosine triphosphase).

Figure 4–1 The Development of the Placenta
Within the first week of cell divisions, the zygote begins to develop five layers of cells, of which the inner three (already described) will become the embryonic and fetal organs. The outer two layers, called here simply outer embryonic membranes, form structures to protect and nourish the embryo (and then the fetus): the placenta and the amniotic sac (see A).[a]

The placenta develops partly from the outer embryonic membranes and partly from the mother's uterine endometrium (lining); it serves as a bed in which maternal and fetal circulatory systems interlock, facilitating transfer of nutrients and oxygen to the fetal blood, and wastes to the maternal blood. The steps in placental development are as follows.

The zygote first affixes itself to the uterine wall; then its outer membranes form projections, or villi, that release enzymes and digest their way into the uterine endometrium. As these elongating villi penetrate the maternal tissue, the zygote follows them, burrowing into the endometrium, until by the end of the second week it is completely implanted there, and its outer membranes have formed about 12 to 14 villi that branch into the maternal endometrium. As they continue invading the endometrium, the villi's enzymes rupture maternal blood vessels, which leak blood into the endometrium's soft tissue, forming tiny lakes, or lacunae (see B). The villi also produce an anticoagulant, which keeps the maternal blood from clotting for as long as the placenta is present there.

By the end of the third week, the embryo already has a closed circulatory system, complete with a pumping heart. A blood vessel of that system loops from the embryo into each of the villi and back again. Only two barriers separate these embryonic blood vessels from pools of maternal blood in which they lie: their own walls, and the membranes of the villi. Thus the embryo's blood vessels can almost be said to bathe in the lacunae of maternal blood. In a sense, that part of the maternal circulatory system that passes through the placenta is an open system; it leaves maternal arteries to circulate in pools, then collects into maternal veins. It never mingles with the embryo's blood, however, for the embryonic circulatory system remains closed. By the time a well-defined placenta has formed, eight weeks have passed, and the embryo is now a fetus.

As the embryo grows into a fetus, it backs *out* of the uterine wall into the uterine cavity, remaining attached to the placenta by a stalk, the umbilical cord. The original outer membranes still surround the fetal portion of the placenta (that is, the villi), the umbilical cord, and the fetus itself. Around the fetus, they balloon out, forming the amniotic sac, which is filled with fluid that bathes, helps nourish, and cushions the fetus until birth (see C).

The umbilical cord has now become a trunk line that connects the growing fetus to the placenta. A major fetal artery runs down this cord into the placenta, and ramifies into branches, one feeding into each of the villi, which are now highly complex structures rooted deeply in the maternal endothelium. Each artery branches into capillaries in its villus, then collects into a fetal vein that returns blood through the main vein in the umbilical cord back to the fetal circulation (see D).

At birth, all of these structures are shed. First the outer membranes of the amniotic sac rupture and the infant is born, still connected to the placenta by the umbilical cord. Then the placenta is released from the uterine wall and expelled, together with all its associated membranes.

[a]The zygote, technically, is called a *blastocyst* by the end of seven days' development. The outer cell layers are the *trophoblastic membranes*—the outermost being the *syncytiotrophoblast* (later the *chorion*) and the inner, the *cellular trophoblast*. These two trophoblastic layers grow the villi that invade the maternal endometrium to form the placenta and also contribute the tissues of the *amniotic sac*.

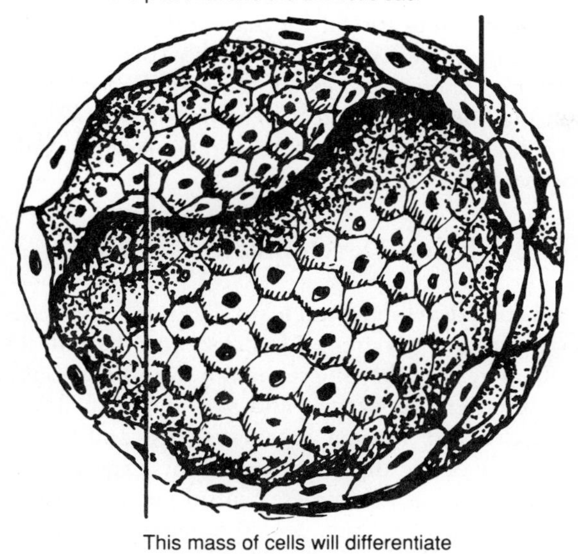

The zygote, still floating free after a few days of cell division.

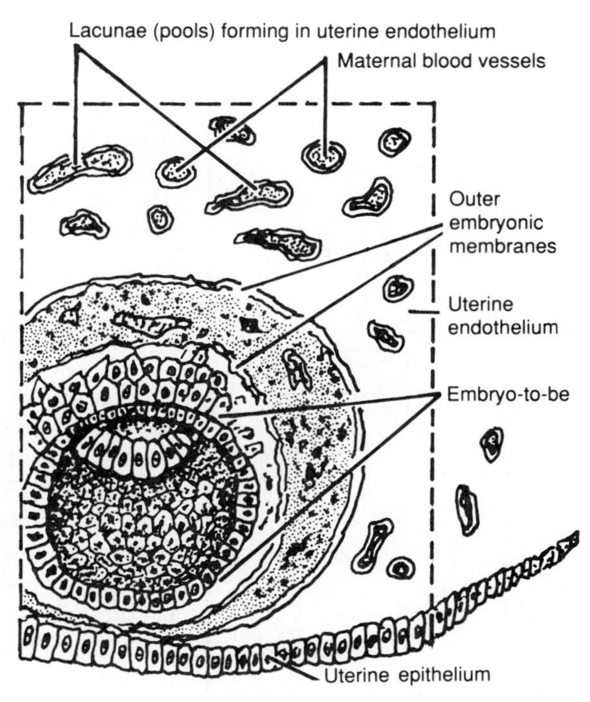

Implantation. A few days later, the embryo-to-be is completely embedded in the uterine wall.

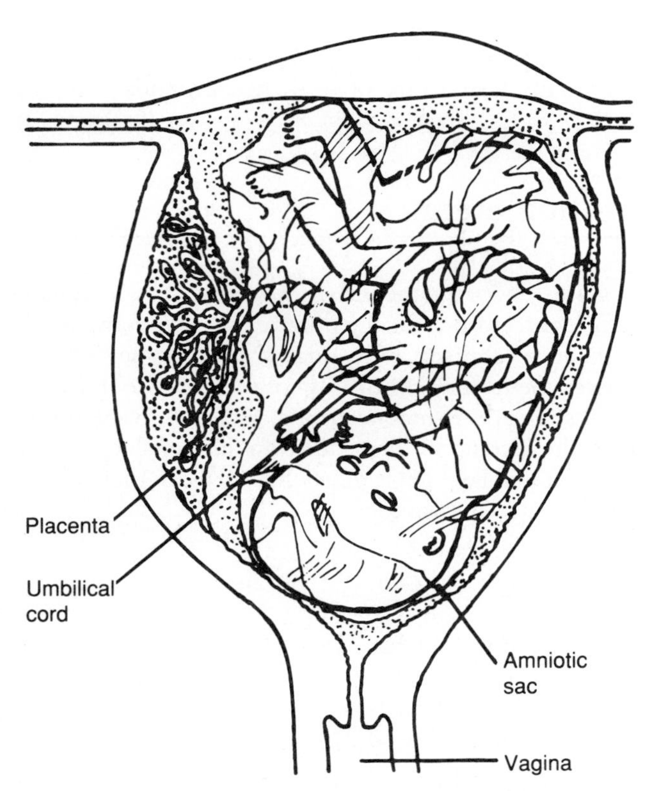

Fully developed placenta and growing fetus. The fetus is shown smaller, and the umbilical cord shorter, than in reality. The circled area is expanded in D.

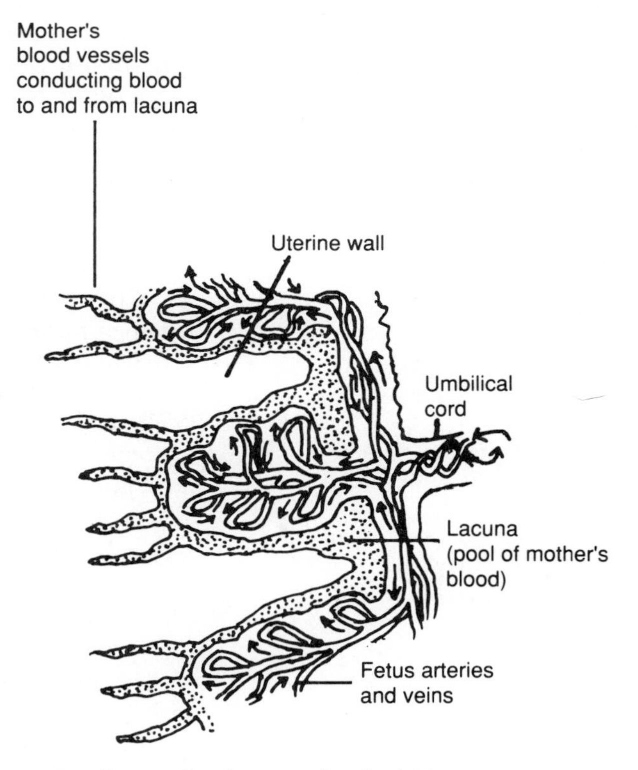

A single villus in the placenta (detail of C).

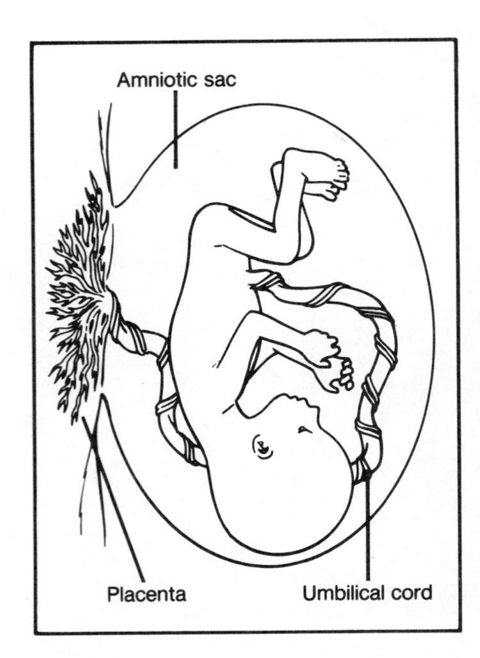

Figure 4–2 The Amniotic Sac
The amniotic sac is filled with fluid to bathe, help nourish, and cushion the infant.

amniotic (am-nee-OTT-ic) **sac** : the "bag of waters" in the uterus, in which the fetus floats.

As mentioned, maternal nutrition is critical for placental growth. Malnutrition limits placental development and results in reduced placental blood flow. This curtails the transfer of energy and essential nutrients to the fetus. Thus maternal malnutrition may negatively affect the fetus in two ways—by making fewer nutrients available for transfer across the placenta and by impairing placental development.

Embryonic and Fetal Growth and Development

The embryo accomplishes amazing developmental feats. The number of cells in the embryo at first doubles approximately every 24 hours; later the rate slows gradually, and during the final ten weeks of pregnancy only one doubling occurs. From the ectoderm, the nervous system and skin begin to develop; from the mesoderm, the muscles and internal organ systems; and from the endoderm, the glands and linings of the digestive, respiratory, and excretory systems. At eight weeks, the embryo has a complete central nervous system, a beating heart, a fully formed digestive system, and the beginnings of facial features. Already, an embryonic tail has formed and almost completely disappeared again, and the fingers and toes are well defined.

The last seven months of pregnancy, the fetal period, bring about tremendous changes. Intensive periods of cell division occur in organ after organ. The amniotic sac fills with fluid to cushion the fetus (see Figure 4–2). As Figure 4–3 describes, the growth of the fetus is phenomenal; the fetus's weight increases from less than a gram to about 3500 grams (7½ pounds). The gestation period, which normally lasts from 38 to 42 weeks, ends with the birth of the infant.

Assessment of Nutrition Status

Assessment of nutrition status in pregnancy is a specialty. The pregnant woman's physiology is different from that of the nonpregnant woman, and it changes from start to finish, so several sets of standards apply to her.

Maternal nutrition assessment may begin by observing the physical appearance of the woman. If she appears malnourished, she may not be eating well, and questions about the extent of her nutrition knowledge, family income, and availability of food will help determine why. Medical, diet, social, and drug histories can provide such information.

Historical Data

Through research, several groups of women have been identified as being more likely than other women to develop complications during pregnancy or to deliver low-birthweight infants. The history-taker should take note of these risk factors, which may indicate a special need for nutrition attention:

- *Age* (under 18 or over 35 years of age).
- *Chronic and acute conditions* (insulin-dependent diabetes, high blood pressure, kidney or respiratory problems, or certain genetic disorders).

- *Prepregnancy weight* (10 percent below or 20 percent above the standard weight for height).
- *Weight gain pattern in previous pregnancies* (abnormal in relation to standards shown in Figure 4–4).
- *Length of time between pregnancies* (intervals of less than one year).
- *Birthweights of previous infants* (both low birthweights and high birthweights).
- *Substance abuse* (cigarette, drug, or alcohol use).
- *Alternative dietary practices and food patterns* (diets that severely restrict kcalories; diets that exclude one or more food groups; diets high in nutrient-poor, high-kcalorie foods; or the use of vitamin-mineral supplements in amounts above the RDA, unless prescribed by a reputable health care provider for a valid reason).
- *Socioeconomic status* (inadequate income).

Should histories or practices that compromise nutrition be revealed, nutrition counseling is in order. Should indications of medical problems arise, further questions should follow.

Figure 4–3 **Stages of Embryonic and Fetal Development**

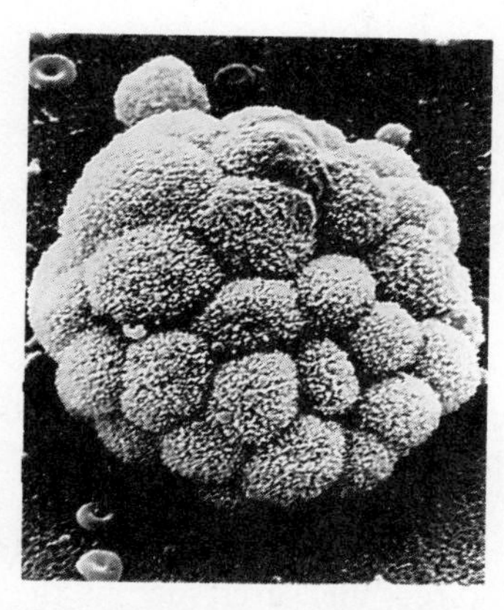

1. A newly fertilized ovum is about the size of the period at the end of this sentence. This zygote, less than one week after fertilization, is not much bigger and is ready for implantation.

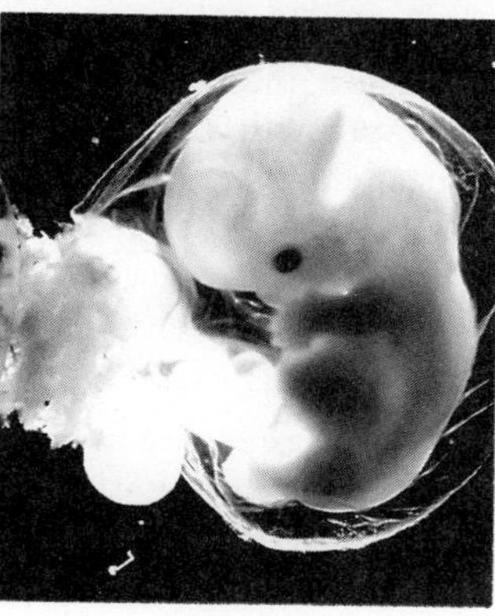

2. An embryo 5 weeks after fertilization is about the size of the capital A that began this sentence; after implantation, the placenta develops and begins to provide nourishment to the developing embryo.

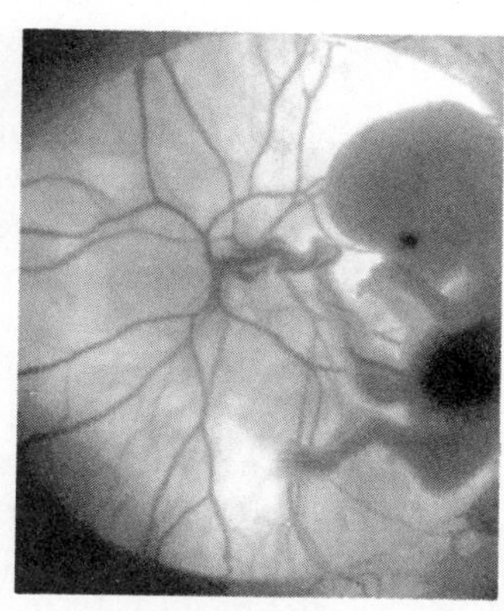

3. A fetus after 11 weeks of development is just over an inch long; notice the umbilical cord and blood vessels connecting the fetus with the placenta.

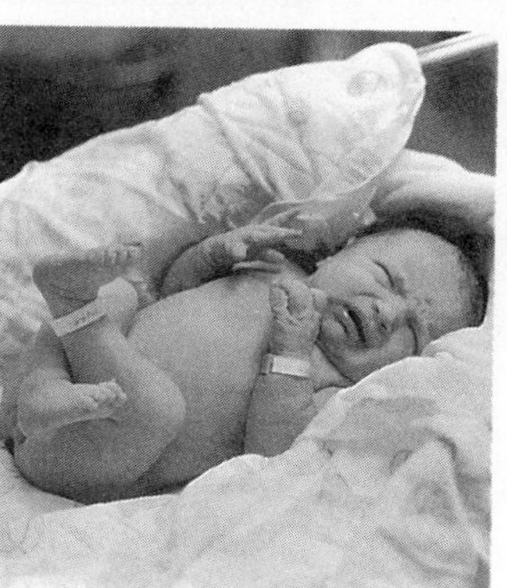

4. A newborn infant after nine months of development measures close to 20 inches in length. From eight weeks to term, this infant grew 20 times longer and 50 times heavier.

gestational diabetes: the appearance of abnormal glucose tolerance during pregnancy, with subsequent return to normal glucose tolerance postpartum.

For example, gestational diabetes requires further attention. Health care providers evaluate a woman's potential for gestational diabetes by checking the following risk factors associated with the condition during the first prenatal examination:[5]

- Previous gestational diabetes.
- History of large infants (9 pounds or more).
- Family history of diabetes.
- Symptoms of diabetes and glycosuria.
- Obesity or excessive weight gain.
- Recurrent urinary tract infections.
- History of spontaneous abortions.
- Previous unexplained stillbirth.

If a woman is at risk for gestational diabetes, laboratory tests (see p. 120) should lead to diagnosis.

Anthropometric Measurements

symphysis-fundus measure: the distance from the junction of the pubic bones on midline in front to the uppermost part of the uterus.

Two of the most important anthropometric measures predictive of the birthweight of a newborn are the mother's prepregnancy weight and the amount and pattern of her weight gain during pregnancy. The normal weight gain pattern is shown in Figure 4–4. Some researchers have suggested that new weight gain standards for pregnancy, based on larger, more diverse groups of women, are desirable.[6] They recommend that future studies used to generate new pregnancy weight gain charts should include adolescents, obese and underweight women, and women who are expecting twins or multiple births. Figure 4–5 presents a chart proposed for use in monitoring maternal weight gains from different starting points.

Assessors should be aware that the pattern of weight gain is even more important than the total weight gain. In the beginning, a 3-pound gain satisfies the needs of the first 3 months, whereas later a 3-pound gain is appropriate for one month, as the weight gain grids show.

The assessor can also gather evidence of fetal and placental growth from the medical record, which may include several indirect measures such as the symphysis-fundus measure, human placental lactogen concentration, and ultrasound measurement. The symphysis-fundus measure indicates uterine (and fetal) growth; the hormone concentration correlates with fetal weight; and the ultrasound measurement accurately determines fetal size.

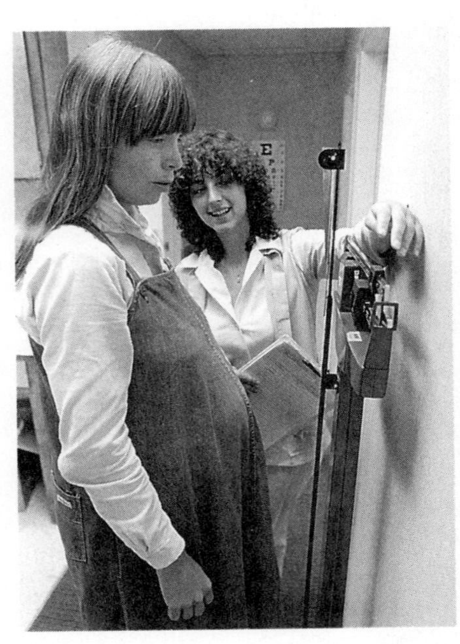

Every prenatal exam includes charting the weight gain.

Physical Examinations

Signs of malnutrition apparent on physical examination of pregnant women are similar to those in other adults (see Chapter 1). Interpretation of laboratory test results, however, refers to several special sets of standards for pregnancy.

Figure 4–4 Nutrition Assessment of the Pregnant Woman: The Standard Prenatal Weight Gain Grid
The standard prenatal weight gain grid plots a pattern of weight gain appropriate during pregnancy for most women.

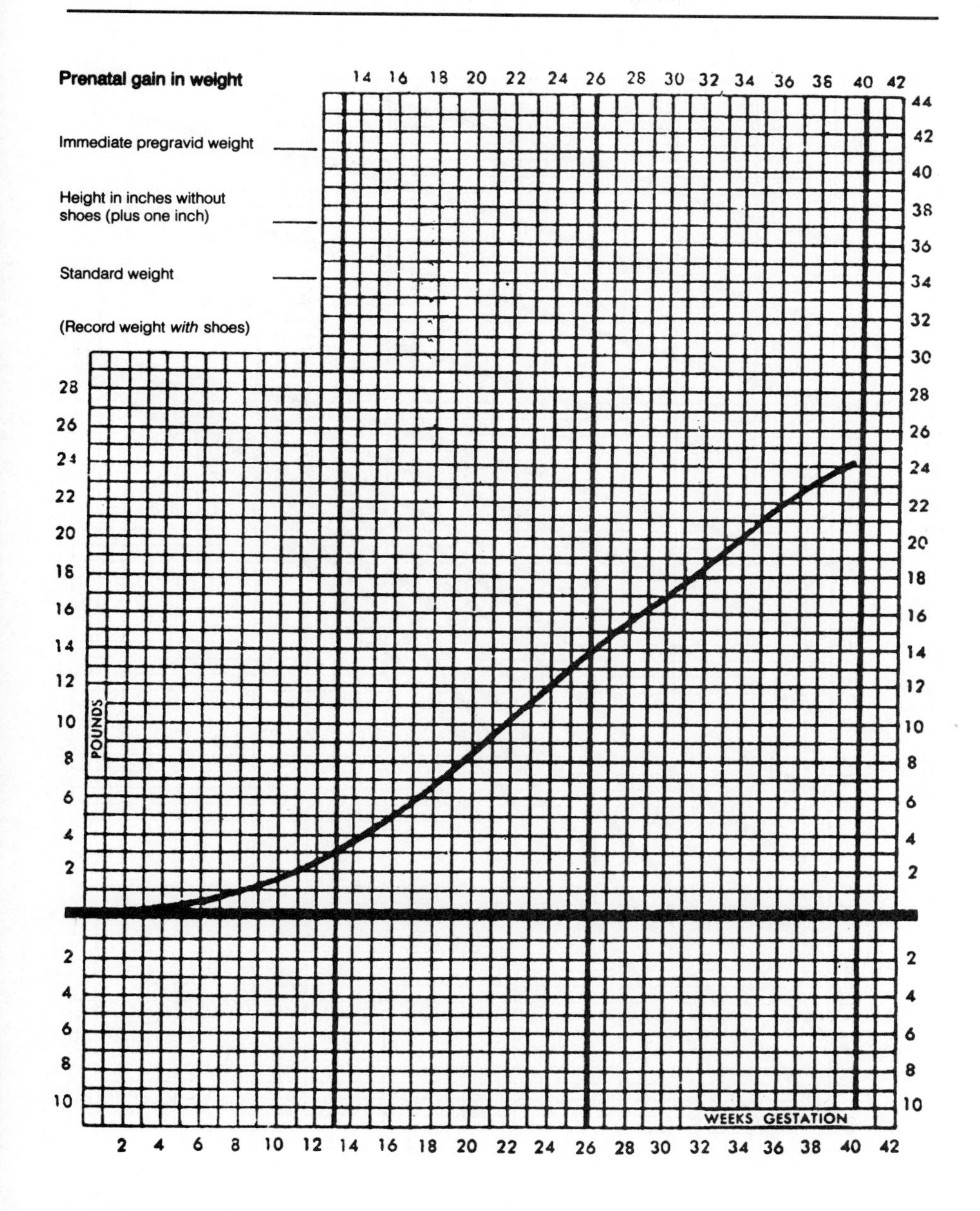

Biochemical Analyses

A health care provider uses a special set of norms to evaluate results of laboratory tests performed on pregnant women. The blood and urine concentrations of many indicators differ for nonpregnant women as Table 4–1 shows. Indices of iron status (see Tables A–13 through A–18 in Appendix A),

Figure 4–5 Nutrition Assessment of the Pregnant Woman: Prenatal Weight Gain Patterns from Different Starting Weights

This chart monitors weight gain during pregnancy considering prepregnancy weight and height.

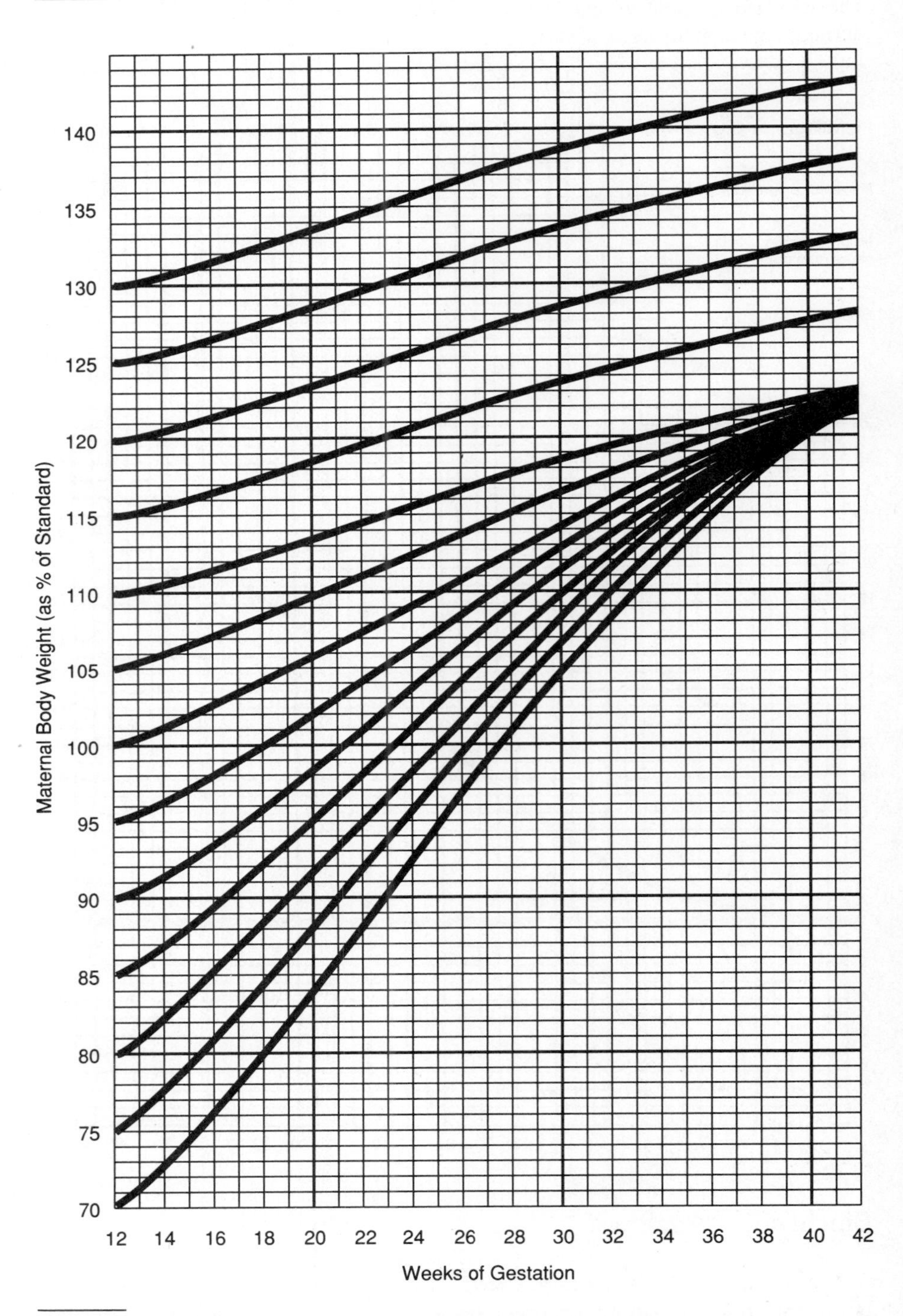

Source: Adapted from P. Rosso, A new chart to monitor weight gain during pregnancy, *American Journal of Clincial Nutrition* 41 (1985): 644–652.

Table 4–1 Nutrition Assessment: Standards for Biochemical Test Results of the Pregnant Woman

Biochemical Test	Minimal Acceptable Values for Women		Deficiency in Pregnancy
	Nonpregnant	*Pregnant*	
Serum protein[a]	≥6.5 g/100 ml	≥6.0 g/100 ml	<6.0 g/100 ml
Serum albumin[b]	≥3.5 g/100 ml	≥3.0 g/100 ml	<3.5 g/100 ml
Serum vitamin A[c]	≥20 μg/100 ml	≥20 μg/100 ml	<20 μg/100 ml
Serum vitamin C[d]	≥0.20 mg/100 ml	≥0.20 mg/100 ml	<0.20 mg/100 ml
Serum folate[e]	≥6.0 ng/ml	≥3.0 ng/ml	<3.0 ng/ml
Hemoglobin[f]	≥12.0 g/100 ml	≥11.0 g/100 ml (2nd trimester)	<9.5 g/100 ml (2nd trimester)
		≥10.5 g/100 ml (3rd trimester)	<9.0 g/100 ml (3rd trimester)
Hematocrit[g]	≥38%	≥35% (2nd trimester)	<30% (2nd trimester)
		≥33% (3rd trimester)	<30% (3rd trimester)

[a]To convert protein values to standard international units (g/L), multiply by 10.
[b]To convert albumin values to standard international units (g/L), multiply by 10.
[c]To convert vitamin A values to standard international units (μmol/L), multiply by 0.03491. Excessively high values may indicate abnormal status or toxicity.
[d]To convert vitamin C values to standard international units (μmol/L), multiply by 56.78.
[e]To convert folate values to standard international units (nmol/L), multiply by 2.266.
[f]To convert hemoglobin values to standard international units (g/L), multiply by 10.
[g]To convert hematocrit values (%) to standard international units, multiply by 0.01.

Source: H. E. Sauberlich, R. P. Dowdy, and J. H. Skala, *Laboratory Tests for the Assessment of Nutritional Status* (Boca Raton, Fla.: CRC Press, 1979); R. H. Aubry, A. Roberts, and V. G. Cuenca, The assessment of maternal nutrition, *Clinics in Perinatology* 2 (1975): 207–219; D. S. Young, Implementation of SI units for clinical laboratory data, *Annals of Internal Medicine* 106 (1987): 114–129.

such as hemoglobin and hematocrit measures, are expected to be lower in pregnancy due to a "physiological anemia" that results from a great increase in the mother's blood volume. In pregnancy, red blood cell number and size increase, but plasma volume increases more, so that the number and volume of cells per milliliter of blood is lower than in the nonpregnant state. At the start of pregnancy, a woman may have a hemoglobin of 14 grams per 100 milliliters and plasma volume of 2 liters, so the total hemoglobin in the blood amounts to 280 grams. At the end of her pregnancy, hemoglobin may have fallen to 10 grams per 100 milliliters, but plasma volume has expanded to 3 liters, so total hemoglobin in the blood is now 300 grams, more than before.[7] Serum protein concentration decreases. Values for some nutrients vary, depending on supplementation (among these are iron, folate, and vitamin D). Values for vitamin A, vitamin E, and serum copper may rise.[8]

The term *anemia* for the so-called physiological anemia of pregnancy is misleading. It is the normal blood profile for pregnancy. It resembles medical anemias only in that the lab values such as hemoglobin and hematocrit are lower than for nonpregnant women.

An edema also occurs during pregnancy that is physiological (meaning expected and normal)—provided that it is not accompanied by indicators of preeclampsia, such as high blood pressure or protein in the urine. This normal edema results from high estrogen, which promotes water retention, and low serum albumin, which lowers osmotic pressure.

preeclampsia: an abnormal condition of pregnancy characterized by edema, increasing hypertension, and protein in the urine (see "Treating Medical Complications," later).

The altered carbohydrate metabolism of a normal pregnancy resembles that of diabetes in a nonpregnant woman. As a result of the large fetal energy demand and the rapid transfer of glucose from mother to fetus, maternal fasting blood glucose concentrations fall. At the same time, maternal tissues place a greater reliance on fat to meet energy needs. Hormonal responses during pregnancy also alter maternal carbohydrate metabolism. Due to a rise in glucocorticoids, estrogen, progesterone, and human placental lactogen, maternal tissues become resistant to insulin. Consequently, with food intake, pregnant women have larger outputs of insulin than nonpregnant women.[9]

Abnormal carbohydrate metabolism (that is, abnormal even for pregnancy) may reflect gestational diabetes. If the history has indicated a risk, the health care provider conducts a fasting or random plasma glucose test (at least two hours after eating). If the fasting level is equal to or greater than 105 milligrams per milliliter, or the random level is equal to or greater than 120, the provider administers a follow-up three-hour glucose tolerance test.[10] A three-hour glucose tolerance test is also administered if routine urinalysis detects glycosuria.

As added insurance against problems with diabetes, all pregnant women have plasma glucose determined, following a glucose load, at 24 to 28 weeks' gestation. Thereafter, at every checkup, the urine is tested for ketones. Ketonuria in conjunction with elevated blood glucose levels may point to diabetes.

Folate needs attention in assessment. Research indicates that the determination of folate status by way of a diet history and possibly laboratory assessment may be more important than previously thought for the pregnant woman, because it is important not only to remedy deficiencies but also to avoid excesses. Folate taken unnecessarily, especially in combination with iron (as occurs in most prenatal supplements), compromises zinc status in women with marginal zinc stores.[11] Pregnancy complications (maternal infection, fetal distress) have been observed in women with high plasma folate and low plasma zinc concentrations.[12] The folate supplements that the pregnant women took were shown to be responsible for the high plasma folate concentrations, and possibly for the inhibition of zinc absorption, which, in turn, may have caused the complications. Zinc deficiency is strongly suspected of causing adverse effects in pregnancy (see Chapter 3).

Therefore, zinc also needs attention. No single biochemical test accurately reflects an individual's zinc status. A minimal assessment of zinc status should therefore include both an estimate of dietary zinc and a measure of serum zinc.[13] A 24-hour urinary zinc excretion measure can provide additional useful information.

During pregnancy, serum zinc concentrations decline. As with iron, however, the decline represents a physiological adaptation to pregnancy due in part to maternal-fetal transfer of zinc and maternal plasma volume expansion.[14] Gestational stage must be considered when evaluating maternal zinc status, since values change as pregnancy progresses.

Unfortunately, pregnancy norms have not been established for all biochemical tests. Still, in recent years, new techniques and methodologies have greatly enhanced early detection and treatment of disease and nutrition disorders. Women with chronic conditions, such as diabetes or heart disease, today can deliver normal, healthy infants, while only a few years ago pregnancy was not even an option for many of them.

How extensively should maternal assessment be pursued? The answer varies, depending on factors such as prior pregnancies, age, and weight, but all pregnancies require a minimal nutrition assessment. If such an evaluation reveals possible problems, then additional tests should be conducted. Table 4–2 offers a hierarchy of priorities.

Table 4–2 Nutrition Assessment of the Pregnant Woman: Tests Used at Three Levels of Approach

	Level of Approach		
	Minimal	*Mid-level*	*In-depth level*
Diet History	Present basic diet: meal patterns, fad or abnormal diets, supplements	The minimal assessment, plus semiquantitative determination of food intake	The mid-level assessment, plus household survey data, diet history, quantitative 24-hour recall
Medical and Socioeconomic History	Obstetrical: Age: parity, interval between pregnancies, previous obstetrical history Medical: Intercurrent diseases and illnesses, drug use, smoking history Family and social: Size of family, whether a "wanted" pregnancy, socioeconomic status	The minimal assessment, plus occupational patterns, utilization of maternity care and family planning services	
Anthropometric Measurements and Physical Examinations	Prepregnancy weight, weight gain pattern during pregnancy, signs and symptoms of gross nutritional deficiencies	The minimal assessment, plus screening for recurrent disease	The mid-level assessment, plus special anthropometric measurements such as fatfold and arm circumference
Biochemical Analyses	Hemoglobin, hematocrit	The minimal assessment, plus blood smear, red blood cell indices, serum iron, test for sickle cell anemia	The mid-level assessment, plus folate and other vitamin levels

Source: Adapted from American Public Health Association, *Nutritional Assessment in Health Programs,* ed. G. Christakis (Washington, D.C.: American Public Health Association, 1974), p. 62.

Energy and Nutrient Needs of Pregnant Women

A woman's nutrient needs during pregnancy are higher than at any other time in her adult life and are greater for certain nutrients than for others, as shown in Figure 4–6. A study of the figure reveals some of the key needs, and the following sections discuss them.

Food Energy and Associated B Vitamins

Energy RDA during pregnancy: +300 kcal/day.

A daily increase of 300 kcalories (during the second and third trimesters) above the allowance for nonpregnant women is recommended.[15] This recommendation represents one of the smallest differences in need between the nonpregnant and pregnant woman. Figure 4–6 confirms this. For women of average size and moderate physical activity, 300 kcalories represents a 15 percent increase

Figure 4–6 Comparison of the Nutrient Needs of Nonpregnant and Pregnant Women (over 25 years old)
The teenage pregnant woman needs even more nutrients than shown here, as Chapter 8 discusses.

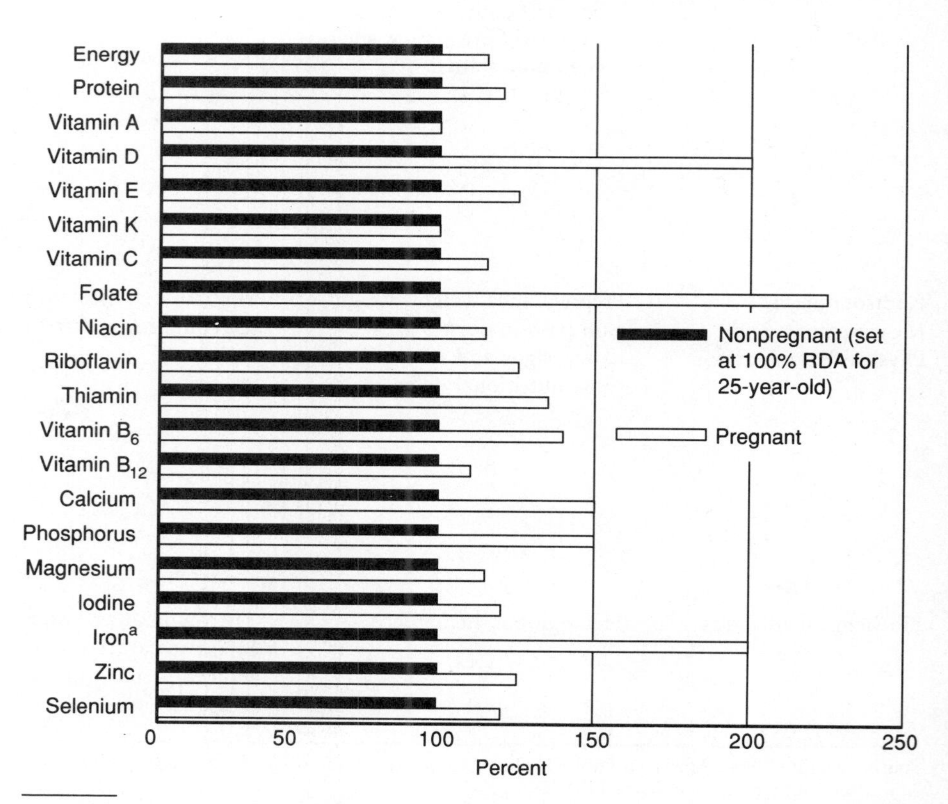

[a]Since the increased requirement for iron cannot be met by typical diets or by the iron stores of most women, daily iron supplements are recommended.

above nonpregnant food energy needs. Pregnant teenagers, underweight women, or physically active women may require more.

For the obese pregnant woman, the emphasis is different. Research suggests that for these women, it is more important to obtain the appropriate distribution of food energy from carbohydrate, fat, and protein than to increase food energy intake.[16]

The source of additional food energy should be high-quality, nutrient-dense foods. Table 4–3 shows the core of a sound food pattern. The NRC recommendations (see Chapter 1) add at least one more fruit or vegetable and two more bread and cereal portions to this. For most women, appropriate food choices include nonfat milk; lean meats, fish, and poultry; eggs; legumes; dark green and other vegetables; citrus and other fruits; and whole-grain breads and cereals. For the pregnant vegan, appropriate food choices include calcium-fortified soy milk, tofu, legumes, nuts, seeds, whole-grain breads and cereals, vegetables (especially dark-green leafy vegetables), and fruits.

Pregnant women do not always eat such nutritious foods. If a woman's diet is already meeting normal nutrient needs at the start of pregnancy, then she can easily adjust it to meet the increased demands; but if the woman has not been eating well, this is definitely the time to begin. High motivation and intensive counseling and practice will help get the pregnancy on the right track (see Practical Point: Diets Tailored to Individual Preferences). Should a pregnant woman's nutrition be poor or should she be at risk for poor nutrition status, help is available to her (see Practical Point: Maternal and Infant Assistance Programs, in Chapter 3).

Added B vitamins are needed in proportion to the added food energy intake, and normally they are obtained from the same foods. Women's needs for thiamin increase early in pregnancy and then remain constant throughout; the RDA committee recommends an added 0.4 milligrams a day above the nonpregnant woman's RDA.[17] The committee also recommends 0.3 milligrams a day of riboflavin and 2 niacin equivalents above the nonpregnant woman's RDA, based on limited evidence of increased needs.[18]

Thiamin RDA during pregnancy: 1.5 mg/day.

Riboflavin RDA during pregnancy: 1.6 mg/day.

Niacin RDA during pregnancy: 17 mg NE/day.

Protein, Carbohydrate, and Vitamin B_6

The pregnant woman's increased need for protein is slightly higher than for food energy. In addition to the 0.8 grams of protein per kilogram appropriate body weight recommended for adults, 10 more grams of protein per day are

Table 4–3 Four Food Group Plan for Pregnant Women

Food Group	Number of Servings	
	Nonpregnant Adult or Obese Pregnant Woman	*Pregnant Woman*
Meat and meat alternates	2	3
Milk and milk products	2	4
Vegetables and fruits	4	4
Breads and cereals	4	4

Note: For pregnant teenagers, the recommended pattern appears in Chapter 8, Table 8–1. For a more detailed summary of serving sizes and food sources, see Appendix C.

Protein RDA during pregnancy: 60 g/day.

recommended during pregnancy. For example, the recommended protein intake for a woman whose appropriate weight is 135 pounds is about 50 grams; an additional 10 grams brings the total daily protein recommendation to about 60 grams. Since the milk allowance for pregnant women is 4 cups a day (32 grams) and the recommended meat/meat alternate intake guarantees at least 28 grams more, the person who builds a diet around the core plan of Table 4–3 will meet the protein recommendation with no difficulty.

▸▸ PRACTICAL POINT

Diets Tailored to Individual Preferences

Perhaps at no other time in life is the motivation to take care of oneself so great as it is during the unique time of pregnancy. The health and well-being of two individuals are inseparable at this time. Many women willingly abstain from smoking and drinking, and pay more attention to their eating habits than ever before. The wise health professional will offer guidance to support this motivation and enhance the chances of an optimal pregnancy and birth.

No two women are alike, and no woman eats the exact diet presented in Table 4–3. One unschooled teenager driven by peer pressure might eat hamburgers, fries, and colas. A busy mother with three children and too many other demands might drink three beers in the evening and miss out on dinner. The newly converted member of a religious cult might be persuaded to give up all but a very few foods. Less extreme examples are more common: one person eats too much meat at the expense of vegetables, whereas another drinks too much milk at the expense of iron-rich foods.

Some women's food practices are not only haphazard but irrational and rigid. Unusual beliefs, superstitions, and dietary practices have surrounded childbirth since the beginning of time. Some women, even today, regardless of age, ethnic group, or income level, may believe that overconsumption or underconsumption of a craved food results in physical or behavioral peculiarities of the infant. For example, some believe that an unsatisfied craving for strawberries can cause a strawberry-shaped birthmark.

Some women develop cravings for, or aversions to, some foods and beverages during pregnancy. These are fairly common, although not well understood. They may arise during pregnancy due to changes in taste and smell sensitivities. One study found that a significant number of women reduced their consumption of coffee, alcoholic beverages, and carbonated beverages during the first half of their pregnancies.[a] Most often, women attributed their reduced coffee consumption to nausea. Their reduced intakes of alcohol and carbonated beverages, however, they ascribed to concern for their infants' health or their own weight gain.

In this same study, the most common food aversions reported were to meat and poultry, a finding consistent with other reports. The most predominant food craving was for ice cream. It has often been hypothesized that food cravings may reflect physiological needs of the mother, fetus, or both, although at present no evidence has proven this to be true. In other words, a woman who craves ice cream or other milk products may not be in need of calcium, but cravings in general may reflect a nutrient-poor diet.

In any case, anyone providing guidance on nutrition to a pregnant woman should listen for clues about dietary practices, beliefs, or superstitions. If any of these appear to be detrimental, warn her that this is so. This may be more easily said than done; it is extremely important not to alienate her. Our suggestions to the diet adviser are as follows.

Identify the problem aspects of the diet based on the historical data. Four examples of diet problems were mentioned in the "Assessment of Nutrition Status" section of this chapter: diets that severely restrict kcalories; diets that exclude one or more food groups; diets high in nutrient-poor, high-kcalorie foods; and the use of self-prescribed vitamin-mineral supplements in amounts above the RDA. Watch for these.

In counseling the woman, support her in all she is doing well, and work within the framework of her beliefs and prejudices. If she is severely restricting kcalories, express approval of her desire to do what she perceives as beneficial, and try to help her realize that adequate food energy is indispensable to her health and the growth of her fetus. Help her select nutrient-dense foods that will not add unneeded body fat but will provide energy enough to support normal growth. If she is omitting food groups, accept this choice, and help her find foods that will meet the need for nutrients normally provided by those groups. For example, a person who eats no meat can use iron-rich foods such as legumes, combined with vitamin C–rich fruits and vegetables, to enhance iron absorption, in place of meat. For the woman whose diet is kcalorie rich and nutrient poor, encourage her to continue enjoying food and to maintain an adequate energy intake but to do so using foods that will provide more vitamins, minerals, and fiber. The person who replaces two soft drinks each day with two glasses of milk significantly improves the nutrient density of an otherwise high-kcalorie, low-nutrient diet. Encourage such a person to eat nutrient-dense snacks such as hard-boiled eggs, yogurt, and fresh fruits and vegetables every day, and provide her with a list of these.

If a woman is taking self-prescribed supplements, for whatever reason, congratulate her for making the effort to do the best she can to obtain all the nutrients she needs. Indicate that while she is correct in thinking she has added nutrient needs, she will do better to obtain her nutrients from foods, and encourage her to take supplements only on the advice of her health care provider.

Support the woman in all that she is doing well during her pregnancy, and without being pushy, try to make her realize why a healthy body and mind are important to her own well-being and to that of her developing fetus. Skillful evaluation of a pregnant woman's diet is essential in determining possible risks due to alternative dietary practices.

As changes that improve her nutrition status gradually become part of her daily routine, the pregnant woman will feel and look better. Her motivation will begin to come from within, and the job of the health counselor will become easier.

▸▸ PRACTICAL POINT
continued

---

[a]E. B. Hook, Dietary cravings and aversions during pregnancy, *American Journal of Clinical Nutrition* 31 (1978): 1355–1362.

Surveys reveal that the diets of U.S. adults exceed the RDA for protein.[19] One survey, for example, found that nonpregnant women between the ages of 19 and 65 had an average protein intake of 64 to 67 grams per day.[20] Because most women already exceed their RDA for protein, they may not need to add to their diets the full 10 grams recommended for pregnancy. In fact, as shown in Chapter 3, excess protein may affect some pregnancies adversely. The pregnant woman does, however, need generous amounts of carbohydrate to spare the protein she eats. To spare protein, the woman should use carbohydrate to supply at least 50 percent of total energy intake each day. For example, in a 2000-kcalorie-per-day intake, this represents about 1000 kcalories of carbohydrate, or about 250 grams. The recommended 4 cups of milk contribute about 50 grams carbohydrate. The other major foods that provide abundant carbohydrate are legumes, fruits, and grains, so generous intakes of these are encouraged.

Recommended carbohydrate intake: at least 50% of energy intake.

Vitamin B_6 needs have been thought to increase in pregnancy for several reasons. For one thing, increased protein is needed, and vitamin B_6 recommendations rise in parallel with protein recommendations. For another, all forms of the vitamin cross the placenta and are concentrated in fetal blood. Furthermore, estrogen's enhanced activity during pregnancy increases the activity of an enzyme (tryptophan oxygenase) that requires a vitamin B_6 coenzyme. During pregnancy, the biochemical tests that reflect vitamin B_6–dependent metabolism yield altered values. These changes can be prevented by giving supplements of 6 to 10 milligrams a day. However, no evidence to date indicates that such supplements provide any therapeutic benefit, or that without them pregnancy complications occur. The RDA committee recommends a conservative, but protective, addition of 0.6 milligrams a day to the woman's normal 1.6 milligrams, sufficient to cover the recommended added protein intake.[21]

Vitamin B_6 RDA during pregnancy: 2.2 mg/day.

Nutrients for Blood Production and Cell Growth

The nutrients required for blood production and for rapid cell proliferation in general are required in greater amounts during pregnancy. New cells are laid down at a tremendous pace as the fetus grows and develops. At the same time, the maternal red blood cell mass expands. All nutrients are important in these processes, but the needs for folate, vitamin B_{12}, iron, and zinc are especially great, due to the key roles of these nutrients in DNA synthesis and the manufacture of red blood cells.

Folate and vitamin B_{12} The additional folate required by the pregnant woman is due to the great increase in her blood volume, the increased urinary excretion of the vitamin, and the rapid growth of the fetus. The RDA for folate more than doubles during pregnancy, increasing from 180 micrograms to 400 micrograms. It is possible to obtain this much folate from a well-selected diet, without supplements. Folate excess from supplements presents a risk to pregnancy as shown on p. 120, so improvement of the diet is the better choice. In addition to being good sources of folate, fruits and vegetables are low in kcalories, rich in other vitamins and minerals, and high in fiber.

Folate RDA during pregnancy: 400 μg/day.

The pregnant woman also has a slightly greater need for the B vitamin that assists folate in the manufacture of red blood cells—vitamin B_{12}. The RDA for vitamin B_{12} for adults is 2 micrograms. During pregnancy, the RDA increases to 2.2 micrograms. Maternal stores can generally meet the needs of pregnancy and because vitamin B_{12} is found almost exclusively in foods of animal origin, meat eaters and lacto-ovo vegetarians are protected from deficiency. Research on hundreds of people eating all-plant diets, however, shows that they all eventually become vitamin B_{12} deficient.[22] People who eat all-plant diets need to eat nutritional yeast grown in a vitamin B_{12}–enriched environment or take a vitamin B_{12} supplement.[23]

Vitamin B_{12} RDA during pregnancy: 2.2 μg/day.

Iron In addition to the usual iron requirements, pregnant women need iron to accommodate the increase in blood, to provide for fetal and placental needs, and to replace blood loss during delivery. During pregnancy, the body conserves iron even more possessively than usual as a result of several changes in iron metabolism. First, menstruation, the major route of iron excretion in women, ceases. Second, absorption of iron increases up to threefold due to a rise in the concentration of the blood's iron-absorbing and carrier protein, transferrin. An additional adjustment is accomplished by the hormones of pregnancy, which raise blood iron by increasing absorption still further and by mobilizing iron from its storage sites in the bone marrow and internal organs. To ensure that enough iron is absorbed to meet the needs of pregnancy, an additional 15 milligrams per day is recommended. Since this recommendation far exceeds the iron contents of typical diets and body stores, most authorities recommend daily supplements for the pregnant woman.

Iron RDA during pregnancy: 30 mg/day.

The best iron sources are meat, fish, and poultry, and thanks to the MFP factor, they also enhance the absorption of iron from nonmeat foods up to threefold. Nonmeat foods to emphasize are legumes, whole grains, dark leafy greens, and dried fruits. Vitamin C in fruits and vegetables can also triple absorption from iron-containing foods eaten at the same meal.

MFP factor: a factor (identity unknown) present in **M**eat, **F**ish, and **P**oultry that enhances the absorption of nonheme iron present in the same foods or in other foods eaten at the same time.

Zinc Zinc is required for DNA and RNA synthesis and, thus, protein synthesis. Over 70 enzymes are now known to require zinc as a cofactor.

Small amounts of zinc must be supplied in the diet of the pregnant woman daily, as the body pool of biologically available zinc appears to be small. The zinc nutrition of many U.S. women may be marginal; a study that measured the zinc content in food samples of 20 adults found an average daily zinc intake of 8.6 milligrams.[24] The zinc RDA for women is 12 milligrams, and for pregnant women, it is 15 milligrams.[25] When dietary zinc intakes of pregnant women were examined, the average intake was 11 milligrams per day, or less than 75 percent of the current RDA.[26]

Zinc RDA during pregnancy: 15 mg/day.

Zinc is most abundant in foods of high protein content, such as shellfish (especially oysters), meats, and liver. Milk, eggs, and whole grains are good sources of zinc when eaten in large quantities, but fiber and phytates in whole grains may render some of the zinc unavailable.

Nutrients for Bone Development

The nutrients involved in building the skeleton are in great demand during pregnancy. Insufficient intakes of calcium, vitamin D, fluoride, or magnesium during pregnancy may have serious consequences for fetal bone and dental development.

Calcium RDA during pregnancy: 1200 mg/day.

Calcium A 50 percent increase in the intake of calcium, from 800 milligrams to 1200 milligrams, is recommended for the pregnant woman. Assuming a customary intake of 800 milligrams a day, this would imply raising the intake by 400 milligrams a day. For women whose prepregnancy intakes are below the RDA, as most are, it may be especially important that mothers' calcium intakes be boosted well above their prepregnancy levels.

However, adaptive physiological changes occur during pregnancy that favor calcium retention. Intestinal absorption of calcium more than doubles early in pregnancy, and the mineral is stored in the mother's bones. At the same time, urinary calcium excretion decreases. During the last trimester, as the fetal bones begin to calcify, a dramatic shift of calcium across the placenta occurs, and the mother's bones give up their stores, accumulated earlier for this purpose, to the fetus. Because of the adaptive mechanisms mentioned above, it is unclear what intake of calcium is really too low to support maternal bone health during pregnancy.

Vitamin D RDA during pregnancy: 10 μg/day.

Vitamin D The relationship between vitamin D and calcium is well known. Vitamin D plays a vital role in calcium absorption and utilization. Maternal vitamin D deficiency will therefore result in abnormal development of the fetal bones. During pregnancy, the RDA increases 5 micrograms for women over age 24, to 10 micrograms per day for women of all ages.[27] Sunlight or vitamin D–fortified milk can provide this amount; supplements are not normally recommended, because of the toxicity risk. The vegan's vitamin D may be obtained from daily exposure to sunlight or else must be included in calcium-fortified soy milk.

Fluoride Formation of the primary teeth begins within the first trimester; for this and for the bones, fluoride is needed. Because there are few food sources of fluoride, a prescription for a supplement that includes fluoride may be desirable for women in areas without fluoridated water.[28]

Magnesium RDA during pregnancy: 320 mg/day.

Magnesium Magnesium assists in building the skeleton, as well as being required for muscle relaxation and energy production. The RDA during pregnancy increases from 280 milligrams per day to 320 milligrams. Many pregnant women in the United States consume diets that are less-than-adequate in magnesium.[29] When pregnant women with low magnesium intakes were given magnesium supplements, the incidence of maternal and fetal complications diminished.[30]

Other Nutrients

Vitamin A RDA during pregnancy: 800 μg RE/day.

The RDA for vitamin A in pregnancy remains at 800 RE a day (the same as for nonpregnant women).[31] While vitamin A is required for fetal growth, the amount available in maternal stores far exceeds the need. As for vitamin E, a pregnant woman's RDA is raised by 2 milligrams above the nonpregnant woman's RDA of 8 milligrams a day. Vitamin E transfer across the placenta occurs mostly late in pregnancy, so in the event of a premature birth, the infant needs supplementation (see Chapter 6). No special recommendation for vitamin K intake is made for pregnancy; usual diets and intestinal flora generally meet the need.

Vitamin E RDA during pregnancy: 10 mg/day.

Plasma vitamin C falls during pregnancy, due to physiological changes, increased needs, or both, and the placenta pumps vitamin C into the fetal blood against a concentration gradient. At term, fetal vitamin C plasma concentration is 50 percent greater than the maternal concentration. Adding 10 milligrams a day to the nonpregnant woman's RDA of 60 milligrams prevents the fall in plasma vitamin C, so this is recommended.[32]

Vitamin C RDA during pregnancy: 70 mg/day.

The only other nutrients for which RDA are given in pregnancy are phosphorus (the recommendation matches that for calcium), iodine (an added 25 micrograms a day are recommended, for a total of 175 micrograms) and selenium (an added 10 milligrams a day are recommended, for a total of 65 milligrams). No special recommendations are made for intakes during pregnancy of the additional nutrients not in the main RDA table; it is assumed that the ranges recommended for adults will cover pregnant women as well.

Phosphorus RDA during pregnancy: 1200 mg/day.

Iodine RDA during pregnancy: 175 μg/day.

Selenium RDA during pregnancy: 65 μg/day.

An adequate nutrient and energy intake throughout pregnancy will help to support the growth and health of both fetus and mother. Dietary intake surrounding the time of birth is also important, as the accompanying Practical Point describes.

▸▸ PRACTICAL POINT

Nutrition at the Time of Giving Birth

When a woman goes into labor, she does not cease to have nutrition needs, nor does her infant. A few pointers regarding her nutrition and the nutrition status of her infant may be helpful.

First of all, her own nourishment is important. She may give birth within two hours or so, in which case she can simply wait to eat again until after the event. However, she may give birth after twenty, or even more than forty hours. Giving birth is an athletic event, and nutritional preparation for it is much like the preparation for competition described in Chapter 8 (see "Foods before Exertion"). She will find the effort easier if she is well nourished and well hydrated at the start. This requires a balance between eating enough food to provide energy during labor, but not too much. As in other stressful events, digestion receives low priority, and vomiting may occur.

If labor proceeds slowly, it is wise to snack periodically, and many health care providers encourage laboring women to carry snacks with them when they go to the birthing center or hospital. After a certain point, the health care provider will proscribe eating, but it still may be permissible and desirable for the woman to sip on ice chips to keep refreshed and hydrated during the heavier stages of labor.

Two notes on the infant's nutrition status are of interest, as well. One has to do with the moment of birth, the other with the time immediately after. As indicated in Chapter 3, probably a greater variation in infant iron status is

▸▸PRACTICAL POINT *continued*

imposed by the way the attending health care professional clamps the umbilical cord than by maternal iron status.[a] When the infant is level with or below the vaginal opening for a few minutes immediately after birth, greater placental blood flow to the infant permits greater iron transfer—up to 40 milligrams of iron, or about one-sixth of the total iron content of a newborn infant. The mode of positioning the infant at birth is controversial and requires careful consideration. Too much blood transfer to the infant can be as critical as too little.

After the umbilical cord is cut, the infant begins to need feeding by mouth for the first time. A newborn infant is born with a small supply of carbohydrate stored as liver glycogen, but this is rapidly depleted, and hypoglycemia can quickly ensue. Once feeding begins, under normal circumstances, the digestive tract easily digests and assimilates dietary carbohydrate to replenish glycogen stores and restore glucose homeostasis. The earlier the infant is fed after birth, the lower the risk of hypoglycemia—and this is important, for hypoglycemia in the newborn can impair central nervous system function. It is desirable for the infant to be put to the breast or fed formula within a half hour or so after birth.

Hypoglycemia defined by blood glucose level:

Term infant:
<30 mg glucose/100 ml (within 72 hr of birth).
<40 mg glucose/100 ml (after 72 hr of birth).

Preterm infant:
<20 mg glucose/100 ml.

[a] A. C. Yao and J. Lind, Effect of gravity on placental transfusion, *Lancet* 2 (1969): 505–508.

Supplements for Pregnant Women

Proper food choices during pregnancy can meet most of a woman's nutrient needs. However, diet alone cannot satisfy her requirements for iron and supplements are recommended during pregnancy. The exact iron amount should be determined by the health care professional administering prenatal care. A woman may make it to the end of pregnancy without developing iron-deficiency anemia, but may bleed excessively at delivery—hence the advisability of a prescribed iron supplement to boost her stores. At birth, an infant needs to have stored enough iron to last three to six months; this iron comes from the mother's iron stores. In order to replenish iron stores depleted by pregnancy, some health care providers recommend continued iron supplementation for two to three months after delivery.

The recommendation that all pregnant and postpartum women take iron supplements is not without opposition among the experts. One critic cites evidence that women given iron supplements may be pushed to the point of macrocytosis—that is, having abnormally large red blood cells.[33] Besides, iron supplements can cause side effects such as indigestion and constipation. The critic suggests that iron deficiency be diagnosed before therapy is initiated, rather than prescribing iron to all pregnant women, some of whom may not need it.

If overall nutrition is suspected to be poor, a health care provider may prescribe a multivitamin-mineral supplement that contains a wide range of vitamins and minerals at physiological levels (particularly vitamins B_6, C, D, and E; folate; pantothenic acid; calcium; magnesium; iron; zinc; copper; and possibly selenium).[34] No special supplementation of any other single nutrient is warranted. Self-medication with individual nutrients seriously threatens pregnancy and should not be undertaken by any woman. In rare cases of nutrient deficiencies, physicians or registered dietitians are the professionals to consult. Isolated reports indicated that multivitamins seem to confer protection against infant neural tube defects in women who had a history of such births, when the multivitamins were taken just prior to, and during, the first few months of pregnancy. Research to confirm this possibility has been inconclusive to date.[35]

Once nutrient criteria are met, then the least expensive product is the best choice. Supplements should be reasonably priced and readily available; cost should not prohibit a woman who needs supplements from using them. In comparison with the costs of malnutrition during pregnancy, a supplement is relatively inexpensive.

The nutrients emphasized here have been, for obvious reasons, those most intensely involved in blood production, other cell growth, and bone growth. However, all nutrients are needed in added quantities during pregnancy, and therefore a diet of mixed whole foods is the best vehicle to supply them. The next task is to obtain the necessary nutrients within a food energy allowance that will support the appropriate weight gain.

Weight Gain during Pregnancy

Authorities recommend a pregnancy weight gain of 22 to 27 pounds (see Table 4–4).[36] Underweight women improve their chances of bearing healthy infants when they gain at least 30 pounds.[37] Adolescents need to gain more weight than older women in order to attain average or optimal infant birthweights.[38] Women who are more than 10 percent above the standard weight for height prior to pregnancy could perhaps gain less, but still should gain between 16 and 20 pounds, depending on prepregnancy weight.[39]

Body composition and nutrition status are important considerations in the estimation of optimal weight gain during pregnancy for the overweight woman. For instance, some women who are 10 percent or more above the

Table 4–4 Weight Gain during Pregnancy

Development	Weight Gain (lb)
Infant at birth	7½
Placenta	1
Increase in mother's blood volume to supply placenta	4
Increase in size of mother's uterus and muscles to support it	2½
Increase in size of mother's breasts	3
Fluid to surround infant in amniotic sac	2
Mother's fat stores	2 to 7
Total	22 to 27

standard weight for height are well nourished, and have dense bones and substantial lean body tissue. Other women who are 10 percent or more above the standard weight for height are poorly nourished; their bodies contain inadequate lean, as well as too much fat tissue. Women in the first group are the ones in whom a 16-pound weight gain during pregnancy may be adequate in terms of fetal outcome, because they have no lean tissue deficit to make up. Women in the second group, however, are probably in need of 20-pound or greater high-quality weight gains, because they need to add lean tissue to their own bodies as well as to support the growth of healthy infants.

If a woman has gained more than the expected amount of weight early in pregnancy, she should not try to diet in the last weeks. Women have been known to gain up to 60 pounds in pregnancy without ill effects. (A *sudden* large weight gain, however, is a danger signal that may indicate the onset of pregnancy-induced hypertension, formerly called toxemia; see "Treating Medical Complications" on p. 134.)

The *pattern* of maternal weight gain is even more important than total weight gain, in terms of maternal nutrition status.[40] The ideal pattern is thought to be about 2 to 4 pounds during the first trimester, and about a pound per week thereafter.

Women who plan to become pregnant, or who are already pregnant, often express concern about the weight gain that accompanies a normal pregnancy. In a society such as ours, where body slimness is frequently equated with youth, vigor, and good looks, many prospective mothers may view the weight gain of pregnancy in a negative light. This is unfortunate. Maternal weight gain during pregnancy correlates closely with infant birthweight.[41] Inadequate weight gain in pregnancy compromises the growth and genetic potential of the fetus.[42] Infant birthweight is in turn a strong predictor of the health and subsequent development of the infant.

Concerned women might find comfort in a reminder that they are *pregnant*, not getting *fat*. The 22 to 27 pounds a pregnant woman is instructed to gain are needed to support the growth of the placenta, uterus, blood, and breasts, as well as an optimally healthy 7½-pound infant and some maternal fat stores to provide energy for labor and lactation.[43] There is little place in the diet for the empty kcalories of sugar, fat, and alcohol, which provide no nutrients to support the growth of these tissues and only contribute to fat accumulation. Obviously, some of the weight the pregnant woman gains is lost at delivery. In the following few weeks, much of the remaining weight is lost as blood volume returns to normal and the fluids accumulated during pregnancy are shed.

For most women, however, not all of the weight gained during pregnancy disappears in the months following the birth. Approximately three out of four women begin their second pregnancy heavier than they did their first.[44] In general, the more weight a woman gains beyond her pregnancy needs, the more weight she is likely to retain. While adequate weight gain supports fetal development, excessive weight gain promotes maternal fat accumulation.

Energy Output and Fitness

Exercise is the energy output component of the energy balance equation; it helps with weight control, so it is included here. Besides, exercise leads to fitness, an important component of health. With this in mind, a few words about the importance of exercise during pregnancy are in order.

It is not unusual today to see pregnant women jogging, swimming, participating in aerobic dance classes, walking energetically, or doing other exercises with the same vigor as their nonpregnant peers. In recent years, there has been increasing interest in the effects of exercise during pregnancy. Is it harmful? Is it beneficial?

Pregnant women can enjoy the benefits of exercise.

Unfortunately, the study of exercise during pregnancy is not without problems. One of the most basic problems is that many variables affect both pregnancy and exercise, making well-controlled studies difficult. Some studies on people have been conducted, but the most reliable data have been obtained from studies on animals. The animals used have four legs rather than two, as well as major metabolic differences from people, so the applicability of these results to human beings is limited.

Despite these limitations and some conflicting results, some basic conclusions seem apparent. Mild exercise during pregnancy is not associated with adverse effects on the fetus or the mother. Most studies involving the effects of strenuous exercise on the outcome of pregnancy have been conducted on women who were highly physically active prior to pregnancy, and these studies report normal fetal outcomes as measured by infant birthweights and conditions at birth.[45]

One study has asked how exercise might affect previously sedentary pregnant women.[46] More than 800 pregnant women were involved. As soon as pregnancy was confirmed, women were asked to begin an exercise program consisting of an easy warm-up on a treadmill or exercise bike, a workout on selected exercise machines, and an aerobic workout on a bicycle. Most of the women did not exercise regularly prior to pregnancy. The women were asked to exercise three days a week; some complied more faithfully than others, and so comparisons were possible. Retrospectively after delivery, the women were divided into groups:

- *Control group:* an average of less than one exercise session completed.
- *Low-exercise group:* an average of 15 exercise sessions completed.
- *Medium-exercise group:* an average of 32 exercise sessions completed.
- *High-exercise group:* an average of 64 exercise sessions completed.

Infants born to mothers in the high-exercise group weighed about 5 ounces (150 grams) more than those born to controls. Other pregnancy outcomes, such as cesarean section incidence and infant condition at birth, were more favorable in the high-exercise group. Many of the women in the study reported that the exercise sessions improved their self-image. Exercise during pregnancy thus appears to be beneficial even for women not previously physically active.

A woman who has not been physically active prior to pregnancy should consult with her physician before beginning an exercise program. Normally, the physician will tell her that exercise that progresses from mild to moderate in small increments is likely to benefit her and her infant. An active, physically fit woman experiencing a normal, healthy pregnancy can more than likely continue exercising throughout her pregnancy, adjusting the intensity and duration as she goes along. Obviously, activities such as skiing, skydiving, contact sports, or games in which the pregnant woman may be hit by a ball should be replaced by safer activities—there are many from which to choose.

For example, a game of tennis played by one partner on each side of the net is safer than a fast-moving doubles game of racquetball.

The American College of Obstetrics and Gynecology has developed the following safety guidelines and criteria for exercise during pregnancy:

- Maternal heart rate should not exceed 140 beats per minute (faster heart rates indicate that oxygen supply is getting short, and oxygen deprivation can impair fetal development).
- No exercise should be performed while lying on the back after the fourth month of pregnancy (the enlarged uterus can obstruct the vena cava and cut off the blood supply to the fetus).
- Energy intake should be adequate to meet the additional energy needs of pregnancy plus the exercise performed (an energy shortage can cause protein deficiency, compromising fetal development).
- No exercise should be performed in hot, humid weather. Avoid overheating. Avoid saunas and hot whirlpools (dehydration threatens fetal development).
- Strenuous activities should not exceed 15 minutes in duration (longer workouts can raise body temperature and induce heat stroke, or raise the heart rate too high for safety).
- The woman should drink liquids liberally before and after exercise to prevent dehydration (normal fluid and electrolyte balance is indispensable to normal fetal development).
- The woman should discontinue exercise at the first sign of discomfort (discomfort may be a sign of impending danger to mother or fetus).

Exercise is beneficial to all pregnant women, but is especially important for those with diabetes, as the next section explains.

Treating Medical Complications

Maternal PKU is discussed in Focal Point 2.

Medical complications of pregnancy can threaten the lives of both mother and fetus. Two such complications merit discussion here due to their relationships with nutrition.

Pregnancy-Induced Hypertension

The normal edema of pregnancy responds to gravity; blood pools in the ankles. The edema of PIH is a generalized edema. This distinction helps with diagnosis.

pregnancy-induced hypertension (PIH): a medical problem in pregnancy that consists of two phases. The first phase, **preeclampsia**, is characterized by increasing hypertension, edema, and protein in the urine. The second phase, **eclampsia**, is characterized by convulsions.

A certain degree of edema is expected in late pregnancy, but it may often be part of the larger cluster of symptoms known as pregnancy-induced hypertension (PIH), formerly called toxemia. PIH is a condition involving high blood pressure and renal problems that requires medical attention. It is important to keep track of maternal blood pressure throughout pregnancy and, if PIH is indicated, to initiate treatment promptly.[47] Both preexisting hypertension and PIH can cause infant death, retarded growth, lung problems, and severe birth defects.

PIH is called the "disease of theories"; its true etiology is unknown, but nutrition-related causes have been suggested for many years. The chief characteristic implicating nutrition in the etiology of PIH is its relatively high incidence in low-income mothers and pregnant teenagers. Women with PIH report consuming diets that are lower in protein and food energy than those of women without PIH. Lack of protein, food energy, calcium, or any combination of these may be involved. Adequate calcium is thought to have a direct effect in preventing high blood pressure.[48]

Epidemiological studies show that a low calcium intake during pregnancy is associated with a high incidence of PIH.[49] Excess dietary sodium has also been suggested as a cause of PIH. To avert PIH, a pregnant woman should consume ample protein-rich foods, including milk. Sodium needs increase during pregnancy, but this does not mean that women should salt their foods; the increased need for sodium will be met simply by eating the added foods needed to meet other nutrient requirements. Sodium restriction is not advised, either, unless a medical condition makes it necessary. Even after the onset of PIH, salt intake should normally not be reduced, and diuretics (to cause sodium excretion) may be harmful. Good nutrition and rest are the basis of treatment.

Diabetes

Diabetes can make pregnancy more difficult than usual. Without management of maternal diabetes, infants may suffer increased mortality and morbidity, congenital abnormalities, and other complications. Gestational diabetes also requires management; 20 to 30 percent of women who have gestational diabetes develop permanent diabetes within five years.[50] Research indicates that the incidence of permanent diabetes in women with previous gestational diabetes is twice as high in those who are 20 percent or more overweight, compared with those who are not overweight. Untreated diabetic women have a 95 to 98 percent infertility rate.[51]

Women with diabetes or gestational diabetes benefit from nutrition counseling. An important aspect of nutrition management is prevention of excessive weight gain, but weight reduction diets are not normally recommended during pregnancy. Women with gestational diabetes should not reduce their carbohydrate intakes but should choose foods rich in complex carbohydrates, such as vegetables and whole-grain breads, and should limit their intakes of concentrated sweets. Optimal protein intake is also important. Dietary recommendations encourage three meals a day plus two snacks, each containing protein, carbohydrate, and moderate fat.

Diet is the cornerstone of treatment for the person with gestational diabetes. If appropriate plasma glucose levels are not maintained, however, insulin therapy may be recommended.

Some complaints such as constipation, heartburn, and nausea also attend pregnancy. These are not major medical problems, but they do cause discomfort, and this chapter's last Practical Point attends to them.

▸▸ **PRACTICAL POINT**

Alleviating Maternal Discomfort

To avoid the most common nutrition-related problems encountered during pregnancy, it helps to be armed with some strategies. Most common conditions are constipation, heartburn, and nausea.

▸▸ PRACTICAL POINT *continued*

An expectant mother may experience constipation as the hormones of pregnancy alter maternal muscle tone and the growing fetus crowds her abdominal organs. A high-fiber diet and a plentiful water intake will help to relieve this condition. Daily exercise may also help alleviate constipation. Laxatives should be used only on the health care provider's orders.

Women often feel heartburn during pregnancy. (Heartburn is a burning sensation in the area of the lower esophagus, near the heart; hence the name.) The condition has nothing to do with the heart itself. The developing fetus puts increasing pressure on the mother's digestive tract, causing a backup of stomach acid, which creates a burning sensation. It may be especially troublesome at night, while the woman is lying down. She may find relief by sleeping with her torso slightly raised. It may also help if she shifts to smaller, more frequent meals than she has been eating. Some health care providers recommend an occasional antacid for heartburn relief.

The nausea of "morning" (actually, any time) sickness seems unavoidable, because it arises from the hormonal changes taking place early in pregnancy, but it can often be alleviated. A strategy some expectant mothers have found effective in quelling nausea is to start the day with a few bites of a soda cracker or other bland carbohydrate food, so that something is in their stomachs before they get out of bed. Some women find that drinking fluids between meals rather than with them is beneficial. The woman with nausea may want to eat small, frequent meals to avoid eating too much at one time.

In Anticipation

Pregnancy for many women is a time of adjustment to major changes. The woman who is expecting to bear an infant is a growing person in more ways than one. Physically and emotionally, her needs are changing. If it is her first infant, she senses that her lifestyle will have to change as she takes on the new responsibility of caring for a child. Ideally, she will be encouraged to develop this sense of responsibility by caring for herself during pregnancy. The expectant mother needs support in thinking of herself as a worthwhile and important person with a new and challenging task that she can and will perform well. She may still be working out her relationship with her mate, and he and she both know that the coming of an infant will affect that relationship profoundly. There is a need for sensitive communication and understanding on both parts in this time of transition.

With all of her planning and adjusting, a woman can still relax and enjoy her pregnancy. With the help and guidance of health professionals, and the support of her family and friends, she can feel healthy and confident as she awaits the birth of her infant. As her uterus expands and that first little kick is felt inside, many earlier apprehensions will give way to joy and satisfaction as she realizes the special bond she has with the life inside her. She has been given the chance to nurture a new life. It is a serious and challenging task, but it is also a rewarding one. As our children grow, so do we as mothers and fathers, discovering strengths and weaknesses, abilities and emotions, and above all, an appreciation for life that we may never have known before.

Chapter 4 Notes

1. H. N. Munro, Placental factors conditioning fetal nutrition and development, *American Journal of Clinical Nutrition* 34 (1981): 756–759.
2. H. N. Munro, S. J. Pilistine, and M. E. Fant, The placenta in nutrition, *Annual Review of Nutrition* 3 (1983): 97–124.
3. R. B. Wilkening and coauthors, Placental transfer as a function of uterine blood flow, *American Journal of Physiology* 242 (1982): H429–H436, as cited by Munro, Pilistine, and Fant, 1983.
4. G. H. Lowrey, *Growth and Development of Children*, 8th ed. (Chicago: Year Book Medical Publishers, 1986), pp. 53–75.
5. I. L. Spratt, Session V: Report of Workshop Chairmen, Summary and Recommendations, *Diabetes Care* 3 (1980): 499–501; C. P. Weiner and M. W. Varner, Nutritional considerations in diabetic pregnancy, *Clinical Nutrition* 3 (1984): 5.
6. P. Rosso, A new chart to monitor weight gain during pregnancy, *American Journal of Clinical Nutrition* 41 (1985): 644–652.
7. T. Lind, Nutrition requirements during pregnancy, *American Journal of Clinical Nutrition* (supplement) 34 (1981): 669–678.
8. J. C. King, Dietary risk patterns during pregnancy, *Nutrition Update* 1 (1983): 205–226.
9. King, 1983.
10. Spratt, 1980.
11. K. Simmer and coauthors, Are iron-folate supplements harmful? *American Journal of Clinical Nutrition* 45 (1987): 122–125.
12. M. D. Mukherjee and coauthors, Maternal zinc, iron, folic acid, and protein nutriture and outcome of human pregnancy, *American Journal of Clinical Nutrition* 40 (1984): 496–507.
13. C. A. Swanson and J. C. King, Zinc and pregnancy outcome, *American Journal of Clinical Nutrition* 46 (1987): 763–771.
14. Swanson and King, 1987.
15. Food and Nutrition Board, *Recommended Dietary Allowances*, 10th ed. (Washington, D.C.: National Academy of Sciences, 1989), pp. 33–34.
16. J. C. King, Obesity in pregnancy, in *Dietary Treatment and Prevention of Obesity*, eds. R. T. Frankle and coeditors (London: John Libbey, 1985), pp. 185–191.
17. Food and Nutrition Board, 1989, p. 128.
18. Food and Nutrition Board, 1989, pp. 134–135, 140.
19. B. H. Dennis and coauthors, Nutrient intakes among selected North American populations in the Lipid Research Clinics prevalence study: Composition of energy intake, *American Journal of Clinical Nutrition* 41 (1985): 312–329.
20. H. Tippett, Nationwide food consumption survey results, *Family Economics Review*, Spring 1980, pp. 3–22.
21. Food and Nutrition Board, 1989, p. 145.
22. I. Chanarin and coauthors, Megaloblastic anemia in a vegetarian Indian community, *Lancet* 2 (1985): 1168–1172, as cited by V. Herbert, Vitamin B_{12}: Plant sources, requirements, and assay, *American Journal of Clinical Nutrition* 48 (1988): 852–858.
23. Herbert, 1988.
24. J. M. Holden, W. R. Wolf, and W. Mertz, Zinc and copper in self-selected diets, *Journal of the American Dietetic Association* 75 (1979): 23–28.
25. Food and Nutrition Board, 1989, pp. 208–209.
26. K. M. Hambidge and coauthors, Zinc nutritional status during pregnancy: A longitudinal study, *American Journal of Clinical Nutrition* 37 (1983): 429–442.
27. Food and Nutrition Board, 1989, p. 96.
28. F. B. Glenn, W. D. Glenn, and R. C. Duncan, Fluoride tablet supplementation during pregnancy for caries immunity: A study of the offspring produced, *American Journal of Obstetrics and Gynecology* 143 (1982): 560–564.
29. K. B. Franz, Magnesium intake during pregnancy, *Magnesium* 6 (1987): 18–27.
30. L. Spatling and G. Spatling, Magnesium supplementation in pregnancy: A double-blind study, *British Journal of Obstetrics and Gynaecology* 95 (1988): 120–125.
31. Food and Nutrition Board, 1989, p. 85.
32. Food and Nutrition Board, 1989, p. 119.
33. D. J. Taylor and T. Lind, Hematological changes during normal pregnancy: Iron induced macrocytosis, *British Journal of Obstetrics and Gynaecology* 83 (1976): 760, as cited by Lind, 1981.
34. V. Newman, R. B. Lyon, and P. O. Anderson, Evaluation of prenatal vitamin-mineral supplements, *Clinical Pharmacy* 6 (1987): 770–777.
35. J. Mulinare and coauthors, Periconceptional use of multivitamins and the occurrence of neural tube defects, *Journal of the American Medical Association,* 260 (1988): 3141–3145.
36. The American College of Obstetrics and Gynecology recommends a pregnancy weight gain of 22 to 27 pounds: Committee on Nutrition, *Nutrition in Maternal Health Care* (Chicago: American College of Obstetricians and Gynecologists, 1974). The National Academy of Sciences recommends a pregnancy weight gain of 24 to 27 pounds; Food and Nutrition Board, Committee on Maternal Nutrition, *Nutrition and the Course of Pregnancy* (Washington, D.C.: National Academy of Sciences, 1970).
37. R. L. Naeye, Weight gain and the outcome of pregnancy, *American Journal of Obstetrics and Gynecology* 135 (1979): 3–9.
38. A. R. Frisancho, J. Matos, and P. Flegel, Maternal nutritional status and adolescent pregnancy outcome, *American Journal of Clinical Nutrition* 38 (1983): 739–746.
39. Naeye, 1979.
40. Pitkin, 1981.
41. A. Gormican, J. Valentine, and E. Satter, Relationships of maternal weight gain, prepregnancy weight, and infant birthweight, *Journal of the American Dietetic Association* 77 (1980): 662–667.
42. C. H. Peckham and R. E. Christianson, The relationship between prepregnancy weight and certain obstetric factors, *American Journal of Obstetrics and Gynecology* 111 (1971): 1.
43. F. E. Hyten and I. Leitch, in *The Physiology of Human Pregnancy* (Oxford: Blackwell Scientific Publications, 1971).
44. G. W. Greene and coauthors, Postpartum weight change: How much of the weight gained in pregnancy will be lost after delivery? *Obstetrics and Gynecology* 71 (1988): 701–707.
45. F. K. Lotgering, R. D. Gilbert, and L. D. Longo, The interactions of exercise and pregnancy: A review, *American Journal of Obstetrics and Gynecology* 149 (1984): 560–568.
46. D. C. Hall and D. A. Kaufmann, Effects of aerobic and strength conditioning on pregnancy outcomes, *American Journal of Obstetrics and Gynecology* 15 (1987): 1199–1203.
47. Blood pressure of 140/90 millimeters of mercury during the second half of pregnancy in a woman who has not previously exhibited hypertension indicates PIH. So does a rise in systolic blood pressure of 30 millimeters or in diastolic blood pressure of 15 millimeters on at least two occasions more than six hours apart. R. J. Worley, Pathophysiology of pregnancy-induced hypertension, *Clinical Obstetrics and Gynecology* 27 (1984): 821–835.

48. J. M. Belizan and J. Villar, The relationship between calcium intake and edema-, proteinuria-, and hypertension-gestosis: An hypothesis, *American Journal of Clinical Nutrition* 33 (1980): 2202–2210.
49. H. M. Linkswiler and coauthors, Protein-induced hypercalciuria, *Federal Proceedings* 40 (1981): 2429–2433.
50. Management of gestational diabetes, *Nutrition and the M.D.*, February 1982, pp. 2–3.
51. M. Wynn and A. Wynn, Effects of nutrition on reproductive capability, *Nutrition and Health* 1 (1983): 165–178.

▶ Focal Point 4

Vegetarian Diets during Vulnerable Times

Eating patterns all along the continuum of dietary choices—from one end, where people eat no foods of animal origin, to the other end, where they eat generous quantities of meat every day—can support or compromise nutritional health. The nutritional quality of the diet depends not on whether a diet includes all plant foods or centers on meat but on whether the diet is based on sound nutrition principles—adequacy, balance, kcalorie control, moderation, and variety.

Both the American Academy of Pediatrics and the American Dietetic Association acknowledge that well-planned vegetarian diets offer nutrition and health benefits to adults in general.[1] Research suggests that adults who eat vegetarian diets reduce their risks of heart disease, hypertension, diabetes, and obesity.[2] This discussion addresses the question whether pregnant and lactating women, infants, and children do equally well.

For purposes of reviewing research, it is important to remember that vegetarian diets, like all diets, encompass a myriad of different eating patterns that reflect the lifestyles and attitudes of the people who choose them. Because vegetarian diets vary in both the types and amounts of foods of animal origin they include, these differences must be considered when evaluating the health status of vegetarians. The accompanying miniglossary defines the types of diets discussed here.

Besides diet, other lifestyle factors must be considered when evaluating research findings. In some instances, membership in certain religious or spiritual groups may influence not only dietary practices but also other lifestyle habits, so as to exert a positive influence on health status. For example, members of the Seventh-Day Adventist faith traditionally follow a lacto-ovo-vegetarian diet and also abstain from tobacco and alcohol use and seek regular medical guidance when pregnant. In contrast, lifestyle habits of other vegetarian groups can negatively affect health. Some vegetarians seek health advice from unorthodox practitioners, and the pregnant women in those groups may avoid contact with medical specialists.

Nutrition experts hold the view that at least in theory, vegetarians at nutritionally vulnerable times such as pregnancy, lactation, infancy, and childhood can easily meet their nutrient needs if they use vegetarian diets that include milk, eggs, and a variety of fruits, vegetables, legumes, and whole-grain foods.[3] However, vegetarian diets that are monotonous and restrictive, such as macrobiotic diets and some vegan diets, can be dangerous during times of rapid growth.[4] This discussion reviews research revealing the nutrition and health advantages of well-planned vegetarian diets and the potential dangers of poorly planned ones for pregnant and lactating women, infants, and children.

Miniglossary of Vegetarian Terms

lacto-ovo-vegetarians: people who include milk or milk products and eggs in, but exclude meat, poultry, fish, and seafood from, their diets.

lactovegetarians: people who include milk or milk products in, but exclude meat, poultry, fish, seafood, and eggs from, their diets.

macrobiotic diet: a diet in which ten stages of dietary restriction lead to gradual elimination of animal products, fruits, and vegetables, at which point the diet is composed only of cereals. Such a diet may be deficient in food energy, protein, iron, zinc, calcium, vitamin B_{12}, vitamin D, and other nutrients as well, depending on the level of restriction.

nonvegetarians: people who abide by no formal restrictions on the eating of any type or group of animal-derived foods; also called **omnivores.**

semivegetarians: people who include some, but not all, groups of animal-derived foods in their diets; they usually exclude meat, and may occasionally include poultry, fish, and seafood; also called **partial vegetarians.**

vegans: people who exclude all animal-derived foods (including meat, poultry, fish, eggs, and dairy products) from their diets; also called **strict vegetarians, pure vegetarians,** or **total vegetarians.**

vegetarians: a general term used to describe people who exclude meat, poultry, fish, or other animal-derived foods from their diets.

Vegetarianism during Pregnancy and Lactation

Like other pregnant women, pregnant vegetarian women are, in general, motivated to do the best they can to ensure successful pregnancies. As long as this motivation leads to practices consistent with health, both mother and fetus benefit. In general, well-planned, liberal vegetarian diets that are adequate in food energy and include milk foods and eggs are compatible with favorable pregnancy outcomes. In contrast, however, despite people's good intentions, severely restricted diets can be inadequate for both mother and fetus in energy, protein, vitamins, or minerals, and thereby can pose significant risks.

As an example of healthy vegetarianism, many vegetarian women willingly eat more liberal diets than usual throughout pregnancy and lactation. In one study, in which researchers evaluated the food choices and nutrient intakes of women during pregnancy and lactation, they found that nonvegetarian women ate more of the nutritious foods they regularly consumed, while vegetarian women added new foods to their diets, usually protein-rich foods and milk foods.[5] All of the vegetarian women drank milk, and most of them ate eggs. All were well-nourished, and their nutrient intakes from diet alone exceeded the RDA for all vitamins and minerals except iron, which was low in both groups. However, more than 90 percent of the women took supplements, and with these included, iron intakes exceeded the RDA. Vegetarian women consumed significantly less protein than nonvegetarians, but both groups exceeded the RDA. In this study, differences in nutrient intakes between vegetarian and nonvegetarian women were insignificant in terms of diet adequacy.

Women who eat exclusively plant-based diets compare less favorably with omnivores during pregnancy. In general, they have lower food energy intakes and are leaner than nonvegetarians, and for pregnant women this may not be desirable.[6] Low prepregnancy weights combined with low weight gains during pregnancy increase the risks of fetal and neonatal death.[7] A typical study showed that infants of very thin mothers had significantly more medical complications (respiratory distress, congenital anomalies, low birthweights, stillbirths, and neonatal deaths) than infants born to normal-weight mothers .[8] Low infant birthweight (less than 2500 grams, or 5½ pounds) is a strong predictor of infant mortality.[9] Low-birthweight infants may be more than twice as likely to die of infections during the first year of life as normal-weight infants.[10]

During lactation, low food energy intakes limit milk production. A study of lactating women found that those who consumed only 1300 kcalories per day were fatigued and, because of inadequate milk production, had to supplement their infants' milk intakes.[11]

The contrast between lacto-ovo-vegetarian and vegan diets can pertain to the protein quality, too, although it need not. Obtaining sufficient protein poses no problem if the diet contains adequate food energy and a variety of protein-containing plant foods, even if it excludes all foods of animal origin. In developed countries, protein-rich plant foods are abundant, and most people who exclude animal-protein foods are aware of the need to eat appropriate plant-protein foods. Pregnant vegetarian women who meet their energy needs by eating ample servings of protein-containing plant foods such as legumes, nuts, and seeds will meet their protein needs as well. Table FP4–1 lists the protein contents of plant foods in descending order and shows that 5 servings of foods on the top half of the table, together with a few servings of each of the other classes of foods listed, can easily meet the RDA for pregnant women (60 grams per day).

Among the vitamins, those of particular concern for vegetarians are vitamin B_{12} and vitamin D. With respect to vitamin B_{12}, again, lacto-ovo-vegetarians have an advantage; they suffer no deficiencies. However, pregnant women who eat exclusively plant-based diets may have inadequate vitamin B_{12} intakes. Because vitamin B_{12} occurs only in foods of animal origin, supplements are necessary to prevent deficiency. Women who have adhered to all-plant diets for many years are especially likely to have low vitamin B_{12} stores. Complications related to vitamin B_{12} deficiency have been reported among nonpregnant British women who eat exclusively plant-based diets.[12] The added demands for the vitamin imposed by pregnancy make it virtually impossible to maintain adequate vitamin B_{12} status without supplementation or the inclusion of a reliable food source of the nutrient.

Vitamin D's best source is sunlight, which promotes synthesis of the vitamin in the skin.[13] Most adults in sunny regions need not make special efforts to obtain vitamin D in food. Darker-skinned people make less vitamin D on limited exposure to the sun. By three hours of exposure, however, vitamin D synthesis in strongly pigmented skin arrives at the same plateau as that at 30 minutes in fair skin.

People who are not outdoors much or who live in northern or predominantly cloudy or smoggy areas are advised to make sure they drink vitamin D-fortified milk, the second best source of the vitamin. Since practically all

Table FP4–1 Protein-Containing Plant Foods

Food	Amount	Protein (grams)
Legumes:		
Soybeans	⅓ c cooked	7
Black beans	½ c cooked	7
Black-eyed peas		
Great northern beans		
Garbanzo beans (chick-peas)		
Kidney beans		
Lentils		
Lima beans		
Navy beans		
Pinto beans		
Tofu (soy cheese)	¼ c	7
Miso	2 tbsp	7
Commercial soy milk	1 c	7
Nuts:[a]	1 to 2 oz	7
Almonds		
Peanuts		
Pistachios		
Cashews		
Walnuts		
Pine nuts		
Pecans		
Nut butters:	2 tbsp	7
Peanut, cashew, sesame (tahini)		
Seeds:[b]	1 to 2 oz	7
Pumpkin, sesame, sunflower		
Breads and cereals	1 slice or ½ c cooked grain, cooked pasta, or dry cereal	3
Starchy vegetables:	½ c	3
Potatoes		
Sweet potatoes		
Corn		
Winter squash (acorn, butternut)		
Green peas		
Other vegetables	½ c cooked or 1 c raw	2

[a]Nuts and nut butters carry significantly more fat and energy with their protein than legumes do. For some people, such as infants, young children, or underweight pregnant women, this may be desirable; for others, it may not be.
[b]Seeds, like nuts, carry more fat with their protein than legumes do.

Source: Compiled from data in P. A. Acosta, A view of vegetarianism by a lacto-ovo vegetarian, in *Nutrition and Vegetarianism: Proceedings of Public Health Nutrition Update,* ed. J. J. B. Anderson (Chapel Hill, N.C.: Health Sciences Consortium, 1981); E. N. Whitney, E. M. N. Hamilton, and S. R. Rolfes, *Understanding Nutrition,* 5th ed. (St. Paul, Minn.: West, 1990).

milk in the United States is fortified, most lactovegetarians are protected from deficiency.

Nutritional osteomalacia, the adult vitamin D-deficiency disease, is practically nonexistent in the United States and Canada, but studies show that the women with the lowest intakes of the vitamin during pregnancy are likely to bear infants with the childhood vitamin D-deficiency disease, rickets.[14] To prevent deficiency, pregnant women who drink no milk are advised to obtain some other reliable source of the vitamin daily—exposure to sunlight, soy milk fortified with vitamin D, or supplements in the amount of 400 IU per day.

As for the minerals, pregnant vegetarian women must make extra efforts to achieve recommended intakes of calcium, iron, and zinc. Vegetarians who exclude milk foods from their diets have lower calcium intakes than do other people. For those who normally consume milk, or who liberalize their diets during pregnancy by adding milk or milk foods, calcium intakes can meet recommendations. People who exclude milk and milk-containing foods are advised to drink calcium-fortified and vitamin D-fortified soy milk. Diets rich in cereals and vegetables (typical vegetarian diets) contain abundant fibers, phytates, and oxalates—compounds that may adversely affect calcium balance even when calcium intakes are adequate.[15] The low calcium intakes of some strict vegetarians, the presence of these dietary calcium binders, and the high calcium needs of pregnancy may impair calcium status among this group of women if reliable sources of calcium are not a regular part of their diets. However, the body may compensate for low calcium intakes by increasing absorption; current recommendations may be unnecessarily high.[16]

Pregnancy imposes a substantial risk of iron deficiency on all women—and especially on those whose diets have been marginal in iron prior to pregnancy. To accommodate increases in maternal blood volume, and to provide iron stores for the fetus, health care providers routinely recommend iron supplementation for all pregnant women. Iron supplementation seems especially appropriate for women whose diets contain only foods of plant origin, which have a low iron bioavailability. One study examined the iron status of pregnant and nonpregnant vegetarian women and found little evidence of deficiency based on hemoglobin, serum iron, total iron-binding capacity, and serum transferrin.[17] However, when other researchers assessed iron status as measured by serum ferritin concentrations, they found it was significantly lower in vegetarians, and especially in vegetarian women, than in nonvegetarians.[18] Although many vegetarian diets contain abundant sources of vitamin C, which enhances iron absorption, apparently this does not fully compensate for the substantially lower bioavailability of iron from vegetable foods.

The richest food sources of zinc include meats, poultry, and fish—foods normally excluded in vegetarian diets. As with calcium, the abundant fibers, oxalates, and phytates in vegetarian diets limit the bioavailability of zinc in foods of plant origin. In short, the potential exists for poor zinc status among vegetarians. Despite this, when researchers compared plasma, hair, and urine zinc concentrations of pregnant women, no differences were found between vegetarians (predominantly lacto-ovo-vegetarians) and nonvegetarians.[19] Dietary zinc intakes were twice as high among vegetarians as nonvegetarians, however. Another study also found no difference in zinc status between lacto-ovo-vegetarian and nonvegetarian pregnant women, even when their dietary zinc intakes were similar.[20] In contrast, other researchers found low dietary zinc intakes and low serum zinc concentrations in vegetarian women, with vegan women's having the lowest serum zinc concentrations.[21] The vegan women's diets contained four times as much dietary fiber as the nonvegetarians' diets, and twice as much as the lactovegetarians' diets; this may partially explain the lower zinc status of the vegan women.

From all of the foregoing, it appears that vegetarianism is perfectly compatible with healthy pregnancy, provided that a woman eats a balanced array of nutritious foods in sufficient quantities. However, a pregnant woman who eats a strict vegetarian diet, who has a low food energy intake, and who does not include fortified soy products or nutrient supplements in her diet increases her risk of a poor pregnancy outcome (having a low-birthweight infant or an infant born with rickets).

Women who have eaten restrictive diets for years prior to pregnancy begin their pregnancies with low nutrient reserves and risk developing overt nutrient deficiencies during pregnancy. In contrast, women who adopt strict vegetarian diets after years of eating liberal nonvegetarian diets often have in their bodies nutrient stores sufficient to protect against deficiency. In either case, most prenatal medical specialists prescribe vitamin-mineral supplements for pregnant women. Vegetarian diets that provide adequate energy; include milk or milk foods; and contain a wide variety of legumes, cereals, fruits, and vegetables favor healthy pregnancy outcomes. Vegetarian diets that exclude all foods of animal origin may require supplementation with vitamin B_{12}, vitamin D, calcium, zinc, and iron, or the addition of foods fortified with these nutrients. Table FP4–2 provides vegetarian eating patterns for pregnant and lactating women.

Table FP4–2 Vegetarian Eating Patterns for Pregnant and Lactating Women

Food Group	Serving Size	Servings/Day
Protein foods:		5
Eggs[a]	1 medium	
Legumes	½ c	
Nuts, seeds	1 oz	
Milk and milk foods[b]		4
Milk or yogurt	1 c	
Cheese	1 oz	
Whole grains:		4
Cereals	½ c	
Bread	1 slice	
Fruits:		3
1 citrus	½ c	
2 other	½ c	
Vegetables:		5
1 dark green	½ c	
4 other	½ c	

[a]Eggs may be replaced in vegan diets with more legumes, nuts, or seeds.
[b]Milk products should be replaced in vegan diets with soy milk fortified with both calcium and vitamin D.

Source: P. A. Acosta, A view of vegetarianism by a lacto-ovo vegetarian, in *Nutrition and Vegetarianism: Proceedings of Public Health Nutrition Update,* ed. J. B. Anderson (Chapel Hill, N.C.: Health Sciences Consertium, 1981), pp. 26–45.

The nutrition status of a woman while pregnant carries over into the time when she is lactating. The only difference so far noted in research between vegetarian and nonvegetarian women in the United States and Canada is that the prevalence and duration of breastfeeding among vegetarian women is higher than among nonvegetarian women.[22] Most vegetarian women enter lactation in good nutrition status, permitting satisfactory growth and health of their infants. The nutrients that require special attention during pregnancy continue to warrant the vegetarian's attention throughout lactation. Diets that support healthy pregnancies are compatible with successful lactation as well.

Vegetarianism during Infancy

The young infant is a lactovegetarian; the infant's primary source of nutrients during the first few months of life is breast milk or infant formula, the ideal food to support health and growth. The infant of a vegetarian mother is no exception. As long as the infant has access to sunlight as a source of vitamin D, and to sufficient quantities of breast milk from a mother who eats an adequate vegetarian or vegan diet, the infant will thrive during the early months. The same is true of infants fed standard or soy-based infant formulas.

Many of today's infants are indoor infants, unfortunately, and for these infants, if they are breastfed, vitamin D is a nutrient of concern. Vitamin D is present in breast milk in insufficient quantities to meet the indoor infant's need. Pediatricians usually recommend vitamin D supplementation for breastfed infants, regardless of the mothers' diets. Infants breastfed by vegan mothers for prolonged periods of time without vitamin D supplementation are at increased risk of rickets.[23]

Vitamin B_{12} deficiency can also afflict breastfed infants of women who consume strict vegetarian diets.[24] In one case, a breastfed infant suffered severe vitamin B_{12} deficiency when the mother showed signs of a marginal blood deficiency.[25] The mother had avoided all foods of animal origin for the previous eight years, but had not herself developed overt anemia, probably because her folate intake was high enough to permit new cell synthesis without the help of vitamin B_{12}. The mother's breast milk, with its low concentration of vitamin B_{12}, was the infant's only source of the vitamin. The infant was in a coma by the time treatment with vitamin B_{12} was initiated.

The speed and severity with which deficiency symptoms developed in this infant emphasizes the important role and rapid use of vitamin B_{12} in growth. Based on observations such as this, it is clear that the infant's need, on a body weight basis, for vitamin B_{12} is much greater than that of the adult.

Infants beyond about four months of age present a greater challenge in terms of meeting nutrient needs by way of vegetarian, and especially vegan, diets. Continued breastfeeding or formula feeding is recommended, but supplementary feedings are necessary to ensure adequate energy and iron intakes. The risks of poor nutrition status in infants increase with weaning and increasing reliance on table foods. Researchers studying growth differences between vegetarian and nonvegetarian infants and children find consistent, significant depressions in growth of vegetarian infants around the time of transition from breast milk to solid foods.[26] Macrobiotic and severely restrictive vegan diets pose the greatest threats to infants' health and nutrition status.

Protein-energy malnutrition and deficiencies of vitamin D, vitamin B_{12}, iron, and calcium have been reported in infants fed macrobiotic and veganlike diets.[27] Vegan and macrobiotic diets that are high in fiber, other complex carbohydrates, and water have a low kcaloric density. Infants in families who eat such diets often receive gruellike mixtures of cereal grains when they are weaned from the breast. Cereals absorb significant amounts of water when cooked, sometimes increasing their volume three to four times. The stomach capacity of infants is limited; when fed gruel mixtures of low kcaloric density, infants cannot consume enough volume to meet their energy needs. This problem can be partially alleviated by providing more nut butters, legumes, dried fruit spreads, and mashed avocado, while limiting the infant's intake of foods with low kcaloric density such as vegetables and gruels.[28] Calcium- and vitamin D-fortified soy milk, vitamin B_{12}-fortified foods or supplements, and the inclusion of vitamin C-containing foods at meals to enhance iron absorption will alleviate other nutrient deficiencies in vegan or macrobiotic infant diets. Table FP4–3 provides diet plans for vegan infants and children.

Vegetarianism during Childhood

As is true with infants, people who feed children vegetarian diets that exclude dairy foods and eggs (vegan diets) must take extreme care to prevent inadequate energy intakes and nutrient deficiencies. Among preschool children

Table FP4–3 Eating Patterns for Vegan Infants and Children

		Average Size of Serving		
Food Group	**Servings/Day**	*6 mo to 1 yr*	*1 yr to 4 yr*	*4 yr to 6 yr*
Soy milk (fortified)[a]	3	1 c	1 c	1 c
Protein foods[b]	2 to 3	1 to 2 oz	1 to 2 oz	2 to 3 oz
Fruits:				
Citrus[c]	2	—	¼ to ½ c	¼ to ½ c
Other	2 or more	1 to 3 oz	1 to 3 oz	
Vegetables:				
Green leafy or deep yellow	1	¼ c	¼ to ½ c	½ c
Other	1	¼ c	¼ to ½ c	½ c
Breads and cereals (whole grain or enriched)	2 to 4	1 slice or 2 oz	1 slice or 2 oz	1 slice or 3 oz
Miscellaneous:				
Brewer's yeast[d]	0 to 1	—	1 tbsp	1 tbsp
Molasses	0 to 1	—	1 tbsp	1 tbsp

[a]Fortified soy milks include soy-based infant formulas.
[b]See Table FP4–1.
[c]Citrus fruits and juices are not recommended for infants less than 9 mo of age.
[d] Nutritional yeast is an inappropriate food for infants. Yeast does not naturally contain vitamin B_{12}. Specially grown yeast or fortified products are available; the amounts of vitamin B_{12} vary.

Source: Data adapted from D. D. Truesdell and P. B. Acosta, Feeding the vegan infant and child, *Journal of the American Dietetic Association* 85 (1985): 837–840.

fed veganlike diets, growth already depressed in infancy remains depressed in childhood.[29] A study that examined the growth and development of three groups of Dutch children who ate three different types of vegetarian diets showed that those who ate a macrobiotic diet were significantly shorter in height and lighter in weight than those eating less restrictive vegetarian diets.[30] As already emphasized, foods of plant origin generally have more bulk, but carry less energy, than foods of animal origin; and while a bulky diet may be advantageous for many adults, it can be detrimental for children, who need energy-dense foods for growth.

The nutrients of greatest concern for children who eat all-plant diets are vitamin B_{12} and vitamin D. Vitamin B_{12} deficiency will invariably occur in unsupplemented vegan diets.[31] The need for supplements of this nutrient when all-plant diets are consumed cannot be overemphasized.

Daily exposure to sunlight or vitamin D supplementation is necessary to prevent rickets in vegan children.[32] Excessive vitamin D is toxic, so the provider of supplements must take care to give the appropriate amount—no more.

Calcium intakes of vegan children tend to be low. Calcium-fortified soy milk offers a reliable source of this nutrient for those who exclude dairy foods. As is true for all children, the iron status of vegetarian children should be checked regularly.

In summary, nutrients that may need attention in lacto-ovo-vegetarian diets are iron and zinc. Vegan diets may also be low in energy, vitamin B_{12}, vitamin D, and calcium. However, diet planners who can recognize the nutrition challenges that vegetarian diets present, who are able to offer competent diet advice, and who are willing to recommend supplements when appropriate, can design vegetarian diets to meet the nutrient needs of people in all stages of life—even the vulnerable stages of pregnancy, lactation, infancy, and childhood.

Focal Point 4 Notes

1. American Academy of Pediatrics, Committee on Nutrition, Nutritional aspects of vegetarianism, health foods, and fad diets, *Pediatrics* 59 (1977): 460–464; American Dietetic Association, Position of the American Dietetic Association: Vegetarian diets—Technical support paper, *Journal of the American Dietetic Association* 88 (1988): 353–355.
2. J. T. Dwyer, Health aspects of vegetarian diets, *American Journal of Clinical Nutrition* 48 (1988): 712–738.
3. C. Jacobs and J. T Dwyer, Vegetarian children: Appropriate and inappropriate diets, *American Journal of Clinical Nutrition* 48 (1988): 811–818.
4. American Academy of Pediatrics, 1977; Jacobs and Dwyer, 1988.
5. D. A. Finley and coauthors, Food choices of vegetarians and nonvegetarians during pregnancy and lactation, *Journal of the American Dietetic Association* 85 (1985): 678–684.
6. F. R. Ellis and V. M. E. Montegriffo, Veganism: Clinical findings and investigations, *American Journal of Clinical Nutrition* 23 (1970): 249–255.
7. R. L. Naeye, Weight gain and outcome of pregnancy, *American Journal of Obstetrics and Gynecology* 135 (1979): 3–9; M. C. Mitchell and E. Lerner, Weight gain and pregnancy outcome in underweight and normal weight women, *Journal of the American Dietetic Association* 89 (1989): 634–638, 641.
8. Mitchell and Lerner, 1989.
9. Mitchell and Lerner, 1989.
10. C. G. Victora and coauthors, Influence of birth weight on mortality from infectious diseases: A case-control study, *Pediatrics* 81 (1988): 807–811.
11. M. R. Thomas and S. Bodily, Maternal anthropometric measurements: Dietary intake and lactation performance of mothers on a weight reduction diet (abstract), *Federation Proceedings* 44 (1985): 1679.
12. M. H. Gleeson and P. S. Graves, Complications of dietary deficiency of vitamin B_{12} in young Caucasians, *Postgraduate Medical Journal* 50 (1974): 462–464.
13. Part of this discussion is adapted with permission from E. N. Whitney, E. M. N. Hamilton, and S. R. Rolfes, *Understand-*

ing Nutrition, 5th ed. (St. Paul, Minn.: West, 1990), pp. 164–168.

14. P. Elinson, L. M. Neustadter, and M. G. Moncman, Nutritional osteomalacia, *American Journal of Diseases of Children* 134 (1980): 427; M. Moncrieff and T. O. Fadahunsi, Congenital rickets due to maternal vitamin D deficiency, *Archives of Disease in Childhood* 49 (1974): 810–811.
15. L. H. Allen, Calcium bioavailability and absorption: A review, *American Journal of Clinical Nutrition* 35 (1982): 783-808.
16. L. W. Turner and E. N. Whitney, Nature versus nurture: The calcium controversy, *Nutrition Clinics,* November-December, 1989.
17. B. M. Anderson, R. S. Gibson, and J. H. Sabry, The iron and zinc status of long-term vegetarian women, *American Journal of Clinical Nutrition* 34 (1981): 1044–1048.
18. A. D. Helman and I. Darnton-Hill, Vitamin and iron status in *new* vegetarians, *American Journal of Clinical Nutrition* 45 (1987): 85–89.
19. J. C. King, T. Stein, and M. Doyle, Effect of vegetarianism on the zinc status of pregnant women, *American Journal of Clinical Nutrition* 34 (1981): 1049–1055.
20. M. J. Abu-Assal and W. J. Craig, The zinc status of pregnant vegetarian women, *Nutrition Reports International* 29 (1984): 485–494.
21. J. H. Freeland-Graves, M. L. Ebangit, and P. W. Bodzy, Zinc and copper content of foods used in vegetarian diets, *Journal of the American Dietetic Association* 77 (1980): 648–654.
22. J. T. Dwyer, Vegetarian diets in pregnancy and lactation: Recent studies of North Americans, *Journal of the Canadian Dietetic Association* 44 (1983): 26–35.
23. M. Rudolf, K. Arulanantham, and R. M. Greenstein, Unsuspected nutritional rickets, *Pediatrics* 66 (1980): 72–76; S. Bachrach, J. Fisher, and J. S. Parks, An outbreak of vitamin D deficiency rickets in a susceptible population, *Pediatrics* 64 (1979): 871-877; D. V. Edidin and coauthors, Resurgence of nutritional rickets associated with breast-feeding and special dietary practices, *Pediatrics* 65 (1980): 232–235.
24. M. C. Higginbottom, L. Sweetman, and W. O. Nyhan, A syndrome of methylmalonic aciduria, homocystinuria, megaloblastic anemia and neurologic abnormalities in a vitamin-B_{12} deficient breast-fed infant of a strict vegetarian, *New England Journal of Medicine* 299 (1978): 317–323; R. Sklar, Nutritional vitamin B_{12} deficiency in a breast-fed infant of a vegan-diet mother, *Clinical Pediatrics,* April 1986, pp. 219–221.
25. Higginbottom, Sweetman, and Nyhan, 1978.
26. J. T. Dwyer and coauthors, Growth in "new" vegetarian preschool children using the Jenss-Bayley curve fitting technique, *American Journal of Clinical Nutrition* 37 (1983): 815-827.
27. E. D. Shinwell and R. Gorodischer, Totally vegetarian diets and infant nutrition, *Pediatrics* 70 (1982): 582–586; P. S. Ward, J. P. Drakeford, and J. Milton, Nutritional rickets in Rastafarian children, *British Medical Journal* 285 (1982): 1242-1243.
28. D. S. Truesdell and P. B. Acosta, Feeding the vegan infant and child, *Journal of the American Dietetic Association* 85 (1985): 837–840.
29. Dwyer and coauthors, 1983.
30. W. A. van Staveren and P. Dagnelie, Food consumption, growth, and development of Dutch children fed on alternative diets, *American Journal of Clinical Nutrition* 48 (1988): 819–821.
31. I. Chanarin and coauthors, Megaloblastic anemia in a vegetarian Indian community, *Lancet* 2 (1985): 852–858.
32. T. B. Sanders, Growth and development of British vegan children, *American Journal of Clinical Nutrition* 48 (1988): 822-825.

Lactation, Breast Milk, and Formula

5

Detail from *Fountain of the Seasons* by Christian Petersen.

Childbirth marks the end of pregnancy and the beginning of a new set of parental responsibilities, decisions, and behaviors. Many of these focus on care of the newborn. Newborns arrive without an instruction manual, and people become parents without any formal training. Parents must gather information from a variety of sources to determine what to do and how and when to do it. Translating this information into action requires making many adjustments and learning a number of new behaviors. This chapter describes the responsibility of choosing what milk to feed an infant, the factors that influence the choice, and the behaviors parents must adopt to carry it out successfully.

Infant Feeding: Breast Milk or Formula?

An infant grows most rapidly during the first four to six months of life. Breast milk and infant formula are the only recommended sources of nutrients during this critical time. Cow's milk is not a health-promoting food for young infants. The American Academy of Pediatrics (AAP) recommends against its introduction into the infant's diet before 6 months, and 12 months is a better time still. The chief reason why the use of cow's milk is contraindicated is that it can cause bleeding in the young infant's gastrointestinal tract. It is also a poor source of iron, so at the same time that it causes blood loss, it does not provide for its replacement. The timing of weaning, the introduction of cow's milk, and the choice of weaning foods for the infant are discussed further in Chapter 6.

In many countries around the world, a woman breastfeeds her newborn without considering the alternatives or consciously making a decision. In other parts of the world, a woman feeds her newborn formula simply because she knows so little about breastfeeding. She may have misconceptions or feel uncomfortable about a process she has never seen or experienced. In both settings, mothers can benefit from knowledge about both alternatives before deciding which best meets their own needs and the needs of their infants.

Appendix F provides a list of nutrition resources.

To learn about infant-feeding practices, a pregnant woman can read at least one of the many books available. Other good sources of information are health care providers and mothers who have successfully breastfed and formula fed their infants. In addition to these resources, a woman examines her personal values and those of the society in which she lives. Each of these factors contributes to the decision-making process. The following discussion examines factors influencing the decision whether to feed an infant breast milk or formula, and the Practical Points later in the chapter offer suggestions as to how to feed, in either case.

Societal Support

wet nurse: a woman who breastfeeds another woman's infant.

Prior to the 18th century, substitutes for breast milk were nonexistent. Human milk was the only source of nourishment for a newborn infant. If a mother could not or did not want to breastfeed her infant, she hired a wet nurse. Deplorable sanitary conditions, as well as strange beliefs held at the time, precluded even the thought of using cow's milk to feed an infant.

Around the mid-1700s, women adopted the practice of feeding infants a mixture of bread and flour soaked in water once the first tooth had erupted. In 1874, an author first wrote of the usefulness of cow's milk when breast milk was unavailable.[1] A major problem at the time was the lack of an appropriate feeding instrument for the infant. (Cow's milk was first fed, logically, by way of a cow's horn.) By the end of the 19th century, the advantages of glass bottles and the need for cleanliness were recognized.

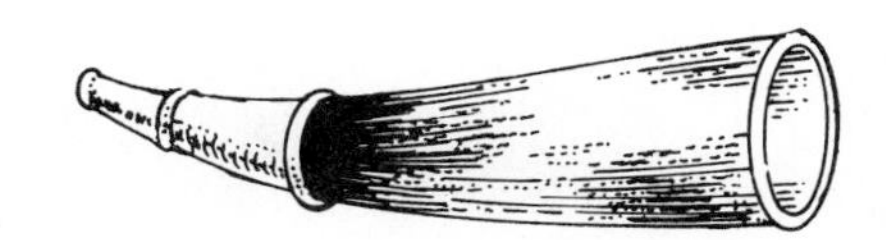

Cow's horn

Formula feeding gained popularity as the "modern," efficient, and practical way to feed an infant at the turn of this century. For the infants of women who were unable to breastfeed or who had died in childbirth, formulas became an alternative. Formula feeding paralleled the rise in hospital births and the associated practice of separating mothers from infants after delivery. Other 20th-century phenomena also influenced the shift from breast milk to formula. One obvious accomplishment of this time was the development of the technology required to analyze breast milk and to create a formula that was similar. Systems to purify water supplies and to control pathogenic organisms allowed formula feeding to develop as a safe method of nourishing infants. Simultaneous trends were seen in the advertising and fashion industries as they grew bolder in glamorizing women's bodies to a point of distorting the natural function of breasts. To view the breasts primarily as sex objects is to forget that their primary function is to provide milk to a suckling infant.

18th century German pewter nursing bottle

Breastfeeding steadily declined until fewer than 20 percent of U.S. mothers of newborns were engaging in it. During the 1970s the trend reversed, until by the 1980s, survey data showed that 60 percent of new mothers were choosing to breastfeed.[2] About half of these women were still breastfeeding at three months, and only one-fourth of those still breastfeeding at three months continued through the first year.[3] Statistics for Canada are somewhat lower and follow a similar trend. As might be expected, the incidence of breastfeeding progressively declines throughout the first year of the infant's life. In developing countries, the initial incidence of breastfeeding is much higher than in developed countries, and the decline over the first year is much smaller.[4]

19th century American glass nurser

Partly because the medical profession endorses breastfeeding as the preferred method of infant feeding, society, too, embraces the old practice of breastfeeding once again as an integral part of motherhood and child care. The increasing incidence of breastfeeding in recent years coincides with the decreasing incidence of formula feeding during the early months of infancy. However, for infants beyond three months of age, formula feeding is on the increase. Specifically, for infants three to five months old, the use of formula has increased 20 percent since 1971.[5]

"There is a reason behind all these things in nature." Aristotle

Although lactation is an automatic physiological process, breastfeeding is a learned behavior. This learning is most successful in a supportive cultural environment. In societies where few women breastfeed, appropriate breastfeeding etiquette remains undefined. A woman faces conflict, confusion, and frustration. Must she retreat to a private place to nurse? What if she cannot find such a place in a public setting? A hungry infant is impatient, and a mother must act quickly. With abundant role models, a consensus defines accepted behaviors, thus offering a nursing mother guidance and confidence. Many public buildings now offer "baby rooms" with tables for changing diapers and comfortable chairs for nursing. These rooms are open to both mothers and fathers, in response to current parenting needs.

lactation: maternal secretion of milk for a suckling offspring.

Another need of parents in today's society is to coordinate work and family. All mothers are working women—many of them with jobs outside of the home. A social system that provides extended, paid maternity leaves, nursing breaks on the job, and at-the-job-site child care promotes breastfeeding as a feasible option for infant feeding.

Nutritional Characteristics of Breast Milk

From the nutrition standpoint, breast milk is the preferred choice for infants. The American Dietetic Association advocates breastfeeding "because of the nutritional and immunologic benefits of human milk and physiological, social, and hygienic benefits of the breastfeeding process for the mother and infant."[6] The Committee on Nutrition of the AAP and the Nutrition Committee of the Canadian Pediatric Society have issued this joint statement: "Breast feeding is strongly recommended for full-term infants, except in the few instances where specific contraindications exist."[7] The recommendations in favor of breast milk arise from its unique nutrient composition and nonnutrient protective factors that promote optimal infant health and development. The AAP also recognizes that the best alternative to breast milk to meet the nutrient needs of infancy is formula—and formula imitates breast milk's composition as closely as possible. Table 5–5, later in this chapter, compares the two.

Because breast milk best meets the human infant's needs, this milk is used as a standard for estimating those needs. The infant's nutrition needs are examined in detail in Chapter 6; the following examination of breast milk composition lays the groundwork for that discussion and reveals how breast milk best serves a newborn infant.

colostrum (co-LAHS-trum): the secretion from the breast before the onset of true lactation; also referred to as the "first milk."

Colostrum During the first two to three days after delivery, before the onset of true lactation, the breasts produce colostrum. Colostrum is a premilk substance that differs from mature breast milk in nutrient composition. Colostrum is lower in fat and energy and higher in protein, sodium, chloride, potassium, sulfur, iodine, zinc, and copper than mature breast milk. During the first two weeks of lactation, the concentrations of proteins and fat-soluble vitamins decrease while lactose, fat, energy, and water-soluble vitamins increase. These changes reflect changing infant needs. Colostrum also contains important protective factors, as described on p. 156.

Carbohydrate The carbohydrate in breast milk and cow's milk–based formula is the disaccharide lactose. Lactose facilitates calcium absorption.[8] Breast milk also contains an amylase enzyme that may facilitate starch digestion.[9] The utility of this enzyme is apparent in view of the tradition that the first food normally fed to infants is some kind of grain food, usually cereal.

renal solute load: a measure of the concentration of all dissolved substances in the urine that result from the feeding of a milk, formula, or diet.

Protein The total protein content of breast milk is lower than that of cow's milk, but this is actually beneficial. The less protein consumed, the more easily the kidney's can excrete the major end product of protein metabolism, urea. The low protein concentration contributes to a low renal solute load. To excrete solutes in excess of the infant's need, the kidneys require extra water. If too little water is available, then the kidneys must concentrate the urine, and their solute-concentrating capacity during infancy is limited. A low renal solute load enables the kidneys to function within their capacity. Moreover, water balance during early infancy is fragile, because infants can afford less water loss before becoming dehydrated. This is a life-threatening condition in the infant. A low renal solute load is therefore beneficial to the infant's immature kidneys and delicate water balance.

The unique protein quality of breast milk also offers benefits. When compared with cow's milk, breast milk contains a greater proportion of the protein alpha-lactalbumin; cow's milk contains a larger proportion of casein. Digestion and absorption of alpha-lactalbumin is efficient, whereas casein produces tough, hard-to-digest curds in the infant's gastrointestinal tract. Alpha-lactalbumin is also richer in sulfur-containing amino acids than is casein. One of these sulfur-containing amino acids, methionine, is an essential amino acid, and another, cysteine, may be essential for premature infants. The enzymes responsible for synthesizing cysteine from methionine are inactive in early infancy.[10] Breast milk contains all the essential amino acids in appropriate amounts.

The amino acid taurine is abundant in breast milk and has gained attention recently in infant nutrition. Researchers are investigating the question of to what extent taurine may be essential for infants. Taurine functions in several organ systems. In the central nervous system and muscles, taurine depresses neural excitation, perhaps by lowering the intracellular calcium concentration.[11] Taurine also functions in the retina, where light causes its release; reduced body pools of taurine are associated with retinal degeneration.[12]

Lipids The total lipid content of breast milk varies considerably, due partly to differences in sampling techniques, but also to individual differences between milks from different women or milk taken from time to time from the same woman. The total fat present in breast milk does not reflect the total fat content of the maternal diet, however, except in severely malnourished women, in whose milk total fat is low. Within the fat of breast milk, fatty acid composition changes in response to maternal diet. For example, carbohydrate-predominant diets (70 percent or more) result in higher concentrations of the saturated fatty acids lauric and myristic, which are synthesized in the mammary gland from carbohydrate metabolites.[13]

The lipids in breast milk, cow's milk, and infant formulas provide the main source of energy in infants' diets. The unique lipid composition of breast milk offers other nutritional advantages as well. The fat in breast milk is more readily and more completely absorbed by the infant than is the fat in cow's milk. The enhanced absorption of breast milk lipids is partially due to the presence of lipases in breast milk that are resistant to acid and therefore can work in the stomach. Another advantage is the unique configuration of the triglycerides upon hydrolysis by the lipases. Figure 5–1 illustrates how triglycerides are hydrolyzed at the 1 and 3 positions of the glycerol molecule, leaving a monoglyceride with the fatty acid attached at the 2 position. In breast milk, palmitic acid occupies the 2 position, whereas in cow's milk, stearic acid is in the 2 position.[14] Monoglycerides with palmitic acid in the 2 position are more efficiently absorbed by the infant than either free fatty acids or monoglycerides with stearic acid in the 2 position.[15]

Triglycerides constitute 98 percent of the lipids present in breast milk, while free fatty acids, cholesterol, and phospholipids make up the remainder.

Figure 5–1 Triglyceride Hydrolysis and Fatty Acid Configuration
Palmitic acid occupies the 2 position in breast milk, whereas stearic acid occupies that position in cow's milk. After hydrolysis, monoglycerides with palmitic acid are more efficiently absorbed than free fatty acids or monoglycerides with stearic acid.

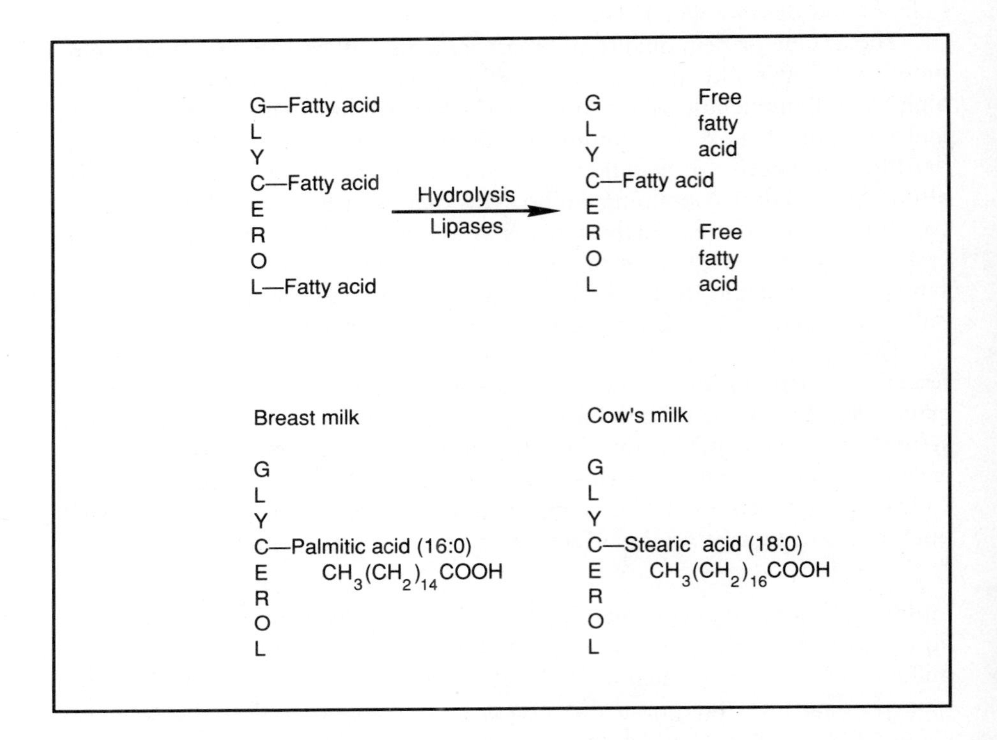

Breast milk contains more than adequate concentrations of the essential fatty acid linoleic acid. Breast milk's cholesterol content of 20 to 30 milligrams per 100 milliliters provides infants with about 200 milligrams per day, about two-thirds of the maximum recommended for adults. The significance of this is still unclear, but considerable interest in the subject continues. One group of researchers speculated that a cholesterol-rich diet during infancy might protect against atherosclerosis in later life.[16] They hypothesized that early high cholesterol consumption might induce the infant to synthesize less cholesterol and break down more of it in later life—responses that might offer protection against atherosclerosis. They tested their hypothesis on male rats by feeding them diets high in cholesterol prior to weaning. These rats had lower serum cholesterol at maturity when compared with rats fed a control diet. Results of further studies, however, have not unambiguously confirmed the first impressions.

Cholesterol in human milk:
20 to 30 mg/100 ml.
Approximately 200 mg/day.
Cholesterol in infant formula:
1 to 3 mg/100 ml.
Approximately 15 mg/day.

The cholesterol content of infant formulas is only about 10 percent of that in breast milk.[17] Many questions remain unanswered concerning exactly how much cholesterol and other lipids are appropriate in the infant's diet. Cholesterol is essential for normal myelination in the central nervous system and is the precursor for synthesis of steroid hormones, but high serum cholesterol concentrations in later life are a risk factor in the development of atherosclrosis. This latter concern may be the basis for formula makers' limiting of cholesterol to only 1 to 3 milligrams per 100 milliliters as of this writing. The AAP discourages restriction of fat during infancy, even for infants at risk of developing atherosclerosis.[18] The question how much cholesterol should be in infant formula remains unresolved.

The energy nutrient composition of breast milk would be as bad for adults, in some ways, as it is good for infants. Breast milk's protein contributes only about 6 percent of its kcalories, whereas protein in the adult diet is supposed to contribute twice that percentage (see Figure 5–2). Its carbohydrate contributes about 40 percent of its kcalories, whereas about 60 percent is recommended for adults. Its fat contributes 55 percent of its kcalories, whereas adults are told to derive less than 30 percent of their food energy from fat. Yet for infants, breast milk is nature's most nearly perfect food, providing an object lesson in the necessity of appreciating that people at different stages of the life span have different nutrient needs.

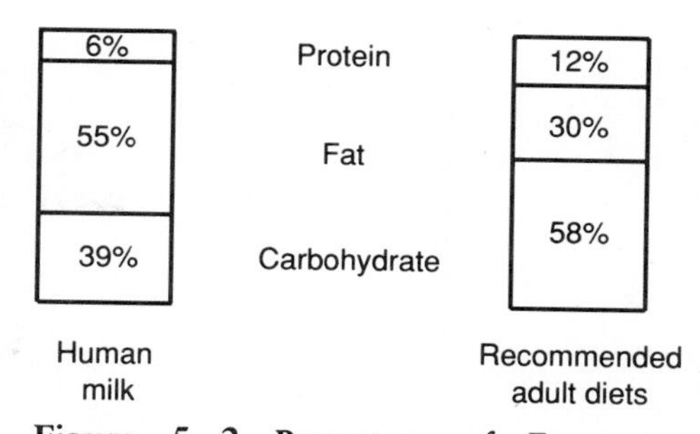

Figure 5–2 Percentage of Energy-Yielding Nutrients in Human Milk and in Recommended Adult Diets
The proportions of energy-yielding nutrients in human breast milk differ from those recommended for adults.

Source: Adapted from Ross Laboratories, Columbus, Ohio, January 1979.

Vitamins and minerals The infant's vitamin and mineral requirements are discussed, nutrient by nutrient, in Chapter 6. In general, breast milk from healthy mothers contains all the infant needs for normal growth, with the exception of vitamin D and fluoride. Until more is known about the vitamin D content of breast milk, supplementation is recommended for the breastfed infant up to the time when infant formula or vitamin D–fortified milk is introduced (see Chapter 6).[19] As for fluoride, its concentration in breast milk is low and varies little, regardless of water supply.[20] Nor does the breastfed infant normally receive much water itself during the first six months of life. For all infants that live in areas of nonfluoridated water, foreseeing that later consumption of water will not deliver fluoride, the AAP favors early introduction of a fluoride supplement—that is, before six months of age.[21]

Breast milk has a low electrolyte content. An advantage of this is the low renal solute load produced, as discussed earlier. Also, breast milk's sodium concentration is low. It is unclear whether a low sodium intake during infancy protects infants from the later development of hypertension, but in adults, an association between high salt intake and high blood pressure does exist. One study of newborn infants investigated the effects of dietary sodium on blood pressure.[22] Infants received either a normal-sodium diet or a low-sodium diet, and their blood pressure was monitored monthly for six months. At the end of that time, the average systolic blood pressure of the low-sodium group was significantly lower than that of the normal-sodium group. Further research in this area is needed before firm conclusions can be drawn, but it appears that the sodium needs of normal, full-term infants are small and are adequately met by breast milk.

The nutritional advantages of breast milk go beyond the nutrient content of the milk. The breastfed infant self-regulates milk consumption, while the formula-fed infant is subject to control by the person offering the bottle. In view of the pervasive concern with childhood obesity, this self-regulation of food intake may be a point in favor of breastfeeding. However, by recognizing and responding to the infant's cues of satiety, parents who feed formula can strengthen their infants' ability to self-regulate food intake.

Protection Conferred by Colostrum and Breast Milk

Parallel to the nutritional benefits of breastfeeding in promoting the infant's health and well-being are the nonnutritional benefits. Evidence for these comes from studies of the milk itself, and from comparative studies of infants fed breast milk and formula.

Protective factors in colostrum and breast milk Colostrum and breast milk are rich in substances that protect the infant from infection. Colostrum has an especially high concentration of antibodies and white blood cells and helps protect the newborn from infections against which the mother has developed immunity. Colostrum may also contain a growth-promoting substance; in newborn rats fed colostrum, intestinal cells grow faster than they do in comparable animals fed mature milk.[23]

immunoglobulins: proteins that are capable of acting as antibodies.

antigens: substances foreign to the body that elicit the formation of antibodies, an inflammation reaction, or both from immune system cells.

secretory IgA: one of several types of antibodies produced by the human immune system; the predominant antibody of breast milk.

bifidus (BIFF-id-us or by-FEED-us) **factors:** factors in colostrum and breast milk that favor the growth, in the infant's intestinal tract, of the "friendly" bacteria *Lactobacillus bifidus* so that other, less desirable intestinal inhabitants will not flourish.

lysozyme: an enzyme that breaks up cell wall structures, including those of bacterial cell walls.

lactoferrin: a factor in breast milk that binds iron and keeps it from supporting the growth of the infant's intestinal bacteria.

The immunoglobulins of breast milk protect the infant's digestive tract against environmental antigens. Most often, the mother and infant are exposed to the same antigens. When the antigens reach the mother's digestive tract, they meet with lymphocytes that become sensitized to the antigens and are then transported to the mammary gland, where secretory IgA production begins. Secretory IgA is the major immunoglobulin of colostrum and breast milk. The infant receives these antibodies with the breast milk. Secretory IgA is resistant to proteolysis, so it survives and functions in the infant's digestive tract. It is thought that secretory IgA, in conjunction with lactoferrin and other factors, accounts for the lower incidence of intestinal infection among breastfed infants compared with formula-fed infants.[24]

Bifidus factors, lipases, lysozyme, and lactoferrin are among the other protective factors in colostrum and breast milk.[25] The bifidus factors favor the growth of the species of bacteria *Lactobacillus bifidus,* which is normally resident in the human gastrointestinal tract. *L. bifidus* in the digestive tract inhibits the growth of potentially harmful *Escherichia coli* and other bacteria, thereby protecting the infant from intestinal infection.[26]

The lipase activity of breast milk also offers the breastfed infant antimicrobial protection. Cow's milk has a high triglyceride and low monoglyceride content, but thanks to its lipases, breast milk is rich in free fatty acids and monoglycerides. One classic study demonstrated that milks with this lipid composition exhibited stronger antiviral activity than milks rich in triglycerides.[27] Based on these results, the researchers suggested that milks used in the manufacture of infant formulas should be fortified with monoglycerides instead of unsaturated vegetable oils, as is the current practice.

Human milk is richer in the enzyme lysozyme than are other milks. Lysozyme breaks apart bacterial cell walls. It is suggested that lysozyme may therefore have a bacteriostatic function in the digestive tract of breastfed infants, although evidence is limited.

Another protein that contributes to the antimicrobial activity of breast milk is lactoferrin. Lactoferrin appears to successfully compete for iron against iron-demanding bacteria in the intestinal tract, thus inhibiting their growth.[28] Also, lactoferrin may be responsible for the high bioavailability of iron from breast milk compared with that from cow's milk, which contains little lactoferrin.[29] The lower bioavailability of iron from infant formulas is compensated for by the manufacture of iron-fortified formulas, which are effective in preventing iron deficiency.

Breast milk also contains zinc-binding proteins that enhance zinc's bioavailability.[30] Zinc's absorption from breast milk is extraordinarily efficient.

An epidermal growth factor that stimulates growth of intestinal cells has been discovered in breast milk.[31] The physiological significance of this growth factor remains to be established. It seems logical that it may promote the replacement of damaged cells, thus helping to keep intact the infant's digestive tract barrier against infection.

Comparative studies Consistent with the discovery of the many anti-infective factors in breast milk is convincing statistical evidence that breast milk does protect infants against infection. In developing countries, where poor sanitation is prevalent and the incidence of infection is high, the infection rate among breastfed infants is low. Failure to breastfeed an infant who lives in a house without piped water and a toilet incurs twice the risk of perinatal mortality as for a breastfed infant living in a house with good sanitation.[32] Similarly, higher hospitalization rates are seen in infants fed formula for the first 18 months of life than in breastfed infants.[33]

In developed countries, findings about the protective effects of breastfeeding are conflicting.[34] Some researchers are skeptical that a protective advantage attends breastfeeding in industrialized nations. Several methodological flaws are common among studies that examine the protective properties of breast milk.[35] One of the most basic inconsistencies concerns the definition of the feeding method itself. For example, researchers often fail to precisely define the term *breastfed*. Infants are categorized as "breastfed" or "bottle fed," without mention of those infants who are partially breastfed, or into what category such infants are placed. Another problem common to infant-feeding studies is controlling for confounding variables such as differences in social class, whether the mother smokes, and family size. If these variables are not controlled for during the course of the study, then the results may be hard to interpret and even harder to compare with those of other studies.

The results of comparative studies may also be confounded by differences in the settings where infants are cared for during their early months. Those who are kept at home are likely to be exposed to fewer pathogens than those that are cared for in day care centers. If the breastfed infants are at home while the formula-fed infants are taken to day care, this difference might account for some of the observed variation in infection rates.

In addition to protection against infection, it is possible that breast milk offers protection against the development of allergies. When breastfed infants were compared with formula-fed infants, those who had been solely breastfed for six months showed a lower incidence of allergic diseases such as asthma and skin rash.[36] This was especially noticeable among those infants with a family history of allergy.

Considerations for Preterm Infants

Preterm infants have special needs, which are discussed in Chapter 6. Researchers have compared three sources of nutrients to meet these needs:

breast milk from mothers of preterm infants, breast milk from mothers of term infants, and formulas designed for preterm infants. They find that preterm milk and special formulas support infant growth better than full-term milk, even though preterm infants tolerate full-term milk well.[37]

A preterm mother's milk supports more rapid infant growth in a preterm infant than the milk of a full-term mother because its composition is more suited to a preterm infant's needs. Formula designed for preterm infants also offers a growth advantage over feeding with mature human milk. The breast milk of preterm mothers differs in some nutrient concentrations and in milk volume from that of term mothers.[38] During early lactation, preterm milk contains higher concentrations of protein and is lower in volume than term milk. The low milk volume is advantageous because preterm infants are unable to consume large quantities of milk per feeding, and the higher protein concentration allows for better growth. The composition of preterm milk closely resembles that of colostrum, the earliest secretion of term lactation. Most authorities agree that preterm milk best meets the specific needs of a preterm infant.

Preterm milk may be an inadequate source of some nutrients, however, based on theoretical estimates of the requirements of preterm infants. Specifically, the calcium and phosphorus in preterm milk may be insufficient to support growth at the rate that would have occurred in utero.[39] The combination of preterm human milk fortified with a preterm supplement supports growth at that rate. In many instances, mixtures of nutrients specifically designed for preterm infants are added to the mother's expressed breast milk and fed to the infant from a bottle.

Medical Considerations

Some medical situations weigh on the question of whether breastfeeding is the appropriate choice. If a woman has a communicable disease that could threaten the infant's health so that they have to be separated, then, of course, she cannot breastfeed. Similarly, a physician may advise a woman with a chronic disease against breastfeeding if it would be a drain on her energy and a strain on her emotional health.

Formula feeding is preferred, and can even be lifesaving, in some medical circumstances. The woman who has a positive antibody test for acquired immune deficiency syndrome (AIDS) is advised to feed her infant formula. The AIDS virus can be transmitted through breast milk and infect the infant.

Infants with specific metabolic disorders, such as PKU, must receive special formulas, such as Lofenelac. In these cases, not only must the infant receive formula instead of breast milk, but the formula must meet exact specifications (see p. 181, "Special Formulas"). Some circumstances, such as lactose intolerance and protein allergies, may appear to warrant formula feeding in preference to breastfeeding. Yet infants with these conditions often do better with mother's milk than with formulas.

If a woman must take medication that is known to harm the infant and that will be secreted in breast milk, she must opt for formula feeding, at least temporarily. Many prescription drugs do not reach nursing infants in sufficient quantities to affect them adversely. Some, however, do. Table 5–1 presents a

Table 5–1 Drugs That Are Contraindicated during Breastfeeding

Drug	Registered Name	Indications	Reported Sign or Symptom in Infant or Effect on Lactation
Amethopterin methotrexate[a]			Possible immune suppression; unknown effect on growth or association with carcinogenesis
Bromocriptine mesylate	Parlodel	Amenorrhea, female infertility, suppression of lactation, Parkinson's disease	Suppression of lactation
Cimetidine hydrochloride[b]	Tagamet	Ulcers	Possible suppression of gastric acidity in infant, inhibition of drug metabolism, and central nervous system stimulation
Clemastine fumarate	Tavist	Allergies	Drowsiness, irritability, refusal to feed, high-pitched cry, neck stiffness
Cyclophosphamide[a]	Cytoxan	Malignancies	Possible immune suppression; unknown effect on growth or association with carcinogenesis
Ergotamine tartrate	Cafergot	Migraine headaches	Vomiting, diarrhea, convulsions (doses used in migraine medications)
Gold salts	Myochrysine	Rheumatoid arthritis	Rash, inflammation of kidney and liver
Methimazole	Tapazole	Hyperthyroidism	Potential for interfering with thyroid function
Phenindione			Hemorrhage
Thiouracil			Decreased thyroid function; does not apply to propylthiouracil

[a] Data not available for other cytotoxic agents.
[b] Drug is concentrated in breast milk.

list of drugs that are contraindicated during breastfeeding and shows their effects on infants and lactation. In addition, drugs containing radioactivity (such as iodine 131) require cessation of breastfeeding for a period of time that depends on the specific drug. (To maintain lactation during times of cessation, the mother should express her milk and discard it.) If a woman must take a medication that is regarded as generally safe, she can minimize any effects by taking the drug while or immediately after the infant nurses. This will produce the lowest amount of drug in the breast milk at the next feeding.[40]

A description of expressing breast milk is given on pp. 176–177.

Alcohol, Illicit Drugs, and Environmental Contaminants

Drug addicts, including alcohol abusers, are capable of consuming such high doses that their infants can become intoxicated or even addicted by way of breast milk; in these cases, formula feeding is preferred. Occasional, moderate consumption of alcohol is compatible with breastfeeding (one cocktail, one glass of wine, or one beer per day). However, even an average of two drinks daily while breastfeeding can diminish an infant's motor development, and indiscriminate or excessive drinking may cause drowsiness, weakness, and slowed growth in infants.[41] Similarly, the use of illicit drugs by a breastfeeding mother can harm her infant. (The effects of such drugs taken during pregnancy were mentioned in Chapter 3.)

Evidence that cocaine enters breastmilk and adversely affects nursing infants comes from reports of a case study involving a two-week-old infant.[42] The infant was healthy at birth, was sent home with the mother at three days of age, and was exclusively breastfed. About two weeks later, the infant was admitted to the hospital emergency room with symptoms of cocaine intoxication—extreme irritability, hypertension, rapid heart rate, tremors, and dilated pupils. The mother had ingested cocaine and breastfed her infant during the four-hour period preceding the infant's symptoms. Analyses of the mother's milk and the infant's urine confirmed the presence of cocaine. Cocaine persisted in the breast milk for 36 hours and in the infant's urine for 60 hours. As long as 72 hours after the last breastfeeding prior to hospitalization, the infant's heart and respiratory rate remained elevated.

Environmental contaminants can also find their way into breast milk and harm an infant. As adaptable and nutritious as human breast milk is, it does not contain a magic filter for contaminants. Substances such as pesticides and toxic wastes may at times be present in breast milk. Whether they are present depends on what the mother has ingested, the quantity consumed, and the interval between ingestion and breastfeeding.

Contaminants are often widespread in the environment; consumed by people in the foods they eat and the air they breathe; stored in the body's fat tissues; and slowly degraded and excreted.[43] Long-term exposure to contaminants leads to a gradual accumulation in the body's fat, including breast milk fat. Lactation is a means of excreting contaminants.

DDT is dichlorodiphenyl trichloroethane.

Prior to the 1972 federal restrictions on the use of the pesticide DDT, several studies reported DDT in the milk of mothers at concentrations greater than the federal government allows in dairy milk meant for human consumption.[44] This may be partly explained by the difference in the DDT excretion rate between women and cows. Lactating women excrete most of their DDT intake, while cows excrete very little.[45] Fortunately, studies did not find that breastfed infants of mothers excreting DDT were less healthy than formula-fed infants.

Other toxic wastes of concern are the PCBs, found in rivers and waterways polluted by industry. PCBs are organic chemicals used in such products as paint and caulking compounds. According to the Committee on Environmental Hazards of the AAP, only women who have eaten large amounts of fish caught in PCB-contaminated rivers, such as the Saint Lawrence Seaway, or who have been directly exposed to PCBs in their occupations need fear breast milk contamination.[46] A woman with concerns about her exposure to toxic substances can obtain information from her local health department.

PCBs are polychlorinated biphenyls.

Effects on the Mother's Body

Breastfeeding may offer medical benefits to the mother as well as to the infant. Some evidence suggests that lactation protects women from later breast cancer.[47] If the correlation between lactation and a reduced breast cancer rate holds up, the causal connection, if any, remains to be elucidated. Perhaps the physical and hormonal events of breastfeeding promote resistance to cancer, but on the other hand, preexisting conditions may be responsible. The imbalances that led to unsuccessful lactation, for example, might also predispose a woman to breast cancer. Of course, in either case, the fear of contracting a life-threatening disease is an unsound basis for deciding to breastfeed.

Some women fear that breastfeeding will cause their breasts to sag. The breasts do swell and become heavy and large immediately after the infant is born, but they may eventually shrink back to their prepregnant size, even when they are producing enough milk to nourish a thriving infant. With proper support, diet, and exercise, breasts return to their former shape and size after weaning, at the latest. Breasts change their shape as the body ages, but breastfeeding does not accelerate this process.

Breastfeeding can actually aid in restoring a woman's body to its nonpregnant state. The infant's suckling stimulates the nerves controlling the muscles of the uterus, causing them to contract. The contractions expel any tissue remaining after the birth, and return the uterus to its normal size.

Milk production also requires energy, facilitating loss of the fat stored during pregnancy. Researchers compared energy intake and weight loss among three groups of women: those who were strictly breastfeeding their infants, those using a combination of breastfeeding and formula feeding, and those who were feeding only formula.[48] The women who were strictly breastfeeding their infants had significantly higher energy intakes, but all three groups experienced similar weight losses over six months' time. Furthermore, the breastfeeding women lost more fat from the hip area than did the other women. A woman who chooses nutrient-dense foods during lactation will experience a gradual weight loss, even though her energy intake is greater than normal.

Breastfeeding does not, however, promote rapid, easy weight loss and should not be expected to do so. Disappointment arising from such false expectations may lead some women to stop nursing their infants earlier than they had originally intended.[49]

Some women find the delayed onset of ovulation during lactation, and the consequent delay of menstruation, a pleasant bonus. (They should not, however, count on it for contraception, as explained on p. 166.) Mothers can continue breastfeeding without harmful effects after menstruation resumes.

Bonding

bonding: a process that occurs immediately after birth, in which a mother forms an affectionate attachment to her infant.

Some breastfeeding proponents argue that breastfeeding encourages bonding. The affection of a mother for her infant depends more on the time spent in close physical contact than on the method of feeding. There is evidence, however, to suggest that mothers allowed early extended contact with their newborns are more likely to breastfeed and to continue to do so for a longer duration than mothers denied contact. This is not to say that breastfeeding mothers bond better than mothers who feed their infants formulas, but that early, prolonged contact facilitates both bonding and breastfeeding.

Controversy surrounds the question of whether early and extended mother-child contact is crucial to bonding.[50] The notion that a bond develops during a sensitive period just after birth that has a profound effect on the parent-child relationship is based on observations of farm animals. In animals, the bond between a mother and her offspring forms within the first few minutes of life. A brief initial contact, even if followed by a separation, is sufficient to establish the relationship, thanks to a variety of visual, auditory, and olfactory cues. Most likely, postpartum hormonal conditions influence maternal acceptance. If mother and offspring are separated during this crucial time, the mother later rejects the offspring. She refuses to nurse and exhibits physically hostile behaviors toward the newborn.

In human beings, mothering behaviors are more complex than can be explained by cues and hormones alone. The actual birth in human beings is preceded by months of maternal thought and anticipation. Many women begin the bonding process during pregnancy. They pat their bellies when they feel a kick, attend childbirth classes, read books, seek advice from friends, and decorate the baby's room. A woman enters motherhood with memories of her own relationship with her mother. She has learned her culture's patterns of mother-child interactions. Whether the infant was planned or wanted also affects the mother's maternal behaviors, as does her relationship with the father.

The notion that events at the time of birth can influence the bonding process has altered birth practices in this country. Earlier this century, mother and newborn were separated at birth and cared for in the maternity and nursery wards, respectively. More recently, there has been a trend toward allowing mother and newborn to lie together in a warm and quiet room, often accompanied by the father. Health care providers encourage breastfeeding mothers to nurse their infants immediately after delivery.

Such family-centered birth experiences may indeed promote loving relationships, but the theory is difficult to prove. As is often the case in research involving human beings, designing methodologically sound and ethical studies is close to impossible.[51] In this instance, the results would require subjective interpretations of significance. If one mother picks up her infant more often than another, who is to determine the significance and effects of that behavior?

Convenience

In a society that embraces microwave ovens and frozen foods, convenience plays a large part in the many decisions parents make about feeding themselves and their children. The lifestyles, attitudes, and habits of each individual dictate what is defined as convenient.

Breastfeeding can be done anywhere, anytime, as long as mother and infant are together. Breast milk is sterile and always at the appropriate temperature. The breastfeeding mother has freedom from sterilizing bottles and mixing formulas. Breastfeeding may not be convenient, however, for a mother whose schedule does not permit her to be with her infant easily at feeding times. A woman who works outside the home may find it less troublesome to feed formula to her infant. Hot cycles on a dishwasher make bottle sterilization easy; and measuring, mixing, and pouring formula is easier than preparing any other meal. Ready-to-pour formulas that do not need mixing, although more expensive, are available for busy mothers who can afford them. Another advantage of formulas is that they allow the father and other family members opportunities to enjoy feeding the infant.

The mother who is motivated to continue breastfeeding when she returns to work outside the home may be able to find ways to do so. She can breastfeed exclusively up to that time; then, depending on her schedule and location, she can continue to breastfeed for some feedings and supplement with formula or expressed breast milk for others. Breastfeeding manuals provide a multitude of hints and suggestions to help her work out the details.

Making the Decision

The decision whether to breastfeed is best made by the parents with the advice of their health care providers. However, quite likely, many friends, relatives, and strangers will voice their opinions. Some will consider breastfeeding old-fashioned and formula feeding more convenient. Others believe the breast best and formulas unnatural. After listening to these people, a woman will ultimately need to seek advice and information from well-informed sources: registered dietitians and other qualified health care providers.

Some women are convinced that breastfeeding offers the best nourishment to their newborns. They are sure within themselves that the suckling of an infant at their breast is a most rewarding experience. Breastfeeding is compatible with their views on mothering and easily fits into their work and home arrangements.

Some women are uncomfortable with breastfeeding; they may attempt to breastfeed, but encounter obstacles in attempting to nurse an infant. Other women find that breastfeeding just does not work with their lifestyles and schedules. These mothers have valid reasons for making their choices, and their feelings need to be honored. Bearing and nurturing an infant involves much more than merely pouring in nutrients.

Many health care providers and dietitians believe breastfeeding is sufficiently important to warrant that every effort be made to do so, even if only for a short time. Of those women who do breastfeed, 25 to 50 percent discontinue within the first month, and 50 to 70 percent discontinue by four months.[52]

This duration is long enough to allow the infant to receive immunological protection and other special advantages of breastfeeding during the most critical first few weeks or months. The mother can then shift to formula, knowing she has given her infant those benefits.

If this discussion appears biased in favor of breastfeeding, it is because breastfeeding offers many benefits to both mother and infant, and every pregnant woman should seriously consider it. Still, as mentioned earlier, there are many valid reasons for not breastfeeding, and formula-fed infants grow and develop into healthy children. After all, the primary goal is to provide optimal nourishment in a relaxed and loving environment.

The responsibilities and decisions continue once a mother decides how she will feed her infant. If she decides to breastfeed, she will need to learn how to select foods to promote optimal milk production. If she decides to feed formula, she will need to learn how to select the appropriate formula and prepare it. The following sections provide the needed information.

Nutrition for the Lactating Mother

While much attention is focused on nutrition for pregnant women, infants, and children, nutrition for lactating women is often neglected. However, the lactating mother has unique nutrient needs and concerns, and her care and feeding deserve special attention. First, a description of the physiology of lactation lays the foundation for understanding the special nutrient needs of a nursing mother and the effects of maternal nutrition on breast milk composition. Then, a discussion of the behavior of breastfeeding builds on that (see Practical Point: How to Breastfeed, at the end of this section).

mammary glands: glands of the female breast that secrete milk.

ducts: narrow tubular vessels that drain the lobes of the mammary gland into the tip of the nipple.

lobes: segments of the mammary gland.

alveoli (al-VEE-oh-lie): the milk-producing cells of the mammary gland; the singular is **alveolus.**

prolactin (pro-LAK-tin): a hormone secreted from the anterior pituitary gland that acts on the mammary glands to initiate and sustain milk production.
pro = promote
lacto = milk

oxytocin (OK-see-TOE-sin): a hormone secreted from the posterior pituitary gland that stimulates the uterus to contract and the mammary glands to eject milk.
oxy = quick
tocin = childbirth

Physiology of Lactation

Lactation is the natural extension of pregnancy—of the mother's body nourishing the infant. The mammary glands secrete milk for this purpose.

Figure 5–3 illustrates breast development from puberty to lactation. The mammary glands, stimulated by estrogen during puberty, develop a system of ducts, lobes, and alveoli. This system remains fairly inactive until pregnancy. During pregnancy, hormones stimulate proliferation of the ducts and alveoli. Estrogen promotes growth and branching of the duct system, while progesterone stimulates development of the alveoli. At the same time, the elevated concentrations of these hormones inhibit the actual secretion of milk. These two hormones return to their basal concentrations within a few days postpartum, thus allowing milk production to begin.[53]

Hormonal support of lactation As illustrated in Figure 5–4, the hormones prolactin and oxytocin finely coordinate lactation. Prolactin is responsible for milk production. Throughout pregnancy, the blood concentration of prolactin increases.[54] With the delivery, its concentration rises substantially in preparation for infant feeding.

Figure 5–3 Breast Development from Puberty to Lactation

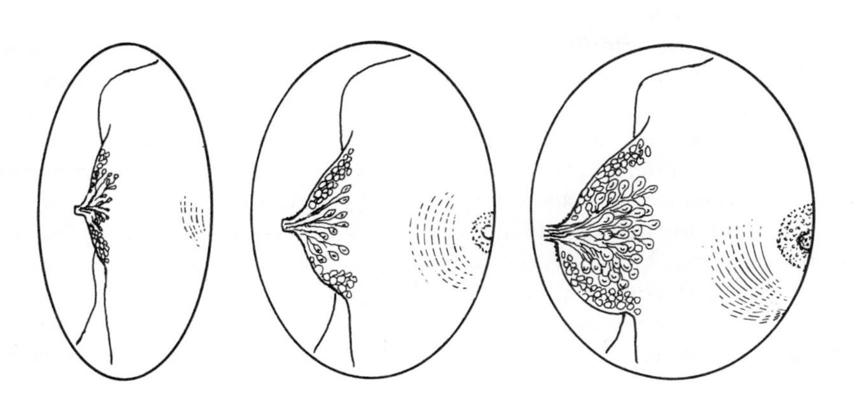

A. During puberty, a system of ducts, lobes, and alveoli develops.

B. This system remains inactive until pregnancy.

C. During pregnancy, growth proliferates, with ductal branching and lobular-alveolar development proceeding at a spectacular rate, yet in an orderly fashion.

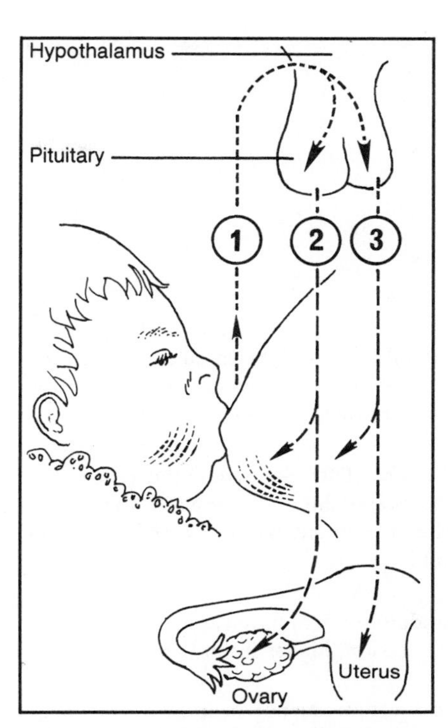

Figure 5–4 Prolactin and Oxytocin Activity

(1) An infant suckling at the breast stimulates the pituitary to release prolactin and oxytocin. Each of these hormones acts on the mammary glands: (2) prolactin encourages milk production, and (3) oxytocin stimulates milk ejection. Each of the hormones also acts on the reproductive organs: (2) prolactin inhibits ovulation, and (3) oxytocin promotes uterus contractions.

High prolactin concentrations signal the release of prolactin-inhibiting hormone, which ensures that prolactin concentrations and milk production do not exceed the need. This is an example of hormone regulation by negative feedback—prolactin is turned off by its own high concentrations.

Superimposed on this, the slow, steady rise in estrogen production during the final trimester of pregnancy requires prolactin-inhibiting hormone to remain inactive, allowing prolactin concentrations to rise in preparation for lactation. Another message to turn off prolactin-inhibiting hormone comes from the infant's suckling. The suckling on the breast signals a demand for milk production (which requires prolactin) and a demand for prolactin-inhibiting hormone to turn off (which permits the release of prolactin). The consequence of these hormonal interactions is that prolactin concentrations remain high and milk manufacture continues as long as the infant is nursing. The infant's demand for milk causes the mammary glands to supply milk.

prolactin-inhibiting hormone: a hormone secreted from the hypothalamus that acts on the anterior pituitary gland to regulate the release of prolactin.

In breastfeeding mothers, concentrations of prolactin are extremely high for the first three months but then decline to near-normal levels, even with continued milk production.[55] In nonbreastfeeding mothers, prolactin concentrations decline to prepregnant levels within the first two to three weeks postpartum.

The hormone oxytocin causes the myoepithelial cells surrounding the alveoli to contract, thus initiating milk ejection into the ducts, known as the let-down reflex. (The mother feels this as a contraction of the breast, followed by a flow of milk and relief of pressure.) After the birth, the progesterone concentration (which was elevated during pregnancy and turned off oxytocin during that time) declines, allowing oxytocin's release. Oxytocin also responds

myoepithelial cells: contractile cells of the mammary glands.

let-down reflex: the reflex that forces milk to the front of the breast when the infant begins to nurse.

to the stretching of the cervix during childbirth, causing two organs to react: the uterus to contract and the mammary glands to eject milk.

An infant's suckling also elicits oxytocin to eject milk from the mammary glands. At first, the stimulus for let-down is the infant's suckling. Later, when the reflex is well established, the sound of the infant's crying may be enough to trigger it. An efficient let-down reflex is essential to successful lactation. Emotional upset, pain, and fatigue may inhibit the let-down reflex. By relaxing and taking care of herself, the nursing mother promotes easy let-down of milk and greatly enhances her chances of successful lactation.

foremilk: the milk released early in a nursing session, the milk at the front of the breast; low in fat, high in nutrients.

draught (DRAFT) **reflex:** the reflex that moves the hindmilk toward the nipple after the infant has drawn off the foremilk.

hindmilk: the milk released late in a nursing session, higher in fat than foremilk.

From early to late in a nursing session, the character of milk changes. The milk released first, known as the foremilk, provides most of the nutrients the infant receives. The draught reflex, which occurs later during a nursing session, draws milk from the hindmost milk-producing glands of the breast after the foremilk has been released. The mother feels this as a tingly sensation within the breast. Hindmilk has a higher fat content than foremilk, allowing infants to become satiated at the end of the breastfeeding session, after they have satisfied their sucking need and received sufficient nutrients from the foremilk.

postpartum amenorrhea: the normal temporary absence of menstrual periods immediately following childbirth.

lactational anovulation: the normal suppression of ovulation during lactation.

Postpartum amenorrhea and contraception Women who breastfeed their infants experience prolonged postpartum amenorrhea. An infant's suckling serves the dual purpose not only of promoting milk production but also of causing lactational anovulation; both of these effects are mediated by the hormone prolactin. As Figure 5–4 depicts, prolactin inhibits the release of the hormones responsible for ovulation. This is beneficial because it allows time for the replenishment of maternal nutrient reserves.

Physically, the new mother's body is still undergoing major changes. Hormones are shifting from a state of pregnancy to one of lactation or nonpregnancy. Some nutrient stores may be low or depleted. A new mother rarely sleeps more than four hours at a time. Even with optimal nourishment and adequate rest, she will not be back to "100 percent" for at least a year. Repeated pregnancies at intervals of less than one year deplete nutrient reserves. Whether contraception is passive or active, it seems best to avoid pregnancy until the mother has had time to readjust to nonpregnancy, restore nutrient banks, and recharge her energy. The reproductive system needs a rest before being called into active duty again.

Chapter 2 discusses the nutrition of a woman prior to pregnancy and the effects of contraception on nutrition.

The uterus has returned to its normal size by about six weeks after delivery. Menstruation usually occurs four to eight weeks postpartum in nonlactating women. Women need not be concerned if menstruation is delayed for three or four months, however, since variation is quite normal. The duration of postpartum amenorrhea is as individual as women themselves. Absent menstrual periods, however, do not protect a woman from pregnancy. An ovum may be released at any time, so to avoid pregnancy, a couple must use some form of contraception.

Maternal nutrition status plays a role in altering plasma prolactin concentration and, therefore, in postpartum amenorrhea and anovulation. When diets of undernourished lactating women are supplemented to provide the needed energy, protein, vitamins, and minerals to support lactation, plasma prolactin concentrations decline.[56] The duration of postpartum amenorrhea is shortened.[57] Thus the length of postpartum amenorrhea depends on both lactation and nutrition. Prolonged lactation and poor nutrition lengthen the period of postpartum amenorrhea and consequently reduce fertility.

The effect of nutrition on fertility is discussed in Chapter 2.

The endocrine events preceding the first menstruation postpartum are rarely like those of an ordinary interpregnancy cycle. Many women do not ovulate before that first menstruation; menstruation results from the degeneration of an ovarian follicle. When ovulation does precede the first menstruation, it is often followed by an incomplete luteal phase of the cycle. These conditions are unlikely to support a pregnancy even if the ovum is fertilized. Fewer than 20 percent of postpartum women have a normal ovulatory cycle prior to their first menstruation.[58] Given a 25 percent probability of conception in a normal menstrual cycle, it is estimated that only 5 percent of lactating women who engage in unprotected intercourse before their first menstruation are likely to conceive. Indeed, the incidence of conception during lactational amenorrhea in developing countries is less than 10 percent. Such statistics seen in populations bespeak the effectiveness of lactation as a contraceptive influence but do not make it acceptable as a contraceptive method for an individual.

Unfortunately, no simple or reliable procedure to detect the onset of fertility has been developed. Many biological factors influence the variability in duration of postpartum infertility. Hormonal responses to the frequency and vigor of sucking are thought to be responsible for much of the variation. Women who frequently nurse their infants have elevated plasma prolactin concentrations. The number of feeding periods, as well as the total time at the breast, also raises plasma prolactin. As mothers begin to supplement their infants' diets and nurse for shorter periods less often, prolactin levels decline and ovarian activity resumes. Most lactating women will start to menstruate and ovulate prior to weaning.

If a lactating mother wants to use the Pill, she is wise to wait until after she has weaned her infant and to use another method of contraception in the meantime. Standard oral contraceptive pills contain estrogen, which reduces milk volume and the protein content of breast milk.[59] The lower milk volume, however, does not appear to impair infant growth or behavior when the mother is well nourished and lactation is well established.[60] For women wishing to use oral contraceptives, progestin-only contraceptive pills probably are preferable, especially if maternal nutrition is inadequate; they do not affect milk volume. The hormones of oral contraceptives reach the infant via the breast milk, but the concentrations are low, and the effects, if any, are probably transient.

End of lactation The woman who wants to stop lactating may be given an injection of estrogen or a large dose of vitamin B_6 to inhibit milk production. Such measures hasten the end of lactation but are not necessary. Without the stimulation of suckling, milk production will eventually stop.

Maternal Nutrition and Breast Milk

During lactation, as during pregnancy, the mother requires sufficient nutrient intakes and stores to support the infant's growth and her own health. If she does not eat well throughout pregnancy and lactation, her health may be compromised; in some instances, to a greater extent than that of her child. In addition, lactation is likely to falter or fail. Ideally, a woman will have consumed high-quality foods throughout pregnancy, and will continue to eat them after she has given birth.

The pattern of nutrient needs for a lactating mother is unique. Depending on the specific nutrient, her need may be less than, equal to, or greater than during pregnancy. Figure 5–5 repeats Figure 4–6 in Chapter 4, adding the nutrient needs of lactating women. Table 5–2 offers a daily food guide to meet these nutrient needs; it is the same as that for pregnant women.

Energy intake and exercise Energy from the maternal diet and tissue reserves provides for both lactation and maternal health and activities. A nursing mother produces approximately 23 ounces of milk in a day during the first 6 months of lactation, and slightly less thereafter.[61] At 23 kcalories per ounce, this quantity represents about 530 kcalories per day. Additional kcalories (5

Figure 5–5 Comparison of Nutrient Needs of Nonpregnant, Pregnant, and Lactating Women
[a]Energy allowance during pregnancy is for 2nd and 3rd trimesters; no additional allowance is provided during the 1st trimester.
[b]The pregnant woman may need to take an iron supplement, as discussed in Chapter 4.

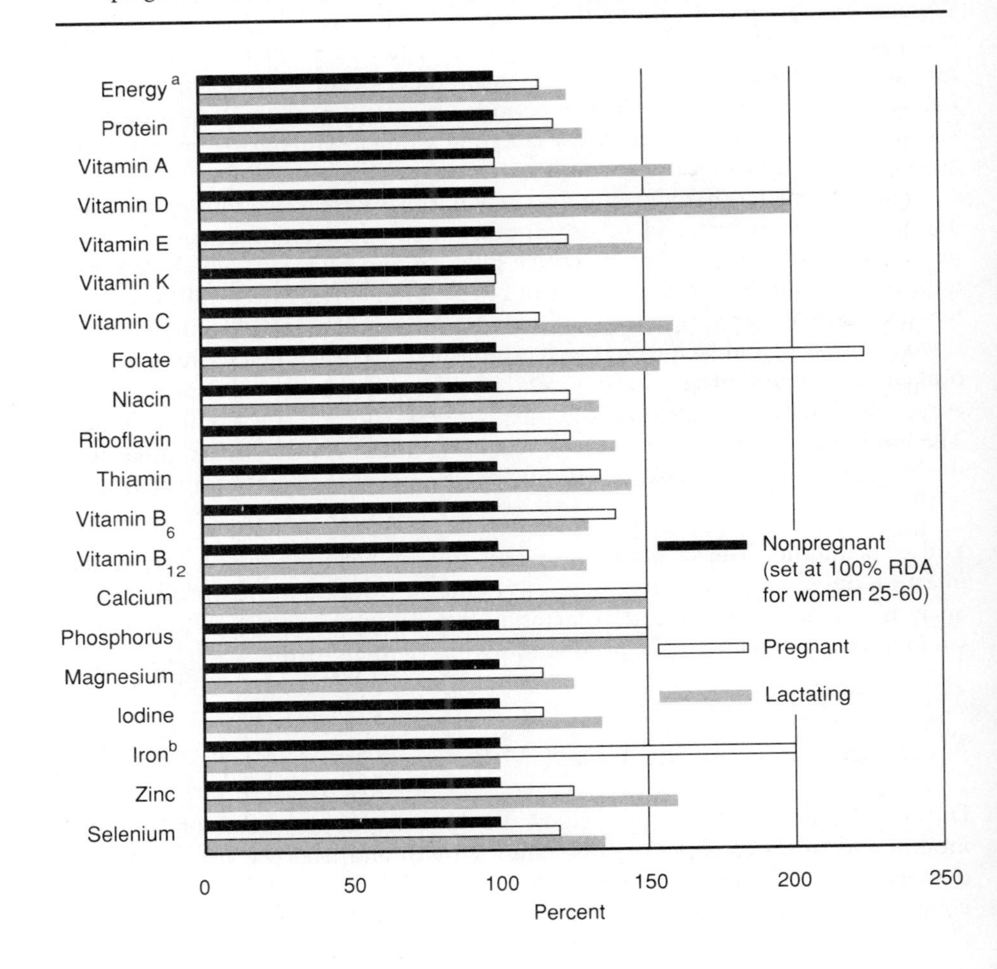

Table 5–2 Four Food Group Plan for Lactating Women

Food Group	Number of Servings[a]	
	Nonpregnant Adult	*Lactating Women*
Meat and meat alternates	2	3
Milk and milk products	2	4
Vegetables and fruits	4	4
Breads and cereals	4	4

[a]See Appendix C for a more detailed summary of serving sizes and food sources.

kcalories per ounce) are required to compensate for the less-than-100-percent efficiency of the mammary glands in converting maternal energy into milk energy. Thus the energy requirement for a lactating woman is about 640 kcalories a day above her nonpregnant need during the first 6 months of lactation, and 510 kcalories per day thereafter.[62] To meet these energy needs, a lactating woman is advised to consume an additional 500 kcalories from foods each day, allowing the fat reserves she accumulated during pregnancy to provide the balance. This allows for gradual weight loss. Many postpartum women, however, want to lose weight rapidly, and so disregard advise to consume recommended energy intakes. Women with energy intakes somewhat below the RDA can successfully breastfeed their infants while losing weight, but at some point a woman's energy intake becomes too low, and her milk volume decreases.[63] Such restrictions are particularly detrimental in the early weeks of lactation, when milk production is still getting established.

Energy RDA during lactation: +500 kcal/ day.

On the energy output side, the lactating mother can and should exercise regularly, as she did before she gave birth. Exercise at the same level as was maintained during pregnancy does not compromise lactation. For the woman who was sedentary during pregnancy, the postpartum period is a good time to begin regular walks or workouts.

Maternal diet has a different influence on each of the nutrients found in breast milk. First and most important is water.

Water Water is the major nutrient in breast milk. Total milk volume varies with infant age, but not with maternal fluid intake, as might be expected.[64] The mother who doesn't drink enough fluids can become dehydrated while lactating. To prevent dehydration, 2 quarts of fluid a day are recommended for the lactating mother. A beneficial habit she may want to adopt is to drink a glass of milk, water, or juice each time she nurses.

Carbohydrate Maternal nutrition has no effect on the carbohydrate in breast milk.[65] The carbohydrate is always lactose, and the concentration is always 39 percent.

Protein Maternal nutrition may have no effect on breast milk's protein content, either. Protein supplementation has been reported to increase milk volume and both increase and decrease the protein concentrations in breast milk. In general, the protein concentration of breast milk in malnourished women is similar to that of well-nourished women. Studies reporting otherwise may have used analytical methods that measured nonprotein nitrogen. Nonprotein nitrogen accounts for 25 percent of the nitrogen in breast milk; if used as an indicator of protein per se, it makes breast milk's protein concentration appear spuriously high.[66] The protein RDA for lactating women is 15 grams per day above nonpregnant requirements during the first 6 months of lactation, and 12 grams per day thereafter. This amount of protein is well within most women's intakes; the average protein intakes of many women exceed their RDA.[67]

Protein RDA during lactation:
65 g/day (1st 6 mo).
62 g/day (2nd 6 mo).

Lipids Maternal dietary intake alters the fatty acid composition of breast milk, but not the total fat concentration or milk volume.[68] Among the fatty acids of interest that respond to diet is the omega-3 fatty acid docosahexaenoic acid (DHA). Its concentration in breast milk increases with maternal consumption of fish oil rich in DHA.[69] This, in turn, raises the infant's consumption of DHA, one of the most abundant structural lipids in the brain. (This is not to encourage the use of fish oil supplements, however; besides using the DHA of human milk, infants can synthesize DHA from linolenic acid.) Cholesterol concentrations in breast milk are unaffected by maternal fat and cholesterol consumption.

omega-3 fatty acid: relatively newly recognized as important in nutrition, a polyunsaturated fatty acid with its endmost double bond three carbons from its methyl (CH_3) end.

Infants' needs for omega-3 fatty acids are discussed in Chapter 6.

Vitamins Breast milk composition may change with maternal dietary excesses and deficiencies of the fat-soluble vitamins, depending on the vitamin. The vitamin A concentration in the milk is maintained at the expense of the mother's stores for as long as they last, and therefore vitamin A deficiency is rare in breastfed infants. However, megadoses of vitamin A can increase the concentration of the vitamin in breast milk to a level that threatens to be toxic. Vitamin D in breast milk is insufficient to meet infants' needs even with adequate maternal dietary intake. Vitamin E in breast milk varies directly with maternal intake and stores. Breast milk vitamin K concentrations are minimal.

Chapter 6 discusses vitamin needs and intakes of infants.

Like the fat-soluble vitamins, water-soluble vitamins in breast milk reflect maternal intakes to varying extents. Marginal deficiencies and daily fluctuations have little, if any, influence on breast milk composition, but severely vitamin deficient mothers produce vitamin-deficient breast milk. For example, lactating women who consume a strict vegetarian diet may produce vitamin B_{12}–deficient milk. Milk concentrations of vitamin C rise with maternal intakes up to 90 milligrams per day. Daily vitamin C intakes greater than 90 milligrams per day do not further raise the vitamin C concentrations in breast milk.[70] The RDA for most of the vitamins during lactation are greater than during pregnancy; the RDA for vitamin D and vitamin K remain at their pregnancy levels during lactation; the RDA for vitamin B_6 and folate during lactation are lower than during pregnancy.[71]

Minerals The calcium and phosphorus RDA are the same during lactation as during pregnancy. The calcium concentration in human milk remains fairly constant, even when maternal calcium intakes are low.[72] However, a deficient calcium intake promotes mobilization of calcium from maternal bone stores.

Thus dietary intake is more critical in preventing maternal bone demineralization than in producing calcium-rich milk.

As with calcium, the phosphorus content of breast milk remains fairly constant, regardless of maternal dietary intake. In general, maternal dietary intake of minerals does not influence the total minerals in milk, although specific amounts of individual minerals may vary as lactation progresses from early months to later months.[73]

Breast milk iron concentration remains fairly constant whether the mother takes an iron supplement or has iron-deficiency anemia. Even when maternal stores and dietary intake are inadequate, the iron that is available reaches the breast milk. (The mother's body is designed to deliver iron to the infant, no matter the cost to her own health.)

Supplements for Lactating Women

The lactating mother may need an iron supplement. Postpartum maternal iron stores are often depleted. During gestation, the fetus takes enough iron to meet its own needs for the first four to six months after birth.[74] In addition, blood losses may have occurred at delivery. That is why the recommendation is made that the woman should continue taking iron supplements during lactation, after the infant's birth. The intent is not to enhance the iron content of her breast milk but to replenish her depleted iron stores. When iron stores become depleted, the symptoms of iron-deficiency anemia (such as weakness, fatigue, and headaches) become evident. A new mother trying to care for her infant cannot do so optimally if she is tired and weak. Her compromised emotional and physical health will eventually undermine her ability to continue producing milk and relaxing sufficiently to nurse successfully.

The health care provider may recommend supplements that contain the full range of vitamins and minerals. However, for most lactating women, iron supplements are all that are necessary, and as always, foods are a better choice for the delivery of nutrients.

The lactating mother needs to pay special attention to her own nutrition at the same time as she is adjusting to the presence of a new, needy, and time-consuming individual in her life. The support of her family and companions can help her succeed at this. In fact, the support of a lactating mother is important enough to warrant its own Practical Point: Care of the Lactating Mother.

▸▸ PRACTICAL POINT

Care of the Lactating Mother

Life is hectic, to say the least, for any new mother, whether she is breastfeeding or formula feeding. The breastfed infant demands especially frequent feedings in the early weeks. The nursing mother may at first feel as if she is doing little else. Finding time to prepare meals for herself and other family members is often difficult in the beginning.

▸▸ PRACTICAL POINT *continued*

Nutrition plays a significant role in successful lactation, affecting both the physical and mental well-being of the mother. Thus it is important that she eat well. In order to do so, she needs help and support. She cannot expect, nor should she be expected, to prepare family meals, breastfeed her infant, attend to other responsibilities, and get the rest she so urgently needs, without help from family members and friends. Her main priorities at first are to feed and care for herself and her newborn infant.

Family members can help by shopping and preparing food. Meals may be quick and simple for a while, but they can still be nutritious. Casseroles, salads, and soups are easy to prepare. They also provide many nutrients, especially when foods such as instant nonfat dry milk, vegetable broth, nuts, seeds, cheese, eggs, and nutritional yeast are added to them. Table 5–3 lists quick and nutritious food suggestions.

Rest is vital to successful lactation and an important contributor to recovery. Giving birth to an infant is an exhausting and stressful experience. Friends and family members can help by caring for the infant for a few hours while mother rests or spends some time away from home. Mothers can benefit by napping when the infant naps.

Once postpartum bleeding has stopped and the mother is feeling stronger, exercise and fresh air are excellent "medicines" to improve her strength, well-being, and self-esteem. Exercise classes that include free baby-sitting are offered throughout the country today. If the mother can afford this luxury, these classes provide an excellent opportunity to do something just for herself, with the added bonus of a break from the infant and the social stimulation of other new mothers' company.

Nursing women are frequently the recipients of unsolicited, "friendly" advice on how to care for themselves and their newborns, including what they should or should not eat during lactation. Aside from creating unnecessary confusion for the mothers, this advice is often incorrect. Each mother and

Table 5–3 Quick and Nutritious Food Suggestions

Foods that derive a large proportion of their kcalories from sugar, fat, or alcohol are not included here, since they are not nutrient dense. This list includes only those foods that require minimal preparation. Don't forget about leftovers from previously prepared meals. Last night's dinner may make a quick, nutritious snack, lunch, or even breakfast.

Nonfat milk and cereal
Cheese and crackers
Cottage cheese and fruit
Yogurt shake (yogurt, juice, and a banana)
Sardines or tuna on crackers
Green pepper stuffed with tuna
Deviled eggs
Peanut butter on bread or crackers
Nuts and raisins or other dried fruit
Fruit and cheese
Raw vegetables and clam, onion, or other dip
Fruit and yogurt

PRACTICAL POINT *continued*

infant combination is unique, and what is true for one nursing mother may not be for another. For example, one mother may find that when she eats garlic, her infant becomes irritable, while another mother has no such experience. Infant reactions to substances in mother's milk are matters that require individual detective work; generalizations are not useful.

In addition to the support that friends and family members offer by helping, a new mother (and her infant) benefit from accurate information on breastfeeding (see Practical Point: How to Breastfeed, for a summary of instructions). When women receive early and repeated postpartum breastfeeding information and support, they breastfeed their infants longer than other breastfeeding women.[a]

Successful lactation requires the support of all those who care. This, plus adequate nutrition, rest, exercise, and fresh air, will do much to enhance the well-being of mother and infant.

[a]S. E. Saunders and J. Carroll, Post-partum breast feeding support: Impact on duration, *Journal of the American Dietetic Association* 88 (1988): 213–215.

PRACTICAL POINT

How to Breastfeed

Most healthy women who want to breastfeed can do so. The mother-to-be may find it reassuring to learn that 95 percent of all women who try are successful. The size and shape of a woman's breasts do not affect her ability to breastfeed an infant.

Newborn infants readily adapt to breastfeeding. In fact, fetuses evince sucking behaviors before birth. Newborns are prepared to suckle immediately after birth and demonstrate a rooting reflex that orients them toward a nipple. Nursing the infant immediately after birth facilitates successful lactation.

rooting reflex: a reflex that causes an infant to turn toward whichever cheek is touched, in search of a nipple.

Beginning at the first feeding, the mother needs to learn how to relax and position herself so that she and the infant will be comfortable. The position must also allow the infant to nurse without obstructing breathing. The mother squeezes the areola, the colored halo around the nipple, between two fingers, slipping enough of it into the infant's mouth to promote good pumping action (see Figure 5–6). If the infant is to successfully milk the mammary glands, the nipple must rest well back on the infant's tongue. The infant's lips and gums pump the areola, thus releasing milk from the mammary glands into the ducts that lie beneath the areola. The sucking and swallowing reflexes work together. The infant's tongue and jaw suck milk from the breast, and the swallow follows. To break the suction, the mother can slip a finger between the infant's mouth and her breast.

areola (ah-REE-oh-la): the colored portion of the mammary gland that surrounds the nipple.

The let-down reflex forces milk to the front of the breast when the infant begins to nurse, allowing the milk to flow. Let-down has to occur for the infant to obtain milk easily, and the mother needs to relax for let-down to occur. This means that at a time when the stress response might be more natural, she must will the relaxation response. Willed relaxation first requires that a person

stress response: the body's response to a physical or psychological threat, mediated by nerves and hormones.

relaxation response: the opposite of the stress response; the normal state of the body.

▸▸ PRACTICAL POINT *continued*

Figure 5–6 Infant's Grasp on Mother's Breast
The mother squeezes the areola, slipping enough of it into the infant's mouth to promote good pumping action. The infant's lips and gums pump the areola, releasing milk from the mammary glands into the milk ducts that lie beneath the areola.

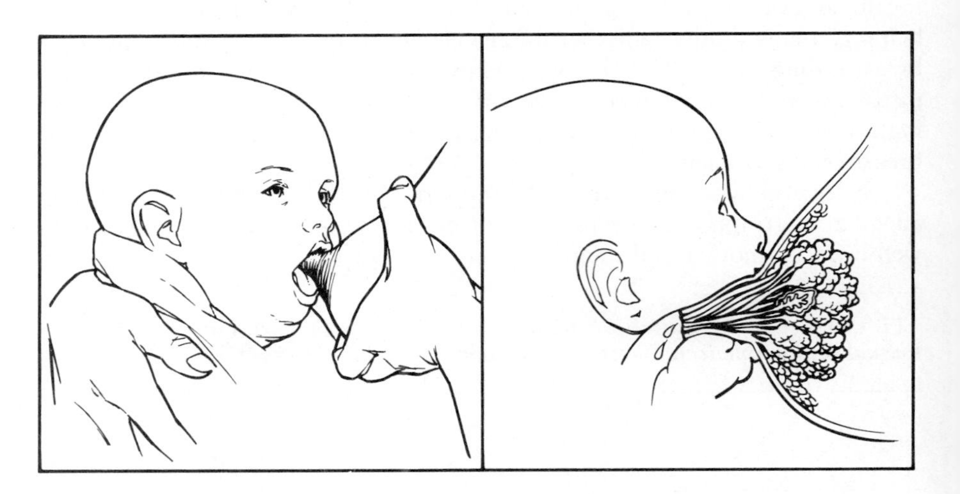

assume a comfortable position in a noninterrupting environment and then maintain a passive attitude toward intervening thoughts. It may take several feeding sessions for the mother to learn how to respond to cries of hunger before she can achieve let-down promptly and fully.

The infant sucks half the milk from the breast within the first two minutes, and 80 to 90 percent of it within four minutes. However, sucking on one breast is encouraged for 8 to 12 minutes. The sucking itself, as well as the complete removal of milk from the breast, stimulates lactation. After ten minutes or so, the mother offers the other breast to finish satisfying the infant's hunger. Nursing sessions start on alternate breasts to ensure that each breast is emptied regularly. This pattern maintains the same supply and demand for each breast, and thus prevents either breast from overfilling. At regular intervals, the mother holds the infant upright to expel any swallowed air, and then offers another chance to nurse.

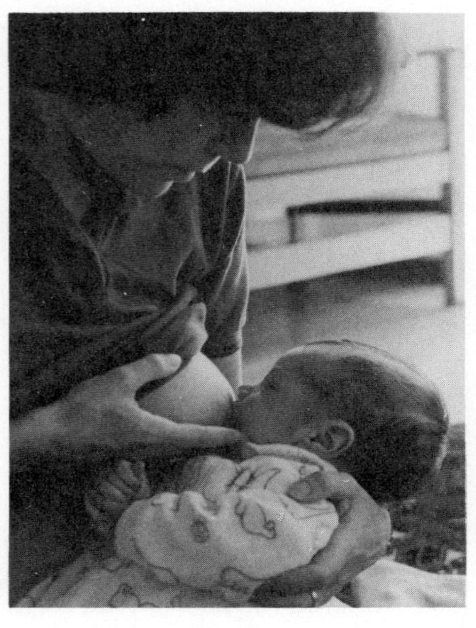

A mother breastfeeds her infant while sitting in a relaxed, comfortable position.

Approximately six feedings a day, when the infant cries with hunger, promote optimal milk production and infant growth. The mother encourages the infant to nurse the full 15 to 20 minutes per feeding. Some infants fall asleep in less time and may need to be aroused to continue feeding. Feeding intervals vary with each infant, but should be at least two hours apart. If they are less than two hours apart, the mother may begin to feel like a human pacifier, and her milk supply, at first unable to meet the demand, may come to exceed it. If a feeding interval exceeds four hours, the mother may need to express some milk to relieve pressure and maintain the demand. Figure 5–7 illustrates methods of expressing milk. Until lactation is well established, the infant should be encouraged to feed regularly and not be allowed to sleep through a feeding.

What if a mother is breastfeeding twins? As far as milk supply is concerned, a woman need have no problem breastfeeding twins. The more milk the infants drink, the more milk the breasts produce. Most mothers of twins agree that the problem lies in finding time to feed two—whether breast milk or formula. Each infant has a unique "hunger clock," and every mother has to work out her own system. Breastfeeding twins is no more difficult than any other task involving twins.

If an infant seems thirsty after a long feed, or prior to the next scheduled feeding, a parent can offer a bottle of water. Sweetened water or supplemental formula feedings in the first two weeks are ill-advised. Sweetened water provides energy, which should come to the infant only from a nutrient-dense source. The full extent of the infant's demand should be communicated to the mother's body by way of sucking, so that the milk supply will increase to meet it.

A mother might want to know if she can skip an occasional feeding, substituting a bottle of formula. To avoid suppressing lactation, the mother will need to express her milk; that way, the breast receives the message to continue producing milk. This may not be necessary once lactation is well established.

A mother who wants to skip one or two feedings daily—for example, if she works outside the home—can substitute formula for those feedings and continue to breastfeed at other feedings. Or the infant can be fed, in a bottle, the breast milk that the mother has expressed and frozen on previous occasions. Breast milk can be kept refrigerated for up to 24 hours or frozen for up to six weeks.

If the mother feels that she is spending immense amounts of time breastfeeding, she should remind herself that the parent who is feeding formula is also spending time sterilizing bottles, preparing formula, and feeding her infant. Breastfeeding will be less time-consuming after the first few weeks. The mother is encouraged to remember the advantages and to enjoy this time with her infant while it lasts.

Most problems associated with breastfeeding can be resolved. Many new mothers experience sore nipples during the initial days of breastfeeding. Sore nipples need to be treated kindly, but nursing can continue. Air and sunlight between feedings help to heal them. Before let-down, the infant must suck hard on the nipple to receive milk. For this reason, a mother will want to nurse on the less-sore breast first. Then, when the milk lets down and is freely flowing, she can switch to the sore nipple. When the fast-flowing, early milk from that sore breast is gone, she can switch back and satisfy the infant's hunger and sucking need.

Engorgement is common before lactation is established; when the schedule is changed, as in weaning; or when a feeding is missed. The breasts become so full and hard that the infant cannot grasp the nipple, and the mother is most uncomfortable. A gentle massage or warming the breasts (with a heating pad or in a shower) helps to initiate let-down and to release some of the accumulated milk. The best solutions are to pump out some of the milk, to use a nipple shield that will help the infant grasp the nipple, and to allow the infant to nurse. The breasts will, in time, get smaller and softer, while still producing ample milk.

A mother can also manage to nurse with an inverted nipple. An inverted nipple folds inward toward the breast when the areola is pressed between two fingers. Pushing the areola toward the chest wall—manually or with a shield—everts the nipple, making it available for infant sucking.

An undrained duct can make a hard, uncomfortable lump in the breast. By massaging the lump while the infant is nursing, the mother can move the milk toward the nipple, where it will join the main supply.

▸▸ PRACTICAL POINT *continued*

engorgement: overfilling of the breasts with milk so that they become swollen and hard.

Figure 5–7 Methods of Expressing Milk

Milk Expression by Hand

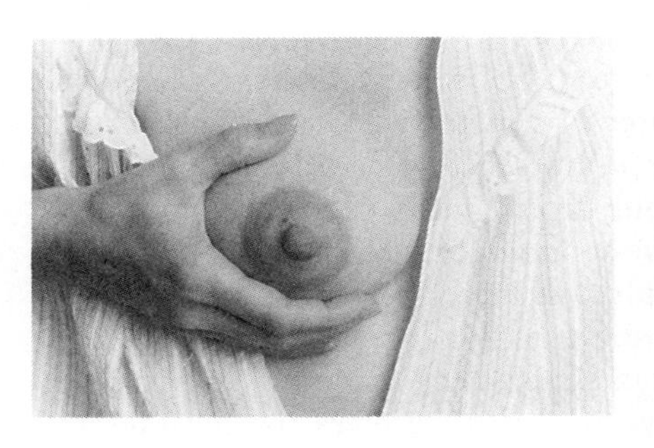

1. The hand is placed on the breast near the chest wall, with the thumb on top and the fingers cupped around and under the breast. The hands gently move toward the nipple.

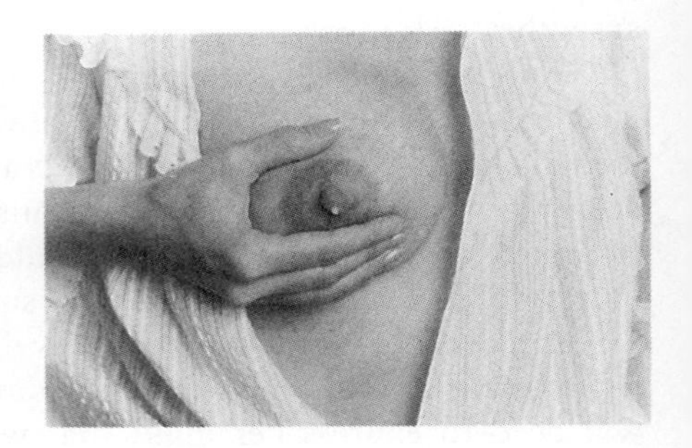

2. With the thumb and forefinger about an inch back from the nipple, the woman presses gently inward toward the chest wall and squeezes the thumb and fingers together. This "push back and squeeze" motion is continued until no more milk comes out. Then the fingers are rotated to another position, and the procedure is repeated.

Manual Breast Pump

The most popular manual (nonelectric) pump is the piston or cylinder pump. Two cylinders fit together, and breast milk is pumped with a piston-type motion into the outer cylinder. The outer cylinder can be used as a baby bottle. The cylinder pump is easy to use and clean. It is small, lightweight, and transportable.

The Loyd-B® pump has a trigger-shaped handle that initiates the suction. Its pumping action, which can be regulated from gentle to strong, makes it one of the most effective of the manual pumps. Although somewhat bulky, the Loyd-B® comes apart and is small enough to transport easily. It is available with two breast shields—one plastic, one glass. The plastic shield fits a regular baby bottle, so the mother can pump directly into a bottle, if desired.

Medela makes a pump that can be operated manually or used in conjunction with Medela's electric pump. The system is designed around the suck, release, relax cycle of the baby's nursing pattern.

Figure 5–7—*Continued*

Electric Breast Pump

An electric pump is the easiest way to express breast milk. Of all pumps, it is most effective at emptying the breast to stimulate milk production. Electric pumps range in price from $100 to $1000. The larger, more expensive models can be rented from some large pharmacies, medical supply companies, and hospitals.

Battery-Powered Breast Pump

The battery-operated pump is the newest addition to the growing line of breast pumps. Two AA batteries supply the power for this small, lightweight, easily cleaned pump. A six-stage suction adjustor allows for individual control, so the pump can be operated with one hand. While this pump does not have the power of the electric pump, the convenience and cost make it worth investigating.

Techniques for Pumping Breast Milk

Learning how to express breast milk effectively will take time, regardless of the method selected. Practice improves efficiency. Effectiveness of pumping may vary due to let-down, stress, and time available for pumping.

One of the best ways to learn the mechanics of pumping is to use the pump on one breast at the same time the baby is nursing from the other. It's easier to master the technique of pumping when milk is flowing than to try to master it without the milk ejection reflex.

1. Massaging the breasts just before pumping helps to increase the quantity of milk collected and to decrease the time needed for collection. To massage the breasts, a woman makes small circular motions with her fingertips starting near the chest wall, and moving toward the areola.
2. She then uses the palms of both her hands to apply gentle, but firm, pressure starting from the outside edge of the breast and working toward the nipple. She massages around each breast several times before attempting to express milk.
3. After massaging the breasts, she moistens the edges of the pump flange with expressed milk or water to lubricate it and to make a better seal with the skin. The nipple should slide along the inside of the top edge of the pump flange to help stimulate the milk ejection reflex and initiate the flow of milk.
4. The flange is held just tightly enough against the breast to make a good seal. Make sure the edge of the flange does not block milk flow from the nipple. Initiate the pumping according to the pump manufacturer's instructions. Continue to pump until one minute after the flow stops. Break the seal by pressing a finger against the breast where it meets the flange. Repeat the pumping process with the other breast.

Source: Reprinted courtesy of Mead Johnson Nutritional Division, Mead Johnson and Company, Evansville, IN 47721.

▸▸ PRACTICAL POINT *continued*

mastitis: inflammation of the breast, most common in women during lactation.

Infection of a breast, known as mastitis, is best managed by *continuing to breastfeed.* By drawing off the milk, the infant helps to relieve pressure in the infected area. The infant is safe, because the infection is between the milk-producing glands, not inside them.

When an infant has colic, the mother may think she should stop drinking cow's milk, believing that its proteins, passed to the infant in her breast milk, may be responsible. However, the theory that cow's milk drunk by the mother causes colic in breastfed infants is entirely unsupported by research.[a]

Most important, if the infant is irritable and wakeful, the mother may fear that her milk supply is inadequate. The small quantity of the infant's bowel movements (which is normal) may suggest to the mother that the infant is underfed, but the health care provider can reassure her that breast milk contains little indigestible material and therefore little waste. The stress of worrying, itself, can inhibit lactation. All infants cry. The mother's ability to relax and set her fears aside will better support lactation than will anxiety about inadequate milk production. However, if she wants to wean the infant to formula, that is an acceptable alternative.

wean: to gradually replace breast milk with infant formula or other foods appropriate to an infant's diet.

To wean an infant, the mother gradually introduces small amounts of formula or milk and solid foods to the infant while she continues to breastfeed. (The type of food or milk and the age of weaning is discussed in the next chapter.) As the infant consumes more solid foods and formula or milk, breast milk consumption will decrease. Because the breasts need time to adjust their supply to the diminishing demand for milk, the key to a comfortable weaning is to do it *gradually.* The less breast milk an infant drinks, the less the mother produces, until finally weaning is complete—that is, the infant is eating solid foods, drinking milk or formula, and not receiving any breast milk.

The mother should allow several weeks for complete weaning. The first step is to replace any one breastfeeding session with formula or milk and solid foods. After a few days, a second breastfeeding session is replaced in the same way. This process continues until all breastfeeding sessions have been replaced. Such a gradual schedule will allow the breasts to adjust with little discomfort until they are no longer producing milk.

---

[a] A. K. C. Leung, Infantile colic, *American Family Physician,* 36 (1987): 153–156.

Formula Feeding

Appendix D provides a table comparing the composition of infant formulas available in the United States.

A woman who breastfeeds for the better part of one year can wean her infant to cow's milk, bypassing the need for infant formula. However, a woman who decides to feed her infant formula from birth, to wean from breast milk to formula after a short time, or to substitute formula for breastfeeding on occasion must select an appropriate infant formula and learn to prepare it. A variety of infant formulas are available, and the selection must be made carefully. After six months, when other foods begin to supply nutrients in significant quantities, then cow's milk can partially replace the formula. However, the continued use of iron-fortified formula throughout the first year of life helps ensure an adequate iron intake. The following discussion offers help with the question of what type of formula is appropriate, and Practical Point: How to Feed Formula appears at the end of this section.

Standard Formulas

Formula makers duplicate human milk as closely as they reasonably can. Not all human milk is the same, though. Breast milk composition varies from one woman to another, from one feeding to another, with the duration of each feeding, and with the duration of lactation. Nevertheless, national and international standards have been established for the nutrient contents of infant formulas. The standard developed by the AAP reflects "human milk taken from well-nourished mothers during the first or second month of lactation, when the infant's growth rate is high."[75] Manufacturers in the United States use this standard to develop their formulas. The Infant Formula Act of 1980 requires that formulas meet a nutrient standard based on the AAP recommendations. Formulas meeting the standard have similar nutrient compositions; small differences are sometimes confusing, but usually unimportant.

Throughout the years, infant formula composition and labeling have reflected the knowledge of the time regarding the nutrient needs of infants. In 1941, labeling regulations pertained only to vitamins A, D, C, and thiamin, plus the minerals calcium, phosphorus, and iron. The remaining vitamins and minerals were considered adequate, because the formulas of the time contained enough milk to provide them. In 1967, as knowledge about the nutrient needs of infants expanded, the AAP proposed minimum, and in some cases maximum, amounts for vitamins and minerals in formulas. As of 1986, labeling regulations specify how nutrient contents and preparation must be listed on labels. The progression of nutrient labeling recommendations for infant formulas and the AAP standard arrived at in 1983 are shown in Table 5–4.

To prepare a standard infant formula, manufacturers start with a nonfat cow's milk base or a mixture of nonfat cow's milk and added whey (if they are creating a whey-predominant formula, for reasons described below). They replace the poorly absorbed butterfat of cow's milk with vegetable oils. They then add lactose, vitamins, and minerals so that the energy content and nutrient distribution closely resemble those of human milk. All standard infant formulas in the United States contain lactose as the principal carbohydrate. Fat is the major source of energy, as well as of essential fatty acids, in infant formulas and human milk. Standard infant formulas provide approximately 20 kcalories per ounce, as does human milk.

whey: the liquid that remains after milk has been coagulated. (The solids are the *curds*.) The term *whey* also refers to the proteins of whey, discounting its liquid; the principal protein of whey is lactalbumin.

The proteins of human milk are whey protein (usually called just *whey*) and casein, in a ratio of 80:20.[76] In contrast, the whey-to-casein ratio of cow's milk is 20:80. Some infant formulas maintain the whey-to-casein ratio of 20:80 (casein-predominant formulas), while others add whey, changing the ratio to 60:40 (whey-predominant formulas). Full-term infants grow equally

casein: the principal protein found in coagulated milk curds, and the principal protein of cow's milk.
caseus = cheese

Table 5–4 Labeling Standards and AAP Recommendations for Formulas

1941[a]	1967	1976	1983 AAP Recommendations (per 100 kcal)
Protein	Protein	Protein	1.8 to 4.5 g
Fat	Fat	Fat	3.3 to 6.0 g[b]
Carbohydrate	Carbohydrate	Carbohydrate	—
Ash			
Vitamin A	Vitamin A	Vitamin A	75 to 225 μg
Vitamin D	Vitamin D	Vitamin D	1.0 to 2.5 μg
	Vitamin E	Vitamin E	0.5 mg tocopherol equivalent
		Vitamin K	4 μg
Vitamin C	Vitamin C	Vitamin C	8 mg
Thiamin	Thiamin	Thiamin	40 μg
	Riboflavin	Riboflavin	60 μg
	Niacin	Niacin	250 μg
	Vitamin B_6	Vitamin B_6	35 μg[c]
	Folate	Folate	4 μg
	Pantothenic acid	Pantothenic acid	300 μg
	Vitamin B_{12}	Vitamin B_{12}	0.15 μg
		Biotin	1.5 μg
		Choline	7.0 mg
		Inositol	4.0 mg
Calcium	Calcium	Calcium	60 mg[d]
Phosphorus	Phosphorus	Phosphorus	30 mg[d]
	Magnesium	Magnesium	6 mg
		Sodium	20 mg (6 mEq[e])
		Potassium	80 mg (14 mEq[e])
		Chloride	55 mg (11 mEq[e])
Iron	Iron	Iron	0.15 mg
	Iodine	Iodine	5 μg
	Copper	Copper	60 μg
		Zinc	0.5 mg
		Manganese	5.0 μg
		Chromium[f]	—
		Cobalt[f]	—
		Molybdenum[f]	—
		Selenium[f]	—
		Fluoride[f]	—

[a] FDA Regulations—1941 Infant Formula Labeling Requirements, from the *Federal Register,* 1941.
[b] The AAP recommends 300 mg of the essential fatty acid linoleic acid.
[c] The vitamin B_6 recommendation provides 15 μg/g protein in formula.
[d] The recommended calcium-to-phosphorus ratio is 1:1 to 2:1.
[e] A milliequivalent (mEq) describes the concentration of electrolytes in a solution.
[f] These trace minerals require further study; recommendations have not been made.

Source: Adapted from H. P. Sarett, The modern infant formula, in *Infant and Child Feeding,* ed. J. T. Bond (New York: Academic Press, 1981), pp. 99–121; American Academy of Pediatrics, Committee on Nutrition, Recommended ranges of nutrients in formulas, in *Pediatric Nutrition Handbook*, 2nd ed., ed. G. B. Forbes (Elk Grove Village, Ill.: American Academy of Pediatrics, 1985), pp. 356–357.

well on either formula. Preterm infants require whey-predominant formulas, which contain an amino acid profile better suited to their metabolic capacities.[77]

In 1984, formula makers began to add the amino acid taurine to infant formulas.[78] Most infant formulas are now supplemented with taurine at a concentration approximating that of human milk.

Further recommendations of the AAP include recognition of interrelationships among nutrients, such as vitamin E and linoleic acid, and vitamin B_6 and protein. For example, the standards require that for each gram of linoleic acid, formulas provide 0.5 milligram alpha-tocopherol.

The AAP also makes specific recommendations regarding iron. All formulas must contain bioavailable iron in amounts approximately equal to that of human milk, which averages about 0.3 milligrams per liter. Formulas fortified with iron contain 6 to 12 milligrams per liter. Both iron-fortified and nonfortified formulas are available, but it is recommended that all formula-fed infants receive iron-fortified formulas by four months. Table 5–5 compares the nutrient composition of human milk, cow's milk, and an infant formula.

Special Formulas

Standard infant formulas are inappropriate for some infants. Special formulas are available that are designed to meet the dietary needs of infants with specific conditions, such as milk intolerance, prematurity, or congenital abnormalities.

Milk intolerance Special soy-protein formulas are available for infants who are unable to tolerate the standard milk-based formulas. Originally developed for infants with milk allergy or lactose intolerance, these formulas are prepared using soy for the protein source, and corn syrup and sucrose instead of lactose. Soy formulas solve the problem of feeding an infant with any of several conditions: a temporary lactase deficiency due to diarrhea, a congenital lactase deficiency, or galactosemia. They are also useful as an alternative to milk-based formulas for vegetarian families. However, soy formulas are often used in situations for which they are inappropriate, such as when an infant fed a standard formula is colicky or regurgitates often. Research does not support any connection between colic in infants and the use of cow's-milk based iron-fortified formulas.[79] The infant's digestive tract adapts enzymatically to the milk it is fed: if not fed cow's milk, it will not produce the enzymes necessary to digest the ingredients that are found in cow's milk. The inappropriate use of soy formulas throughout infancy may make the later transition to cow's milk more difficult than if milk-based formula had been used.

Soy-based formulas include Prosobee, Nursoy, Isomil, Isomil SF, and Soyalac.

colic: a condition in infants characterized by three or more hours of crying a day, three or more days a week for no identifiable reason such as pain, hunger, or sickness.

Some infants with milk allergies are also allergic to soy protein. An infant with multiple food allergies or chronic diarrhea may require a hydrolysate formula. In a hydrolysate formula, the casein is hydrolyzed to amino acids and peptides to permit easier absorption. Some of these formulas also replace the long-chain triglycerides with medium-chain triglycerides, for infants with impaired fat absorption. Hydrolysate formulas are expensive and taste unappealing, but for the infant unable to tolerate milk- or soy-based formulas, they are indispensable.

Casein hydrolysate formulas include Nutramigen and Pregestimil.

Table 5–5 Comparison of Human Milk, Cow's Milk, and Infant Formula

Nutrient (per 100 ml)	Human Milk	Cow's Milk	Commercial Formula[a]
Macronutrients			
Energy (kcal)	64	66	67
Protein (g)	0.9	3.4	1.5
Fat (g)	3.4	3.7	3.7
Carbohydrate (g)	6.6	4.9	7.1
Minerals			
Sodium (mg)	17	58	20
Potassium (mg)	55	138	68
Chloride (mg)	43	103	43
Calcium (mg)	26	125	47
Phosphorus (mg)	14	96	35
Magnesium (mg)	4	12	5
Iron (mg)	0.5	0.5	1.2[b]
Zinc (mg)	0.2	0.4	0.5
Copper (mg)	0.04	0.01	0.06
Vitamins			
Vitamin A (IU)	190	103	225
Thiamin (μg)	16	44	63
Riboflavin (μg)	36	175	110
Vitamin B_6 (μg)	10	64	41
Niacin (μg)	159	93	700
Pantothenic acid (μg)	198	352	277
Biotin (μg)	1	4	1.4
Folate (μg)	5	5	9
Vitamin B_{12} (μg)	0.04	0.42	0.14
Vitamin C (mg)	4.6	1.2	5.6
Vitamin D (IU)	2.2	3.4	41
Vitamin E (IU)	0.2	0.04	1.7
Vitamin K (μg)	1.5	6.0	5.7
Inositol (μg)	37	17	3
Choline (μg)	6	20	10

[a]Values represent the average for three major commercial products: (1) Similac, Ross Laboratories, (2) Enfamil, Mead-Johnson Laboratories, and (3) SMA, Wyeth Laboratories.
[b]Value represents formulas with iron fortification. The value for unfortified formula is 0.1 mg.

Source: Adapted with permission from K. J. Motil, Breast-feeding: Public health and clinical overview, in *Pediatric Nutrition*, eds. R. J. Grand, J. L. Sutphen, and W. H. Dietz (Stoneham, Mass.: Butterworths, 1987) pp. 251–263.

Preterm infants Preterm infants, especially very-low-birthweight infants (less than 1500 grams), often have limited digestive abilities. For this reason, formulas designed for preterm infants contain a mixture of lactose and glucose polymers, and a blend of medium-chain triglycerides and unsaturated long-chain triglycerides.[80] Preterm infants also have greater nutrient needs than full-term infants. For this reason, formulas for preterm infants generally have higher protein, mineral, and vitamin concentrations than standard formulas. They also have an energy density of 24 kcalories per ounce, greater than the standard 20 kcalories per ounce used in regular infant formulas.

The whey-to-casein ratio in formulas for preterm infants is adjusted to 60:40, to approximate that in human milk. Whey protein is preferred because it is higher in cysteine (in the paired form cystine) than is casein protein. Preterm infants may lack the hepatic enzyme cystathionase needed to convert methionine to cysteine.[81] Soy formulas are not recommended for preterm infants.

Congenital disorders Other formulas are available for infants with inborn errors of metabolism who cannot metabolize specific amino acids. These formulas purposely lack one or more nutrients and are called *incomplete* formulas. For this reason, they are not appropriate for other infants. Figure 5–8 illustrates the process of choosing a formula, and the accompanying

Figure 5–8 Choosing a Formula
[a]Manufacturers design soy-based formulas for infants with milk sensitivities—whether lactose intolerance of soy allergy. These formulas use corn syrup or sucrose in place of lactose. See Appendix D for more details on formula composition.

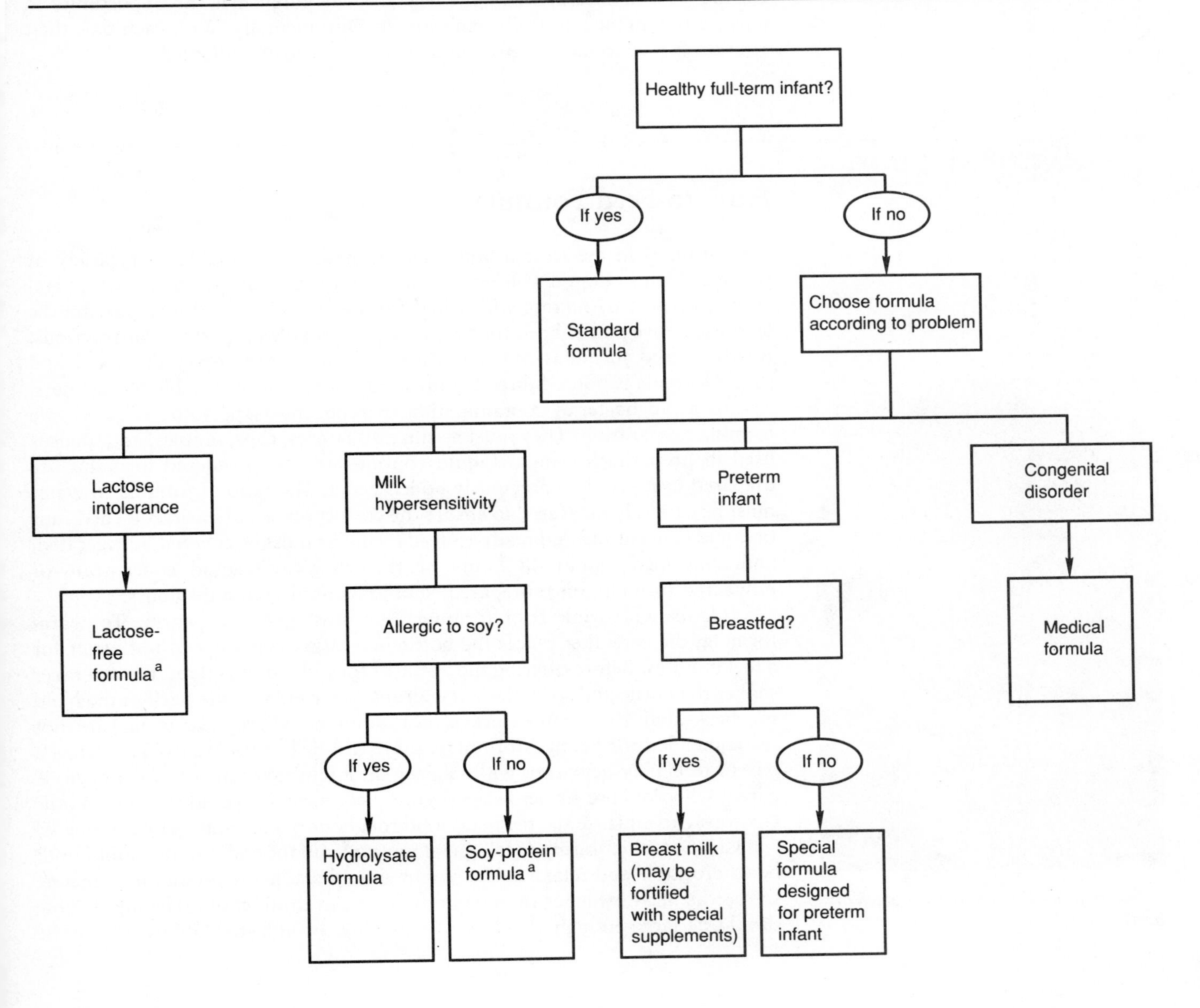

Practical Point: How to Feed Formula provides directions for the feeding process itself.

Infants born in technologically advanced countries are fortunate to have a diverse array of formulas available to them. Not too long ago, death was the inevitable outcome for infants without access to breast milk. Recent changes in government regulations of formula manufacturing ensure protection for formula-fed infants. Mothers feeding their infants formula can feel confident that modern infant formulas offer a safe, nutritionally sound alternative to breast milk.

The initial care of a healthy newborn involves such simple tasks as the parents' providing clean diapers, warm clothing, a place to sleep, nourishment, and love. With each interaction, the parent and infant relationship develops. During each of several feedings every day, the parent has the opportunity to nurture the infant physically, emotionally, and mentally. With each day, their relationship continues to grow and change, just as the infant does.

▸▸ PRACTICAL POINT

How to Feed Formula

Formulas in the United States and Canada are available in a variety of physical forms. Liquid concentrate formulas are relatively inexpensive and easy to prepare by mixing with equal parts water. Powdered formulas are the least expensive and lightest for travel. Labels provide preparation instructions. Ready-to-feed formulas are the easiest and most expensive; premixed and sterile formula is poured directly into sterile bottles or disposable bottle liners.

To avoid bacterial contamination, parents must apply the rules of safe formula preparation. They must sterilize all bottles, caps, nipples, and utensils used in preparing formula. Liquid concentrate and powdered formulas are prepared using cooled, previously boiled water. The ratio of formula to water must be carefully measured to ensure the correct nutrient density. Opened cans of liquid concentrate and ready-to-feed formulas must be covered, refrigerated, and consumed within 48 hours or thrown away. Liquid concentrate or powdered formula, once prepared, should be used within 24 hours.

Infants will drink cold formula, but most prefer it warm. To warm formula, the caretaker places the bottle in a larger container of hot water for a few minutes. Before offering the bottle of formula to the infant, the caretaker shakes the bottle and sprinkles a few drops of formula on the back of the hand to check that the temperature is not too hot. Microwave ovens are not recommended for heating formulas; they tend to heat unevenly. The drops a parent feels may be warm, while the sip an infant takes may cause a burn. A parent who does use a microwave oven to heat formula should shake the bottle vigorously to equalize the temperature throughout the formula before testing it.

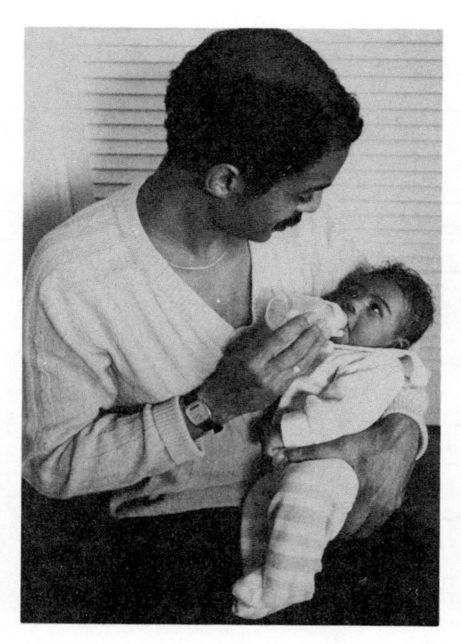

A father feeds his infant formula from a bottle.

Close contact during feeding is important. Infant and parent should both be comfortable and relaxed. The parent should cradle the infant on an incline so that its head is higher than its body, so it can drink easily. The nipple hole should be large enough to allow one swallow of milk to flow each time the infant sucks; if it is too small, it should be enlarged with a sterile needle; if it is too large, it should be replaced. The parent tilts the bottle so that the nipple is full of formula, not air, while the infant is sucking.

Now and then, the parent should hold the infant upright and give a gentle pat on the back to help eliminate any bubbles of air. Infants generally feed for about 15 minutes and should not be forced to empty the bottle. Like making children clean their plates, forcing the bottle promotes obesity. Formula left in a bottle after feeding should be discarded.

The feeding schedule can vary, but at first it is best if adjusted to the infant's expressed hunger needs, within reason. Some infants need to feed more frequently than others. Most infants enjoy bottles; the sucking provides stimulation and satisfaction, as well as nutrients. Infants cannot be allowed to sleep with bottles, however, because of the potential damage to developing teeth. A bedtime bottle may be the most wanted, but it must be firmly denied.

▸▸ PRACTICAL POINT *continued*

Nursing bottle syndrome is discussed in Focal Point 5.

Chapter 5 Notes

1. M. Underwood, as cited by T. E. Cone, History of infant and child feeding: From the earliest years through the development of scientific concepts, in *Infant and Child Feeding*, eds. J. T. Bond and coeditors (New York: Academic Press, 1981), pp. 3–34.
2. G. A. Martinez, Trends in breastfeeding in the United States, in *Report of the Surgeon General's Workshop on Breastfeeding and Human Lactation*, HHS publication no. (HRS-D-MC) 84–2 (Washington, D.C.: Government Printing Office, 1984).
3. United Nations Children's Fund (UNICEF), *The State of the World's Children 1988* (New York: Oxford University Press, 1988), pp. 61–67.
4. United Nations Children's Fund (UNICEF), 1988.
5. S. Fomon, Reflections on infant feeding in the 1970s and 1980s, *American Journal of Clinical Nutrition* 46 (1987): 171–182.
6. American Dietetic Association, Position of the American Dietetic Association: Promotion of breastfeeding, *Journal of the American Dietetic Association* 86 (1986): 1580–1585.
7. American Academy of Pediatrics, Committee on Nutrition, and the Nutrition Committee of the Canadian Pediatric Society, Breastfeeding: A commentary in celebration of the International Year of the Child, *Pediatrics* 62 (1978): 591–601.
8. E. M. E. Poskitt, Infant feeding: A review, *Human Nutrition: Applied Nutrition* 37A (1983): 271–286.
9. T. Lindberg and G. Skude, Amylase in human milk, *Pediatrics* 70 (1982): 235–238.
10. Poskitt, 1983.
11. K. C. Hayes and J. A. Sturman, Taurine in metabolism, *Annual Review of Nutrition* 1 (1981): 401–425.
12. T. A. Picone, Taurine update: Metabolism and function, *Nutrition Today*, July–August 1987, pp. 16–20.
13. M. W. Borschel and coauthors, Fatty acid composition of mature human milk of Egyptian and American women, *American Journal of Clinical Nutrition* 44 (1986): 330–335.
14. K. Brostrøm, Human milk and infant formulas: Nutritional and immunological characteristics, in *Textbook of Pediatric Nutrition*, ed. R. M. Suskind (New York: Raven Press, 1981), pp. 41–64.
15. Poskitt, 1983.
16. R. Reiser and Z. Sidelman, Control of serum cholesterol homeostasis by cholesterol in the milk of the suckling rat, *Journal of Nutrition* 102 (1972): 1009–1016, as cited in Cholesterol and the Reiser hypothesis, *Journal of Nutrition Education*, March 1983, p. 27.
17. D. W. Spady, Infant nutrition, *Journal of the Canadian Dietetic Association* 38 (1977): 34–41.
18. American Academy of Pediatrics, Committee on Nutrition, Toward a prudent diet for children, *Pediatrics* 71 (1983): 78–79.
19. American Academy of Pediatrics, Committee on Nutrition, *Pediatric Nutrition Handbook*, 2nd ed. (Elk Grove Village, Ill.: American Academy of Pediatrics, 1985), pp. 37–48.
20. S. Esala, E. Vuori, and A. Helle, Effect of maternal fluorine intake on breast fluorine content, *British Journal of Nutrition* 48 (1982): 201–204; O. B. Dirks and coauthors, Total and free ionic fluoride in human and cow's milk as determined by gas-liquid chromatography and the fluoride electrode, *Caries Research* 8 (1974): 181–186; J. Ekstrand and coauthors, Distribution of fluoride to human breast milk, *Caries Research* 18 (1984): 93–95.
21. American Academy of Pediatrics, 1985.
22. A. Hofman, A. Hazebroek, and H. A. Valkenburg, A randomized trial of sodium intake and blood pressure in newborn infants, *Journal of the American Medical Association* 250 (1983): 370–373.
23. C. L. Berseth, L. M. Lichtenberger, and F. H. Morriss, Comparison of the gastrointestinal growth-promoting effects of rat colostrum and mature milk in newborn rats in vivo, *American Journal of Clinical Nutrition* 37 (1983): 52–60.
24. M. G. Kovar and coauthors, Review of the epidemiologic evidence for an association between infant feeding and infant health, *Pediatrics* 74 (1984): 615–638.
25. B. Lonnerdal, Biochemistry and physiological function of human milk proteins, *American Journal of Clinical Nutrition* 42 (1985): 1299–1317; J. K. Welsh, I. J. Skurrie, and J. T. May, Use of Semliki Forest virus to identify lipid-mediated an-

tiviral activity and anti-alphavirus immunoglobulin A in human milk, *Infection and Immunity,* February 1978, pp. 395–401.
26. Lonnerdal, 1985.
27. Welsh, Skurrie, and May, 1978.
28. J. J. Bullen, H. J. Rogers, and L. Leigh, Iron-binding proteins in milk and resistance to *Escherichia coli* infection in infants, *British Medical Journal* 1 (1972): 69–75, as cited by Lonnerdal, 1985.
29. J. A. McMillan, S. A. Landaw, and F. A. Oski, Iron sufficiency in breast-fed infants and the availability of iron from human milk, *Pediatrics* 58 (1976): 686–691.
30. C. D. Eckhart, Isolation of a protein from human milk that enhances zinc absorption in humans, *Biochemical and Biophysical Research Communications* 130 (1985): 264–269.
31. G. Carpenter, Epidermal growth factor is a major growth-promoting agent in human milk, *Science* 210 (1980): 198–199.
32. J. P. Habicht, J. DaVanzo, and W. P. Butz, Mother's milk and sewage: Their interactive effects on infant mortality, *Pediatrics* 81 (1988): 456–461.
33. Y. Chen, S. Yu, and W. Li, Artificial feeding and hospitalization in the first 18 months of life, *Pediatrics* 81 (1988): 58–62.
34. U. M. Saarinen, Prolonged breast feeding as prophylaxis for recurrent otitis media, *Acta Paediatrica Scandinavica* 71 (1982): 567–571; R. K. Chandra, Prospective studies of the effect of breast feeding on incidence of infection and allergy, *Acta Paediatrica Scandinavica* 68 (1979): 685–689; A. L. Frank and coauthors, Breast-feeding and respiratory virus infection, *Pediatrics* 70 (1982): 239–245; A. H. Cushing and L. Anderson, Diarrhea in breast-fed and non-breast-fed infants, *Pediatrics* 70 (1982): 921–925, as cited by H. Bauchner, J. M. Leventhal, and E. D. Shapiro, Studies of breast-feeding and infections: How good is the evidence? *Journal of the American Medical Association* 256 (1986): 887–892.
35. Bauchner, Leventhal, and Shapiro, 1986.
36. U. M. Saarinen and coauthors, Prolonged breast-feeding as prophylaxis for atopic disease, *Lancet* 2 (1979): 163–166.
37. S. J. Gross, Growth and biochemical response of preterm infants fed human milk or modified infant formula, *New England Journal of Medicine* 308 (1983): 237–241.
38. D. M. Anderson and coauthors, Length of gestation and nutritional composition of human milk, *American Journal of Clinical Nutrition* 37 (1983): 810–814; J. A. Lemons and coauthors, Differences in the composition of preterm and term human milk during early lactation, *Pediatric Research* 16 (1982): 113–117.
39. M. S. Brady and coauthors, Formulas and human milk for premature infants: A review and update, *Journal of the American Dietetic Association* 81 (1982): 547–552.
40. W. A. Bowes, The effect of medications on the lactating mother and her infant, *Clinical Obstetrics and Gynecology* 23 (1980): 1073–1080.
41. R. E. Little and coauthors, Maternal alcohol use during breast-feeding and infant mental and motor development at one year, *New England Journal of Medicine* 321 (1989): 425–430; American Academy of Pediatrics, Committee on Drugs, The transfer of drugs and other chemicals into human breastmilk, *Pediatrics* 72 (1983): 375–384.
42. I. J. Chasnoff, D. E. Lewis, and L. Squires, cocaine intoxication in a breast-fed infant, *Pediatrics* 80 (1987): 836–838.
43. W. J. Rogan, A. Bagniewska, and T. Damstra, Pollutants in breast milk, *New England Journal of Medicine* 302 (1980): 1450–1453.
44. Rogan, Bagniewska, and Damstra, 1980.
45. J. A. Knowles, Drugs in milk, *Ross Timesaver* 21 (1972): 28–32.
46. American Academy of Pediatrics, Committee on Environmental Hazards, PCB in breastmilk, *Pediatrics* 62 (1978): 407.
47. T. Byers and coauthors, Lactation and breast cancer: Evidence for a negative association in premenopausal women, *American Journal of Epidemiology* 121 (1985): 664–674.
48. M. Brewer, M. R. Bates, and L. P. Vannoy, Postpartum changes in maternal weight and body fat depots in lactating vs. nonlactating women, *American Journal of Clinical Nutrition* 49 (1989): 259–265.
49. A. M. Ferris and coauthors, Biological and sociocultural determinants of successful lactation among women in eastern Connecticut, *Journal of the American Dietetic Association* 87 (1987): 316–321.
50. B. J. Myers, Mother-infant bonding: The status of this critical-period hypothesis, *Developmental Review* 4 (1984): 240–274.
51. Myers, 1984.
52. S. E. Saunders and J. Carroll, Post-partum breast feeding support: Impact on duration, *Journal of the American Dietetic Association* 88 (1988): 213–215.
53. R. A. Lawrence, Human lactation as a physiologic process, in *Report of the Surgeon General's Workshop on Breastfeeding and Human Lactation,* HHS publication no. (HRS-D-MC) 84–2 (Washington, D.C.: Government Printing Office, 1984).
54. A. S. McNeilly, Effects of lactation on fertility, *British Medical Bulletin* 35 (1979): 151–154.
55. Lawrence, 1984.
56. P. G. Lunn and coauthors, The effect of improved nutrition on plasma prolactin concentrations and postpartum infertility in lactating Gambian women, *American Journal of Clinical Nutrition* 39 (1984): 227–235.
57. H. Delgado, Nutrition, lactation, and postpartum amenorrhea, *American Journal of Clinical Nutrition* 31 (1978): 322–327.
58. R. V. Short, Breast feeding, *Scientific American* 250 (1984): 35–41.
59. American Academy of Pediatrics, Committee on Drugs, The transfer of drugs and other chemicals into human breast milk, *Pediatrics* 72 (1983): 375–381; Task force on oral contraceptives of the WHO Special Programme of Research, Effects of hormonal contraceptives on milk volume and infant growth, *Development and Research Training in Human Reproduction,* vol. 30, December 1984, pp. 505–522.
60. Task force on oral contraceptives, 1984; Task force on oral contraceptives of the WHO Special Programme of Research, Long-term follow-up of children breast-fed by mothers using oral contraceptives, *Development and Research Training in Human Reproduction,* vol. 34, November 1986, pp. 443–457.
61. Food and Nutrition Board, *Recommended Dietary Allowances,* 10th ed. (Washington, D.C.: National Academy of Sciences, 1989), pp. 34–35.
62. Food and Nutrition Board, 1989, pp. 34–35.
63. N. F. Butte and coauthors, Effect of maternal diet and body composition on lactational performance, *American Journal of Clinical Nutrition* 39 (1984): 296–306.
64. American Academy of Pediatrics, Committee on Nutrition, Nutrition and lactation, *Pediatrics* 68 (1981): 435–443.
65. B. Lonnerdal, Critical review: Effects of maternal dietary intake on human milk composition, *Journal of Nutrition* 116 (1986): 499–513.
66. Lonnerdal, 1986.
67. B. B. Peterkin, Women's diets: 1977–1985, *Journal of Nutrition Education* 18 (1986): 251–257.
68. Lonnerdal, 1986.

69. W. S. Harris, W. E. Connor, and S. Lindsey, Will dietary ω-3 fatty acids change the composition of human milk? *American Journal of Clinical Nutrition* 40 (1984): 780–785.
70. L. O. Byerley and A. Kirksey, Effects of different levels of vitamin C intake on the vitamin C concentration in human milk and the vitamin C intakes of breast-fed infants, *American Journal of Clinical Nutrition* 81 (1985): 665–671.
71. Food and Nutrition Board, 1989.
72. L. H. Allen, Calcium bioavailability and absorption: A review, *American Journal of Clinical Nutrition* 35 (1982): 783–808.
73. D. B. Jelliffe, Unique properties of human milk, *Journal of Reproductive Medicine* 14 (1975): 133.
74. American Academy of Pediatrics, 1981.
75. Brostrøm, 1981.
76. Brostrøm, 1981.
77. D. Wink, Getting through the maze of infant formulas, *American Journal of Nursing* 4 (1985): 388–392.
78. Picone, 1987.
79. B. Taubman, Parental counseling compared with elimination of cow's milk or soy milk protein for treatment of infant colic syndrome: A randomized trial, *Pediatrics* 81 (1988): 756–761; D. W. Thomas and coauthors, Infantile colic and type of milk feeding, *American Journal of Diseases of Children* 141 (1987): 451–453.
80. M. S. Brady and coauthors, Specialized formulas and feedings for infants with malabsorption or formula intolerance, *Journal of the American Dietetic Association* 2 (1986): 191–200.
81. O. G. Brooke, Nutritional requirements of low and very low birthweight infants, in *Annual Review of Nutrition,* eds. R. E. Olson, E. Beutler, and H. P. Broquist (Palo Alto, Calif.: Annual Reviews, 1987), pp. 91–116.

▶ Focal Point 5

Dental Health

Teeth begin to develop in the fetus before birth, erupt during the first year, and serve their owners thereafter for a lifetime. It seems appropriate to present the relationships between nutrition and oral health early in the chronology of the life span, because the care parents deliver early in their children's lives can make a lifelong difference to their children's dental health.* As one authority put it, "The best time to start practicing good oral hygiene was yesterday. The next best time is today."†

The mouth is the normal passageway for all foods and beverages entering the body. Its parts work to prepare foods for their journey through the digestive system. The tongue senses the flavors that encourage or discourage food consumption. The teeth break large pieces of food into smaller ones, and saliva blends with these pieces to ease swallowing. The teeth also contribute to diction and facial appearance, and the gums support the teeth.

saliva: the secretion of the salivary glands.

periodontal health: health of the tissues surrounding and supporting the teeth.

alveolar (al-VEE-oh-lar) **bone:** the part of the jawbone that forms the sockets of the teeth.

Nutrition and diet are important to dental and periodontal health. Conversely, oral health is important to nutrition. Due to both dental and periodontal disease, almost half of U.S. adults over age 65 have no teeth at all.[1] Tooth loss and alveolar bone resorption change the contour of the jaw, impairing chewing ability and the fit of dentures. The loss of a tooth can reduce chewing efficiency, thereby creating difficulty in making food ready for swallowing. Dentures, even when they are comfortable, well-designed, and well-maintained, are less effective than natural teeth.

As surprising as it may sound, tooth loss can even be fatal. Circumstantial evidence points to toothlessness as a cause of death: a large majority of adult choking victims are denture wearers, and choking correlates with absence of teeth. Missing teeth or improperly fitting dentures reduce chewing efficiency, and this results in the attempt to swallow dangerously large pieces of food.[2]

People with advanced gum disease, tooth loss, and ill-fitting dentures tend to select soft foods over fibrous, sometimes more nutritious foods. Foods such as corn on the cob, apples, and hard rolls that are difficult to chew are swallowed mostly unchewed, or avoided altogether. If they are replaced by creamed corn, applesauce, and rice, then nutrition status may not be greatly affected, but when food groups are avoided and variety is limited, nutrient deficiencies follow.

Since this discussion is about the connections of nutrition with oral health, it speaks little about dental care and oral hygiene, merely acknowledging their high priority. This discussion is intended to answer the questions:

---

*This discussion is adapted from S. R. Rolfes and E. N. Whitney, Say cheese and smile: The nutrition and oral health picture, *Nutrition Clinics*, December 1987 (available from J. B. Lippincott, Route 3 Box 20B, Hagerstown, MD 21740).

†H. Hopkins, editorial director of *FDA Consumer*.

- How does nutrition before birth affect tooth development?
- How do nutrition, food, and eating patterns throughout life affect the health of the teeth and gums?

It concludes with a set of recommendations for the person interested in applying the answers.

Tooth Development

Primary tooth development in human beings begins between two and three months in utero. By the last trimester of gestation, permanent teeth are forming. Of the ultimate 52 primary and permanent teeth that human beings form, 32 have begun to develop during gestation.[3] Maternal nutrition during pregnancy therefore profoundly influences the development of the teeth. Maternal nutrients must supply the preeruptive teeth with the building materials needed to develop in the proper sequence.

Like other tissues, the tissues in the mouth develop in stages. When nutrition insults occur during critical stages of their growth, the damage that results is irreversible. For example, defects in dentin or enamel formation cannot be corrected at any time after the critical stage.

To a great extent, heredity determines the potential arrangement of teeth, their eruption time, the tooth pattern and bite, the pits and fissures on the tooth surface, and their resistance to decay. Nutrition is one of several factors that help to determine the extent to which these potentials are realized. Chemical insults during pregnancy, including nutrient deficiencies during fetal development, can impair the development of the mouth structures. Severe nutrient deficiencies are not the only cause of abnormalities, but they can result in malformations that will make eating, chewing, and swallowing difficult. Subtle nutrient deficiencies during tooth development can reduce tooth size, interfere with tooth formation, delay the time of tooth eruption, and increase susceptibility of the teeth to caries.

Figure FP5–1 shows the anatomy of a tooth. The cells responsible for creating it are odontoblasts (dentin-forming cells) and ameloblasts (enamel-forming cells). The dentin interior and the outer enamel shell of the tooth are built on protein matrices that are subsequently mineralized—primarily with calcium, magnesium, and phosphorus. For dentin, the protein foundation is a collagen matrix, which requires a variety of substances, including vitamin C, for proper formation. The protein matrix for enamel is keratin, which depends in part on vitamin A for its synthesis. If protein or either vitamin is deficient during tooth development, then an imperfect matrix is laid down, and even with successful mineralization, the final structure will be imperfect. Likewise, if the protein matrix is normal but mineralization is not, then the tooth will be poorly formed. Table FP5–1 summarizes some of the effects of nutrient deficiencies on dental development.

The table shows the effects of deficiencies of all of the nutrients just mentioned—protein, the minerals that serve as building materials, and the vitamins that assist in the building of the tooth. In most instances, the effects are easily explained. As would be expected, protein deficiency makes teeth

primary teeth: the first set of 20 teeth that are eventually replaced by permanent teeth; also called **deciduous** or **baby teeth.**

permanent teeth: the final set of 32 teeth that replace the primary teeth.

dentin: the main tissue of a tooth surrounding the pulp.
dens = tooth

enamel: the hard, white, dense substance made up mainly of calcium and phosphorus that covers the crown of a tooth. Enamel is the hardest substance in the human body.

caries (KARE-eez): gradual decay and disintegration of a tooth.
carius = rottenness

odontoblasts: cells from which dentin is formed.

ameloblasts: cells from which tooth enamel is formed.

Figure FP5–1 The Anatomy of a Tooth

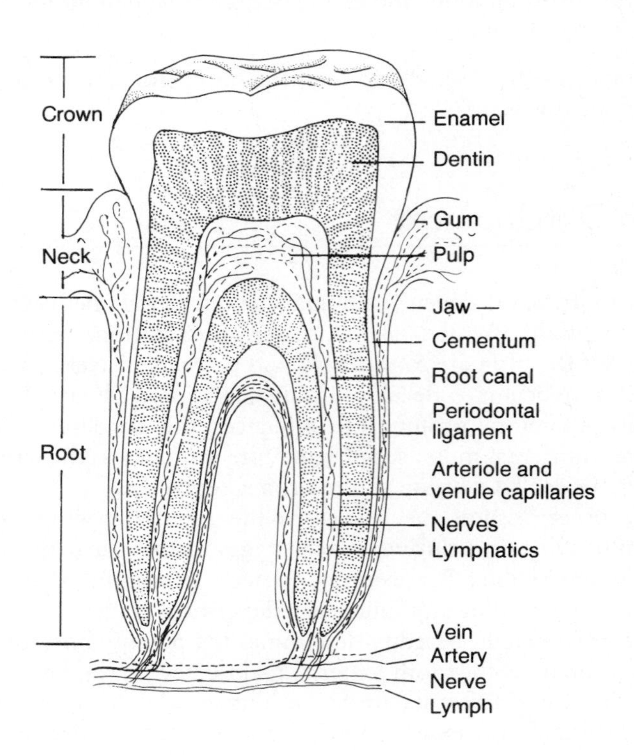

Table FP5–1 Nutrient Deficiencies Affecting Tooth Development

Nutrient Deficiency	Effect on Tooth Development
Protein	Small, irregularly shaped teeth, delayed eruption, high caries susceptibility
Vitamin C	Disturbance of collagen matrix of dentin
Vitamin A	Disturbance of keratin matrix of enamel
Vitamin D	Poor calcification, pitting, striations
Calcium	Poor calcification
Phosphorus	Poor calcification
Magnesium	Enamel hypoplasia
Iron	High caries susceptibility
Zinc	High caries susceptibility
Fluoride	High caries susceptibility

Source: Adapted from H. M. Leicester, Nutrition and the tooth, *Journal of the American Dental Association* 52 (1956): 284–289; A. E. Nizel, Preventing dental caries: The nutritional factors, *Pediatric Clinics of North America* 24 (1977): 141–155; J. H. Shaw and E. A. Sweeney, Oral health, in *Nutritional Support of Medical Practice*, eds. H. A. Schneider, C. E. Anderson, and D. B. Coursin (Philadelphia: Harper & Row, 1983).

smaller, slower to erupt, and more irregularly shaped than normal.[4] The way in which it makes them susceptible to caries appears to be by reducing salivary flow.[5] Even when fetal and neonatal protein deficiency is later corrected with an adequate diet, reduced salivary flow persists, reflecting an irreversible effect of early protein deficiency on gland function.

Iron deficiency appears to act in the same way. Prenatal iron deficiency, even if marginal, decreases salivary flow and salivary protein content in children.[6] These salivary factors correlate with a high incidence of caries. Suboptimal zinc status prior to tooth eruption is also associated with an increase in dental caries.[7] The last nutrient mentioned in Table FP5–1 is fluoride, known to be important in converting the basic tooth crystal, hydroxyapatite, to the more decay-resistant crystal, fluorapatite. Altogether, nutrition during tooth development clearly has a major influence on future dental health.

hydroxyapatite (high-drox-ee-APP-ah- tite): the major calcium-containing crystal of bones and teeth. See also **fluorapatite.**

fluorapatite (floor-APP-ah-tite): the stabilized form of bone and tooth crystal (hydroxyapatite), in which fluoride replaces the hydroxy groups of the hydroxyapatite.

Adequate maternal nutrition during pregnancy, with respect to all of these nutrients, is important to permit the optimal development of the child's teeth. Some controversy attends one question, however: whether pregnant women would benefit their children's teeth by taking prenatal fluoride supplements. The Food and Drug Administration prohibits manufacturers from claiming that fluoride supplements will prevent dental caries in infants of women who take supplements during pregnancy. Some researchers believe that because permanent teeth mineralize postnatally, these teeth receive little benefit from prenatal supplements. Others point out that fluoride supplementation is most protective when at least part of the preeruptive phase of tooth development is included. If fluoride exposure during preeruptive tooth development is necessary, then prenatal supplementation may indeed be important to protecting primary teeth.

Fluoride does pass through the placenta to the fetus, but whether the placenta can defend against excess fluoride is questionable. Fluoride supplements are therefore not recommended for pregnant women who drink fluoridated water. However, supplementation is ordinarily regarded as safe, and some physicians and dentists do prescribe prenatal fluoride supplements to women who live in communities without fluoridated water. One long-term, well-controlled study reported that children of mothers taking prenatal fluoride supplements (2.2 milligrams sodium fluoride, which delivers 1 milligram fluoride) had, at the ages of five and six, teeth that were virtually immune to caries.[8] Their teeth contained greater concentrations of fluoride than those of children whose mothers had used just fluoridated water, and still greater concentrations than the teeth of children whose mothers had used nonfluoridated water during pregnancy.

The 20 primary teeth (sometimes referred to as baby teeth) begin to erupt at around four months of age and continue erupting through the third year of life. Tooth development continues to depend on systemic nutrition—supplied via the vascular system—until the final tooth erupts at around age 13. The formation and mineralization of teeth continues to depend on all of the nutrients named in Table FP5–1.

At the same time, as childhood progresses, the effect of nutrition on the teeth becomes increasingly more environmental than systemic; that is, the presence of food in the mouth increasingly affects the health of the teeth.

Dental Caries

Dental caries is an infectious oral disease that attacks the structure of the teeth. It is a pervasive health problem affecting 95 percent of the population.

The relationship between teeth, food, and caries development is complex and is complicated further by unique individual factors, such as the hormonal and immunological milieux, that affect susceptibility to caries. Additional considerations are the behaviors and lifestyles that influence food selection, eating habits, and oral hygiene.

Caries develops as the result of the metabolism of fuels by microorganisms that reside in plaque on the surface of the teeth. These microorganisms consume carbohydrates, producing organic acids, such as lactic acid and pyruvic acid, as wastes. These acids cause the pH in the plaque and saliva to fall, and this leads to demineralization of the basic crystal of the enamel, hydroxyapatite. Calcium and phosphorus dissociate from the hydroxyapatite crystals and diffuse into the plaque. Fortunately, salivary fluids dilute and salivary proteins buffer the contents of the mouth, returning the pH to neutral. This results in plaque that is at neutral pH and supersaturated with calcium and phosphorus. A reverse flow of the calcium and phosphorus back into the enamel—that is, a remineralization of the enamel—can now occur. Until recently, caries development was considered to be a continuing demineralization process. Now it is viewed more as a dynamic process—one of alternating phases of demineralization and remineralization. When the net result is demineralization, caries develops.

microorganisms: small living bodies, such as bacteria, not perceptible to the naked eye.

plaque (PLACK): a sticky, colorless cluster of microorganisms, protein, and polysaccharides that adheres to teeth and gums. Plaque contributes to dental caries and periodontal disease. When calcium combines with the plaque and hardens, it becomes **tartar**.

organic acids: any organic compound containing one or more acid (carboxyl) groups (for example, lactic and pyruvic acids).

buffer: a substance capable of neutralizing both acids and bases.

cariogenic: conducive to caries formation.

Research conducted using animals reveals that at least two main ingredients are required to make dental caries: microorganisms and carbohydrates. Without microorganisms, there is no caries; that is, decay does not develop in a germ-free mouth, even with a cariogenic diet. Likewise, without a carbohydrate source, caries does not develop. Teeth remain caries-free when carbohydrate is fed via a tube into the gastrointestinal tract, even when the mouth is infected with microorganisms.

This discussion of caries development begins with the microorganisms that inhabit the mouth, the saliva that influences the oral environment, and the protective role of fluoride. The focus later shifts to the diet, the cariogenicity of foods, and the person's eating habits.

Microorganisms

The principal dental plaque-forming, caries-producing microorganism is *Streptococcus mutans,* although other microorganisms have been shown to cause caries. *S. mutans* is found in caries, adheres readily to tooth surfaces, and uses carbohydrate from food to produce acid. Research to develop a vaccine or oral antibiotics against dental caries is focusing on this organism.

In addition to acids, *S. mutans* produces the sticky polysaccharides glucan and fructan from sucrose and other carbohydrates. These polysaccharides allow the microorganisms to adhere to the smooth enamel surfaces, creating the clusters of plaque. As the bacteria continue to metabolize carbohydrates, the acids become concentrated at the site, and a carious lesion begins.

S. mutans bacteria are not found in the mouths of infants prior to tooth eruption.[9] Nor are they evident in people who have lost their teeth and do not wear dentures. The bacteria appear shortly after tooth eruption begins, and studies suggest that parents or other caretakers infect the infant's mouth with them.

A relationship is evident between the microbial infection of mothers and that of their children. Mothers with high concentrations of *S. mutans* have children with high concentrations, and mothers with low *S. mutans* concentrations have children with low concentrations. In one study, a preventive program for mothers was implemented that involved diet counseling, professional dental care and instruction, and fluoride treatment.[10] These preventive measures were not offered to their children, but the children were compared with the children of a control group. In both groups, the percentage of infected children increased with age, but it was lower in the experimental group. Only 16 percent of the children in the preventive group developed caries, compared with 43 percent in the control group. Prevention techniques reduced the *S. mutans* population in the mothers as well as in their children, resulting in fewer carious teeth.

Saliva

Most people fail to fully appreciate the complexities and contributions of saliva to their oral health. Fluids secreted by the salivary glands vary according to stimulation, age, sex, time of day, diet, diseases, and drug intake. Approximately 1 liter of saliva is secreted in a day in response to the chewing of food. Salivary flow during other times of the day is small, and during sleep, it is minimal. One of the major actions of salivary fluids is protection against dental caries. Saliva dilutes acid and normalizes pH, as mentioned earlier; rinses the mouth; provides minerals; and exerts antibacterial activity.[11] The power to elicit secretion of saliva is one of the factors that determines the cariogenicity of foods, discussed later.

When salivary flow is reduced, the oral environment is less able to defend against caries. Reduced salivary flow may occur as a symptom of a disease, a side effect of medication, or in response to radiation therapy. Fasting also reduces salivary flow. Salivary flow progressively decreases during a 300-kcalorie, 3-liter liquid fast.[12] Researchers note an increase in the rate of plaque formation during fasting. This effect is not simply due to the lack of chewing stimulation, but might be explained by general dehydration.

Fluoride

The effect of prenatal fluoride has already been noted; debated until recently, its importance now appears to be supported by research. The importance of postnatal fluoride, on the other hand, has long been known. Numerous studies have shown that when fluoride is added to the water supply, the children in the community have fewer dental caries than children who drink nonfluoridated water. Children provided with optimally fluoridated water from birth have 50 to 70 percent fewer caries than otherwise expected.

Water fluoridation is the most effective, least expensive way to provide dental care to everyone. It protects the poor, the uninformed, and people who simply do not practice regular preventive measures or seek professional care. However, one-third of the U.S. population is not receiving fluoride because water fluoridation has not been adopted by local communities or the private water companies that serve them.

The National Research Council of the National Academy of Sciences recommends fluoridation of drinking water to approximately 1 part fluoride per million parts of water (1 ppm, which is the same as 1 milligram per liter). Water with 1 ppm fluoride offers the greatest caries protection at virtually no risk of fluorosis. Liquid fluoride supplements are available by prescription, and supplementation is recommended when the natural fluoride concentration in water is below 0.7 ppm. Table FP5–2 lists the American Dental Association's recommended supplement dosages by age.

fluorosis (flur-OH-sis): mottling of tooth enamel caused by excess fluoride.

All food and water supplies naturally contain variable amounts of fluoride in trace quantities.[13] About half of the U.S. population has access to water with an optimal fluoride concentration. Foods are not a major source of fluoride. The fluoride content of foods that are processed with fluoridated water, however, is higher than that of the same foods processed with fluoride-free water. The effect of water fluoridation on the food chain is becoming evident.

As would be expected, the fluoride content of beverages is also higher when they are processed with fluoridated water. The Food and Drug Administration has set limits on the natural and added fluoride content for domestic and imported bottled water. Most teas contain appreciable natural fluoride (contributing about 0.1 milligram fluoride per cup), even when brewed in fluoride-free water. An exception is herbal tea, popularly accepted as a caffeine-free alternative, which has negligible fluoride.

Excess fluoride causes dental fluorosis, a developmental imperfection of the tooth surface. At doses of 2 ppm, the teeth appear extremely white; at doses greater than 4 ppm, brown stains appear. (Stains on the teeth are also produced by other factors. When taken prenatally or during the first eight years of life, tetracycline stains teeth. Like the fluoride stains, tetracycline stains are permanent but do not weaken the tooth structure.) While the brown stains of fluorosis are cosmetically unattractive, dental fluorosis does not threaten health. Studies confirm that drinking water fluoridated to recommended levels poses no adverse health effects.[14]

The *preeruptive* maturation stage of tooth development is the critical time for *systemic* fluoride to offer its benefits in making the tooth resistant to caries throughout life.[15] This stage begins before birth and ends when the last molar

Table FP5–2 American Dental Association's Recommended Fluoride Supplement Dosages for Low-Fluoride Areas

Age	Dosage
0 to 2 yr	0.25 mg/day
2 to 3 yr	0.50 mg/day
3 to 10+ yr	1.00 mg/day

Source: Adapted from Effect of fluoride on dental health, *Nutrition and the M.D.*, December 1980, pp. 3–4.

erupts. It is during this time that the calcium and phosphate in the enamel are combining into hydroxyapatite, and systemic fluoride can convert it into fluorapatite—a combination of calcium fluoride and calcium phosphate. The benefit of this conversion is, as mentioned, that fluorapatite is more resistant than hydroxyapatite to the acid demineralization process that initiates dental caries.

The *posteruptive* maturation phase of tooth development is when *topical* fluoride makes the tooth resistant to caries.[16] Immediately after the tooth erupts and for the following two to three years, the outer enamel surface is immature. This is an ideal time to expose the teeth to the protection of fluoride, because they can take up minerals. (They are also more prone to decay if exposed to harmful substances.) Topical fluoride produces calcium fluoride and fluorapatite compounds. Such fluoride application also disrupts the normal growth and activity of dental plaque bacteria.[17]

Fluoridated drinking water offers both systemic benefits to developing teeth and topical benefits to those teeth already present. Fluoride in the drinking water washes over the teeth during their development, enabling them to incorporate fluoride continuously into their crystals.

Topical fluoride can partially compensate for long periods of enamel formation without fluoride, but this enamel is not as resistant to decay—it contains less fluorapatite. Even after teeth are formed, it is ideal to have fluoride continuously present in the oral environment. Teeth continue to exchange materials with the surrounding fluid all the time. This is why fluoridation of water is preferable to topical fluoride. For children who do not drink fluoridated water, fluoride tablets or drops are an effective method of providing both topical and systemic benefits.

The rate of dental caries in the general population is declining, with major credit going to water fluoridation. Even in communities without fluoridated water, the prevalence of caries is declining. This may be due to the increase of fluoride in the food chain, as already mentioned, because the use of fluoridated water in food processing is becoming increasingly common.[18]

Gum Disease

Although caries is declining, gum disease, or periodontal disease, still poses a large threat to most people, affecting over half of adults over age 45.[19] Gum disease is preventable with diligent oral hygiene, but if left untreated, it leads to bleeding gums, loosening of the teeth, and eventual loss of teeth.

Systemic nutrition may influence periodontal health by way of the immune system, bone metabolism, collagen formation, and epithelial tissue function. Consider, for example, that the oral epithelium has a rapid cell turnover rate, replacing cells every three to seven days. Any stress that compromises this ability to regenerate weakens the defense against microorganisms and, therefore, against gum disease.

The progress of gum disease can be slow and unnoticeable. It may first become evident when gums bleed while a person is brushing teeth. The same plaque that causes dental caries is the major initiating factor in gum disease. The plaque on tooth surfaces collects calcium salts, hardens, and turns into deposits of calculus, or tartar. The gums surrounding the tooth's root become

calculus: the general term for any abnormal concentration of mineral salts, also referred to as **tartar**.

tartar: calcium salts, mucin, and bacteria deposits found on the teeth and gums; also called **calculus**.

inflamed and infected. If the infection progresses, resorption of the bone below the tooth begins, causing the tooth to lose its anchor.

Many factors contribute to gum disease. The primary cause, of course, is poor oral hygiene, but any irritation of the gums weakens their resistance to infection. Stresses such as bad tooth alignment, tooth loss, and tooth grinding can contribute to periodontal disease development.

The extent of bone resorption may reflect dietary inadequacies. The rate of alveolar bone loss is slowed by a calcium intake of 1000 to 1500 milligrams per day.[20] One study reported that alveolar bone loss in clients receiving a calcium and vitamin D supplement was 36 percent less than in clients receiving a placebo.[21] As noted earlier, alveolar bone resorption causes the teeth to lose their anchor.

Another nutrition connection, not often encountered but worth mentioning, is vitamin C, which has long been associated with the integrity of the gums. Gum deterioration is a classic clinical symptom of acute vitamin C deficiency. The effects of subclinical vitamin C deficiency are less well documented, but one study reports that subclinical vitamin C deficiency does influence the early stages of gingival inflammation even under conditions of sustained oral hygiene.[22] Gingival bleeding and inflammation varied directly with changes in vitamin C intake and serum concentrations. Vitamin C depletion did not affect other dental measurements observed, such as plaque accumulation.

gingiva (jin-JYE-va or JIN-jih-va): the tissue surrounding the necks of the teeth and supporting bone; also called the **gums**. Gingival inflammation is known as **gingivitis** (jin-jih-VYE-tis).

Foods and Eating Habits to Foster Oral Health

Healthful eating habits from the nutrition standpoint are not necessarily healthful eating habits from the dental standpoint. A selection of foods may provide all the nutrients in adequate amounts to support overall health, but still may promote caries development. Of course, all meals should be followed by proper oral hygiene, but realistically, this does not always happen. The question of what foods are most and least cariogenic is therefore of interest.

The American Dental Association is trying to develop a rating system for the cariogenicity of foods. Most likely, it will be based on the key factor that results from all the characteristics of a food working together—namely, the amount of acid a food produces in plaque. Guidelines based on cariogenicity may eventually find their way to food labels. In Switzerland, foods that pass the acid-plaque test are labeled with a smiley-faced tooth to signify that the product is "safe for teeth." This positive labeling system encourages consumers to purchase such items for between-meal snacks.

Prime among the relevant characteristics of foods is their carbohydrate content. Carbohydrates are the fuel source for bacteria—carbohydrates of many kinds, not just refined table sugar (sucrose). Honey, molasses, brown sugar, glucose, fructose, and starches all have a strong cariogenic potential. However, sugar alcohols, which are used as sugar substitutes, either are noncariogenic or have extremely low cariogenicity potential. Some may actually have anticariogenic effects. In experimental studies in rats, partial or total substitution of xylitol for dietary sucrose results in caries reduction.[23] The effect is greater than just the displacement of sucrose—xylitol seems to have a therapeutic effect against caries. Rinsing with a xylitol solution after a sucrose-containing meal reduces the cariogenicity of the diet. Xylitol stimulates salivary flow, increases pH, maintains a high pH, and resists microbial metabolism.

Other sugar substitutes, such as saccharin, aspartame, and cyclamate, are thought to be protective against caries simply because they are not metabolized to acids. However, one study concluded that saccharin actually inhibited caries in rats.[24] Rats fed a saccharin-supplemented diet developed fewer caries than rats fed the same diet without supplementation or with aspartame supplementation. Offsetting this effect of saccharin are other health risks, though, so its use should be moderate.

In addition to the presence of carbohydrate in foods, the retention of those foods in the mouth is critical.[25] Foods that stay in the mouth for a long time yield acid for a long time. Sticky foods are retained on tooth surfaces longer and present a greater risk than foods that are readily cleared from the mouth. For that reason, the sugar in a soft drink is less significant than that in caramels, pastries, or jelly. By the same token, the sugar in a sticky food such as dried fruit is more detrimental than its quantity alone would suggest.

Stickiness is not the only factor affecting food retention. Curiously, sugar speeds up the clearance rate of starchy foods from the mouth. Starchy foods with a high sugar content are removed more rapidly and lower the pH of the plaque for a shorter time than do starchy foods with less sugar.[26]

The sugar content of a food and the amount of acid produced from that sugar do not always parallel the amount of enamel dissolved. Some foods, such as citrus fruits and carbonated beverages, contain acids of their own, and these acids can act directly on the tooth enamel. These dietary acids are strong enough to depress the pH below the point at which bacterial enzymes are active, so no new acid is formed, but the acid already present is strong enough to significantly dissolve enamel.[27]

Interestingly, a high sugar concentration can also depress bacterial growth and activity. Foods with high concentrations of sugar (candies) rapidly leave the mouth and destroy less enamel than do foods with less sugar in combination with starch (breads and cookies).[28] This effect is not simply explained by the stickiness of the foods. A variety of other factors, including fat and salt content, also influence food clearance.[29] Thus to predict which foods will be cariogenic is not as easy as might be expected. Researchers often isolate one dietary factor to determine the extent of its effect on caries development, but when they do so, the usefulness of the findings is limited, because people eat meals that contain multiple dietary factors. Quite often, the findings of such research are variable and inconsistent. Researchers lack standardized reference foods and methods for assessing cariogenic potential. To establish a cariogenicity rating for a food is to rely on many questionable assumptions. Nevertheless, pieces are being collected and analyzed in the hope of assembling a puzzle in which, someday, they will all fit.

The cariogenicity of a food depends on its chemical composition and physical form. The chemical composition of a food includes not only the type of carbohydrate but also the content of dietary acids, calcium, phosphorus, and fluoride. In addition, the food's ability to stimulate salivary flow is considered. The physical form of a food affects its retention in the mouth.

Some high-fiber carbohydrate foods are examples of anticariogenic foods. In particular, raw vegetables such as celery and carrots are sometimes called "detergent" foods. Their crisp and crunchy texture serves as a mechanical cleanser, removing food particles from teeth. They do not stick to the teeth, and they require vigorous, thorough chewing, which stimulates salivary flow. Increased saliva flow helps to clear the food from the mouth and buffer the plaque acid. The person wishing to minimize caries formation could munch on these types of foods at the end of any meal, if brushing were not feasible. Rinsing the mouth with water also helps, of course.

Apples offer an interesting contrast. They are recommended by some as a good food to eat at the end of a meal to help prevent caries, and they have been called "nature's toothbrush" because they stimulate salivary juices and "brush" the surfaces of the teeth. However, while it is true that they stimulate salivary flow, they also liberate sugar after they have been crushed by the teeth; the sugar contributes to acid formation, which soon offsets the buffering effect of the saliva. One study measured the pH changes that occurred when either apples or peanuts were consumed after a lump of sugar.[30] In some individuals, apples caused the pH to fall even lower than the sugar had already done. In contrast, peanuts raised the pH that had been lowered by the sugar. Apples, then, may offer saliva-stimulating benefits, but they still may lower pH, and they do not brush teeth clean at potential caries sites. This example illustrates how foods may have both caries-promoting and caries-preventing effects. The best foods to eat at the end of a meal are those that have a saliva-stimulating effect and do *not* depress pH.

Another food of interest is cheese. Cheese is a powerful saliva stimulant, and its proteins also buffer pH in the mouth. Even when eaten immediately after sugary foods, cheese raises plaque pH.[31] A piece of cheese eaten at the end of a meal may therefore reduce the cariogenicity of the meal. A further contribution cheese makes to dental health is its high calcium and phosphorus content.

So far, these factors have been mentioned: the quantity of carbohydrate, the nature of the carbohydrate, its context (such as the stickiness of the food, or acid, fiber, or protein present in it), and its saliva-stimulating effect. Another concern is the *frequency* of its consumption. Carbohydrate eaten between meals poses a greater risk of dental caries than does carbohydrate eaten with meals. Bacteria produce acid for 20 to 30 minutes after an exposure to sugar. So, if a person were to eat three pieces of candy at one time, the teeth would be exposed to approximately 30 minutes of acid demineralization. If that person were to eat three pieces of candy at half-hour intervals, the time of exposure to acid would increase to 90 minutes. Likewise, slowly sipping a sugar-sweetened soft drink between meals may be more harmful than drinking the entire soda at mealtime.

An extreme effect of prolonged tooth exposure to carbohydrate is seen in infants who are put to bed sucking on a bottle of formula, milk, or fruit juice, or who use such a bottle as a pacifier for extended periods of time. They experience extensive and rapid loss of tooth material. Prolonged sucking on such a bottle bathes the upper teeth for long periods in a carbohydrate-rich fluid. (The tongue covers and protects most of the lower teeth, although they, too, may be affected.) Salivary flow, which normally cleanses the mouth and neutralizes the acid, diminishes as the child falls asleep. The result is decayed teeth (nursing bottle syndrome). This syndrome has also been reported in

nursing bottle syndrome: extensive tooth decay due to prolonged tooth contact with formula, milk, fruit juice, or other carbohydrate-rich liquid offered to an infant in a bottle.

breastfed infants offered the breast for extended times. To prevent it, children should not be given a bottle as a pacifier at bedtime. If a bottle is given, it should be filled with water. In fact, a wise mother would offer her infant water after each feeding to rinse the mouth.

It makes sense to select foods with dental health as well as nutrition in mind. Of course, it is always best to brush and floss the teeth, or at least to rinse the mouth, after eating meals and snacks, but there is no harm in applying some knowledge of the relative cariogenicity of foods as well. For example, the person who likes raisins might be better advised to eat a carrot-raisin salad or raisin muffins with meals after which toothbrushing will follow rather than to eat raisins between meals and let them stick to the teeth. Recommendations with respect to foods approved as snacks are provided in Table FP5–3.

Teeth can last a lifetime with proper care. They do not have to loosen and fall out, even with old age. It is evident that diet and nutrition can promote dental health throughout life. The same balanced diet that promotes general health can also contribute to sound dental health, provided that, as the American Dental Association recommends, consumers control the frequency with which they eat cariogenic foods, especially when they cannot brush their teeth immediately afterwards. The following guidelines are offered to maximize protection against dental caries and gum disease. Parents can help their children establish good dental habits and health by:

Table FP5–3 Dietary Recommendations for Controlling Dental Caries

Food Group	Low Cariogenicity: Use When Teeth Cannot Be Brushed Immediately	High Cariogenicity: Do Not Use Unless Followed by Prompt and Thorough Dental Hygiene
Dairy	Milk, cheese, plain yogurt	Chocolate milk, ice cream, ice milk, milk shakes, fruited yogurts, eggnog
Meat/meat alternates	Meat, fish, poultry, eggs, legumes	Peanut butter with added sugar, luncheon meats with added sugar, meats with sugared glazes
Fruit	Fresh fruit, packed in water or juice	Dried fruit, fruit packed in syrup, jams, jellies, preserves, fruit juices and drinks
Vegetable	Most vegetables	Candied sweet potatoes, glazed carrots
Bread/cereal	Popcorn, soda crackers, toast, hard rolls, pretzels, potato chips, corn chips, pizza	Cookies, sweet rolls, pies, cakes, ready-to-eat sweetened cereals as a between-meal snack
Other	Sugarless gum, coffee or tea without sugar	Sugared soft drinks, candy, fudge, caramels, honey, sugars, syrups

- Not allowing their infants to sleep with bottles of carbohydrate-rich liquids.
- Watching for hidden sugars in foods they provide their children; using low-sugar or sugar-free products whenever possible.
- Restricting sweet treats to mealtimes.
- Encouraging their children to practice oral hygiene after eating between-meal snacks.
- Limiting the duration of time their children's teeth are exposed to adhesive foods.
- Encouraging their children to brush and floss daily, and visit a dentist for regular checkups.
- Encouraging their children to rinse with water after eating, if brushing and flossing are not possible.
- Allowing children to drink fluoridated water, and providing fluoride supplements when such water is not available.
- Providing their children with a balanced diet composed of a variety of foods that will maintain an adequate nutrition status.
- Encouraging their children to eat foods rich in calcium and phosphorus.
- Providing their children with a variety of firm, fibrous foods that will stimulate gingival tissues, alveolar bone, and salivary glands.

These measures will serve personal dental health well. For the benefit of the younger generation, it is also important to make efforts to improve the social context so that it will better support their dental health. Unfortunately, much of the effort of the food industry is not directed toward this goal. Printed advertisements and television commercials compete to attract public attention to new items that delight the taste buds but threaten the teeth. The average television-watching child sees over 21,000 commercials each year; approximately half of those commercials are for foods and beverages, most of which contain sugar.[32] Cereals, candy, and gum lead the list of kinds of foods advertised on the nation's airwaves. Cookies, crackers, desserts, and soft drinks follow close behind. At the bottom of the list are vegetables, citrus fruits and juices, and cheese.

Food companies spend billions of dollars on television advertising. In essence, they try to encourage consumption of the very foods that health experts warn us not to indulge in. The effect of advertising is seen in children's influence on food selections. The tantalizing messages about high-sugar (and therefore low-nutrient) foods take unfair advantage of an impressionable, nutritionally naive audience.

Take a moment to consider this naive audience. Children receive about 70 hours a year learning about foods by way of television commercials—information that is almost invariably misleading. Compare that to the number of hours parents, teachers, and dentists spend each year providing children with sound nutrition and dental care information. Television commercials do not offer nutrition education, nor do they warn of problems certain foods pose to health. The Netherlands requires that commercials of sugary foods show an insignia of toothpaste being applied to a toothbrush during the last few seconds of the ad.

Consumers can influence television commercials. When the surgeon general warned of health problems associated with smoking, antismoking commercials were aired until cigarette advertising was eventually banned. The advertising of sweet foods corrupts good nutrition and dental habits. Parents and health professionals need to continue making efforts to pressure the industry to respond to their concern for healthy teeth.

A poster in a dental office reminds clients, "There is nothing the dentist can do that will overcome what the patient will not do." Professional dental care, in other words, augments, but does not replace, personal dental hygiene. Learning and practicing good dental hygiene habits early in life will serve a child through adulthood. Parents will want to teach their children how to brush with fluoridated toothpaste and gently floss regularly to remove plaque. When water fluoridation is not available, parents can provide fluoride supplements, lozenges, and rinses. Beyond these strategies, parents can be assertive in helping their children to resist social influences that pull the wrong way, be conscientious about providing an adequate diet, and be faithful in encouraging eating habits consistent with dental health.

Focal Point 5 Notes

1. Dentistry at the crossroads: The future is uncertain, the challenges are many, *American Journal of Public Health* 72 (1982): 653–654.
2. C. A. Geissler and J. F. Bates, The nutritional effects of tooth loss, *American Journal of Clinical Nutrition* 39 (1984): 478–489.
3. F. B. Glenn, W. D. Glenn, and R. C. Duncan, Fluoride tablet supplementation during pregnancy for caries immunity: A study of the offspring produced, *American Journal of Obstetrics and Gynecology* 143 (1982): 560–564.
4. M. C. Alfano, Effect of diet and malnutrition during development on subsequent resistance to oral disease, in *National Symposium on Dental Nutrition*, ed. S. Wei (Iowa City: University of Iowa Press, 1979), p. 23, as cited by M. C. Alfano, Nutrition, sweeteners, and dental caries, *Food Technology*, January 1980, pp. 70–74.
5. L. Menaker and J. M. Navia, Effect of undernutrition during the perinatal period on caries development in the rat: V. Changes in whole saliva volume and protein content, *Journal of Dental Research* 53 (1974): 592, as cited by Alfano, 1980.
6. M. C. Alfano, J. Sintes, and D. P. DePaola, Effect of marginal dietary iron deficiency during development on caries susceptibility in rats, *Journal of Dental Research* 58 (special issue A, 1979): 422, as cited by Alfano, 1980.
7. Increased dental caries in young rats suckled by zinc-deficient rats, *Nutrition Reviews* 37 (1979): 232–233.
8. Glenn, Glenn, and Duncan, 1982.
9. *Streptococcus mutans* and human caries, *Nutrition Reviews* 36 (1987): 107–109.
10. Relation of caries prevention in mothers to the infection of their children's mouths, *Nutrition Reviews* 41 (1983): 341–342.
11. I. D. Mandel, Relation of saliva and plaque to caries, *Journal of Dental Research* 53 (1974): 246–266.
12. I. Johansson, T. Ericson, and L. Steen, Studies of the effect of diet on saliva secretion and caries development: The effect of fasting on saliva composition of female subjects, *Journal of Nutrition* 114 (1984): 2010–2020.
13. G. S. Rao, Dietary intake and bioavailability of fluoride, *Annual Review of Nutrition* 4 (1984): 115–136.
14. V. L. Richmond, Thirty years of fluoridation: A review, *American Journal of Clinical Nutrition* 41 (1985): 129–138.
15. A. E. Nizel, Preventing dental caries: The nutritional factors, *Pediatric Clinics of North America* 24 (1977): 141–155.
16. Nizel, 1977.
17. Nizel, 1977.
18. D. H. Leverett, Fluorides in the changing prevalence of dental caries, *Science* 217 (1982): 26–30.
19. Dentistry at the crossroads, 1982.
20. Diet, nutrition, and oral health: A rational approach for the dental practice, *Journal of the American Dental Association* 109 (1984): 20–32.
21. K. E. Wical and P. Brussee, Effects of a calcium and vitamin D supplement on alveolar ridge resorption in immediate denture patients, *Journal of Prosthetic Dentistry* 41 (1979): 4–11.
22. P. J. Leggott and coauthors, The effect of controlled ascorbic acid depletion and supplementation on periodontal health, *Journal of Periodontology* 57 (1986): 480–485.
23. K. K. Makinen and A. Scheinin, Xylitol and dental caries, *Annual Review of Nutrition* 2 (1982): 133–150.
24. J. M. Tanzer and A. M. Slee, Saccharin inhibits tooth decay in laboratory models, *Journal of the American Dental Association* 106 (1983): 331–333.
25. B. G. Bibby and coauthors, Oral food clearance and the pH of plaque and saliva, *Journal of the American Dental Association* 112 (1986): 333–337.
26. Bibby and coauthors, 1986.
27. B. G. Bibby and S. A. Mundorff, Enamel demineralization by snack foods, *Journal of Dental Research* 54 (1975): 461–470.

28. Bibby and Mundorff, 1975.
29. Bibby and Mundorff, 1975.
30. D. A. M. Geddes and coauthors, Apples, salted peanuts, and plaque pH, *British Dental Journal* 142 (1977): 317–319.
31. A. J. Rugg-Gunn and coauthors, The effect of different meal patterns upon plaque pH in human subjects, *British Dental Journal* 139 (1975): 351–356.
32. R. B. Choate, Selling cavities—U.S. style. Address presented at the American Dental Association Council on Dental Health meeting, Miami Beach, Florida, 11 October 1977.

Nutrition during Infancy

6

Father and Son by Paul T. Granlund.

The first year of life is a time of phenomenal growth and development. To attain full potential, the infant requires an abundant supply of nutrients. The infant's high nutrient needs and developing maturity define the foods most appropriate for each stage of the first year. Nutrition during that year is the focus of this chapter.

Growth, Development, and Assessment

Physical growth and development involve not only the progressive increase in size of a living being but also the changes that accompany this increase. They depend on a variety of interrelated factors, of which nutrition is but one. The presence or absence and combination of these factors influence how growth and development progress.

Development is "the progress of an egg to the adult state." In many ways, growth and development go hand in hand: as something grows, it also develops. But development is broader than growth both on the microscopic level and in terms of qualitative anatomical and physiological changes. On the macroscopic level, development refers to attainment of motor and sensory skills and psychological attributes, as well as to anatomical and physiological changes.[1]

Growth and development are not uniform. Each body system has its own unique schedule of growth and development, varying in rate, pattern, and duration. Assessment of nutrition status monitors growth and development to confirm that they are proceeding as expected.

Growth and Development

An infant grows faster during the first year of life than ever again, as Figure 6–1 shows. The infant's birthweight doubles by about four months of age and triples by one year. (If an adult, starting at 150 pounds, were to do this, the person's weight would increase to 450 pounds in a single year!) This tremendous growth is a composite of the differing growth patterns of all the internal organs.

The course of development in the first year is remarkable. Externally, an observer can note that at birth the infant can hardly see, and cannot roll over; at a year the infant can crawl and is beginning to walk and talk. At birth it can only suck; at a year it can hold a spoon and feed itself. Internal physiological changes parallel the external ones. The internal changes, especially the development of the gastrointestinal tract and kidneys, are of particular relevance to nutrition. These internal changes enable the infant to handle more and more complex foodstuffs, from breast milk or formula alone at the start of the year to foods from all food groups at the end. A later section, "Readiness of the GI Tract," shows the significance of these developmental changes for food choices and feeding approaches.

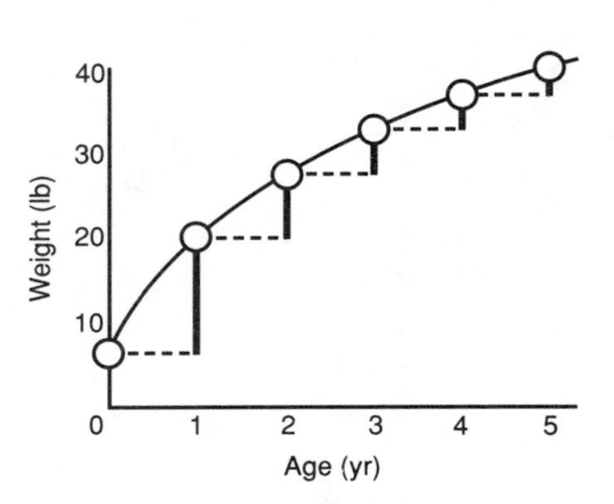

Figure 6–1 Weight Gain of Human Infants in their First Five Years of Life An infant grows faster during the first year of life than ever again.

Assessment of Nutrition Status

The infant's nutritional health at birth reflects influences experienced during pregnancy, including the adequacy of the mother's nutrition. The main parameters looked at in evaluating the newborn's health are weight, length, and Apgar score. The parameters of interest throughout infancy and childhood are those related to growth. The first years are busy ones in these terms, involving many changes.

Apgar score: a system of scoring an infant's physical condition right after birth, based on heart rate, respiration rate, color, muscle tone, and responses to stimuli.

Historical data Thorough history information can reveal medical conditions, eating behaviors, and other factors such as the physical, social, and emotional environments that influence the infant's nutrition status. Some of the environmental factors that indicate a failure to thrive (FTT) infant include marital strife, financial instability, and an overcrowded or substandard home. Comments from caretakers that the infant is "temperamental and difficult to care for" and observations from the health care providers that the infant is "irritable and unresponsive" are additional reasons to suspect FTT.

failure to thrive (FTT): failure of an infant or child to develop mentally and physically.

Failure to thrive may be accompanied by medical illness, but most often it reflects a lack of parenting. A fine line separates FTT and child abuse. When health care providers suspect FTT, the infant requires further evaluation and medical observation.

A complete diet history for infants includes information about:

- *Type of feeding.* Is the infant breastfed, formula fed, or both? If formula fed, what kind of formula?
- *Quantity and/or frequency of feeding.* If on formula, how much does the infant drink each day? If breastfed, how frequently does the infant nurse, and how long do feedings last?
- *Vitamin and mineral supplements.* Is the infant given supplements? If so, which ones, and how much of each?
- *Solid food intake.* Is the infant offered solid foods? If so, at what age were solids introduced, which ones does the infant eat, and how often? Are solid foods of the commercial type or home prepared?
- *Feeding behavior.* Does the infant exhibit unusual or abnormal feeding behaviors, such as food aversions?
- *Alternative dietary practices.* Does the infant's family omit foods or food groups?
- *Food allergies.* Does the infant appear to be allergic to specific foods?

Counseling and further evaluation (such as a detailed dietary analysis) are in order if the assessor suspects that dietary inadequacies or other conditions exist that adversely affect the infant's health.

Anthropometric measurement: Infant birthweight Birthweight is most often used as an indicator of an infant's probable future health status; it is convenient and easy to determine, and its significance is universally understood. Infants of less than 2500 grams (5½ pounds) and 1500 grams (3½ pounds) are defined as being of low birthweight and very low birthweight, respectively. More precise standards are now available, however. During the last two decades, standards have been developed that enable health care providers to correlate infants' birthweights with their gestational ages at birth. This is desirable because different birthweights are appropriate for infants of different gestational ages, and because newborn morbidity and mortality correlate both with birthweight and with gestational age.

To evaluate birthweight this way, one must first determine gestational age. In the past, gestational age has been estimated on the basis of the mother's menstrual history, consistent with clinical findings. With the availability of more precise measurements of fetal growth, including physical characteristics and neurological development, more accurate estimates of gestational age are now possible. Using these, the precise classification of newborn infants assigns infants to categories based first on gestational age, then on birthweight.[2] The dividing line between preterm and term infants is drawn at 38 weeks; the line between term and post-term infants, at 42 weeks.

large for gestational age (LGA): infants whose weight for gestational age falls above the 90th percentile.

appropriate for gestational age (AGA): infants whose weight for gestational age falls between the 10th and 90th percentiles.

small for gestational age (SGA): infants whose weight for gestational age falls below the 10th percentile.

Within each gestational age group, three subgroups of infants are defined by birthweight. Those above the 90th percentile in weight for their gestational age category are *large for gestational age*, those between the 10th and 90th percentiles are *appropriate for gestational age*, and those below the 10th percentile are *small for gestational age*. In this manner, nine groups of newborn infants are defined. When all newborns are classified at birth using this system, health care providers can easily recognize those infants who must be watched closely and can predict the types of morbidity likely to occur. Infants who are large for gestational age have different problems from those who are small for gestational age. For example, a large-for-gestational-age term or post-term infant would be more likely to be traumatized during delivery, whereas a small-for-gestational-age term or preterm infant would be more likely to suffer from respiratory distress.

The distinction between preterm infants and small-for-gestational-age infants is important in terms of health and development. Preterm infants are born before their gestational development is complete. They may be small, but those who are appropriate in size and weight for gestational age do catch up in growth to their full-term peers. In contrast, small-for-gestational-age infants have experienced fetal growth retardation and do not catch up as well. They reach only about the 25th percentile for height, and some fail to attain the same mental ability as normal-birthweight children.[3] Fetal growth retardation may also have long-term effects on development and immune function.[4]

Although birthweight for gestational age is a useful statistic, plain birthweight is often the only one available. In discussions of the research literature that follow, understand *low birthweight* to refer to an infant with a birthweight below 5½ pounds (2500 grams), unless otherwise indicated.

Infants may lose up to 10 percent of their birthweight during the first few days of life without cause for alarm. Health care providers become concerned when weight loss continues beyond 10 days or the lost weight is not regained to achieve birthweight by 3 weeks. Such observations would indicate FTT.

Anthropometric measurements: Infant growth The growth of infants directly reflects their nutritional health, and is a parameter used to assess their nutrition status. Nutrient deficiencies and excesses, especially in early infancy, can have long-term, irreversible effects.

For infants and children younger than three, health care professionals may often use special equipment to measure length. The assessor lays the barefoot infant on a measuring board that has a fixed headboard and movable footboard attached at right angles to the surface (see Figure 6–2). Often it takes two people to obtain an accurate measurement: one to hold the infant's head against the headboard and keep the legs straight, and the other person to

Figure 6–2 Nutrition Assessment of the Infant: How to Measure Length
An infant is measured lying down on a measuring board with a fixed headboard and movable footboard. Note that two people are needed to measure the infant's length.

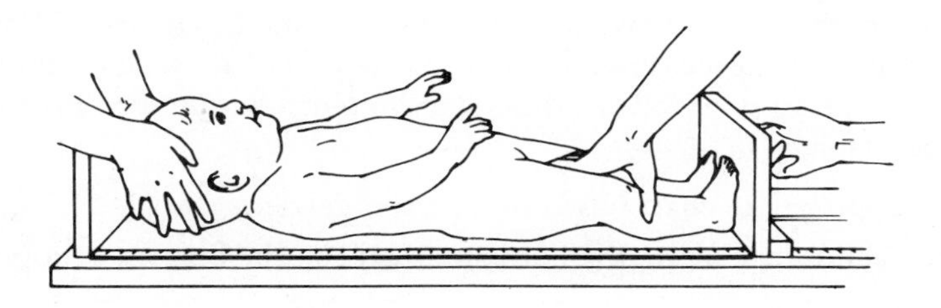

Source: Reprinted with permision of Ross Laboratories, Columbus, OH 43216.

do the measuring. This method provides the most accurate measure possible, but many health care providers use a less exacting strategy. They may simply hold the infant straight with its head against the crib headboard or other vertical support, then mark the blanket with a chalk or pen at the infant's heel, and then measure the distance from the headboard to the mark. Even more informally and less accurately, with the infant lying on a flat surface, they may stretch a nonstretch measuring tape along the side of the infant from the top of the head to the heel of the foot.

Special beam balance and electronic scales are available to measure infants' weights (see Figure 6–3). Their design allows for infants to lie or sit on the scales. Weighing infants naked, without diapers, is standard procedure.

Assessors may also measure head circumference, to confirm that growth is proceeding normally or to help detect protein-energy malnutrition and evaluate the extent of its impact on brain size. The assessor places a nonstretchable tape so as to encircle the largest part of the infant's or child's head: just above the eyebrow ridges, just above the point where the ears attach, and around the occipital prominence at the back of the head. For accuracy in recording, the measurer immediately notes the measure in either inches or centimeters.

Some assessors routinely measure head circumferences in children under three years of age. Other assessors do so only if they have reason to believe that a child has been severely malnourished during infancy. Malnutrition severe enough to cause a detectable reduction in head circumference is seldom seen in developed countries but is not uncommon among the poor in developing countries. The brain grows rapidly during early infancy, and researchers believe that malnourished children with small head circumferences will have fewer brain cells.[5]

Growth retardation indicated by measurements below standard for height, weight, or head circumference is an important sign of poor nutrition status. The standards almost universally used for infants and children are the growth charts presented in this and the next chapter. Standard charts compare weight to age, height to age, and weight to height; ideally, height and weight are in roughly the same percentile. Although individual growth patterns may vary, in general, a child's growth curve will stay at about the same percentile

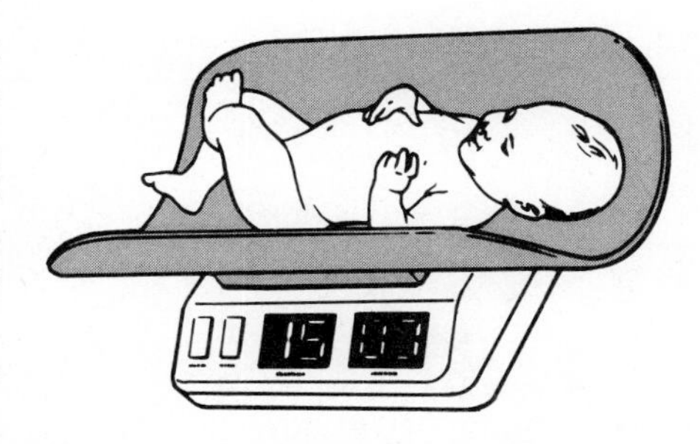

Figure 6–3 Nutrition Assessment of the Infant: How to Measure Weight
Infants sit or lie down on scales that are designed to hold them while they are being weighed.

throughout childhood. In children whose growth has been retarded at first, nutrition rehabilitation will ideally induce height and weight to increase to higher percentiles. In overweight children, the goal is for weight to remain stable as height increases, until weight becomes appropriate for height.

To evaluate growth in infants, as long as length is measured lying down, an assessor uses the charts shown in Figures 6–4 (A and B) and 6–5 (A and B). As soon as height can be measured standing, the assessor can switch to the charts of Chapter 7. The assessor follows these steps to plot a weight measurement on a percentile graph:

- Select the appropriate chart based on age and gender.
- Locate the child's age along the horizontal axis on the bottom or top of the chart.
- Locate the child's weight in pounds or kilograms along the vertical axis on the lower left or right side of the chart.
- Mark the chart where the age and weight lines intersect, and read off the percentile.

To assess length, height, or head circumference, the assessor follows the same procedure, using the appropriate figure. Head circumference percentile should be similar to the child's weight and height percentiles.

With length, weight, and head circumference measures plotted on growth percentile charts, a skilled clinician can begin to interpret the data. Percentile charts divide the measures of a population into 100 equal divisions. Thus half of the population falls above the 50th percentile, and half falls below. The use of percentile measures allows for comparisons among people of the same age and gender. For example, a six-month-old female infant whose weight is at the 75th percentile weighs more than 75 percent of the female infants her age. Infants whose weights fall below the 5th percentile are defined as FTT infants and those below the 10th percentile are suspect for FTT. Such infants require medical evaluation and care.

Physical examinations and biochemical analyses Clues to an infant's nutrition status can be identified by examining the infant for physical signs of malnutrition. Table 1–9 in Chapter 1 lists some general physical signs of malnutrition valid for all ages, and Table A–8 in Appendix A lists the signs of specific vitamin, mineral, and other imbalances. If the assessor suspects a nutrient deficiency or excess, the next step is to perform the appropriate laboratory tests.

Iron deficiency does not usually develop in infants prior to six months of age thanks to the iron they accumulate during gestation, but it becomes the most common nutrient deficiency in infants beyond six months of age. After six months, therefore, assessment of iron status is advisable. At a minimum, laboratory assessment of hemoglobin or hematocrit is suggested. Tables A–17 and A–18 in Appendix A show deficient and acceptable values for infants and children.

Other useful laboratory tests of nutrition status for infants include serum protein and serum albumin, when protein malnutrition is suspected; and serum vitamin A, when vitamin A deficiency is suspected. Tables 6–1 and 6–2 show deficient and acceptable serum concentrations for these nutrients.

Figure 6–4A Nutrition Assessment from Birth to 36 Months: Length and Weight for Age—Girls

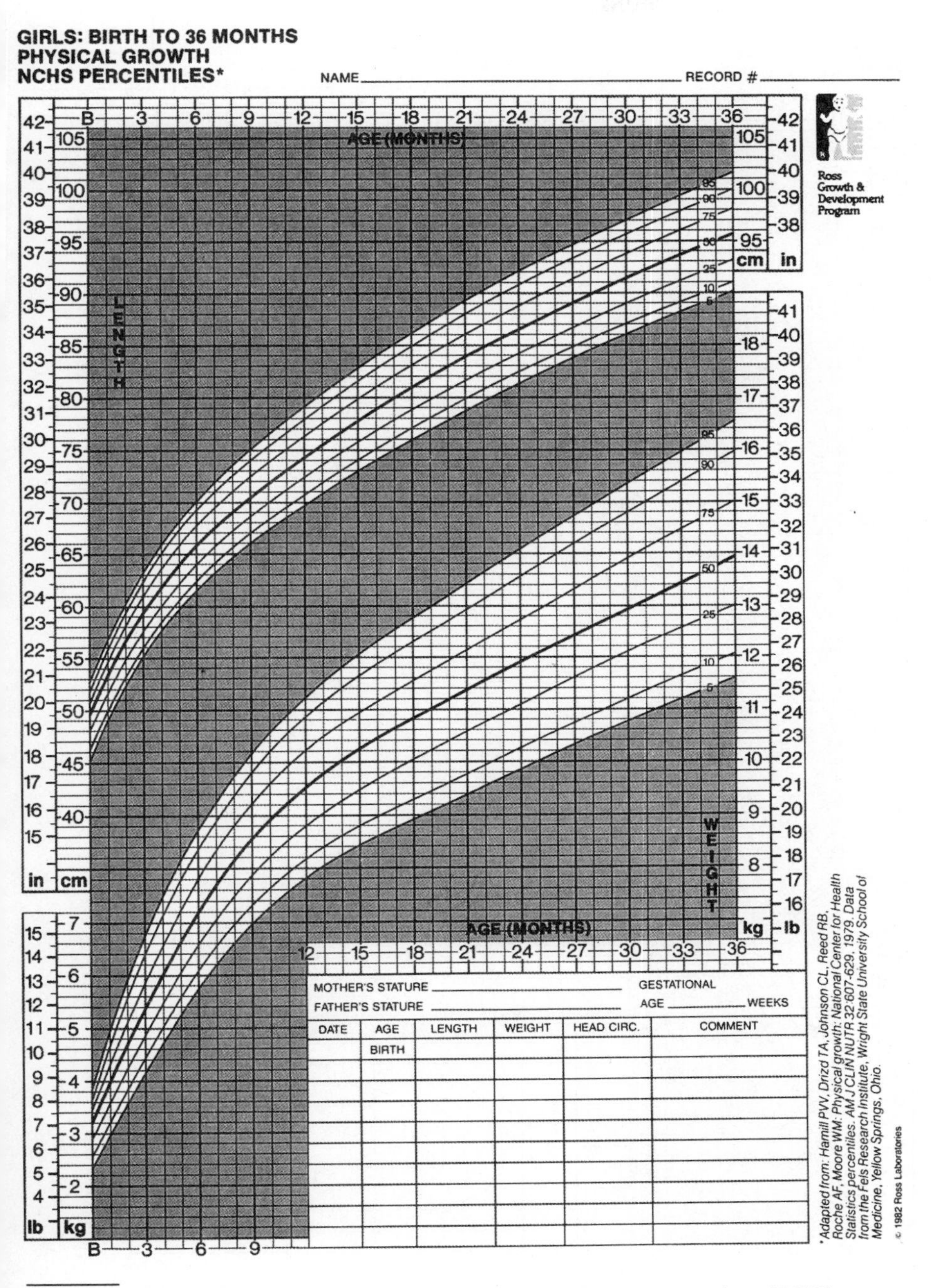

Source: Used with permission of Ross Laboratories, Columbus, OH 43216, from NCHS Growth Charts, © 1982 Ross Laboratories.

Figure 6–4B Nutrition Assessment from Birth to 36 Months: Length and Weight for Age—Boys

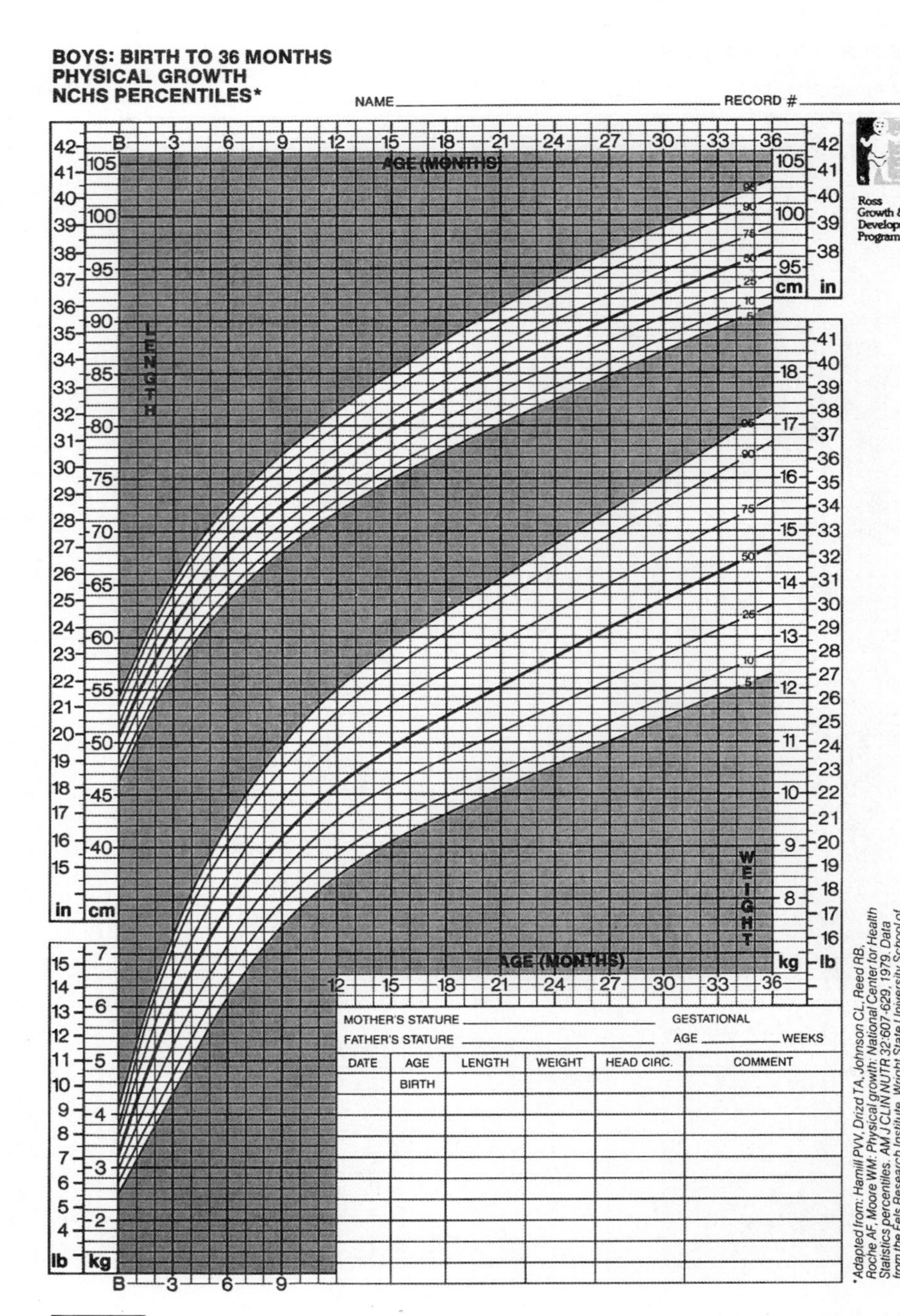

Source: Used with permission of Ross Laboratories, Columbus, OH 43216, from NCHS Growth Charts, © 1982 Ross Laboratories.

Figure 6–5A Nutrition Assessment from Birth to 36 Months: Head Circumference for Age and Weight for Length—Girls

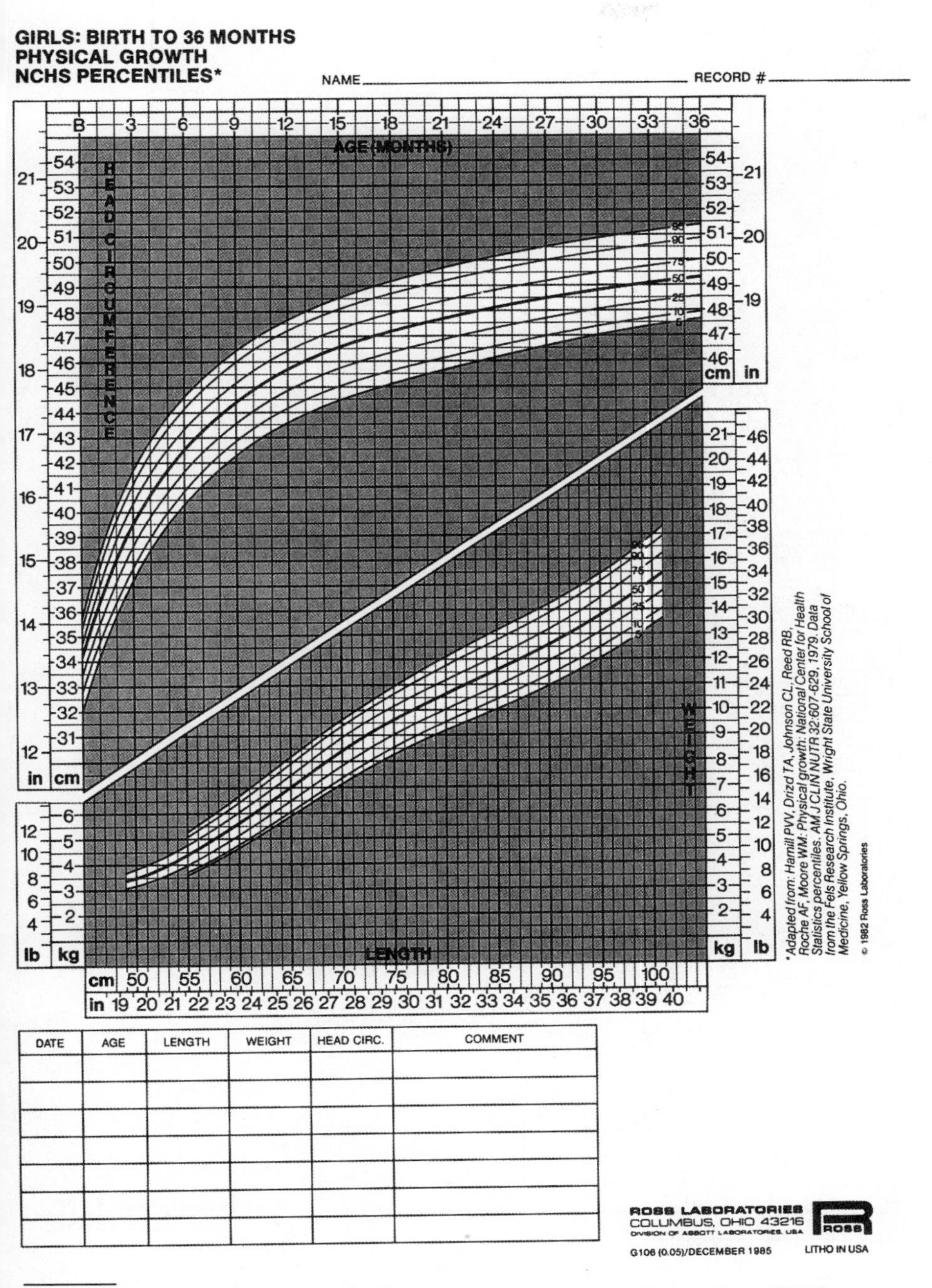

Source: Used with permission of Ross Laboratories, Columbus, OH 43216, from NCHS Growth Charts, © 1982 Ross Laboratories.

Figure 6–5B Nutrition Assessment from Birth to 36 Months: Head Circumference for Age and Weight for Length—Boys

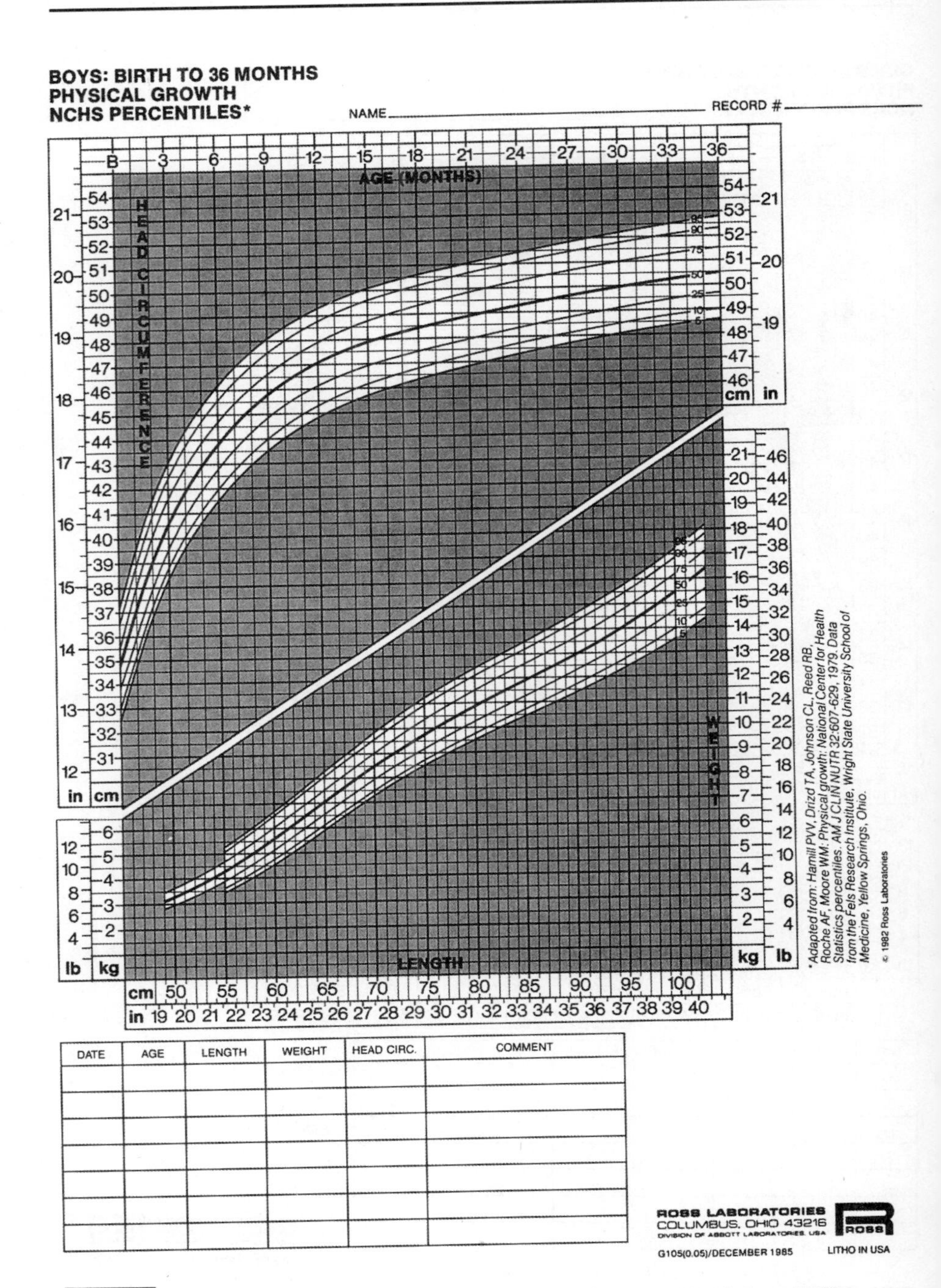

Source: Used with permission of Ross Laboratories, Columbus, OH 43216, from NCHS Growth Charts, © 1982 Ross Laboratories.

Table 6–1 Nutrition Assessment of the Infant, 0 to 11 Months: Standards for Serum Protein and Albumin

Indicator	Deficient	Acceptable
Serum Total Protein (g/dL)[a]	[b]	5.0 or >
Serum Albumin (g/dL)[a]	[b]	2.5 or >

[a]To convert to standard international units (g/L), multiply by 10.
[b]The reference provided no line to demarcate deficiency, but the lower the value, the more likely deficiency is.

Source: Adapted with permission form A. Grant and S. DeHoog, *Nutritional Assessment and Support,* 3rd ed. (Seattle, Wash.: Anne Grant and Susan DeHoog, 1985).

Table 6–2 Nutrition Assessment of the Infant: Standards for Serum Vitamin A

Indicator	Age	Deficient	Acceptable
Serum Vitamin A (μg/dL)[a]	0 to 5 months	<10	20 or >
Serum Vitamin A (μg/dL)[a]	5 months to 1 year	<20	30 or >

[a]To convert μg/dL to standard international units (μmol/L), multiply by 0.03491.

Source: Adapted with permission from A. Grant and S. DeHoog, *Nutritional Assessment and Support,* 3rd ed. (Seattle, Wash.: Anne Grant and Susan DeHoog, 1985).

Vitamin and mineral deficiencies other than those just mentioned are seldom seen in small infants. Their assessment in children is described in the next chapter.

Keep in mind as you read the following section that although PEM and FTT have been described separately, they can overlap.[6] In both cases, the infants are malnourished and appear lethargic but their environments may differ.

The family of an infant with PEM lacks adequate food, income, housing, sanitation, health care, and education. In short, the family lives in poverty. The infant usually suffers an infection and falls into a vicious malnutrition cycle.

The family of an FTT infant lives with financial instability, in overcrowded or substandard housing, and with emotional discord. Often, the father is absent and the mother feels overwhelmed. Described as "sickly," the infant may not even have a medical disease. Such an infant simply does not receive an adequate energy intake. This may be because food intake is limited or because the infant refuses to eat or vomits upon eating. In short, the family lives in strife and the infant lacks nourishment—both nutrients from food and love from parents.

Malnutrition in Infants

Malnutrition is not common in U.S. infants, but it is common worldwide, and world hunger is a subject of concern to all. Malnutrition does occur in the

United States, and the placement of this topic here provides some basis for understanding nutrient needs during infancy, the topic of the section that follows this one.

Malnutrition in early infancy can take many forms, from protein-energy malnutrition (PEM) to effects of individual vitamin and mineral deficiencies and toxicities. These forms are described here, with an emphasis on the kinds of malnutrition most prevalent worldwide and most likely to be observed in the United States.

Protein-Energy Malnutrition

As is true of malnutrition during pregnancy, much of what is known about malnutrition during infancy comes from studies of animals. One question of critical importance asks whether the effects of malnutrition in infancy or early childhood leave a permanent mark, or can be rectified by subsequent good nutrition. The answer may be either, depending on the time and nature of the deprivation. One study examined the effects of supplying unlimited food to animals that had suffered undernutrition at different ages.[7] Rats malnourished during the first three weeks of life grow more slowly than normal, and continue to grow slowly even when provided an unlimited food supply. They do not catch up in growth to their well-nourished peers, and they become small adults. In contrast, rats well nourished until the 9th week of life and then undernourished from the 9th to the 12th weeks grow rapidly when given unlimited food after that. Their mean weight actually surpasses that of rats that are never malnourished. In both instances, the rats experience three weeks of nutrient deprivation, yet one group's growth potential is never achieved. The critical difference is in the timing of the deprivation; the first occurred during a critical period, and the second did not (recall Figure 3–1 in Chapter 3).

Human beings differ from rats in many ways, of course, and one of the ways most relevant to studies of malnutrition is that human beings are more mature at birth. Rats, however, grow faster and mature earlier than human beings do. Rats are weaned and fully independent of their dams at three weeks, whereas human infants are not weaned until after several months and remain dependent for many years.

Because each body system has a particular schedule of growth, the time of deprivation has different effects on each system. The growth of the brain illustrates this concept. The timing of brain cell replication varies among different species, but in general, it occurs most rapidly just prior to, or immediately following, birth. In human beings, brain cell number rises rapidly (as determined by increase in DNA content) until birth and more slowly until six months of age.[8] The brain stops its cell replication earlier than other organs do. Achievement of total adult brain weight occurs at approximately two years of age.

In one study of rats, malnutrition during the period of active brain cell division reduced the total brain cell number finally achieved.[9] Based on this information, researchers have speculated that if the response to malnutrition in the human brain is similar to that in the rat brain, then the critical postnatal period of cell division would be the first six months of life. The researchers then compared well-nourished infants who died in accidents at various times within the first year of life with infants who died of severe malnutrition at the same times. They examined both brain protein, which reflects the total mass of cells in the brain, and brain DNA, which reflects the number of cells. (The ratio of protein to DNA reflects cell size.) Brain cell sizes were normal in both groups, but cell numbers were significantly reduced in the brains of the malnourished infants.[10] Severe early malnutrition, then, can curtail the normal increase in brain cell number in human beings.

Economic and social environment can modify the effects of early malnutrition. The intellectual capacities of young boys who had suffered severe malnutrition (protein deprivation, energy deprivation, or a combination of the two) during their first two years of life were compared with those of a matched group of boys who had never been malnourished. The malnourished boys came from differing social and economic backgrounds, and the effects of those backgrounds were compared. Malnutrition in the context of an unfavorable social and economic environment resulted in intellectual impairment.[11] The effect of malnutrition on intellectual development was negligible, however, when that malnutrition occurred in a generally favorable social and economic environment. This study was unique in that it examined the effects of malnutrition during infancy while controlling for living conditions.

Another study examined the effects of severe malnutrition during infancy on intellectual functions as well as on subsequent physical growth, but did not control for social and economic variables.[12] The study did, however, produce some interesting results based on a 15-year follow-up of 20 severely malnourished infants. Over the 15 years, evidence of gross, irreversible intellectual impairment was collected. This impairment did not disappear, despite improved nutrition and environmental conditions; it appeared to be permanent. Over the years, with improved nutrition, the difference in mean height between the malnourished and well-nourished groups diminished, but the difference in mean head circumference increased. Evidently, a reduced head circumference reflects suboptimal brain growth and tends to be a permanent effect of severe malnutrition.

These studies indicate that early malnutrition can retard both mental development and physical growth. The extent of growth retardation and mental impairment depends on the timing, severity, and duration of the malnutrition. Physical growth may in some cases be less vulnerable than mental development to nutritional insult. Except for those malnourished the most severely at the earliest ages, children have an amazing power to return to their predicted growth curves with adequate nutrition—that is, they can catch up. In fact, they show astonishingly rapid rates of catch-up growth. The rate of catch-up growth is influenced by the nature of the initial deficit and the composition of the rehabilitation diet.[13] Catch-up velocity in height can reach four times the normal velocity for chronological age (see Figure 6–6).[14]

catch-up growth: the acceleration in growth that occurs when a period of growth retardation ends and favorable conditions are restored. It is a self-correcting response that, at best, restores the individual to his or her original growth channel.

Vitamin Deficiencies

Vitamin and mineral deficiencies, identifiable as such, are not commonly reported in infants, although generalized malnutrition is common and, of course, entails vitamin and mineral deficiencies. However, specific vitamin and mineral deficiencies are beginning to develop in infancy whenever the cluster of

Figure 6–6 Catch-up Growth in a Child Following Two Periods of Starvation
The graph shows that growth in height nearly ceased during periods of starvation and that growth accelerated markedly as soon as the child began eating adequate food.

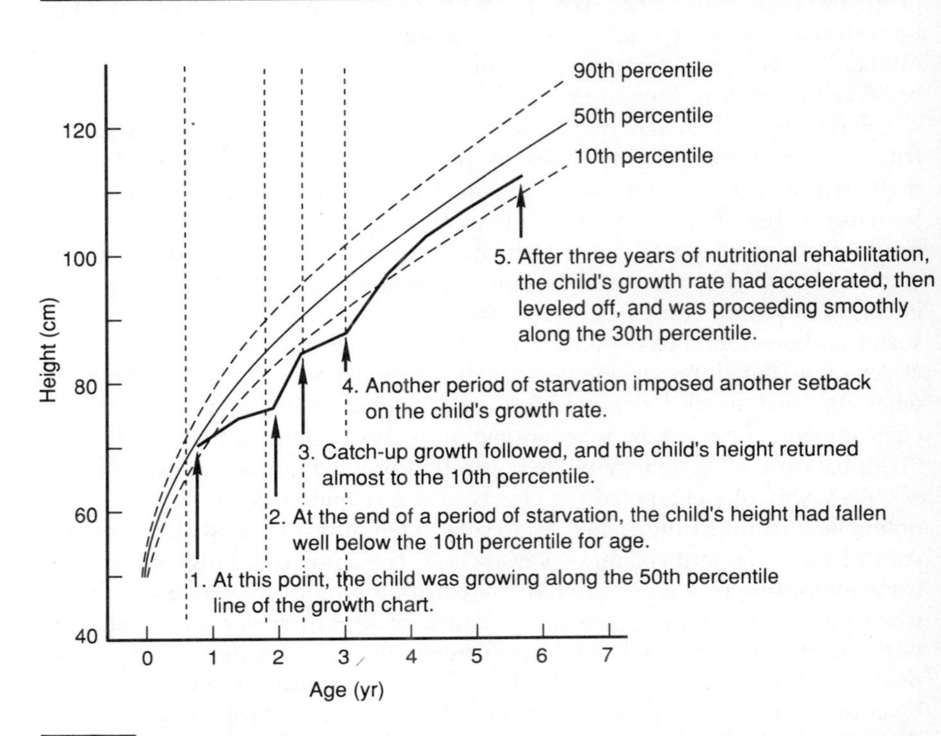

Source: Adapted from A. Prader, J.M. Tanner, and G. A. von Harnack, Catch-up growth following illness or starvation: An example of developmental canalization in man, *Journal of Pediatrics* 62 (1963): 646–659.

environmental factors that predispose people to malnutrition is present—the cluster consisting of poverty, lack of education, disease, and hunger. By the time infants have grown to two or three years or older, all of the vitamin and mineral deficiency diseases that can affect children are recognizable: nutritional blindness, rickets, scurvy, iron-deficiency anemia, and the others described in the next chapter. Typically, they all took root in conditions surrounding the family of the infant.

In the United States and other developed countries, few vitamin A deficiencies begin to develop during infancy. They occur only in infants with impaired fat absorption or in those receiving milk that is not fortified with vitamin A. The earliest clinical sign of vitamin A deficiency is impaired dark adaptation in the retina, which is difficult to detect in infants. As the deficiency progresses, failure to thrive, apathy, anemia, dry skin, and corneal changes appear.

In developing countries, vitamin A deficiencies far more commonly do begin to develop in infancy, though they take an especially terrible toll after infancy. Chapter 7 describes the extent and severity of this worldwide problem.

Vitamin D deficiency may develop in infancy but becomes apparent largely after the first year, when children begin to use their bones to support their bodies. When they sit up, their ribs can become malformed; when they stand, their legs can become bowed. Rickets is discussed further in Chapter 7.

Plasma vitamin E concentration in newborn infants is about one-half that of adults; that of low-birthweight infants is even lower.[15] The smaller the infant, the lower the vitamin E levels at birth; most premature and low-birthweight infants are deficient in vitamin E.[16] Full-term, breastfed infants attain adult values of the vitamin soon after birth, as do infants who consume adequate quantities of commercial infant formula.[17]

Deficiencies of vitamins E and K are not seen in infants except in cases of nutritional or medical mismanagement. The need for vitamin E correlates directly with the amount of polyunsaturated fatty acids (PUFA) in the diet. One study showed that infants fed a commercial infant formula with the standard 12 milligrams of iron per quart and a high PUFA-to-vitamin-E ratio suffered hemolytic anemia.[18] To improve the PUFA-to-vitamin-E ratio, formula manufacturers have reduced the level of PUFA and raised the vitamin E content of their products. As for vitamin K, newborns, and especially premature infants, are susceptible to a deficiency and the resultant occurrence of hemorrhagic disease, because they have sterile intestines and few clotting factors. This can easily be prevented: in many states, the administration of a dose of vitamin K to every newborn is required by law (see "Supplements for Infants," later in this chapter).

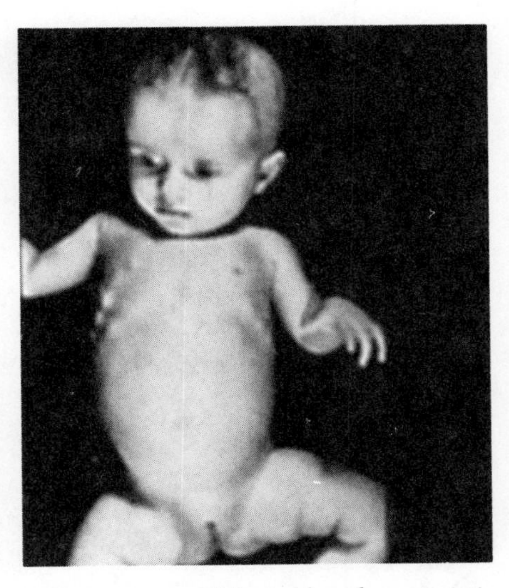

Infant scurvy. This is the characteristic "scorbutic pose," with legs bent and thighs rotated open. The infant's joints are painful, and she will cry if made to move.

A deficiency of vitamin B_{12} is rare in infants, because they are usually born with stores sufficient for the first year of life, and then receive additional amounts from breast milk or infant formula.[19] Thus vitamin B_{12} deficiency does not occur in infants who are breastfed by women with adequate serum vitamin B_{12}. However, vitamin B_{12} deficiency has been observed in infants of women who consume strict vegetarian diets, as described in Focal Point 4.

Vitamin C deficiency is uncommon in the United States, but scurvy does occur in infants fed exclusively cow's milk for the first 6 to 12 months of life. Symptoms of scurvy in infants include anorexia, diarrhea, failure to gain weight, irritability, and increased susceptibility to infection.[20] As the disease worsens, hemorrhages under the skin and failure of spontaneous leg movements occur. Dramatic improvement of symptoms is seen with the administration of 25 milligrams of vitamin C four times a day.

Iron Deficiency

Among mineral deficiencies, iron deficiency stands out as the most common one affecting infants. Zinc deficiency also occurs, but is usually concealed within the broader condition of protein-energy malnutrition. Others, as mentioned, are beginning to develop but first become full-blown and diagnosable, for the most part, in children. These are described in Chapter 7.

Iron deficiency is common in young children throughout the world and ranks as the number one nutrition problem among infants and children in the United States.[21] It becomes most common in children between the ages of six months and three years, for two reasons. Their rapid growth early in this period drains their iron stores; also, cow's milk often becomes a major source of their food energy.[22] Within the past decade or so, the prevalence of anemia among low-income children in the United States has declined, thanks in large part to the positive impact of public health programs.[23]

Evidence of a positive relationship between iron status and behavior has been observed in animals. Lactating mice were fed an iron-deficient diet for 21 days in order to reduce total iron in the brains of their offspring.[24] Then the iron-deficient offspring were placed in an adverse, novel environment. They responded with less alertness and exploratory behavior (reared on their hind legs and stood immobile) less frequently than iron-sufficient mice. Further studies indicate that behavioral effects such as these persist over the long term.

The question of whether iron deficiencies also influence infant and child behaviors has been extensively explored in recent years. Iron-deficient infants and children are known to display symptoms of irritability, anorexia, and poor weight gain.[25] Many studies also reveal that iron deficiencies commonly correlate with behavioral disturbances in infants and that they tend to persist.[26]

Before further discussion of the results of studies in young human beings, a mention of the problems involved in studies of iron deficiency and infant behavior is in order. Conflicting results of studies that seem on the surface to be similar in design are partially attributable to some of these problems. Researchers have yet to agree on the exact criteria by which to define iron deficiency, oftentimes using different biochemical indexes and cutoff points. Researchers relying on only one measure of iron status risk error in the classification of individuals being studied.[27]

A point of clarification is also needed. Iron deficiency and anemia are not one and the same, though they often go hand in hand. Infants may be iron deficient without being anemic. The term *iron deficiency* refers to depleted body iron stores, without regard to the degree of depletion or to the presence of anemia.[28] The term *anemia* refers to the hematologic state resulting from a severe deficiency. In the case of iron-deficiency anemia, body iron stores are severely depleted.

Infant research entails all of the many problems associated with human research, and in addition, specific problems of its own. Researchers encounter difficulty when measuring infants' behaviors and intelligence, because their responses are so limited. Language as a component of intelligence is easily measured later, in an older child, but not in an infant. Despite such shortcomings, tests such as the Bayley Infant Scale of Mental and Motor Development are available to researchers who study infant behavior. The Bayley scale attempts to establish norms for certain behaviors that emerge during infancy, such as reaching for toys or responding to voices, and to assign levels of mental development based on these standards.

Research into the effects of iron deficiency on infant behavior has proceeded despite these pitfalls. In general, the results suggest that iron deficiency and its subsequent resolution do influence mental development. In one study, iron-deficient, anemic infants were treated with either an intramuscular dose of iron or a placebo.[29] Those treated with iron demonstrated significant improvement on the mental portion of the Bayley test. Another study showed significant differences in test performances between iron-deficient, anemic infants and iron-sufficient infants. With iron treatment, the iron-deficient, anemic infants' test scores improved. A study of iron-deficient, nonanemic infants and iron-sufficient infants found no statistically significant differences in mental test performances between the two groups.[30] The iron-deficient infants, however, did show significant test performance improvement after iron treatment. Also, iron-deficient infants, with or without anemia, tend to score lower on mental development tests.[31] In summary:

- Iron-repletion therapy given to iron-deficient infants results in an improvement in performance on the Bayley Infant Scale of Mental and Motor Development.
- The low test performance in iron-deficient infants cannot be attributed to specific mental abilities and skills at this time.
- No evidence exists that points to an association between iron deficiency and delayed motor development.

One more observation of iron-deficient infants deserves mention. It seems that at least one researcher has stated that the most noticeable behavioral characteristic of anemic infants is that they are unhappier (more irritable and apathetic) than nonanemic infants.[32] The researcher concludes that an iron-deficient infant is a happiness-deficient infant.

Now, how to account for the effects of iron deficiency on behavior? The brain contains significant quantities of iron, much of it in iron-dependent enzymes that participate in the synthesis and catabolism of neurotransmitters such as norepinephrine and serotonin.[33] The breakdown of these neurotransmitters to their excretion products requires the iron-dependent enzyme monoamine oxidase. When monoamine oxidase activity is depressed, the rate of breakdown of these neurotransmitters is reduced. In fact, iron-deficient children excrete excessive amounts of norepinephrine in their urine, an abnormality unique to the anemia of iron deficiency and reversible with iron therapy.[34] Thus researchers theorize that the behavioral abnormalities seen in iron-deficient children may be secondary to abnormal amounts of neurotransmitters in the central nervous system caused by the reduced activity of monoamine oxidase.

Because the critical period for brain development in the human infant occurs within the first six months, iron deficiency during early infancy may impair central nervous system development and functioning. Such a deficiency may be evidenced in behavioral disturbances, and the damage may be irreversible.

The length of this discussion on iron accentuates the importance of this nutrient to the infant, especially to the older infant. In view of the fact that iron deficiency is the most common nutrient deficiency of infants and children in the United States, Canada, and developing countries as well, knowledge about this nutrient and the practical application of this knowledge should be a high priority for all health professionals involved in the care of young children. The continuing problem of iron deficiency through childhood, and the related problem of lead poisoning, receive attention in Chapter 7.

Food sources of iron for the infant are discussed in the "Feeding Infants" section of this chapter.

Energy and Nutrient Needs of Infants

That nutrient deficiencies can and do occur in infancy confirms the crucial role adequate nutrition plays in normal growth and development. The rapid growth and metabolism of the infant demand an ample supply of *all* the nutrients, but the energy-yielding nutrients and those vitamins and minerals critical to the growth process have special importance during infancy. This section emphasizes those nutrients that scientific research has thus far deemed to play the largest roles in growth and development.

Before any discussion of nutrient needs for the infant can begin, it must be acknowledged that the information on which nutrient recommendations for infants are based is incomplete at this point. Much remains to be learned, even about protein and energy. Estimates of vitamin and mineral needs for the infant are based on even more uncertainty. Despite this, health care providers must have standards for nutrient adequacy on which timing of introduction of supplemental foods for the infant can be based. Because of large variations in growth rate, activity level, size, metabolic rate, environment, and other factors among infants, it is impossible to establish a single standard applicable to all infants. For this reason, recommendations are often expressed as ranges. The standard (RDA or RNI) used in establishing nutrient allowances for infants is the average amount of nutrients consumed by thriving infants breastfed by healthy, well-nourished mothers. Therefore, much of the discussion of infant nutrient needs that follows is a discussion of the nutrient contents of breast milk.[35]

The selected reference value for mean breast milk volume for the first six months of life is 750 ml/day.

During the first year of life, the infant's need for most nutrients in proportion to body weight is about double that of the adult; for some nutrients it is more than double. Figure 6–7 shows this to be true by comparing a three-month-old infant's needs with those of an adult male. The sections that follow describe the normal infant's needs, nutrient by nutrient, and the bases on which requirements for individual nutrients are established. After them is a section on the premature infant's needs.

Water

One of the most essential nutrients for the infant, as for anyone, is water. The younger the infant, the greater the percentage of body weight is water. The water in an infant's body is easily lost, because compared with the water in an adult's body, a larger percentage of an infant's body water is located in the interstitial (extracellular) and vascular spaces. During early infancy, breast milk or infant formula normally provides enough water for a healthy infant to replace water losses from the skin, lungs, feces, and urine. Under normal conditions, the American Academy of Pediatrics (AAP) recommends little or no water be offered to infants before the introduction of solid foods, because its volume may displace needed milk.[36] However, conditions that cause fluid loss without replacement, such as sweating in hot weather, may justify offering additional water. Adults must remember that infants may cry for thirst as well as hunger. When water is needed, allow infants to drink it until their thirst is quenched.

Dehydration in an infant, such as that caused by diarrhea or vomiting, can be a serious medical emergency. Do not attempt to remediate dehydration without medical advice. All infants need supplemental water once they are eating solid foods. Foods with a high protein content, such as meats and eggs, impose a high renal solute load on the kidneys. Additional water is needed to ease the burden on the kidneys. The infant's ability to concentrate urine does not reach adult capacity until around a year of age. Without supplemental water, the kidneys are stressed, and dehydration becomes a threat. Foods such as fruits and vegetables present the

Figure 6–7 Nutrient Needs of a Three-Month-Old Infant and an Adult Male Compared on the Basis of Body Weight
The man's needs are set at 100% RDA.

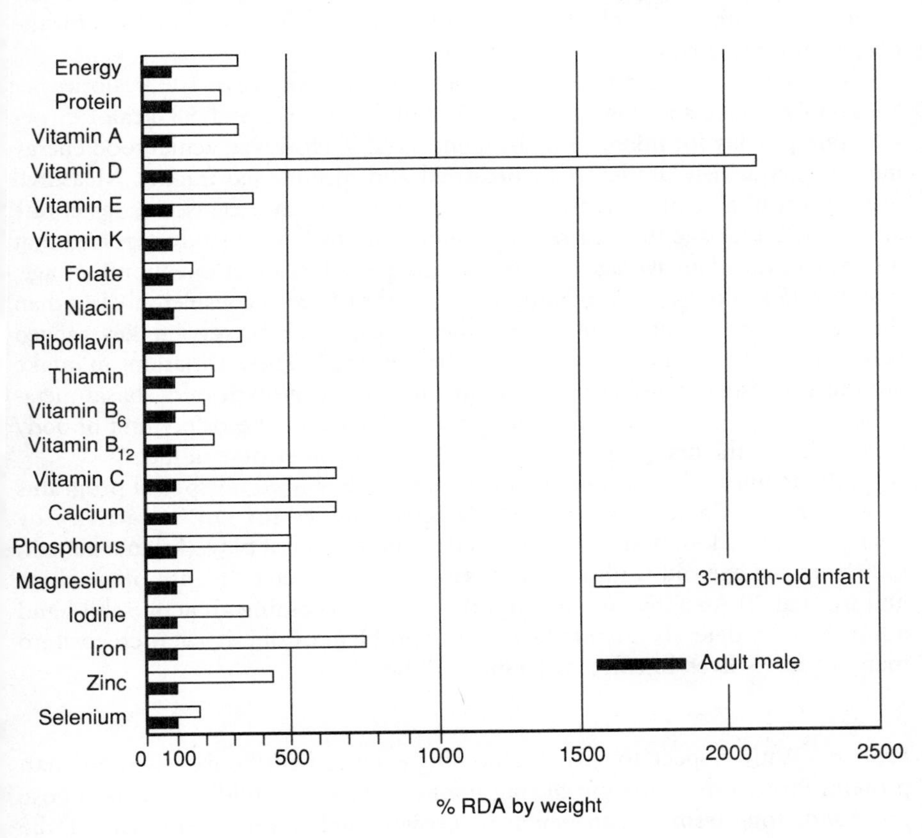

Figure 6—8 Estimated Energy Requirements of Infants during the First Year
Energy needs for the infant during the first year vary according to age and weight. As the infant gets older, the growth rate diminishes, and the activity level increases.

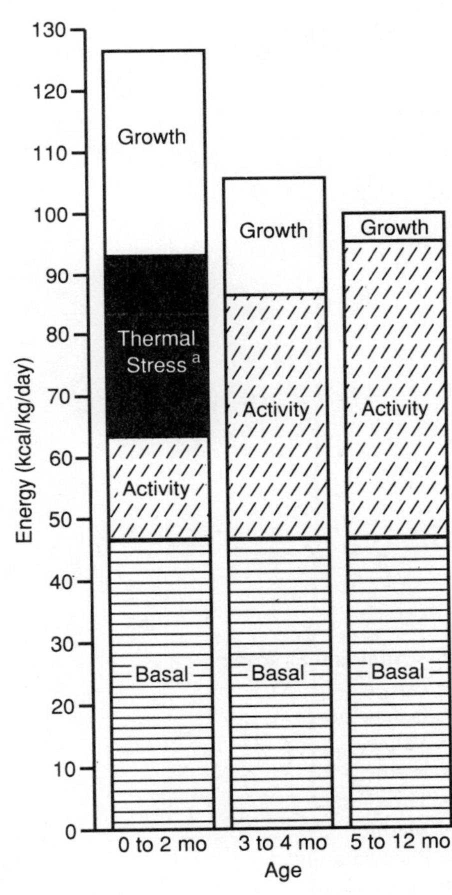

[a]The newborn infant's thermoregulatory system requires extra energy at first to adapt from womb temperature to room temperature.

Source: Adapted from data in E. M. Widdowson, Nutrition in *Scientific Foundation of Pediatrics,* 2nd ed. (Baltimore: University Park Press, 1981), pp. 41–53, as cited by M. Gracey and F. Falkner, *Nutritional Needs and Assessment of Normal Growth* (New York: Raven Press, 1985), pp. 23–40.

kidneys with a low renal solute load. In addition, water satisfies fluid needs without providing kcalories. Many adults today would no doubt be healthier had they learned early to quench their thirst with water.

Energy Intake and Activity

Energy needs per unit of body weight peak during the first year of life due to the infant's rapid basal metabolic rate and growth. This may not be obvious, for a newborn requires only about 650 kcalories per day, whereas most adults require at least 2000 kcalories per day, but to examine the energy needs per unit of body weight is to see the great differences. Infants require about 100 kcalories per kilogram body weight per day; most adults require fewer than 40. A 170-pound adult who tried to eat like an infant, then, would have to ingest over 7000 kcalories a day.

The infant uses energy for two major purposes—to support basal metabolism, including the metabolism necessary for growth, and to support activity. The proportion between these components of energy expenditure changes throughout the first year. Figure 6–8 compares the infant's uses of energy at three different periods within the first year. Note how the infant's daily energy needs for growth are greatest during the early months of life, then decline by the infant's first birthday, as the energy needs for activity rise. Note also that basal metabolic needs do not change; the energy requirement for basal metabolism is about double that of the adult on a per body weight basis for the healthy infant or child.[37]

Energy RDA for infants:
108 kcal/kg (0 to 6 mo).
98 kcal/kg (6 to 12 mo).

The National Research Council recommends an average of 108 kcalories per kilogram per day for infants up to six months of age, and 98 kcalories per kilogram per day for infants 6 to 12 months old.[38] However, actual food energy intakes vary widely among both breastfed and formula-fed infants. Measured food energy intakes of breastfed infants in North America decrease at a faster rate and at an earlier age than current recommendations.[39] One study reported that infants consumed an average of 110 kcalories per kilogram at one month of age, but only 85 kcalories per kilogram by six months of age— considerably less than the current recommendation. In another study, food energy intakes ranged between 52 and 152 kcalories per kilogram per day.[40] These variations in intake are not surprising, considering differences in activity, growth rates, basal metabolic rates, age, and weight. The fact remains that energy needs per unit of body weight during the first year are greater than at any other time of life.

Infants normally are spontaneously active; they need no special programs or equipment. Caretakers can provide opportunities for normal activity by giving them freedom of movement and the stimulation of play. (Infants benefit, too, from the age-old tradition of making sure they have "plenty of sunshine and fresh air.") An active infant is on the way to becoming an active child and adult; activity defends the infant from the problem of obesity that creeps into many people's lives as early as young childhood.

Protein With respect to growth, no single nutrient is more essential than protein. Protein malnutrition during infancy and early childhood can impose profound, long-term impairments on growth and development. All of the body's cells and most of its fluids contain protein; it is the basic building material of the body's tissues. Nine amino acids are essential for growth, and the absence of any one of them can result in stunted growth.[41] Other amino acids are conditionally essential. That is, an amino acid may be essential for an infant under certain conditions, such as illness, prematurity, or inborn errors of metabolism. For example, premature infants and some term infants require cystine, tyrosine, and taurine.[42]

Essential amino acids:

- Leucine.
- Isoleucine.
- Valine.
- Lysine.
- Threonine.
- Methionine.
- Phenylalanine.
- Tryptophan.
- Histidine.

Conditionally essential amino acids:

- Cystine.
- Tyrosine.
- Taurine.

Dietary protein is needed daily to replace daily losses of amino acids. Knowledge of the protein needs of infants and children has come from dietary surveys, studies of nitrogen balance and creatinine excretion, and measurements of growth at different ages. An adequate protein intake is defined as one that supplies all of the essential amino acids in amounts sufficient for maintenance and growth; this definition considers protein quality as well as quantity.[43] The protein foodstuffs in the normal one-year-old infant's diet in North America consist of human milk or infant formula; meat, fish, and eggs; and some foods of grain and vegetable origin. These foods supply all of the essential amino acids in reasonably high concentrations relative to their food energy contents.

Focal Point 2 discusses the unique nutrient needs of infants with inborn errors of metabolism.

Dietary allowances for protein for infants are the same as the amounts of protein provided by the quantities of milk that ensure adequate rates of growth.[44] Infants consuming 150 to 200 milliliters of breast milk per kilogram body weight per day receive adequate protein.

150 to 200 ml breast milk contains between 1.7 and 2.2 g protein.

Protein recommendations are designed for healthy, full-term infants. Infection, premature birth, illness, and genetic factors all increase protein requirements. Protein requirements increase with an insufficient food energy intake, which results in the use of dietary protein for energy.

100 ml is approximately equal to 3½ oz or just under ½ c.

Protein contributes a small percentage of the energy in breast milk—about 6 to 7 percent. Since breast milk in the quantity normally consumed provides approximately 2 grams protein per kilogram body weight per day for the average 7½-pound infant, the AAP recommends that infant formulas provide a comparable amount: a minimum of 1.8 grams protein (that is, 7.2 kcalories) per 100 kcalories, with a protein efficiency ratio equal to or greater than that of casein.

Protein RDA for infants:
2.2 g/kg (0 to 6 mo).
1.6 g/kg (6 to 12 mo).

Protein in breast milk:
- 6% of kcal.
- 1.1 g/100 ml.

Infants consume approximately 750 ml/day. Therefore, protein intake = 8.25 g protein/day.

Excess dietary protein can cause problems, especially in the small infant, producing a buildup of amino acids in the blood. This stresses the kidneys and liver, which have to metabolize and excrete these amino acids. Signs of protein overload, including acidosis, dehydration, diarrhea, elevated blood ammonia, elevated blood urea, and fever, have been observed in infants with protein intakes greater than 6 grams per kilogram per day.[45]

750 ml is approximately equal to 25 oz or about 3 c.

protein efficiency ratio (PER): a measure of protein quality assessed by determining how well a given protein supports weight gain in laboratory animals.

Carbohydrate Carbohydrate provides a readily available source of energy for all human beings and, in doing so, spares protein from being used for energy. The brain relies almost exclusively on carbohydrate for its energy. The percentages of carbohydrate kcalories in the diets of infants range between 35 and 55 percent. Human milk provides about 39 percent of its energy from carbohydrate; infant formula, about 42 percent. Figure 6–9 illustrates the contribution each energy nutrient makes toward total kcalories in both human milk and standard infant formula.

In most instances, lactose is by far the major carbohydrate consumed by infants. Lactose makes up about 90 percent of the total carbohydrate in human milk, and contributes most or all of the carbohydrate in standard infant formulas.[46]

Figure 6–9 Percentages of Energy-Yielding Nutrients in Human Milk and Infant Formula
The proportions of energy-yielding nutrients in human breast milk and formula differ slightly.

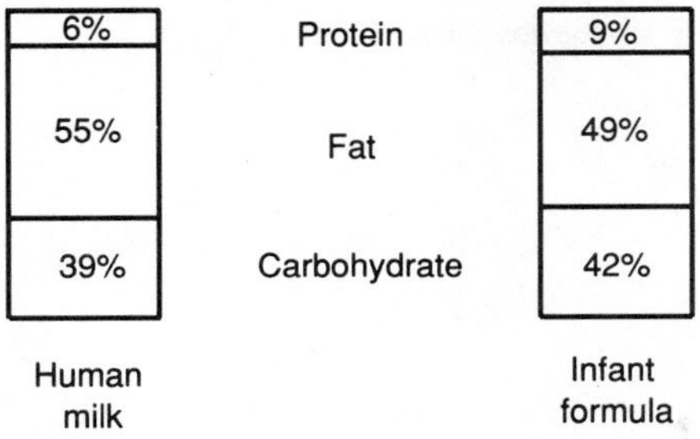

Source: Ross Laboratories, Columbus, Ohio, January 1979.

Fat Fat is important in developing the central nervous system and maintaining body temperature. Fat in milk imparts flavor and satiety and serves as a vehicle for the absorption of the fat-soluble vitamins. Fat provides about 55 percent of the kcalories in human milk, and this percentage is considered ideal for infants. As Chapter 5 mentioned, approximately 98 percent of the fat is in the form of triglycerides; the remainder is phospholipids, cholesterol, and free fatty acids.

Human milk is a relatively rich source of cholesterol. Formula manufacturers replace cow's milk fat with vegetable oils, thus lowering the cholesterol content of infant formulas. This raises the question of how much cholesterol infants need. Clinical studies show no adverse effects of low-cholesterol formulas, but further research is needed regarding the long-term effects.

Linoleic acid, which is as essential for infants as it is for adults, contributes about 7 percent of the total kcalories in human milk; in infant formula, about 10 percent. Infants develop deficiency symptoms, such as dermatitis and failure to thrive, when linoleic acid provides less than 0.1 percent of the total daily

Cholesterol in human milk: 20 to 30 mg/100 ml.

Breastfed infants consume approximately 200 mg cholesterol/day.

energy intake, but such deficiencies are rare.[47] Linoleic acid deficiencies occur when premature infants with low body fat stores are fed fat-deficient diets.[48]

Another essential fatty acid, linolenic acid, is a member of the omega-3 fatty acid family. Linolenic acid has been the longest known of the omega-3 fatty acids, but two others, eicosapentaenoic acid (EPA) and docosahexaenoic acid (DHA), have gained worldwide attention in the last decade or so. The emphasis in research on the roles of omega-3 fatty acids has been on the prevention of heart disease, but interest is now turning to their other roles, including those in growth and development. The retina and the brain are rich in DHA. About half of the DHA accumulates in the brain before birth and half, after birth, emphasizing the importance of lipids during pregnancy and lactation.[49]

Human milk contains linoleic acid, linolenic acid, EPA, and DHA; infant formula contains significant quantities of only linoleic acid.[50] Some researchers suggest that including the other fatty acids in infant formula may be desirable for optimum development. Because breast-fed infants have more DHA in their bodies than formula-fed infants, research is under way to determine the physiological significance of this difference.

Vitamins

The extraordinary growth of an infant during the first year of life may not be obvious, because each day brings only gradual changes. Yet internally, the metabolic machinery is fast at work creating new body parts—with the assistance of the vitamins.

retinol: the active form of vitamin A found in milk.

Vitamin A RDA for infants:
375 μg/day (0 to 12 mo).

Vitamin A The average retinol content of human milk is about 50 micrograms per 100 milliliters. Assuming a milk intake of 750 milliliters per day, the average intake of vitamin A for the infant is 375 micrograms per day.[51]

Vitamin D RDA for infants:
7.5 μg/day (0 to 6 mo).
10 μg/day (6 to 12 mo).

Vitamin D Vitamin D is similar in structure, metabolism, and mechanism of action to steroid hormones, and the active form to which it is converted in the body is actually a hormone.[52] Nevertheless, for convenience and historical reasons, it is still called vitamin D.[53] A major function of vitamin D is to promote calcium absorption and transport.

From recent research, breast milk appears unsuitable as a standard from which to estimate the vitamin D needs of infants; it provides less than they need, and the difference is made up by sunlight. In case breastfed infants are deprived of sufficient sunlight, they require supplemental vitamin D (see the section on supplements for infants, later in this chapter).[54]

The question has been raised whether breast milk may contain more vitamin D than early analyses have shown. Researchers have traditionally used the lipid fraction of breast milk to determine its vitamin D content. An unconfirmed report claims that the protein whey fraction of breast milk revealed vitamin D sulfate activity—that is, that an active vitamin D compound existed in the water portion of the milk.[55] More recent reports indicate, however, that vitamin D sulfate levels in human milk are indeed negligible, confirming that the milk truly does not supply enough vitamin D to meet infant needs.[56]

Vitamin E Human breast milk contains 1.3 to 3.3 milligrams vitamin E per liter and is assumed to provide an adequate intake for nursing infants. Infant formula contains about 10 milligrams vitamin E per liter.

Vitamin E RDA for infants:
3 mg/day (0 to 6 mo).
4 mg/day (6 to 12 mo).

Vitamin K The newborn infant presents a unique case when it comes to vitamin K nutrition.[57] At birth, the intestinal tract of the newborn is sterile and thus lacks vitamin K–producing bacteria. At the same time, plasma prothrombin concentrations fall, to reduce the likelihood of fatal blood clotting during the stress of birth.[58] Prothrombin synthesis depends on vitamin K, and prothrombin concentrations climb back to adult levels as milk is consumed and the intestinal tract gradually develops a population of vitamin K–producing bacteria. The AAP therefore recommends that a single dose of vitamin K be given to infants at birth; see "Supplements for Infants," later in this chapter.

Vitamin K RDA for infants:
5 μg/day (0 to 6 mo).
10 μg/day (6 to 12 mo).

The B vitamins If infants consume adequate amounts of infant formula or breast milk from healthy women, then their daily requirements for the B vitamins will be met. Normal-weight infants may experience deficiencies if requirements increase—for example, due to severe injury or malabsorption. Premature infants may also experience deficiencies, because they consume small quantities of breast milk or formula; they may not receive adequate amounts of the B vitamins without supplementation.

Two of the B vitamins, vitamin B_{12} and folate, deserve special mention for their crucial roles in growth and development. Vitamin B_{12} is required to enable folate to perform its role in DNA synthesis and thus to support growth. The decreased DNA synthesis that results from vitamin B_{12} deficiency causes all replicating cells in the body to have a megaloblastic (giant cell) appearance. This is usually observed in blood cells and is called megaloblastic anemia.[59]

Vitamin B_{12} RDA for infants:
0.3 μg/day (0 to 6 mo).
0.5 μg/day (6 to 12 mo).

Folate is required for the synthesis of DNA and is dependent on the presence of vitamin B_{12} to properly fulfill its role in this process. Folate deficiency is the most common cause of megaloblastic anemia in infants and children. At birth, an infant's serum folate is three times the maternal concentration, but infant stores are limited, and the rapid growth of the infant depletes these stores. By two weeks of age, serum folate falls below adult values and stays there for several months.[60]

Folate RDA for infants:
25 μg/day (0 to 6 mo).
35 μg/day (6 to 12 mo).

Daily outputs of vitamin B_{12} in breast milk range from 0.2 to 0.8 micrograms.[61] The folate needs of infants are adequately met by breast milk or infant formula, but not by goat's milk, which is notoriously low in its folate content. Human milk and cow's milk contain about 50 micrograms of folate per liter.

Vitamin C Based on an average daily milk output of 750 milliliters, the daily vitamin C output in human milk ranges between 23 and 75 milligrams, depending on the dietary intake of the mother.[62] Infant formulas in the United States contain 55 milligrams of vitamin C per liter.[63]

Vitamin C RDA for infants:
30 mg/day (0 to 6 mo).
35 mg/day (6 to 12 mo).

Minerals

The minerals are as actively involved in the growth and development of infants as are the vitamins. All are essential, and each serves a unique function. Those discussed here are particularly important to infant nutrition.

Calcium Not surprisingly, calcium needs are great during infancy, a period of rapid skeletal growth. During infancy, the calcium content of the body increases faster with respect to body size than at any other time in the life span. By nature's design, milk is both the best source of calcium and the main component of the infant's diet. Human breast milk contains about 300 milligrams of easily absorbed calcium per liter (two-thirds absorption efficiency, or 200 milligrams).[64] Standard infant formula contains more calcium (400 to 500 milligrams per liter), but it is less readily absorbed (less than one-half retention).[65] The full-term infant's calcium needs are fully met by either breastfeeding or formula feeding.

Calcium RDA for infants:
400 mg/day (0 to 6 mo).
600 mg/day (6 to 12 mo).

The idea persists that excessive dietary phosphorus impairs calcium absorption.[66] To the contrary, the ratio of calcium to phosphorus in the diet appears not to play as important a role in calcium absorption as once thought.[67] The calcium-to-phosphorus ratio of human milk is 2:1; the ratio of calcium to phosphorus in commercial infant formulas ranges between 1.3:1 and 1.5:1, amounts compatible with the AAP recommendations (shown in Table 5–4 of Chapter 5).[68]

Sodium (estimated minimum for infants):
120 mg/day (0 to 6 mo).
200 mg/day (6 to 12 mo).

Sodium Breast milk provides 7 milliequivalents of sodium per liter, while infant formulas contain between 7 and 11 milliequivalents per liter.[69]* This provides an infant with an average intake of 120 milligrams per day. Sodium intakes of infants who are breastfed or fed formula are within the recommended range.

Potassium (estimated minimum for infants):
500 mg/day (0 to 6 mo).
700 mg/day (6 to 12 mo).

Potassium The infant's potassium requirement for growth exceeds that of sodium, as reflected in the sodium-to-potassium ratio of 0.6:1 in breast milk.[70] Infant formulas contain slightly more than the potassium content of breast milk. Consistent with the infant's limited renal concentrating capacity, upper limits for the electrolyte content of infant formula have been set by the AAP. If the blood concentration of potassium rises three to four times above normal, the heart stops beating. Years ago, a breastfed infant died when given potassium supplements for colic.

Iron The rapid growth of infants and children imposes large iron needs on them. Many infants' diets and most toddlers' diets contain a marginal supply

*A milliequivalent is the amount of a substance that contains the same number of charges as 1 milligram of hydrogen. The number of milliequivalents is a more useful measure than milligrams or grams when considering ions, because what is of interest is the number of charges present. If two solutions contain the same number of milliequivalents, they contain the same number of charges, and exert the same amount of osmotic pressure.

of iron. The combination of high need and low intake makes deficiency a likely consequence, and iron deficiencies are common among infants and children, as a previous section attests.

During gestation, the fetus receives high priority when it comes to available iron, often at the expense of the mother. In fact, iron stores of newborn infants, based on serum ferritin concentrations, show negligible differences between infants of iron-deficient, anemic mothers and infants of iron-sufficient mothers.[71] The full-term, newborn infant arrives endowed with enough iron to last for at least the first four to six months, assuming that breast milk or iron-fortified formula also is provided.

The infant continues to receive high priority for available iron during lactation. Poor iron status of the mother does not reduce the iron concentration of her breast milk.[72] Human breast milk contains relatively small amounts of iron (0.5 milligrams per liter), but this iron has a high bioavailability (50 percent).[73] In contrast, iron-fortified infant formula contains about 12 milligrams of iron per liter, but this iron has low bioavailability (about 4 percent). At about four to six months, though, the infant begins to need still more iron than stores plus breast milk or iron-fortified formula can provide. Iron deficiency becomes possible, and in fact likely, unless iron-fortified cereal or iron supplements are introduced (see later sections).[74]

Iron RDA for infants:
6 mg/day (0 to 6 mo).
10 mg/day (6 to 12 mo).

Zinc Zinc requirements of infants are not well defined, but a deficiency of this nutrient causes impaired growth. It appears that growth velocity is the main determinant of infant zinc requirements.[75] Thus zinc needs are highest during early infancy and decline with advancing age. The zinc content of human breast milk declines as lactation progresses, ranging from 1.5 milligrams per liter in the first half year to 1.0 milligram per liter during the second half year.[76] Despite this marked decline in breast milk zinc, breastfed infants' zinc intakes remain adequate, as long as maternal zinc intakes are adequate.[77] Availability of zinc from cow's milk is lower than from human milk, and even lower from soy-based formulas. Thus standard infant formulas are supplemented with zinc and soy-based formulas, with even more.[78]

Zinc RDA for infants:
5 mg/day (0 to 12 mo).

Supplements for Infants

The AAP recommends that a single dose of vitamin K be given to infants at birth.[79] In many states, this preventive dose of vitamin K is required by law.

In addition, pediatricians routinely prescribe supplements containing vitamin D, iron, and fluoride. Table 6–3 shows the timing of introductions and provides options.

Vitamin K:
Intramuscular dose: 0.5 to 1.0 mg.
Oral dose: 1.0 to 2.0 mg.

The nutrient needs discussed to this point apply to full-term, healthy infants. Preterm, low-birthweight infants present special nutrition concerns.

Nutrient Needs of Preterm Infants

The terms *preterm* and *premature* were introduced on p. 31. They are used interchangeably to refer to a shortened gestation period, and they imply

Table 6–3 Supplements for Full-Term Infants

	Vitamin D[a]	Iron[b]	Fluoride[c]
Breastfed infants:			
Birth to 6 mo	√		√
6 mo to 1 yr	√	√	√
Formula-fed infants:			
Birth to 6 mo			√
6 mo to 1 yr		√	√

[a]Vitamin D supplements are recommended for breastfed infants for as long as breast milk is the major milk the infant consumes. Once infant formula or vitamin D–fortified cow's milk replaces breast milk in the infant's diet, vitamin D supplements are no longer needed.
[b]Most pediatricians recommend the use of iron-fortified formula from birth for formula-fed infants, although some infants are fed noniron-fortified formula for the first few months. The Committee on Nutrition of the American Academy of Pediatrics recommends the use of iron-fortified infant formula by four months of age for formula-fed infants. Iron-fortified infant cereal is a reliable source of iron for both breastfed and formula-fed infants during the second half of the first year. The iron-fortified infant cereal contains about 0.45 mg iron per 100 g or ½ c.
[c]The use of fluoride supplements for infants less than six months of age is controversial. The Committee on Nutrition of the American Academy of Pediatrics recommends initiating fluoride supplements for breastfed infants, formula-fed infants who receive ready-to-use formulas (these are prepared with water low in fluoride), or those who receive formula mixed with water that contains little or no fluoride (less than 0.3 ppm) shortly after birth. (For safety and consistency, infant formulas contain no fluoride.) The committee acknowledges that fluoride supplementation could be initiated at six months of age, however.

Source: Adapted from American Academy of Pediatrics, Committee on Nutrition, Vitamin and mineral supplement needs of normal children in the United States, in *Pediatric Nutrition Handbook,* 2nd ed., eds. G. B. Forbes and C. W. Woodruff (Elk Grove Village, Ill.: American Academy of Pediatrics, 1985), pp. 37–48.

incomplete fetal development, or immaturity, of many body systems. The preterm infant faces physical independence before the growth of some of its organs and body tissues is complete. The rate of weight gain in the fetus is greater during the last trimester of gestation than at any other time in the life span.[80] Therefore, a preterm infant is most often a low-birthweight infant as well. With a premature birth, the infant is forced to endure the time of maximal growth without the continued nutritional support of the placenta.

The last trimester of gestation is also a time of building nutrient stores. Being born with limited nutrient stores intensifies the precarious situation for the infant. Further compromising the nutrition status of preterm infants is their metabolic immaturity. Nutrient absorption, especially that of fat and calcium, from the immature gastrointestinal tract is impaired.[81] The immature brain and liver are vulnerable to damage from high plasma concentrations of certain amino acids, and immature renal function makes these high concentrations likely. In short, preterm, low-birthweight infants are perfect candidates for nutrient imbalances. For these reasons, premature infants require special dietary and medical attention.

Few guidelines are available concerning the preterm infant's nutrient requirements. Authorities on infants' and children's nutrition disagree as to how fast the preterm infant should grow and what the body composition should be. In one study, formula-fed preterm infants accumulated more fat than breastfed preterm infants.[82] Whether this is of significance in terms of future growth is not known.

What *is* known about low-birthweight infants, and especially very-low-birthweight infants, is that nutrient deficiencies appear during the first days of life. Deficiencies of vitamin E, folate, other B vitamins, iron, and calcium are among them.[83]

For the premature infant, the attainment of adequate vitamin E status is difficult. Many of these infants are so small that ingesting adequate breast milk or formula orally is impossible at first. In addition, their immature digestive systems often fail to absorb fat efficiently, and some do not respond to the administration of a water-miscible form of vitamin E, either. They may require supplementation by way of parenteral solutions until oral feeding is possible.[84] Vitamin E deficiency in premature infants is associated with hemolytic anemia. Vitamin E deficiency in older infants, children, and adults produces a shortening of the red blood cell life span in the absence of anemia.[85]

parenteral nutrition: the delivery of nutrients through a vein, bypassing the intestines. In contrast, **enteral** refers to delivering nutrients into the intestines.
para = opposite
enteron = intestine

One group of researchers has questioned whether the low plasma vitamin E of low-birthweight infants represents a normal range for these infants, or whether a true deficiency state exists.[86] Based on several measures of vitamin E status, they found that the antioxidant protective role of vitamin E is best attained at plasma concentrations close to adult values. They concluded, therefore, that true vitamin E deficiency exists for most low-birthweight infants at birth.

hemolytic anemia: anemia characterized by breakage of the red blood cells, with resultant low hemoglobin levels and an increased production of immature red blood cells.
heme = blood
lysis = to break

Premature, low-birthweight infants have small, readily depleted reserves of folate. At birth, plasma folate in both premature and full-term infants is higher than in adults, but only temporarily. Plasma folate concentrations decline in both groups of infants, but in the premature infants, the decline is more rapid. Infants with the lowest birthweights experience the greatest declines.[87] A maintenance dose of 50 to 100 micrograms per day is recommended to prevent anemia.[88] As mentioned earlier, premature infants may also experience deficiencies of other B vitamins, because they consume small quantities of breast milk or formula; they may require supplementation.

Low-birthweight infants have a greater requirement for dietary iron than do full-term normal-weight infants.[89] Low-birthweight infants have less stored iron on a weight basis to begin with, and their more rapid growth rate exhausts their iron stores sooner. For these reasons, breastfed low-birthweight infants are prescribed ferrous sulfate drops at a dose of 2 to 3 milligrams per kilogram body weight not to exceed 15 milligrams per day, beginning at two months of age.[90] Preterm infants who receive iron-fortified formula do not require supplementation.

Infants who are born eight to ten weeks prior to term have acquired only about 30 percent as much calcium as full-term infants, so their calcium requirements are high.[91] They miss out on the normal mineralization of bone that takes place in the uterus during the last trimester of gestation. Their inadequate bone mineralization may result in the metabolic bone disease referred to as osteopenia, or the rickets of prematurity. The probability of this condition's occurring varies directly with the infant's weight; the smaller the infant, the greater the risk of bone disease.

osteopenia: a metabolic bone disease common in preterm infants; also called **rickets of prematurity**.

The calcium and phosphorus needs of the preterm infant are not fully met by human breast milk, due to the milk's low content of the minerals, the poor absorptive capacity of the immature digestive tract, and the preterm infant's small intake. Formulas designed for the preterm infant contain higher concentrations of calcium and phosphorus than standard formulas. Preterm infants receive these formulas, or human milk supplemented with calcium and phosphorus, to ensure adequate intakes of these minerals.

Disagreement abounds regarding the best method of feeding the low-birthweight infant. Once the infant can accept enteral feedings, the question at issue is which type of feeding best serves the nutrition needs of the infant: breast milk or formula. Each offers unique advantages, and in fact, they are often used in combination.

Breast milk does provide protection against infection, but its primary advantage lies in its composition. The composition of breast milk, with its readily digestible fat, low renal solute load, and unique protein profile, makes it particularly suitable to the immature intestine, kidneys, and liver. Despite the obvious advantages, breast milk, even the premature infant's own mother's milk, is sometimes less than ideal. Formulas designed for preterm infants offer the advantages of known composition and precise measurement of intake.

Feeding Infants

Infant nutrition and feeding recommendations are much like the infant: constantly changing—in a word, dynamic. Historically, the timing of weaning to cow's milk and of introducing solid foods to the infant's diet has ranged from a few weeks to three years, and is currently recommended between four and six months of age.[92] On the surface, ever-changing infant feeding recommendations and practices are frustrating and confusing for parents and health care providers alike, but their significance goes much deeper. Such changes occur because new knowledge is being gained about the crucial importance of nutrition during infancy. As discussed earlier, undernutrition in the first year of life can have devastating, long-lasting effects. Infancy sets the stage for eating habits that affect nutrition status and health for a lifetime.

All newborns begin life receiving breast milk, formula, or a combination as their only source of food. This continues to be the sole food source until solid foods are introduced. Even then, breast milk and formula still contribute significantly to infants' diets, providing about half of the day's energy intake. At some point, cow's milk can begin to replace breast milk or formula, but not too soon.

Introducing Cow's Milk

The timing of the introduction of whole cow's milk into the infant's diet remains a controversial issue. Many pediatricians recommend the use of breast milk or iron-fortified formula throughout the first year. The AAP currently recommends the introduction of whole cow's milk at any time after six months, or as soon as at least one-third of the infant's energy intake is from a balanced mixture of cereal, fruits, vegetables, and other foods.[93]

At no time before six months should infants be given whole cow's milk. As mentioned earlier, in some infants—particularly those younger than six months of age—its consumption is associated with intestinal bleeding and iron deficiency.[94] Also, whole cow's milk is a poor source of iron. Furthermore, the young infant's stomach absorbs whole proteins, and those in unmodified cow's milk can cause allergies if absorbed whole.[95] Because milk makes significant nutrient contributions to infants' diets and because a limited number of nutritionally comparable foods are available, it is especially important to avoid cow's milk allergy if at all possible.

Introducing Solid Foods

The high nutrient needs of infancy must be met by breast milk or formula only, at first, and then by a limited diet to which foods are gradually added. Infants only gradually develop the ability to chew, swallow, and digest the wide variety of foods available to adults. The caretaker's selection of appropriate foods at the appropriate stages of development is prerequisite to the infant's optimal growth and health.

When to begin Few issues of infant feeding have been debated more than the question of when it is most appropriate to introduce solid foods into the infant's diet. As is true with all aspects of infant feeding, recommendations for the timing of the introduction of solid foods have changed repeatedly over the years. At the beginning of this century, solid foods were not introduced until ten months of age or older.[96] In the 1950s, the AAP suggested that iron-containing foods be introduced during the third month of life.[97] In the 1970s, most infants received solid foods by six weeks of age.[98] Currently, the AAP recommends the introduction of solid foods between four and six months of age.

A term sometimes used to describe any nonmilk food given to an infant is the German word **beikost** (BY-cost).

The choice of this later time to introduce solid foods accompanied the resurgence of breastfeeding and was probably in the interest of postponing weaning. Studies show a difference in the timing of solid food introduction between breastfed and formula-fed infants: the latter receive solid foods earlier.[99] Survey data indicate that over half of formula-fed infants receive solid foods between two and three months of age, compared with one-fourth of breastfed infants.[100]

In practice, the timing of the introduction of solid foods to infant diets should not depend on whether the infant is breastfed or formula fed, or even on the infant's chronological age. It should depend on the individual infant's nutrient needs and developmental readiness. These factors vary from infant to infant due to differences in growth rates, activity, and environmental conditions.

The main purpose of introducing solid foods to infants is to provide nutrients that are no longer supplied adequately by breast milk or formula alone. Solid foods provide an alternative source of nutrients when the needs of the infant exceed what can be provided by a volume of fluid that the infant can easily consume. However, the foods chosen must be foods that the infant is developmentally capable of handling and metabolizing. Just as it is unrealistic to expect a one-year-old to read and write or ride a bicycle, so it is unrealistic to expect an infant to handle adult foods, or even baby foods, before the physiological capacity exists.

Not only the nutrient needs but also the energy needs of some infants become too great, by four to six months of age, for breast milk or formula alone to provide.[101] At about four months of age for some infants, and no later than six months of age for all infants, foods other than breast milk or formula should be making significant energy contributions to the infant diet. It is important, though, that foods not displace milk, for milk remains the infant's most important source of nutrients. Foods should enter the diet only as supplemental foods given in addition to breast milk or formula—not as replacements for either.

Research findings substantiate the statement that for infants less than four months of age, solid food represents a replacement of milk rather than an addition to the diet. A study found that when breastfed infants less than four months of age were given solid foods, nursing frequency declined; when solid foods were introduced later than four months of age, nursing frequency either remained the same or increased.[102] The impact of an early introduction of solid foods was seen in lower growth rates of the infants who received solid foods early. The infant's immature digestive capacities explain this finding: solid foods limited the amount of milk they could consume, and thereby, their growth. Early introduction of solid foods may also interfere with lactation, rather than appropriately supplementing it. These findings strengthen the argument for delaying solids until sometime between four and six months of age.

Breast milk and formula thus remain the most efficient vehicles for delivering both energy and most nutrients throughout the first year, but by the end of the year, infants need to have become accustomed to eating foods as well. As their energy needs increase, their intakes of nonmilk foods can increase steadily. Breast milk or formula consumption can decrease even more gradually, so that throughout the first year of life, milk foods continue to be their major energy and nutrient source.

From the foregoing discussion, it is obvious that a variety of factors supports the recommendation to delay the timing of introduction of solid foods to infants until four to six months. The GI tract's development further explains the basis for this recommendation.

Readiness of the GI tract The exact time to introduce solid foods depends on the developmental readiness of the child. The type and amount of food an infant can swallow, digest, and absorb at any given time varies according to the infant's neuromuscular maturity and control of the GI tract.

The full-term, newborn infant is capable of sucking and swallowing at birth, although somewhat inefficiently. Within a few days, however, the infant is better able to coordinate sucking, swallowing, and breathing, so this process becomes more efficient. Until about four months of age, all solid objects entering the infant's mouth are pushed out through it by the action of the tongue; this is how the infant manages to obtain milk from the breast or bottle: sucking it in, drawing the milk out of it, and then pushing it out. This action is called the extrusion reflex; it also prevents foods from being swallowed for the first four to six months of age. Only after the extrusion reflex disappears is the infant able to push food to the back of the mouth for swallowing. Food in the back of the mouth touches the soft palate of the roof of the mouth. This stimulus triggers the swallowing reflex. This is why the nipple has to be well inside the infant's mouth when breastfeeding, and why a spoon has to be well inside when spoon feeding.

extrusion reflex: the reflex that causes an infant to push food out, rather than swallowing it.

Spoon feeding of solid foods promotes proper swallowing movements. In contrast, the use of an infant feeder, which allows an infant to suck pureed food from a bottlelike container, discourages the maturation of swallowing movements. Sometime after about four months of age, the infant is able to sit alone and has greater control of head movement than previously. Thus the infant can sit in a high chair and is capable of giving signs of satiety, such as turning the head away when satisfied.

The main role of the stomach is to physically and chemically prepare foods for digestion. As Figure 6–10 illustrates, the infant's stomach is much smaller, shaped differently, and functionally immature compared with the older child's stomach. During the newborn period, foods move more slowly and are mixed less effectively in the stomach than they are in later infancy. The small capacity and prolonged emptying time of the infant's stomach, together with the large energy requirements of the infant, necessitate that foods consumed be of high energy density. Human milk or infant formula help compensate for the stomach's limited capacities during early infancy.

The concentrations of gastric secretions, such as hydrochloric acid and pepsin, rise throughout infancy. Early on, the low acidity of the infant's stomach prevents some types, as well as large amounts, of food from being digested.[103] This explains why the large, tough curds of casein in cow's milk do not readily break down. Similarly, enzymatic secretions of the pancreas and small intestine are low in early infancy, so the digestion of solid foods is inefficient.

The limited capacity of the infant's renal system during the first few months of life further justifies delaying solid food introduction. Water should be offered to the infant regularly once solid food is added to the diet, to compensate for the higher renal solute load.

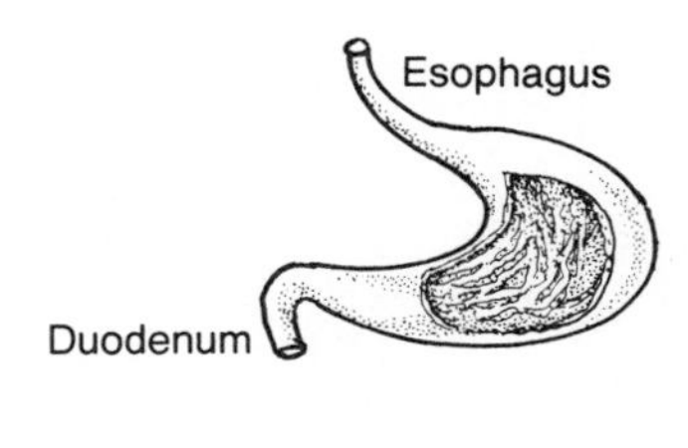

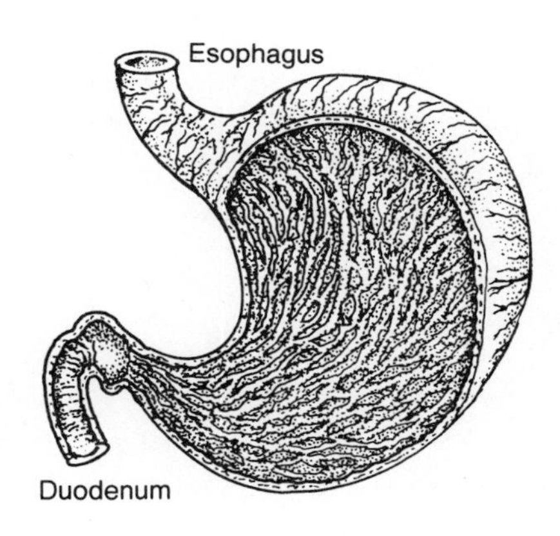

Figure 6–10 Infant and Child Stomachs
The stomach of the normal newborn is more horizontal than that of the older child. The stomach becomes curved and upright as the child assumes a more erect posture. The subdivisions of fundus, body, and antrum are not well defined for the first year of life. The vigorous mixing action of the antrum, which is most involved with emptying of solids, is apparent in older children but is minimal in the first months of infancy. The differences in stomach anatomy is one reason why solid foods are digested less readily at this time.

What to introduce first While energy needs are increasing with steady growth and increasing activity, needs for other nutrients are increasing as well. Prominent among nutrients demanding attention at this time is iron. Even though the infant is born with ample supplies of iron, body iron must double as birthweight triples in the first year of life.[104] Iron absorption is minimal during early infancy, when iron stores are large. Therefore, no advantage is gained by supplementing the diets of full-term infants with iron during the first four months of life. By four to six months of age, however, iron stores have diminished, and additional sources of iron beyond breast milk or formula are needed. For the breastfed infant, iron-fortified cereals are a desirable source of supplemental iron by *six* months of age. For the formula-fed infant, iron-fortified formula, iron-fortified cereal, or a combination of the two are recommended by *four* months of age.[105]

Iron-fortified infant cereal is advantageous as the first solid food, for several reasons. The needed iron is provided in a convenient, economical form. Cereals readily mix with fluids, so they are of a consistency easily swallowed by the infant. They provide energy, calcium, phosphorus, and other vitamins and minerals as well as iron, and most infants accept and tolerate them well.

Food allergies and intolerances The appropriate timing and type of food introduced may also facilitate or prevent the onset of an allergic response. In early infancy, the mucosal barrier of the small intestine is permeable to large molecules.[106] This is advantageous, because it allows maternal antibodies to cross the barrier intact, affording the infant protection against infection. If, however, inappropriate protein sources, such as those in unmodified cow's milk and some solid foods, are fed, undigested proteins may become antigens and induce allergies to these foods.[107] Symptoms of food allergy include nausea, vomiting, abdominal discomfort, respiratory disturbances, and skin rashes.

Chapter 7, which deals with children's nutrition, discusses allergies and intolerances in detail; this chapter focuses on prevention and early detection of allergy. To prevent allergy, and to facilitate its prompt identification should it occur, experts recommend the introduction of single-ingredient foods, one at a time, in small portions, allowing four to five days before introducing the next new food. The gradual introduction of single-ingredient foods permits identification of problem foods. Rice cereal, barley cereal, or other single-ingredient infant cereals are appropriate first foods. Mixed cereals and combination foods can be offered once sensitivity to specific foods has been ruled out.

The introduction of plain, unmodified cow's milk to an infant's diet should be made after six months, for reasons already mentioned (p. 230), and in most infants it should then cause no problems. However, adverse reactions may occur in some infants due to either milk-protein hypersensitivity (milk allergy) or to lactose intolerance. The incidence of milk-protein hypersensitivity in infants is estimated to range from 0.4 to 7.5 percent. The actual incidence is difficult to pinpoint because at present, no satisfactory, generally accepted way of making the diagnosis exists. Many infants with milk-protein hypersensitivity outgrow it by early childhood. As for lactose intolerance, it occurs when an infant is unable to completely digest lactose, the carbohydrate in milk. It is very rare in infants; the onset is usually much later, at about year four or so. For either milk allergy or lactose intolerance, soy formulas and other special foods are necessary, as described in Chapter 7.

Other than milk, the foods most often implicated in food allergies in infants include citrus fruits, soy protein, egg whites, and wheat.[108] Some pediatricians recommend delaying the introduction of these foods until around nine months of age to coincide with more complete GI tract development.

Aside from true food allergies, infants may exhibit adverse reactions to foods due to ingredients in them such as bacterial toxins, monosodium glutamate, or the natural laxative in prunes. Adverse reactions may also be caused by digestive tract disorders, such as obstructions or injuries, or inborn errors of metabolism. In some cases it is not possible or necessary to say whether an adverse food reaction is due to allergy or something else, but avoidance of the food, at least for a while, is indicated.

Infants frequently outgrow food allergies and other adverse reactions to foods within a short time. Avoidance of offending foods, therefore, should not have to be permanent unless a physician confirms the presence of a food allergy.

Infant foods Infant foods should be selected to provide variety, balance, and moderation. Commercial baby foods offer a wide variety of palatable, nutritious foods, in a safe and convenient form. Home-made infant foods can be as nutritious as commercially prepared ones, as long as the preparer minimizes nutrient losses during preparation. Ingredients for home-made baby food recipes should be fresh, whole foods without added salt, sugar, or seasonings. Pureed food can be frozen in ice cube trays, providing convenient-sized blocks of food that can be thawed, warmed, and fed to the infant. The preparer should take precautions to guard against food contamination or infection; hands and equipment must be clean.

A wide variety of commercial infant foods is available today. The foods fall into eight groups: cereals, fruits, fruit juices, vegetables, meats, meat and vegetable combinations, yogurts, and desserts. Most are available either in dry form (to which liquid is added) or ready-prepared in jars. In response to consumer concerns during the 1970s, manufacturers have reduced or eliminated salt, sugar, and other additives in commercial infant foods.[109] All infant food ingredients are listed on the label in descending order by weight. In the United States, FDA regulations require that certain nutrients be listed on the label, as shown in Figure 6–11.

Iron-fortified infant cereal is usually recommended as the first solid food for the infant.[110] Infant cereal provides needed iron and energy and is

Figure 6–11 Nutrition Information on a Baby Food Label and Cap

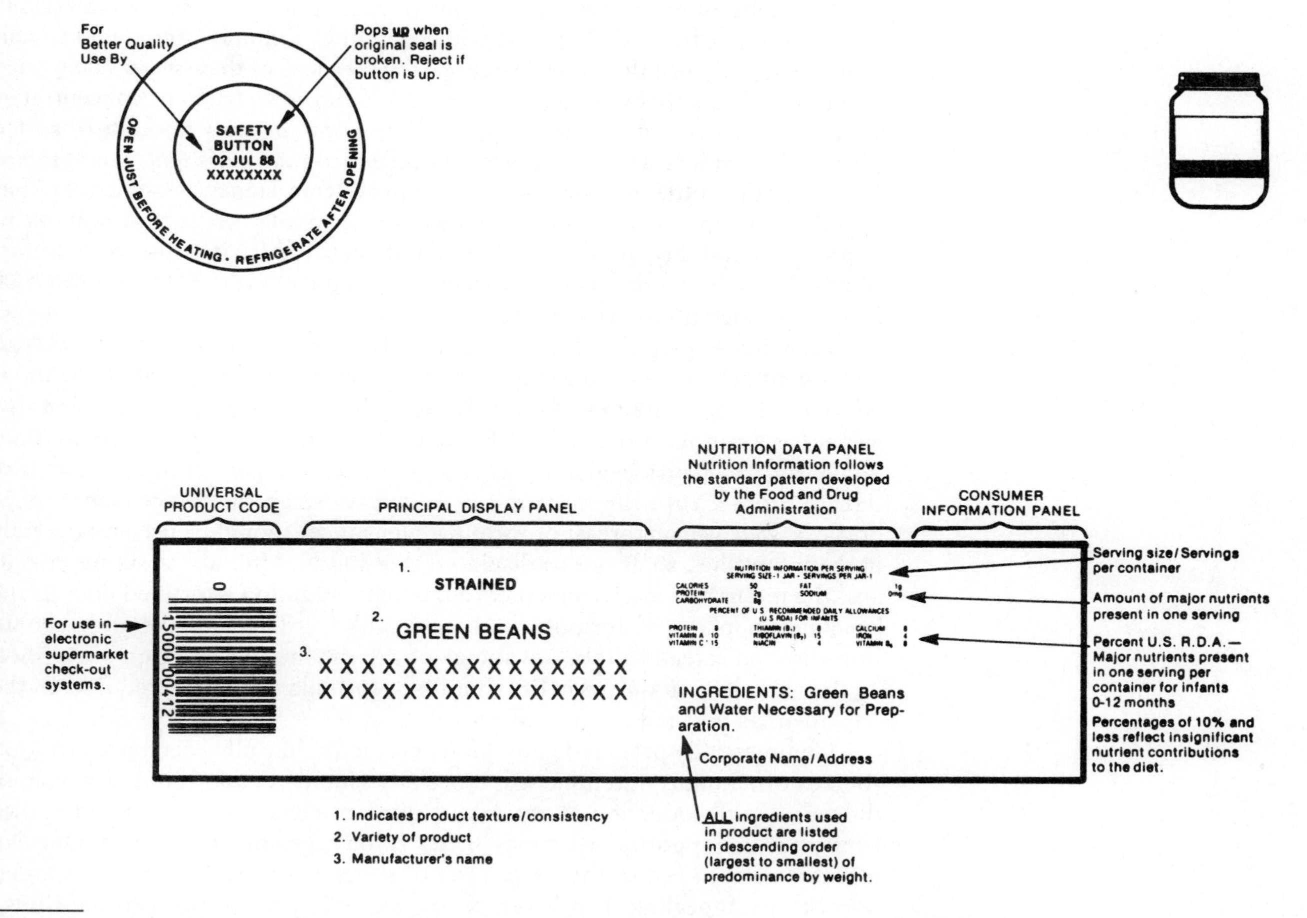

Note: Nutrient information may vary among manufacturers due to crop variety, formulations, and analytic methods.

Source: American Council on Science and Health, *Baby Foods* (New York: American Council on Science and Health, 1987).

convenient, economical, and well tolerated by infants. Rice cereal is usually the first cereal introduced, because it is the least allergenic.[111] Infant cereals are available as single-grain or mixed varieties. Single-grain cereals are rice, barley, and oatmeal. Mixed cereals include mixed-grain varieties and cereals mixed with fruit. Ready-prepared cereals in jars usually contain fruit, as do the dehydrated, canned varieties. Dry cereals are prepared by adding water, breast milk, or formula to obtain the desired consistency, which should be almost liquid at first. Cereals and other solids should be fed by spoon and offered once a day in small quantities (1 to 2 tablespoons) until the infant becomes accustomed to the food and spoon.

The foods given to the infant at first contain mostly nonheme iron rather than the heme iron found primarily in meat. Heme iron is well absorbed, regardless of the nature of the meal. In contrast, the absorption of nonheme iron is influenced by the presence of other foods consumed at about the same time. For example, vitamin C greatly enhances the absorption of nonheme iron. Once the infant is consuming cereals, the parent or other caretaker should begin selecting vitamin C–rich foods to go with them, to promote optimal iron absorption.

Fruits are often the second food introduced in infant diets. Commercially prepared infant fruits are fortified with vitamin C and may provide up to half of the infant RDA of this vitamin per serving. Because of their sweet taste, most infants like fruits. Baby food fruits are made from fresh fruits or concentrates. Some of the more tart, acidic, prepared fruits contain small amounts of added sugar, while others, such as apples and pears, contain no sugar. Some infant fruits contain a little modified starch to improve consistency. Modified starches include corn, tapioca, potato, or wheat starch treated in such a way as to improve palatability, product stability, and shelf life.[112] The use of modified starches in baby foods has been extensively studied by the FDA and the AAP and is considered safe as practiced.[113]

Fruit juices prepared for infants are also fortified with vitamin C. Fruit juices contain no added sugar and are prepared from pressed whole fruits or concentrates. Several brands of regular adult fruit juices are prepared similarly, and infants can have these, also. Fruit juices are useful both to increase fluid intake and to provide vitamin C. An infant can begin drinking fruit juice directly from a cup; this circumvents having to wean the infant from a juice bottle as well as from breast or formula later. Juices should be used moderately in the infant diet, so as not to displace other foods. After about six months of age, the infant can receive one meal a day with vitamin C–fortified juice as the beverage in place of formula or breast milk.[114] The meal should contain iron-fortified cereal, to take advantage of the enhancement of iron absorption by the vitamin C in the juice. Breast milk or formula can be offered later in the day, then, as a snack in place of juice.

Commercially prepared baby food vegetables, like all vegetables, are good sources of vitamins and minerals, especially vitamin A and the B vitamins. In the past, the introduction of vegetables in infant diets usually followed that of fruits, but the opposite order may better favor vegetable acceptance. Once an infant is accustomed to the sweet taste of fruits, the tastes of some vegetables may be less appealing than if vegetables had been offered prior to the fruits.

Commercial infant yogurts have a lower sugar content than adult fruit-flavored yogurts and are good sources of calcium and protein. Plain or reduced-sugar varieties of regular yogurt are also acceptable for infants. Yogurt is a popular food for infants and is often introduced around eight months of age.

Meats are generally introduced between eight and ten months of age. Meats for infants are prepared with broth. They are excellent sources of protein, B vitamins, and iron. A suitable choice for the first meat food to offer is a meat-and-vegetable combination, which contains less protein per serving than single-ingredient meats. In most cases, protein from meat sources is not critical to an infant's nutrition, because adequate protein is available from breast milk, formula, and cereal. Meats may be less readily accepted than other foods at first, but this is no cause for concern. As long as the infant is consuming iron-fortified cereal and breast milk or formula daily, the iron and protein contribution of meats is dispensable.

Children in vegetarian families should be encouraged to eat iron-fortified infant cereals well into the second year of life. Legumes and whole-grain foods can be added to their diets in place of meat. Children who eat no foods of animal origin will require vitamin B_{12} supplements, as well as calcium- and vitamin D–fortified food sources or supplements.

Most infants do not need commercial infant desserts. They are intended as energy supplements for infants whose energy needs are greater than those supplied by a regular diet of milk, cereal, fruits, vegetables, and meats, such as underweight and seriously ill infants. They provide more kcalories and fewer nutrients than other infant foods.

Feeding an infant appropriately supports sound nutrition and health, and eating habits acquired during infancy and childhood influence the overall food attitudes of the individual throughout life. The nurturing of an infant, however, involves more than nutrition. In light of infants' developmental and nutrient needs, and in the face of their often contrary and willful behavior, a few feeding guidelines may be helpful (see Practical Point: Mealtimes with Infants). Those who care for infants are responsible for providing not only nutritious milk, foods, and water, but also a safe, loving, secure environment in which the infants may grow and develop.

▸▸ PRACTICAL POINT

Mealtimes with Infants

The following are several problem situations that may be encountered when feeding older infants, with some suggestions for handling them:

- *He stands and plays at the table instead of eating.* Don't let him. This is unacceptable behavior and should be firmly discouraged. Put him down, and let him wait until later to eat again. Be consistent and firm, not punitive. If he is really hungry, he will soon learn to sit still while eating. An infant's appetite is less keen at a year than at eight months, and his energy needs are relatively lower. An infant will get enough to eat if he lets his own hunger be his guide.

▸▸ PRACTICAL POINT *continued*

- ▸ *She wants to poke her fingers into her food.* Let her. She has much to learn from feeling the texture of her food. When she knows all about it, she'll naturally graduate to the use of a spoon.
- ▸ *He wants to manage the spoon himself, but can't handle it.* Let him try. As he masters it, withdraw gradually until he is feeding himself competently. During the second half of the first year, an infant can learn to feed himself and is most strongly motivated to do so. He will spill, of course, but he'll grow out of it soon enough.
- ▸ *She prefers sweets—candy and sugary confections—to foods containing more nutrients.* Human beings of all races and cultures have a natural inborn preference for sweet-tasting foods. Limit them strictly. If they are kept in the house, keep them out of sight. There is no room in an infant's daily 1000 kcalories for the kcalories from sweets, except occasionally.

These recommendations reflect a spirit of tolerance that serves the best interest of the child emotionally as well as physically. The wise parent of a one-year-old offers nutrition and love together.

Source: Adapted with permission from E. M. N. Hamilton, E. N. Whitney, and F. S. Sizer, *Nutrition: Concepts and Controversies,* 4th ed. (St. Paul, Minn.: West, 1988), pp. 450–451.

Chapter 6 Notes

1. D. Sinclair, *Human Growth after Birth,* 4th ed. (New York: Oxford University Press, 1985), pp. 1–22.
2. F. C. Battaglia and L. O. Lubchenco, A practical classification of newborn infants by birthweight and gestational age, *Journal of Pediatrics* 71 (1967): 159.
3. J. M. Tanner, Growth before birth, in *Foetus into Man: Physical Growth from Conception to Maturity* (London: Open Books Publishing, 1978), p. 46.
4. R. Fancourt and coauthors, Follow-up study of small-for-date babies, *British Medical Journal* 1 (1976): 1435–1437; C. Neumann, E. R. Stiehm, and M. Swenseid, Longitudinal study of immune function in intrauterine growth retarded infants (abstract), as cited in *Federation Proceedings* 39 (1980): 888.
5. M. B. Stoch and P. M. Smythe, 15-year developmental study on effects of severe undernutrition during infancy on subsequent physical growth and intellectual functioning, *Archives of Disease in Childhood* 51 (1976): 327–336.
6. J. M. Rathbun and K. E. Peterson, Nutrition in failure to thrive, in *Pediatric Nutrition,* R. J. Grand, J. L. Sutphen, and W. H. Dietz, Jr., eds. (Boston: Butterworths, 1987), pp. 627–643; American Academy of Pediatrics, Committee on Nutrition, *Pediatric Nutrition Handbook,* 2nd ed. (Elk Grove Village, Ill.: American Academy of Pediatrics, 1985), pp. 6, 198–199; S. J. Fomon, *Infant Nutrition,* 2nd ed. (Philadelphia: W. B. Saunders, 1974), pp. 81–84.
7. E. M. Widdowson, Early nutrition and later development, in *Diet and Bodily Constitution,* eds. G. E. W. Wolstenholme and M. O'Connor (Boston: Little, Brown, 1964), pp. 3–11; R. A. McCance, Some effects of undernutrition, *Journal of Pediatrics* 65 (1964): 1008–1014.
8. M. Winick, Changes in nucleic acid and protein content of the human brain during growth, *Pediatric Research* 2 (1968): 352–355.
9. M. Winick and A. Noble, Cellular response in rats during malnutrition at various ages, *Journal of Nutrition* 89 (1966): 300–306.
10. M. Winick and P. Rosso, The effect of severe early malnutrition on cellular growth of human brain, *Pediatric Research* 3 (1969): 181–184.
11. S. A. Richardson, The relation of severe malnutrition in infancy to the intelligence of school children with differing life histories, *Pediatric Research* 10 (1976): 57–61.
12. Stoch and Smythe, 1976.
13. A. Ashworth and D. J. Millward, Catch-up growth in children, *Nutrition Reviews* 44 (1986): 157–163.
14. A. Prader, J. M. Tanner, and G. A. von Harnack, Catch-up growth following illness or starvation: An example of developmental canalization in man, *Journal of Pediatrics* 62 (1963): 646–659.
15. J. G. Bieri, Vitamin E, in *Present Knowledge in Nutrition,* 5th ed. (Washington, D.C.: Nutrition Foundation, 1984), pp. 226–240.
16. G. R. Gutcher, W. J. Raynor, and P. M. Farrell, An evaluation of vitamin E status in premature infants, *American Journal of Clinical Nutrition* 40 (1984): 1078–1089.
17. F. A. Oski, Vitamin E in infant nutrition, in *Textbook of Pediatric Nutrition,* ed. R. M. Suskind (New York: Raven Press, 1981), pp. 145–151.
18. M. L. Williams and coauthors, Role of dietary iron and fat on vitamin E defi-

ciency anemia of infancy, *New England Journal of Medicine* 292 (1975): 887–890.

19. V. Herbert, Nutritional anemias of childhood—Folate, B_{12}: The megablastic anemias, in *Textbook of Pediatric Nutrition,* ed. R. M. Suskind (New York: Raven Press, 1981), pp. 133–144.
20. H. L. Greene, Disorders of the water-soluble vitamin B-complex and vitamin C, in *Textbook of Pediatric Nutrition,* ed. R. M. Suskind (New York: Raven Press, 1981), pp. 113–131.
21. F. A. Oski and J. A. Stockman, Anemia due to inadequate iron sources or poor iron utilization, *Pediatric Clinics of North America* 27 (1980): 237–252.
22. American Academy of Pediatrics, 1985, pp. 215–220.
23. R. Yip and coauthors, Declining incidence of anemia among low-income children in the United States, *Journal of the American Medical Association* 258 (1987): 1619–1623.
24. J. Weinberg, Behavioral and physiological effects of early iron deficiency in the rat, in *Iron Deficiency: Brain Biochemistry and Behavior,* eds. E. Pollitt and R. L. Leibel (New York: Raven Press, 1982), pp. 93–123.
25. Oski and Stockman, 1980.
26. E. Pollitt and coauthors, Iron deficiency and behavioral development in infants and preschool children, *American Journal of Clinical Nutrition* 43 (1986): 555–565.
27. T. Walker, Infancy: Mental and nutritional development, *American Journal of Clinical Nutrition* (supplement) 50 (1989): 655–661.
28. E. Pollitt and R. L. Leibel, Iron deficiency and behavior, *Journal of Pediatrics* 88 (1976): 372–381.
29. F. A. Oski and A. S. Honig, The effects of therapy on the developmental scores of iron-deficient infants, *Pediatrics* 92 (1978): 21–25; T. Walter, J. Kovalskys, and A. Stekel, Effect of mild iron deficiency on infant mental development scores, *Journal of Pediatrics* 68 (1983): 828–838.
30. F. A. Oski and coauthors, Effect of iron therapy on behavior performance in nonanemic, iron-deficient infants, *Pediatrics* 71 (1983): 877–880.
31. Pollitt and coauthors, 1986.
32. Walter, Kovalskys, and Stekel, 1983.
33. R. L. Leibel, Behavioral and biochemical correlates of iron deficiency, *Journal of the American Dietetic Association* 71 (1977): 398–404.
34. M. L. Voorhees and coauthors, Iron deficiency anemia and increased urinary norepinephrine excretion, *Journal of Pediatrics* 86 (1975): 542–547.
35. J. A. Olson, Recommended dietary intakes (RDI) of vitamin A in humans, *American Journal of Clinical Nutrition* 45 (1987): 704–716.
36. American Academy of Pediatrics, 1985, p. 31.
37. G. H. Lowrey, *Growth and Development of Children,* 8th ed. (Chicago: Year Book Medical Publishers, 1986), pp. 383–410.
38. Food and Nutrition Board, *Recommended Dietary Allowances,* 10th ed. (Washington, D.C.: National Academy Press, 1989), p. 35.
39. R. G. Whitehead and coauthors, A critical analysis of measured food energy intakes during infancy and early childhood in comparison with current international recommendations, *Journal of Human Nutrition* 35 (1981): 339–348.
40. R. A. Stewart, *Infant and Child Feeding* (New York: Academic Press, 1981), pp. 123–133.
41. Food and Nutrition Board, 1989, pp. 56–58.
42. J. A. Sturman, D. K. Rassin, and G. E. Gaull, Minireview: Taurine in development, *Life Science* 21 (1977): 1–21.
43. Food and Nutrition Board, 1989, pp. 52–77.
44. Food and Nutrition Board, 1989, pp. 62–64.
45. L. A. Barness, Nutritional requirements of the full-term neonate, in *Textbook of Pediatric Nutrition,* ed. R. M. Suskind (New York: Raven Press, 1981), pp. 21–28.
46. K. Brostrøm, Human milk and infant formulas: Nutritional and immunological characteristics, in *Textbook of Pediatric Nutrition,* ed. R. M. Suskind (New York: Raven Press, 1981), pp. 41–64.
47. Brostrøm, 1981.
48. Z. Friedman and coauthors, Rapid onset of essential fatty acid deficiency in the newborn, *Pediatrics* 58 (1976): 640–649.
49. A. P. Simopoulos, Omega-3 fatty acids in growth and development and in health and disease, *Nutrition Today,* March–April 1988, pp. 10–19.
50. M. A. Crawford and coauthors, Structural lipids and their polyenoic constituents in human milk, in *Dietary Lipids and Postnatal Development,* eds. C. Galli, G. Jacini, and A. Pecile (New York: Raven Press, 1973), p. 41, as cited by Simopoulos, 1988.
51. Olson, 1987.
52. H. L. Henry and A. W. Norman, Vitamin D: Metabolism and biological actions, *Annual Review of Nutrition* 4 (1984): pp. 493–520.
53. R. H. Herman, Disorders of fat-soluble vitamins A, D, E, and K, in *Textbook of Pediatric Nutrition,* ed. R. M. Suskind (New York: Raven Press, 1981), pp. 65–111.
54. American Academy of Pediatrics, 1985, pp. 37–48.
55. D. R. Lakdawala and E. M. Widdowson, Vitamin-D in human milk, *Lancet* 1 (1977): 167–168.
56. L. E. Reeve, R. W. Chesney, and H. F. DeLuca, Vitamin D of human milk: Identification of biologically active forms, *American Journal of Clinical Nutrition* 36 (1982): 122–126.
57. J. A. Olson, Recommended dietary intakes (RDI) of vitamin K in humans, *American Journal of Clinical Nutrition* 45 (1987): 687–692.
58. Herman, 1981.
59. Herbert, 1981.
60. V. Herbert, Recommended dietary intakes (RDI) of folate in humans, *American Journal of Clinical Nutrition* 45 (1987): 661–670.
61. V. Herbert, Recommended dietary intakes (RDI) of vitamin B-12 in humans, *American Journal of Clinical Nutrition* 45 (1987): 671–678.
62. Food and Nutrition Board, 1989, p. 119; L. Salmenpera, Vitamin C nutrition during prolonged lactation: Optimal in infants while marginal in some mothers, *American Journal of Clinical Nutrition* 40 (1984): 1050–1056.
63. Brostrøm, 1981.
64. Food and Nutrition Board, 1989, p. 180.
65. Food and Nutrition Board, 1989, p. 180.
66. C. R. Paterson, Calcium requirements in man: A critical review, *Postgraduate Medical Journal* 45 (1978): 244–248, as cited by K. Brostrøm, 1981.
67. H. Spencer and coauthors, Effect of high phosphorus intake on calcium and phosphorus metabolism in man, *Journal of Nutrition* 86 (1965): 125; H. Spencer and coauthors, Effect of phosphorus on the absorption of calcium and on the calcium balance in man, *Journal of Nutrition* 108 (1978): 447; World Health Organization, Calcium requirements, Technical Report Series, 230 (Geneva: World Health Organization, 1962), pp. 16–18; D. M. Heg-

sted, Mineral intake and bone loss, *Federation Proceedings* 26 (1967): 1947, as cited by H. Spencer, Minerals and mineral interactions in human beings, *Journal of the American Dietetic Association* 86 (1986): 864–867.

68. Brostrøm, 1981.
69. Food and Nutrition Board, 1989, p. 254.
70. Brostrøm, 1981.
71. E. D. Rios and coauthors, Relationship of maternal and infant iron stores as assessed by determination of plasma ferritin, *Pediatrics* 55 (1975): 694–699.
72. B. Lonnerdal, C. L. Keen, and L. S. Hurley, Iron, copper, zinc, and manganese in milk, *Annual Review of Nutrition* 1 (1981): 149–174.
73. G. H. Johnson, F. A. Purvis, and R. D. Wallace, What nutrients do our infants really get? *Nutrition Today,* July–August 1981, pp. 4–10, 23–26.
74. American Academy of Pediatrics, 1985, pp. 213–220.
75. N. F. Krebs and K. M. Hambridge, Zinc requirements and zinc intakes of breastfed infants, *American Journal of Clinical Nutrition* 43 (1986): 288–292.
76. Food and Nutrition Board, 1989, pp. 209–210.
77. Krebs and Hambridge, 1986.
78. Brostrøm, 1981.
79. American Academy of Pediatrics, 1985, p. 40.
80. O. G. Brooke, Nutrition in the preterm infant, *Lancet* 1 (1983): 514–515.
81. O. G. Brooke, Nutritional requirements of low and very low birthweight infants, *Annual Review of Nutrition* 7 (1987): 91–116.
82. B. Reichman and coauthors, Diet, fat accretion, and growth in premature infants, *New England Journal of Medicine* 305 (1981): 1495–1500.
83. O. G. Brooke, Nutrition in the preterm infant, *Lancet* 1 (1983): 514–515; J. Senterre and J. Rigo, Nutritional requirements of low-birthweight infants, in *Nutritional Needs and Assessment of Normal Growth,* eds. M. Gracey and F. Falkner (New York: Raven Press, 1985), pp. 45–59.
84. Bieri, 1984.
85. P. J. Leonard and M. S. Losowsky, Effect of alpha-tocopherol administration on red cell survival in vitamin E–deficient human subjects, *American Journal of Clinical Nutrition* (1971): 388–393.
86. Gutcher, Raynor, and Farrell, 1984.
87. M. S. Rodriguez, A conspectus of research on folacin requirements of man, in *Nutritional Requirements of Man: A Conspectus of Research* ed. M. I. Irwin (Washington, D.C.: The Nutrition Foundation, 1980), pp. 397–489.
88. E. E. Ziegler, R. L. Biga, and S. J. Fomon, Nutritional requirements of the premature infant, in *Textbook of Pediatric Nutrition,* ed. R. M. Suskind (New York: Raven Press, 1981), pp. 29–39.
89. P. R. Dallman, M. A. Siimes, and A. Stekel, Iron deficiency in infancy and childhood, *American Journal of Clinical Nutrition* 33 (1980): 86–118.
90. American Academy of Pediatrics, 1985, p. 216.
91. M. I. Irwin and E. W. Kienholz, A conspectus of research on calcium requirements of man, in *Nutritional Requirements of Man: A Conspectus of Research* (Washington, D.C.: The Nutrition Foundation, 1980), pp. 135–211.
92. D. W. Spady, Infant nutrition, *Journal of the Canadian Dietetic Association* 38 (1977): 34–40; American Academy of Pediatrics, 1985, p. 30.
93. American Academy of Pediatrics, Committee on Nutrition, The use of whole cow's milk in infancy, *Pediatrics* 72 (1983): 253–255.
94. J. F. Wilson, M. E. Lahey, and D. C. Heiner, Studies on iron metabolism, V. Further observations on cow's milk-induced gastrointestinal bleeding in infants with iron-deficiency anemia, *Journal of Pediatrics* 84 (1974): 335–344.
95. American Academy of Pediatrics, 1985, p. 34.
96. D. M. Paige, Infant growth and nutrition, *Clinical Nutrition* 2 (1983): 14–18.
97. R. A. Stewart, Supplementary foods: Their nutritional role in infant feeding, in *Infant and Child Feeding,* eds. J. T. Bond and coeditors (New York: Academic Press, 1981), pp. 123–133.
98. S. J. Fomon, Reflections on infant feeding in the 1970's and 1980's, *American Journal of Clinical Nutrition* 46 (1987): 171–182.
99. D. W. Marlin, M. F. Picciano, and E. C. Livant, Infant feeding practices, *Journal of the American Dietetic Association* 77 (1980): 668–675; D. L. Yeung and coauthors, Infant feeding practices, *Nutrition Reports International* 23 (1981): 249–260.
100. Fomon, 1987.
101. E. M. E. Poskitt, Infant feeding: A review, *Human Nutrition: Applied Nutrition* 37A (1983): 271–286.
102. S. A. Quandt, The effect of beikost on the diet of breastfed infants, *Journal of the American Dietetic Association* 84 (1984): 47–51.
103. The gastrointestinal tract: Development and nutrition, in *Dynamics of Infant Physiology and Nutrition,* ed. L. J. Filer (Bloomfield, N.J.: Health Learning Systems, 1982), pp. 1–16.
104. Dallman, Siimes, and Stekel, 1980.
105. American Academy of Pediatrics, 1985, p. 42.
106. W. A. Walker, Antigen handling by the gut, *Archives of Disease in Children* 53 (1978): 527–531.
107. American Academy of Pediatrics, 1985, pp. 16–27.
108. American Council on Science and Health, *Baby Foods* (Summit, N.J.: American Council on Science and Health, 1987).
109. American Council on Science and Health, 1987.
110. D. L. Yeung, *Infant Nutrition: A Study of Feeding Practices and Growth from Birth to 18 Months* (Ottawa, Canada: Canadian Public Health Association, 1983), pp. 132–149.
111. Yeung, 1983.
112. American Council on Science and Health, 1987.
113. H. M. Barry, Addressing confusion over role of modified starches, *Food Engineering* 58 (1986): 56–57.
114. C. Briggs, Recent developments in infant feeding and nutrition, in *Nutrition Update,* vol. 1, eds. J. Weininger and G. M. Briggs (New York: Wiley, 1983), pp. 227–261.

▶ Focal Point 6

Nutrition Care of Sick Infants and Children

The emphasis of this text is on wellness and the prevention of disease, but even children who are generally in good health get sick on occasion. This discussion examines some of the common symptoms that most children experience at one time or another—infections and fever, diarrhea, and constipation. It then continues with a look at the special needs of any child who requires hospitalization.*

A healthy, well-nourished child can easily slip into poor nutrition status with an illness. Serious illness can affect a child's nutrition status in several ways. An illness can alter:

- Appetite.
- Chewing and swallowing abilities.
- Digestion and absorption.
- Metabolism.
- Excretion.

Any or all of these effects can occur during illness and compromise nutrition status. Competent medical care includes attention to nutrition.

Infections and Fever

Each day, the body confronts an environment teeming with disease-causing organisms. The body's remarkable capacity to survive such an environment is a tribute to its immune system. The immune system has no central organ of control, but rather depends on various organs and white blood cells. Their interactions and secretions defend the body against infectious organisms, such as bacteria and viruses. Figure FP6–1 summarizes the action of the immune system; the Miniglossary defines related terms. The immune system recognizes infectious organisms and destroys or otherwise neutralizes them. When organisms do manage to penetrate the immune defenses, an infection, or an even more damaging disease, develops. Still, for every successful penetration of foreign organisms, the immune system averts thousands of attempts.

---

*Parts of this discussion have been adapted with permission from E. N. Whitney, C. B. Cataldo, and S. R. Rolfes, *Understanding Normal and Clinical Nutrition*, 2nd ed. (St. Paul, Minn.: West, 1987).

Figure FP6–1 The Immune System

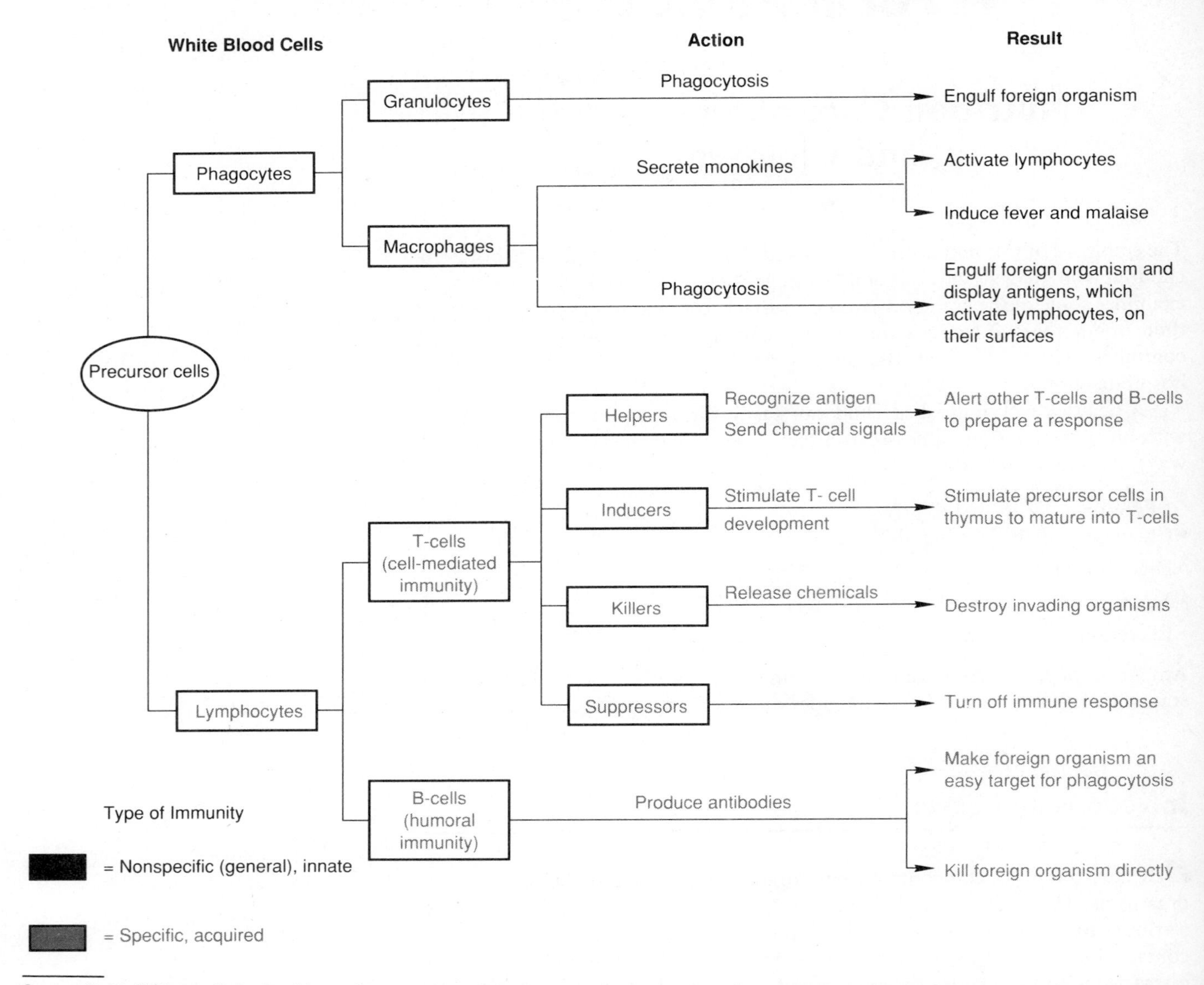

Source: E. N. Whitney, C. B. Cataldo, and S. R. Rolfes, *Understanding Normal and Clinical Nutrition*, 2nd ed. (St. Paul, Minn.: West, 1987). Used with permission.

Malnutrition alters immune system components in ways that compromise their function, thus impairing the defense against infecting organisms. It is little wonder that malnourished children develop more infections than well-nourished children. Infection is a major cause of mortality and morbidity in children with protein-energy malnutrition. A vicious cycle develops in which malnutrition reduces resistance to infection, and infection further aggravates malnutrition. This synergistic relationship between malnutrition and infection threatens a child's survival.

synergism: the effect of two factors operating together in such a way that their combined actions are greater than the sum of the actions of the two considered separately.

Miniglossary of Immunity Terms

acquired immunity: immunity directed at specific organisms (also called **specific immunity**). The lymphocytes mediate this type of immunity, which depends on prior exposure, recognition, and reactions to invading organisms. Two types of specific immunity are **cell-mediated immunity** and **humoral immunity.**

antibody: a protein produced by the B-cells in response to invasion of the body by a foreign protein.

antigen: a substance foreign to the body that elicits the formation of antibodies or an inflammation reaction from immune system cells.

cell-mediated immunity: immunity conferred by the actual reaction of T-cells to an invading organism.

granulocytes: a type of phagocyte.

helper T-cells: the T-lymphocytes capable of recognizing antigens and alerting other T-cells to prepare to mount a response.

humoral immunity: immunity conferred by antibodies secreted by B-cells and carried to the invaded area by way of the body fluids.

immune system: the body's natural defense system against foreign materials.

immunity: the body's ability to recognize and eliminate foreign materials.

inducer T-cells: T-lymphocytes that stimulate precursor cells in the thymus to develop into mature T-cells.

innate immunity: immunity directed at foreign organisms in general (also called **nonspecific immunity**). The skin, mucous membranes, and phagocytes are a part of this type of immunity.

killer T-cells: T-lymphocytes that release chemicals that can destroy an invading organism. Also called **cytoxic T-cells.**

lymphocytes: white blood cells that originate from precursors in the bone marrow and mature into two distinct types of lymphocytes—B-cells and T-cells—upon their release.

macrophage: the type of phagocyte that secretes *monokines* (see below).

monokines: various proteins that are secreted by phagocytes and that help mediate the immune response, such as interferon and interleukin-1.

nonspecific immunity: see *innate immunity.*

phagocytes: cells that have the ability to engulf and destroy foreign materials.

phagocytosis (FAG-oh-sigh-TOE-sis): the process by which some cells (phagocytes) engulf and destroy foreign materials.
phagein = to eat
kytos = cell
osis = intensive

precursor cell: a simple cell that matures. Precursor white blood cells are capable of maturing into three different types of cells.

specific immunity: see *acquired immunity.*

suppressor T-cells: T-lymphocytes that slow down the immune response and eventually turn it off.

fever: an increase of body temperature of more than 1° F above normal (98.6° F).

An infection generally progresses as follows. Disease-causing organisms invade the body, overcoming initial immune defenses. A few days after exposure to these infective organisms, symptoms begin to appear, with fever developing shortly thereafter.

People fear fever and rush to treat it because it accompanies many dangerous diseases, but the fever itself may actually assist the immune system.[1] Clearly, fever stresses the body—it raises the heart rate and increases the tissues' demand for oxygen. High temperatures (over 104 degrees Fahrenheit or, in some cases, lower) can cause convulsions and demand medical attention not only to control the fever, but to determine and treat the underlying condition. Moderate temperatures (between 102 degrees Fahrenheit and 104 degrees Fahrenheit) may be cause for concern and require a consultation with a health care provider. Generally, though, a mild fever should be allowed to do its job of assisting the immune system.[2] Fever causes the T-cells of the immune system to proliferate, thus enhancing the system's activity.[3] When researchers suppress fever experimentally, sick animals are more likely to die from infection than when fever is allowed to run its course. Furthermore, animals kept continuously at fever temperatures resist viruses better than nonfeverish ones.

After the onset of fever, catabolic changes begin. The body loses nitrogen (from protein catabolism) and intracellular electrolytes. These losses are great enough to result in negative balances. If diarrhea and vomiting accompany the infection, the body's losses are even greater.

The scientific name for the centigrade temperature scale is Celsius. To convert from centigrade to Fahrenheit (and vice versa), use these equations:

$$t_F = 9/5\ t_C + 32.$$
$$t_C = 5/9\ (t_F - 32).$$

Fever is a major factor in determining the energy needs of an infected child. The basal metabolic rate increases (roughly 10 to 13 percent for each 1 degree centigrade; 7 percent for each 1 degree Fahrenheit) as the temperature rises above normal (98.6 degrees Fahrenheit, or 37 degrees centigrade). Additional food energy may be needed to support physical activity—for example, if a child is restless, coughing, crying, or the like.

sodium salicylate: a compound used as an analgesic and antipyretic; aspirin is acetylsalicylic acid.

interleukin-1: a protein released by the immune system that mediates many responses to infection.

Ironically, during this time of heightened energy needs, appetite diminishes. Infection-related anorexia aggravates negative nutrient balances and contributes to weight loss. Researchers have questioned whether this anorexia is a result of the fever itself.[4] When they blocked the fever of infected rats with sodium salicylate, appetite remained depressed. Additional investigation revealed that interleukin-1, a protein released by the immune system in response to infection, both induces fever and suppresses food intake.

The immune system functions best when protein status is optimal. The protein requirements of a child with an infection depend primarily on total energy consumption. At a minimum, protein intake should meet the RDA for age, weight, and gender. With fever, protein allowances increase above the RDA by 10 percent per 1 degree centigrade (5.5 percent per 1 degree Fahrenheit) of fever.

All the nutrients appear to play some role in the functioning of the immune system. Changes in immune function have been associated with deficiencies of folate, iron, and zinc, as well as with excesses of vitamin E and essential fatty acids. Furthermore, clinical reports suggest that changes in immune function accompany deficiencies of vitamins A, B_6, B_{12}, and C; deficiencies of pantothenic acid; and elevated blood cholesterol levels. Studies in animals link the concentrations of many vitamins, minerals, trace elements, amino acids, fatty acids, and cholesterol with immunological changes. Table FP6–1 lists nutrients reported to influence immune function, along with the effect of each.

Table FP6–1 Effects of Selected Nutrients on the Immune System

Nutrient Deficiency	Effects
Vitamin A	Depletion of T-lymphocytes;[a] increased frequency and severity of some infections;[a] increased incidence of infections (human beings)[b]
Vitamin B_6	Depressed cell-mediated and humoral immunity;[a] reduced antibody responses to vaccines (human beings)
Folate	Impaired response to skin tests; lymph tissue atrophy;[a] reduced numbers of white blood cells;[a] impaired cell-mediated and humoral immunity[a]
Pantothenic acid	Depressed antibody responses (human beings)
Vitamin B_{12}	Some reduction of phagocytosis by granulocytes (human beings)
Vitamin E[c]	Depressed antibody responses;[a] impaired response to skin tests[a]
Iron	Atrophy of lymph tissue; impaired response to skin tests; defective phagocytosis (human beings)
Zinc	Atrophy of lymph tissues; abnormalities in cell-mediated and humoral immunity (human beings)
Individual amino acids	Impaired humoral immunity[a]

[a]Information is from animal studies.
[b]Some reports support this finding, but more information is needed.
[c]Vitamin E excess can cause inhibition of multiple immune functions in human beings.

Source: E. N. Whitney, C. B. Cataldo, and S. R. Rolfes, *Understanding Normal and Clinical Nutrition,* 2nd ed. (St. Paul, Minn.: 1987). Used with permission.

Little information is available regarding specific vitamin and mineral requirements during infection. The need for B vitamins increases with increasing energy and protein intakes, but remains consistent with the RDA. Foods can generally cover the vitamin and electrolyte losses incurred during catabolism. Children who have lost their appetites or are feeling nauseated may prefer liquids to solid foods. Liquids are an acceptable alternative to solid foods, provided that the caretaker selects those that offer energy, vitamins, and minerals. Drinking liquids also prevents dehydration, of course.

Children with infections may develop a type of anemia referred to as the anemia of infection. At the onset of the infection, the blood concentration of iron rapidly declines as iron moves into the liver for storage. This shift of iron from the blood to the liver helps to fight the infection by making the body's iron unavailable for the infecting bacteria, which require iron to perform their metabolic functions.[5] However, iron in storage is also unavailable for hemoglobin synthesis, so anemia results. This type of anemia is the body's normal physiological response to infection and does not respond to iron, folate, or vitamin B_{12} supplements, nor to any other dietary or medical treatment. Instead, the situation corrects itself; iron returns to the blood from the liver as the infection resolves.

anemia of infection: a condition in which iron moves from the blood to the liver to help fight infection, resulting in a decline in hemoglobin synthesis.

Maintaining fluid balance is of particular concern during an infection. In some infectious diseases, especially those that involve vomiting, diarrhea, or considerable sweating, fluid needs may be as high as 3 to 4 liters per day. If the child does not drink fluids at this rate, dehydration can quickly develop. On the other hand, the child may retain water due to the hormonal changes that accompany fever. In the rare case that a child's fluid intake is excessive, water intoxication can develop. Table FP6–2 lists the symptoms associated with dehydration and water intoxication.

dehydration: loss of too much fluid from the body.

water intoxication: the condition in which body water content is too high.

Physicians' primary concerns for children with infections are to identify and eliminate the infecting organisms. Quite often, they prescribe antibiotics, but these can interfere with nutrient absorption, thus compromising nutrition status (see Table A–1 of Appendix A).

When the infection ends, a well-balanced diet best restores the body to its normal status. The time it takes to reach a positive balance and the duration of positive balance depend on the extent of nutrient deficiencies, the quality of the diet, and the quantity of food intake. A well-balanced diet that restores the nutrient reserves also helps to defend against future infections. In some cases, physicians may prescribe a multivitamin-mineral supplement to augment nutrient intake.

Quite often, a child with a fever does not require medical attention, but may benefit from tender loving care. The child's caretaker can comfort the child by:[6]

- Helping the child's body maintain its own body temperature by keeping the room temperature moderate and bed coverings to a minimum.
- Sponging the child with lukewarm water to increase heat loss by evaporation.
- Providing the child with plenty of fluids.
- Providing the child with acetaminophen if a physician recommends drug intervention.

acetaminophen: an antipyretic, analgesic drug used to reduce fever or relieve pain.

Pediatricians recommend acetaminophen instead of aspirin because of aspirin's association with Reye's syndrome, a rare disease that primarily affects children and adolescents, generally following flu or chicken pox. Symptoms begin with tiredness and vomiting, progress to permanent brain damage, and result in death in 20 to 30% of the cases.[7]

Fever is the body's signal that something is wrong and it is trying to defend itself. In many cases, the body is successful without medical attention, but infants with any degree of fever and older children with fevers of 103 degrees Fahrenheit or greater require medical attention. In addition, fevers that go away and recur, that persist for more than 72 hours, or that accompany a rash or marked irritability and confusion require consultation with a physician.

Table FP6–2 Dehydration and Water Intoxication Symptoms

Dehydration Symptoms	Water Intoxication Symptoms
Thirst	Low plasma sodium concentrations
Muscle cramps	Headache
Weakness, fatigue, exhaustion	Muscular weakness and fatigue
Delirium	Lack of concentration, poor memory, delirium
Death	Loss of appetite
	Seizure
	Death

Diarrhea

Diarrhea is characterized by frequent, loose, watery stools. Such bowel movements indicate that the chyme has moved too quickly for the intestines to absorb enough fluids from it, or that the chyme has drawn water from the cells lining the intestinal tract. In both cases, the result is the same—extensive fluid and electrolyte losses. If diarrhea continues without treatment, an infant or young child can quickly become dehydrated and malnourished. The smaller and younger the child, the more dramatic are the effects.

Nearly every child suffers from diarrhea at one time or another. Many times an acute case of diarrhea develops and remits in 24 to 48 hours. Well-nourished children with acute diarrhea can usually endure the uncomfortable symptoms without medical treatment. Caretakers can support children during such episodes of diarrhea by eliminating food irritants from the diet and offering clear liquids. Fruit juices aggravate diarrhea and are therefore inappropriate beverages to offer. Children may enjoy such clear liquids as gelatin dessert, carbonated beverages, and Popsicles, but these treats fall short of correcting for dehydration. Their low electrolyte content and high osmolality make them unsuitable for rehydration therapy.[8]

Nutrient reserves of well-nourished children protect them from the detrimental effects of diarrhea for a short while. The availability of medical treatment and high-quality food in industrialized countries offers children the opportunity to recover, both from fluid and electrolyte losses and from growth losses.

The story is quite different in developing countries. Acute diarrhea in a malnourished child threatens life and requires immediate medical attention. In developing countries, more children suffer from malnutrition, and acute diarrhea seriously threatens their tenuous nutrition status. With repeated episodes of diarrhea falling close together, these children's recovery time becomes limited. Without full restoration of nutrients, children are progressively less able to defend against future infections.

Diarrhea is the most common cause of dehydration and malnutrition among children in developing countries.[9] Millions of children die from the complications of diarrhea each year. With reduced dietary intake, impaired intestinal absorption, and the increased nutrient requirements that accompany diarrhea, malnutrition is inevitable.

Children with diarrhea eat less, and therefore their energy intakes are low, averaging between 15 and 50 percent less than their usual intakes.[10] Factors interfering with food intake include anorexia, nausea, and vomiting. In addition, parents or health care workers may withhold food in an effort to resolve the diarrhea. (The wisdom of such a practice is discussed in upcoming paragraphs.)

Maldigestion and malabsorption accompany diarrhea. Normally, hormones and nerves orchestrate the digestive and absorptive processes by signaling several organs to respond at the appropriate times with contractions that move the intestinal contents along, secretions that dismantle nutrients into absorbable molecules, and receptors that transport these molecules into the body. When the contents of the intestine pass too rapidly, digestion is incomplete. When intestinal cells that allow the transport of nutrients into the

Diarrhea that results from an accelerated movement of fluids and electrolytes from the intestinal capillaries into the lumen of the intestine is called **secretory diarrhea**. When unabsorbed water and electrolytes cause diarrhea by increasing the osmolality of the intestinal contents, then **osmotic diarrhea** exists.

The term **acute** describes diseases or conditions that develop rapidly, have severe symptoms, and are of short duration. A disease or condition that develops slowly, shows little change, and lasts a long time is said to be **chronic**. Severe, chronic diarrhea is often called **intractable diarrhea.**
acutus = sharp
chronos = time

body are damaged, absorption is limited. Absorption of protein, carbohydrate, and fat in children with diarrhea can be 10 to 30 percent less than in healthy children.[11] On the positive side, 70 to 90 percent of these nutrients *are* getting absorbed, a fact that emphasizes the value of feeding children with diarrhea. Without food intake, pancreatic function and intestinal cell production and maturation remain low, thus limiting the supply of enzymes for digestion and the surface area for absorption.

Clinicians focus on the most appropriate way to minimize and replace nutrient losses incurred by diarrhea. Controversy surrounds the question of whether to withhold or provide nutrients during episodes of diarrhea. The advantages of both delayed feeding and continued feeding are worthy of consideration.

The traditional practice of withholding food is based on the premise that the bowel needs to rest and that ingested food is malabsorbed. Malabsorption results from the rapid intestinal transit time and damage to the intestinal mucosa. Injuries to the intestine diminish mucosal surface area, alter villus structure, and lower enzyme concentrations. Lactose intolerance is a common, usually temporary, consequence of diarrhea. The intestinal cells that produce lactase are located on the delicate fringe of microvilli that form the brush border of the intestinal villi. Any condition (such as diarrhea) that damages the brush border can lead to lactose intolerance. For this reason, dairy products are reintroduced into the diet gradually following diarrhea.

The obvious consequence of malabsorption is the loss of potential nutrients, but unabsorbed nutrients in the intestine present other complications as well. For one, they have an osmotic effect, drawing water and electrolytes into the gut; this can intensify the diarrhea beyond that caused by the original infection.[12] Unabsorbed nutrients in the intestine may also bind with bile acids, thus preventing their normal conservation via enterohepatic circulation. (Figure FP6–2 illustrates the loss of body materials via the GI tract in diarrhea.) Such losses deplete the bile acid pool and can contribute to steatorrhea.

The recycling of nutrients through the intestine and liver is known as **enterohepatic circulation.**
enteron = intestine
hepat = liver

steatorrhea (stee-ah-toe-REE-ah): fatty diarrhea characteristic of fat malabsorption; stools are foamy, greasy, and malodorous.

Diminished absorption capacity is not the only consequence of damaged intestinal cells. Absorption of whole proteins is another concern. Some researchers speculate that ingestion of whole proteins during acute diarrhea may induce food sensitivities; however, evidence of this possibility is lacking.[13]

Even considering the problems of malabsorption, the argument in favor of feeding the child appears to weigh more heavily. For well-nourished children, short-term fasting may be appropriate therapy, but for malnourished children, fasting throughout the course of diarrhea can be devastating. Consider that the annual prevalence rate of diarrhea in children under the age of three in Bangladesh is 55 days a year.[14] To fast for close to two months a year is to lose a significant percentage of a year's nutrient intakes. It is unreasonable to expect that the diet during times without diarrhea could replace losses incurred by such prolonged fasting. Health care providers argue that "suboptimal absorption of some food is preferable to no malabsorption of no food."[15]

oral rehydration therapy (ORT): the administration of a simple solution of sugar, salt, and water taken by mouth, to treat dehydration caused by diarrhea.

The potential for accelerated deterioration of nutrition status demands rapid replacement of fluids and nutrients.[16] Health care workers around the world are treating diarrhea with oral rehydration therapy (ORT). Table FP6–3 lists the World Health Organization (WHO) standards for ORT formulas. The components of ORT formulas provide needed energy and electrolytes. Except

Figure FP6–2 Loss of Body Materials via GI Tract in Diarrhea

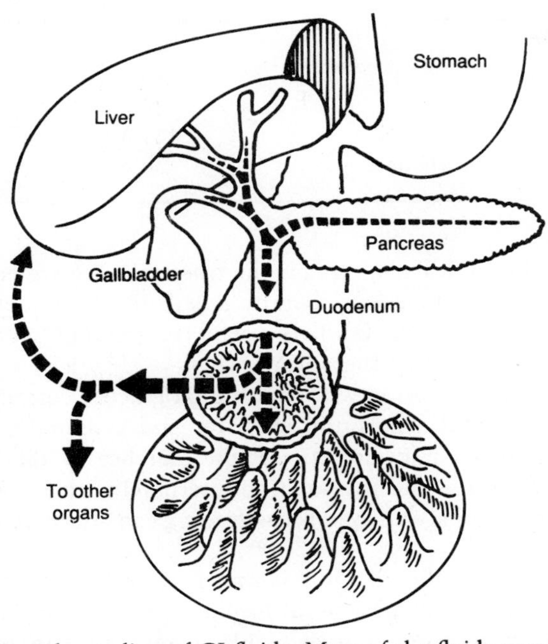

1. Bile and pancreatic juices enter the duodenum through the common bile duct.
2. The intestinal walls contribute additional juices.
3. Most of these fluids are absorbed into the body.
4. Some fluids leave the body with the feces.

A. Normal recycling of GI fluids. Most of the fluids secreted into the GI tract are returned to the circulation; only a little fluid is excreted.

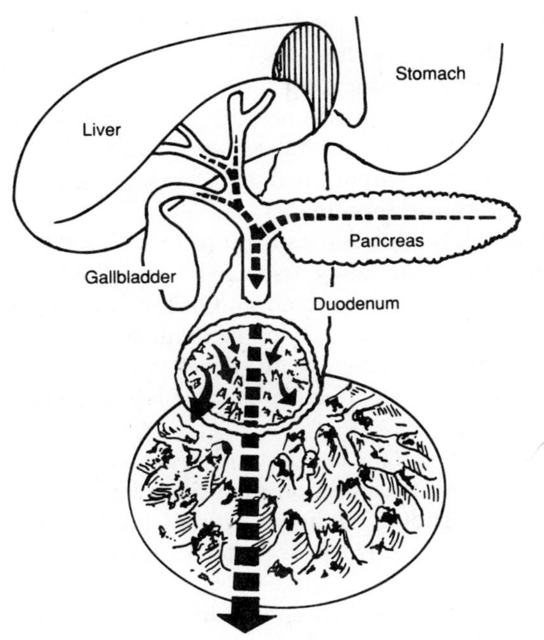

1. Bile and pancreatic juices enter the duodenum as usual.
2. The intestinal walls are damaged. They lose not only the normal intestinal secretions but much additional body fluid, blood and whole cells.
3. Little or no absorption takes place across the intestinal wall.
4. Massive amounts of fluids leave the body.

B. Losses of body fluids in diarrhea. In diarrhea there is little or no recycling of GI fluids into the body. Also, fluids, including blood, and eroded tissue are lost from damaged villi at an accelerated rate.

Table FP6–3 World Health Organization ORT Formula
The World Health Organization (WHO) recommends this ORT formula.

Nutrient	Concentration
Sodium	90 mmol/l
Chloride	80 mmol/l
Potassium	20 mmol/l
Glucose	111 mmol/l
Citrate tribasic or bicarbonate	30 mmol/l

intravenous (IV) fluid therapy: the administration of nutrient solutions through a vein.

A simple ORT recipe: 1 c boiling water.
2 tsp sugar.
A pinch of salt.

in cases of severe dehydration, ORT can replace the traditional treatment of intravenous (IV) fluid therapy. Intravenous therapy is still most valuable in treating severely dehydrated children; ORT is useful in treating mild to moderate cases and following initial IV therapy.[17]

ORT provides an electrolyte solution with an optimal glucose concentration, which favors rapid intestinal absorption of water and sodium.[18] This solution reverses dehydration but may not correct the diarrhea. If diarrhea continues, the child receives water and the solution alternately. Infants with diarrhea usually tolerate breast milk well, and breastfeeding can alternate with supplements of the solution.

Perhaps the most significant value of ORT is that it is oral—it does not require hospitalization, as with IV or parenteral feedings. In addition, if the specific WHO recommended solution is unavailable, caretakers can make other suitable ORT solutions. A mother who lives miles from the nearest pharmacy or clinic and who does not have the resources to purchase medicines anyway can prepare a solution to refeed her dehydrated infant. A properly prepared rice powder solution facilitates the absorption of electrolytes and water, as well as limiting the duration of diarrhea. The primary role of health care providers, then, becomes one of educating those who care for children. The people of a community must learn how to prepare a rehydration solution from ingredients available locally. They must learn to measure ingredients carefully and to use sanitary water. Parents will also want to learn how to recognize diarrhea and dehydration symptoms.

The success of ORT in developing countries is central to WHO's effort to counter the dehydration and death commonly associated with diarrhea. Yet in developed countries, physicians are reluctant to adopt ORT, recommending IV fluids instead.[19] In some cases, the practice of giving IV fluids serves as the only justification to hospitalize a child. However, treating well-nourished children in developed countries with ORT is as effective as, and less expensive and invasive than, IV therapy.

Once rehydrated, children can resume eating foods. At first, they tolerate frequent, small meals best. If food consumption intensifies the diarrhea and threatens dehydration, then they should again go without food temporarily. Clinical observation and the child's willingness are often the best determinants of whether to offer food.

The nutrient needs of an undernourished child with diarrhea are exceptionally high. Convalescence time must include food consumption of greater quantity and higher quality than normal to replenish nutrient losses and to allow for catch-up growth. The nutrient intake must compensate for impaired intestinal absorption, support a raised metabolic rate, repair damaged tissue, and make up for growth losses. A conservative goal during convalescence is to provide at least 25 percent more energy than the average requirement for healthy children.[20] A protein intake twice as high as the recommendation for healthy children covers the catabolic and malabsorption losses incurred and the inefficient use of protein when energy intake or absorption is inadequate. These recommendations serve as guidelines and should be adjusted according to the child's growth response.

Constipation

The consequences of constipation are far less life-threatening than those of diarrhea. Each child's digestive tract responds to food uniquely, with its own rhythm. When a child receives the signal to defecate and ignores it (as active children having fun will do), the signal may not return for several hours. During this time, the intestine continues withdrawing water from the fecal matter, so that when the child does take time to defecate, the bowel movement is dry and hard. Bowel movements that are hard and passed with difficulty, discomfort, or pain define constipation. The amount of time that has elapsed since the previous bowel movement is irrelevant. In the case of painful bowel movements, a parent will want to consult with a physician in order to rule out the presence of organic disease.

constipation: the condition of having painful or difficult bowel movements (elapsed time between movements is not relevant).

Some fibers attract water into the digestive tract, thus softening the stools and preventing or relieving constipation. For this reason, increased fluid intake should accompany increased fiber intake. Wheat bran is one of the most effective stool-softening fibers, although convincing a child to eat bran regularly can be a challenge. One group of pediatricians recommends 1 to 2 quarts of popped popcorn per day to relieve constipation.[21] They found this "treatment" to soften stools and increase stool volume, thus providing an enjoyable and inexpensive solution to the problem of constipation. Fluids also help to relieve constipation by augmenting stool weight and softness. Children may prefer fruit juices to water, and these are acceptable alternatives.

The Hospitalized Child

At times, children may require hospitalization. Health care providers encounter unique problems when feeding children in the hospital. To effectively solve these problems, they need an understanding of the concepts underlying diet therapy and child development. This discussion does not examine the specific dietary treatments of diseases, but does recognize a child's developmental needs.

To work effectively with an infant or child, a health care provider must work effectively with the family and community. Family members must feel comfortable and be able to communicate openly with hospital staff. Their participation in the child's care helps family members to feel needed and helps the child to recover.

Mealtimes

Much of the discussion in Chapter 7 on feeding children applies to hospitalized children as well. Hospitalized children require careful attention to ensure that their nutrient needs are met. Health care providers and parents must also be sensitive to the child's emotional needs. Pointers from people experienced in working with hospitalized children include:

- Notice the child's posture. Body language can indicate fear, pain, or discomfort.
- Touch the child often and lovingly. Touch communicates more than words.
- Allow the child to choose what foods to eat as much as possible. If permissible, foods brought from outside the hospital can help stimulate appetite.
- Encourage the child to eat the food; putting it in front of a child is not enough. Notice the quantities and types of food not eaten.
- Stay with the child during the meal, or make sure a caring person is present. The child will eat and digest food better if someone is nearby to soothe anxieties and loneliness.
- Encourage the child to eat the most needed foods first.This ensures that the child will receive valuable nutrients before becoming too full to complete the meal.
- Allow children to eat with other children, if possible. They will enjoy mealtimes more, accept more food, and eat for a longer period.
- Avoid painful procedures near mealtimes. The stress of pain or fear shuts down digestion and turns off interest in food.
- Serve small servings of well-liked foods.
- Serve foods attractively, selecting a colorful variety of foods. Cut foods into different shapes. Arrange foods in patterns, such as faces or vehicles.

Even though its effects may not be immediately obvious, nutrition care contributes importantly to a child's recovery. Providing nourishment is sometimes not as simple as serving a meal. Special circumstances may require providing nourishment through alternative methods.

Tube Feedings

Tube feedings nourish children who have functioning digestive tracts but are unable to orally ingest enough nutrients to meet their needs. This may be due to a physical problem that impairs chewing or swallowing, lack of appetite, coma, or intense nutrient requirements. Physicians determine the most appropriate formula and feeding route for each case, after completing a thorough nutrition assessment (see Appendix A).

At first glance, tube feedings may appear to be a horrible experience. A closer inspection of the procedure reveals that tube feedings offer lifesaving nutrition in many critical situations. When health care providers select the correct size and type of tube and prepare the child for the procedure, there is little discomfort. They can explain the tube insertion and feeding procedure to a child by using dolls or stuffed animals. They can alleviate parental discomfort in handling the child or fears of causing pain or dislodging the feeding tube by showing parents how to hold and move the child.

For infants and young children, especially small feeding tubes minimize discomfort and interference with the airways. Removal of the feeding tubes after each feeding frees the infant's airways between feedings and reduces the risk that formula will back up into the esophagus and enter the lungs. The feeding route for infants is usually from the mouth to the stomach, rather than from the nose (as is common for older children and adults). This route allows infants to breathe more easily, because infants breathe through their noses, not their mouths.

An infant's stomach is small, and gastric emptying is slow. For these reasons, the quantity of formula in a tube feeding must be small. If too much formula empties into the stomach at one time, complications can quickly arise. Health care workers who continually monitor the formula's concentration, infusion rate, and volume help to ensure tolerance and avoid problems.

Children are not simply growing; they are also developing. They learn new skills as they grow. Health care workers must always be aware of the developmental age of a child. Self-feeding skills missed at the appropriate age may be difficult to learn at a later age. Thus infants fed by tube feedings should have partial feedings by bottle, if possible. Using a pacifier during a tube feeding helps to maintain an association of sucking and swallowing with eating and fullness. Likewise, older infants must learn to use and accept food by spoon, even when primary nourishment is from tube feedings. A therapist can help stimulate appropriate development when problems occur in cases of long-term tube feedings.

At times, physicians use infusion pumps to administer tube feedings to young children. Such a procedure requires additional precautions. The bright lights, interesting sounds, and many controls of the pump stimulate a child's curiosity. To ensure safety to the child and to the pump, health care workers must position the pump a safe distance from the child's bed.

Adolescents are more likely to physically tolerate and adapt to tube feedings than are younger children. However, their social and psychological development may interfere with their acceptance of tube feedings. The prospect of tubes feeding them can horrify teenagers, particularly those overly concerned with appearances. These suggestions might help:

- Encourage teens to be as active as possible.
- Allow teens to dress in their own clothes and to bring their favorite personal items from home, when possible.
- Explain the tube-feeding procedure and its importance.
- Include teens in decisions. Allow them to arrange daily schedules and help with feedings, when possible.
- Encourage time with friends and favorite activities.

Health care workers best serve teenagers by being available to listen to their fears and their problems.

Tube feedings offer an alternative when children are unable to eat regular meals. When the digestive tract cannot handle meals or tube feedings, IV nutrition must be employed.

Intravenous Nutrition

Intravenous nutrition is required when the digestive system is not functioning. Many of the concerns that arise with tube feedings also arise when infants and children receive IV feedings. For example, it is important to involve parents in the child's care, develop the child's skills at the appropriate age, maintain a safe distance between the child and the infusion pump, and ease the child's and family's fears with proper instructions.

To feed an infant intravenously poses additional problems. The particular nutrient needs of each infant, especially of premature infants, are difficult to determine. To feed a mixture of water, glucose, electrolytes, amino acids, fats, vitamins, and trace elements that will adequately promote growth without overloading an immature body is a challenge. Such mixtures may omit substances that naturally occur in foods and that infants require, but that are still unknown.

The infant's renal system is immature and cannot adjust to changes in the blood's composition as rapidly as can an adult's renal system. Furthermore, an infant's body contains a larger percentage of fluid, so fluctuations cause major problems. Delivering too much or too little of any solution constituent directly into an infant's vein leads to immediate imbalances. Too much fluid can quickly stress the infant's immature renal system; too little can quickly cause dehydration. Too much glucose can rapidly lead to hyperglycemia, with severe consequences; ceasing a feeding or dislodging the tube too rapidly can precipitate hypoglycemia. The list of such possible complications could go on for virtually every substance found in the IV solution. Health care workers must explore all possible sources of any complications that may arise.

The care of children in times of sickness requires careful attention to their nutrition needs. To feed children is to provide them with much more than nutrients and fluids alone. Foods carry both a physical and an emotional comfort. Chicken noodle soup does offer fluids, some energy, a little protein, and a variety of vitamins and minerals, but its healing power also comes from the caretaker's concern about the child. Children given tender loving care recover more quickly than those deprived of it. As Dr. F. W. Peabody of Harvard University said, "The secret of the care of the patient is in the caring for the patient."

Focal Point 6 Notes

1. M. S. Kramer, L. Naimark, and D. G. Leduc, Parental fever phobia and its correlates, *Pediatrics* 75 (1985): 1110–1113.
2. H. D. Jampel and coauthors, Fever and immunoregulation: III. Hyperthermia augments the primary in vitro humoral immune response, *Journal of Experimental Medicine* 157 (1983): 1229–1238.
3. E. Atkins, Fever: The old and the new, *Journal of Infectious Diseases* 149 (1984): 339–348.
4. D. O. McCarthy, M. J. Kluger, and A. J. Vander, Suppression of food intake during infection: Is interleukin-1 involved? *American Journal of Clinical Nutrition* 42 (1985): 1179–1182.
5. American Academy of Pediatrics, Nutrition and infection, in *Pediatric Nutrition Handbook*, eds. G. B. Forbes and C. W. Woodruff (Elk Grove Village, Ill.: American Academy of Pediatrics, 1985), pp. 267–273.
6. A. Hecht, Fever: What to do—and what not to do—when the heat is on, *FDA Consumer*, November 1985, pp. 16–18.
7. Reye syndrome: New research, regulation, *FDA Consumer*, March 1986, pp. 2–30.
8. J. D. Snyder, From Pedialyte to Popsicles: A look at oral rehydration therapy used in the United States and Canada, *American Journal of Clinical Nutrition* 35 (1982): 157–161; Z. Weizman, Cola drinks and rehydration in acute diarrhea (letter), *New*

England Journal of Medicine 315 (1986): 768.

9. L. M. Roberson, A. J. McLaughlin, and J. K. Lund, Promoting oral rehydration therapy for acute diarrhea, *Journal of the American Dietetic Association* 87 (1987): 496–497.
10. National Research Council, Commission on Life Sciences, Food and Nutrition Board, Committee on International Nutrition Programs, Subcommittee on Nutrition and Diarrheal Diseases Control, *Nutritional Management of Acute Diarrhea in Infants and Children* (Washington, D.C.: National Academy Press, 1985).
11. National Research Council, 1985.
12. K. H. Brown and W. C. MacLean, Jr., Nutritional management of acute diarrhea: An appraisal of the alternatives, *Pediatrics* 73 (1984): 119–125.
13. Brown and MacLean, 1984.
14. R. E. Black and coauthors, Longitudinal studies of infectious diseases and physical growth of children in rural Bangladesh: Incidences of diarrhea and association with known pathogens, *American Journal of Epidemiology* 115 (1982): 315–324, as cited by Brown and MacLean, 1984.
15. Brown and MacLean, 1984.
16. National Research Council, 1985.
17. T. I. Bhutta, Oral rehydration for diarrhea (letter), *New England Journal of Medicine* 307 (1982): 952.
18. Roberson, McLaughlin, and Lund, 1987.
19. M. Santosham and coauthors, Oral rehydration therapy of infantile diarrhea: A controlled study of well-nourished children hospitalized in the United States and Panama, *New England Journal of Medicine* 306 (1982): 1070–1076.
20. National Research Council, 1985.
21. D. Chen and B. Sullivan, Constipation (letter), *Pediatrics* 77 (1986): 933.

Nutrition during Childhood

7

Children by Paul T. Granlund.

The quantities of nutrients a child needs continue to change throughout the growing years. Individual variables such as genetic constitution, rate of growth, gender, and previous nutrition and health status continue to influence the body's requirements. When nutrient supplies fall short of needs, growth, health, and behavior are all affected.

To deliver nutrients in the form of meals and snacks that are nutritious and delicious to children is challenging for their caretakers. Children develop likes and dislikes without regard to their nutrient needs, and they are easily influenced by peers, the media, and their taste buds. Yet, of a food's many qualities, its nutritional quality has the greatest impact on a child's health. Childhood is the time of continuing to develop food habits that were rooted in infancy—habits that will be carried into the future. This chapter explores the world of nutrition for preschool and school-age children.

Growth, Development, and Assessment

The growth of children slows after the first year, but over the 10 to 12 years of childhood, its cumulative effects are dramatic. Many developmental changes take place to transform the one-year-old child into a child of 12, ready to start the teen years. As with infants, growth is the parameter always monitored as an index of the nutrition status of children.

Growth and Development

2½ inches is approximately 5 cm.

5 lb is 2.2 kg.

After having grown all of 10 inches during the first dramatic year of life, a child grows approximately 5 inches in height between the ages of one and two.[1] Thereafter, the rate slows to about 2½ inches per year until adolescence, when this steady rate of increase in height rises abruptly and markedly. Like growth in height, weight gain settles into a similar pattern, averaging out as an annual increase of approximately 5 to 6 pounds per year until the onset of adolescence.

Increases in height and weight are only two of the many dramatic changes growing children experience. At age one, children can stand alone and are beginning to toddle; by two, they can walk and are learning to run; by three, they can jump and are climbing with confidence.[2] Bone and muscle tissue increase in both mass and density, making these new accomplishments possible.

The development of fetal bones into those of an adult involves a series of well-coordinated dynamic processes. Bone tissue is continually remodeled throughout life, but the turnover of skeletal tissue varies with age. Each bone's outer shell of cortical bone grows by adding new tissue on its outer surfaces and resorbing tissue from its inner surfaces. Meanwhile, the inner tissue, composed of spongy, or trabecular, bone, constantly rearranges itself in response to the stresses of weight and exercise. During the growing years, formation exceeds resorption, and calcium balance is positive.

Skeletal muscles grow by increasing both muscle fiber size and number. Muscle mass is reflected in the amount of creatinine excreted in the urine, since muscle tissue is the site of creatinine production (see "Assessment of Nutrition

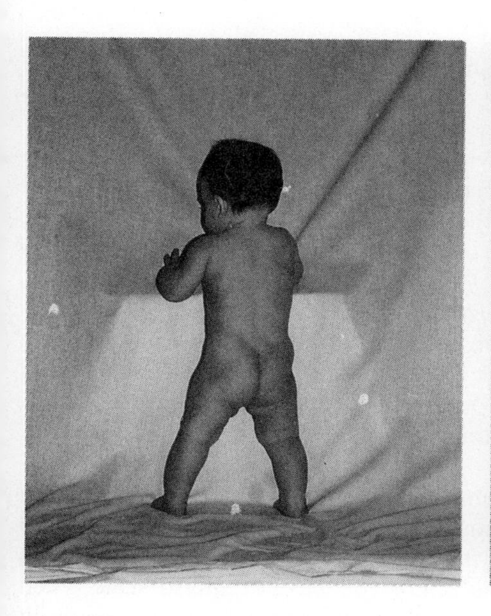

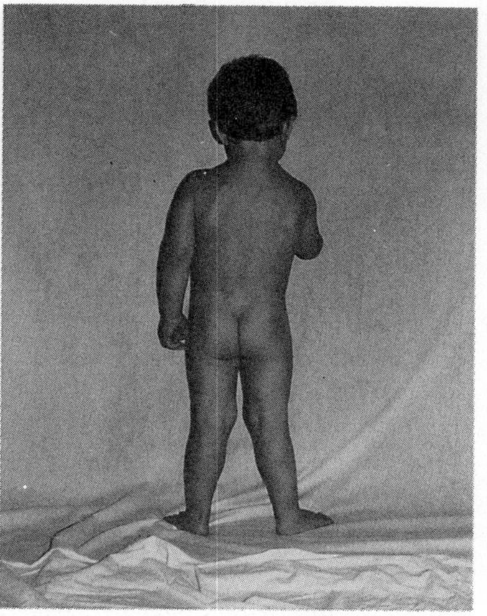

One-year-old and two-year-old shown for comparison of body shape. The two-year-old has lost much of the baby fat; the muscles (especially in the back, buttocks, and legs) have firmed and strengthened, and the leg bones have lengthened.

Status"). The lengthening of bones and development of muscles are reflected outwardly as the obvious signs of a child's growth. Other tissues such as connective tissues, teeth, body fat, skin, and the nervous system are also growing.

As children enter the second year of life, although growth slows, development continues to progress rapidly. The brain and central nervous system mature at a tremendous pace, as evidenced by increasing muscle control, coordination, and the ability to perform new skills. By the age of two, most of the primary teeth have erupted, and control of the jaw muscles is voluntary. During the second year of life, children can handle significantly more types of foods than they could at age one.[3]

At the same time, after a year of age, children who as infants were more than willing to taste anything and everything, whether it was food or not, become assertive and selective about what they will ingest and how they will do it. This behavior reflects children's psychological development.

Assessment of Nutrition Status

Nutrition status assessment of children relies on the same four approaches as are used for infants and adults. A fifth approach—observation of behavior—is considered useful by some assessors, and is discussed here as part of the physical examination. The section "Nutrition and Behavior," later in the chapter, shows the ways in which nutrition affects behavior, and the ways in which it does not.

Historical data A child's nutrition status often correlates with the economic status of the family. The socioeconomic history, including the extent of participation in food assistance programs, can reveal useful information regarding the child's food intake. Questions similar to the ones listed in the history section in Chapter 6 will provide clues about a child's nutrition status.

Unlike infants, young children often snack between meals. Snacks can have a major influence on a child's nutrient intake, so it is important to ascertain which foods a child snacks on.

Diet history information can provide clues to children's iron and calcium status, two nutrients that are of special concern for growing children. If the assessor suspects that one or both of these nutrients is lacking in a child's diet, nutrition counseling for the parents, and further evaluation of the child, is appropriate.

As discussed in Focal Point 7, overweight children often become overweight adults. Diet history information can identify eating habits that promote obesity. At the other extreme, the diet history can help identify children with less than adequate energy intakes—whether due to poverty or to parent-imposed food restrictions that undermine the child's health.

Anthropometric measurements As soon as they can stand upright, children can be weighed on the same beam balance used for adults. Assessors can also measure their standing height, but they must use care in doing so. It is especially important never to use the flimsy rod on a standard scale when measuring children's height, for the errors that arise in doing so are greater in proportion to a child's small height than to an adult's. The best way to measure height is with the child's back against a flat wall alongside an affixed, nonstretchable measuring tape or stick (see Figure 7–1). The child stands erect, without shoes, with heels together. The child's line of sight should be horizontal, with the heels, buttocks, shoulders, and head touching the wall. The assessor uses a block, book, or other inflexible object to ensure that the top of the head is measured at an exact right angle to the wall. The assessor carefully checks the height measurement and immediately records the result in either inches or centimeters. Such a practice prevents misplacing or forgetting the measurement. To evaluate growth in children, an assessor uses the charts shown in Figures 7–2 (A and B) and 7–3 (A and B).

Anthropometric measures on children can throw light not only on their growth and development but also on the probability of their becoming obese—a likelihood in our society. Weight gains in children that exceed those appropriate for height reflect eating in excess of energy need. In children, the growth charts may provide the first clue to obesity; they are most useful in assessing a child according to that individual child's own growth and development.

Childhood obesity is the subject of Focal Point 7.

Another index used to evaluate childhood obesity is the Eid index, named for the person who developed it. This index assumes the child is at the age appropriate for height and then asks if the weight is appropriate for that age. The assessor measures the child's height, then turns to the height-for-age chart and notes the age at which the child's height would be at the 50th percentile. The assessor then looks to see what weight would be appropriate for that age (the weight at the 50th percentile for that age on the weight-for-age chart). The assessor then calculates the child's actual weight as a percentage of this weight. The formula is:

$$\frac{\text{Actual weight}}{\text{50th percentile weight for the age at which the child's height is in the 50th percentile}} \times 100$$

Figure 7–1 Nutrition Assessment of the Child: How to Measure Height
Height is measured most accurately when the child stands against a flat wall to which a measuring tape has been affixed. The movable bar or board that comes in contact with the child's head should be at an unvarying 90° angle to the wall.

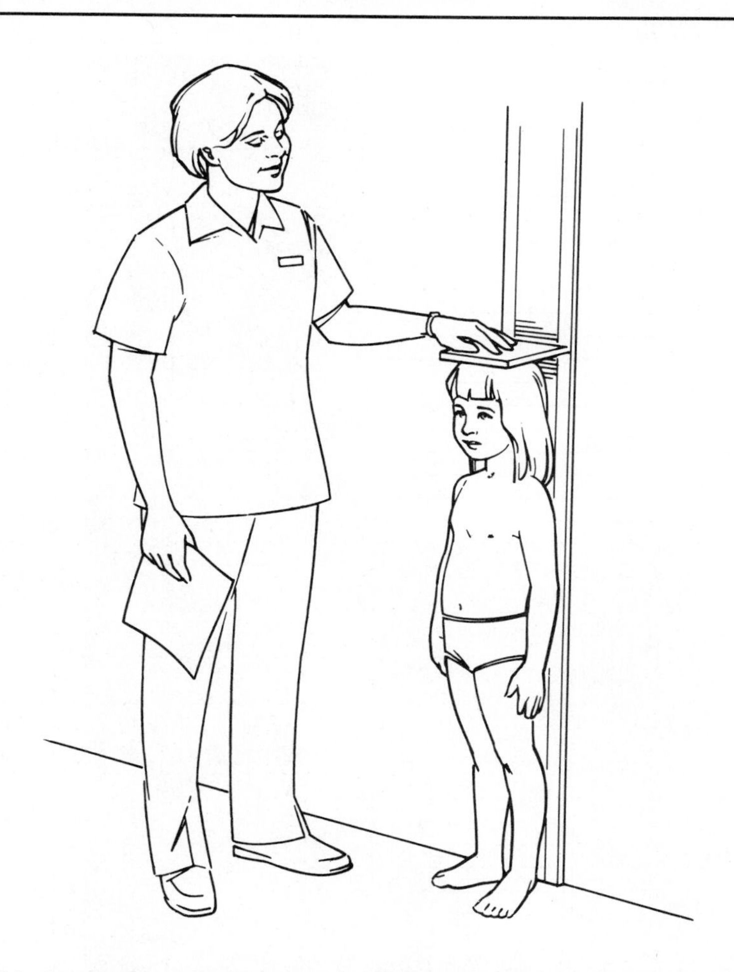

The Eid index permits comparison of children at similar developmental stages, regardless of age.[4]

Another index, the developmental index, considers that growth rates vary in any one child over time, and also differ among different children. It also recognizes that *normal* growth involves weight gain; weight gain, even in an obese child, is fine as long as height and lean body mass are gained faster than fat. The developmental index calculates a weight *change* adjusted for atypical changes in height and permits the assessor to check whether the observed change in weight is appropriate for the actual change in height. The index is expressed as a ratio of expected to actual changes in height. The adjusted weight is:

$$\text{Actual weight change} - \frac{\text{actual height change}}{\text{expected height change}} \times \text{expected weight change}$$

Figure 7–2A Nutrition Assessment from 2 to 18 Years: Height and Weight for Age—Girls

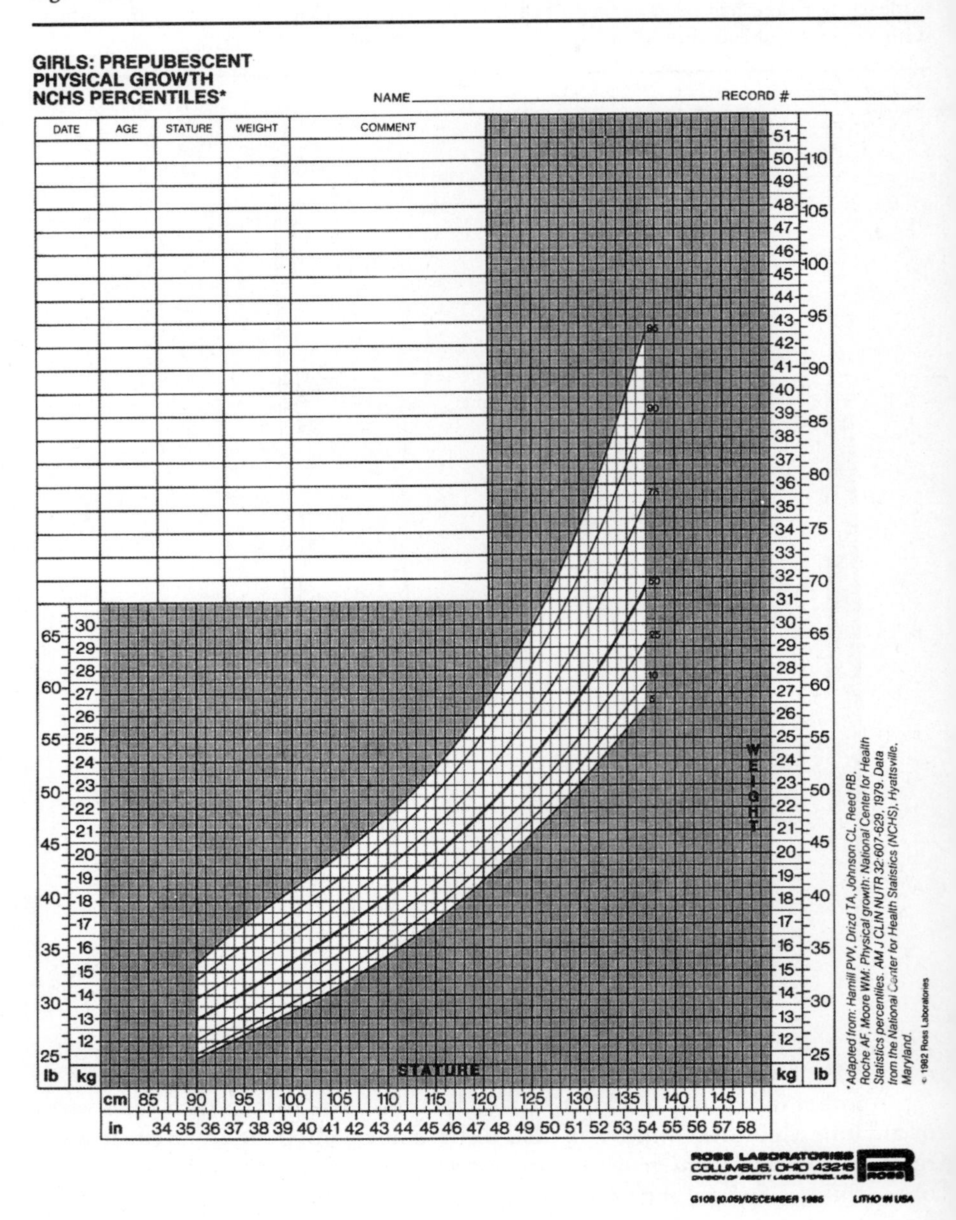

Source: Used with permission of Ross Laboratories, Columbus, OH 43216, from NCHS Growth Charts, ©1982 Ross Laboratories.

A careful visual assessment of a child can determine whether these measures are even required. Often, the best clinical assessment of obesity is the trained eye. It is quite likely that those who are visually identified as obese are also obese by other criteria as described in Focal Point 7. When a child does not appear obese, or if the assessment is questionable, both weight-for-height (as

Figure 7–2B **Nutrition Assessment from 2 to 18 Years: Height and Weight for Age—Boys**

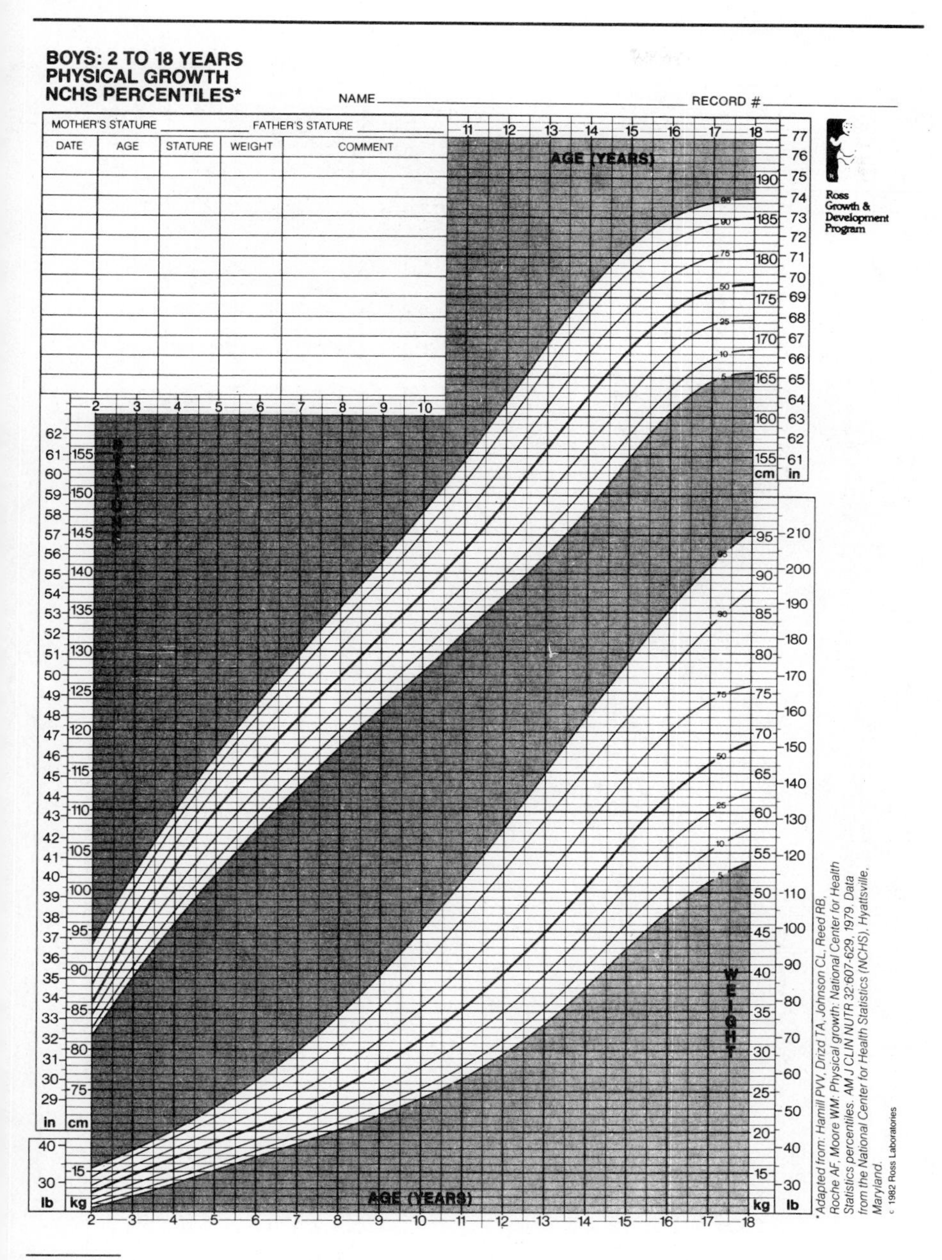

Source: Used with permission of Ross Laboratories, Columbus, OH 43216, from NCHS Growth Charts, ©1982 Ross Laboratories.

shown in Figures 7–2 A and B and 7–3 A and B) and triceps fatfold measurements (as in Appendix A, Table A-5) should be used to confirm the diagnosis.

Protein-energy malnutrition (PEM) is detectable by means of anthropometric measures. If height for age is 85 percent or less of the median height for

Figure 7–3A Nutrition Assessment during Prepubescence: Weight for Height—Girls

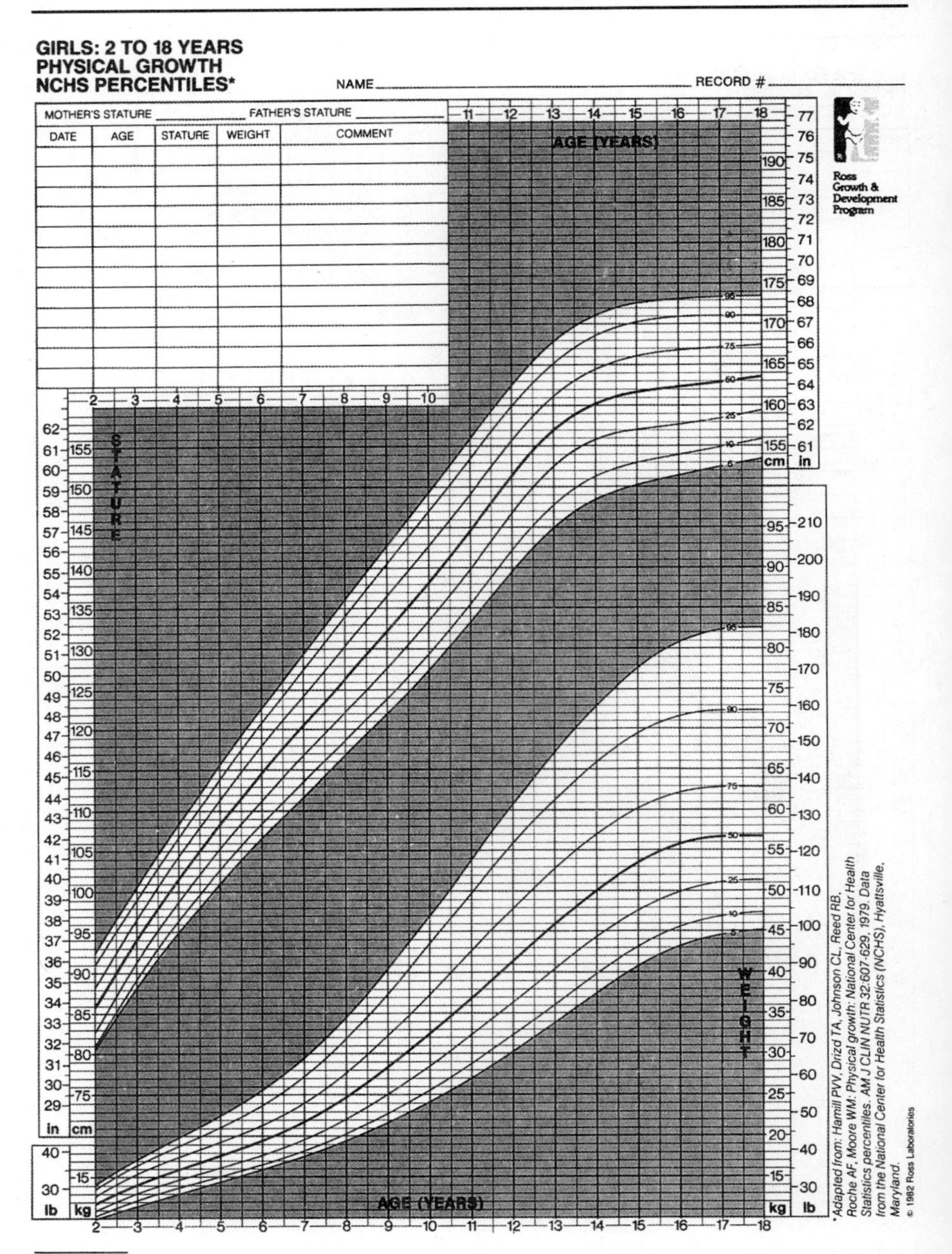

Source: Used with permission of Ross Laboratories, Columbus, OH 43216, from NCHS Growth Charts, ©1982 Ross Laboratories.

age, this indicates stunted growth of a degree consistent with effects of severe PEM. If weight for age is 70 percent or less of the median weight for age, this indicates wasting consistent with severe PEM.[5] Physical examination and biochemical testing can help pinpoint the type of PEM (marasmus or kwashiorkor) so that appropriate therapy can be instituted.

Figure 7–3B Nutrition Assessment during Prepubescence: Weight for Height—Boys

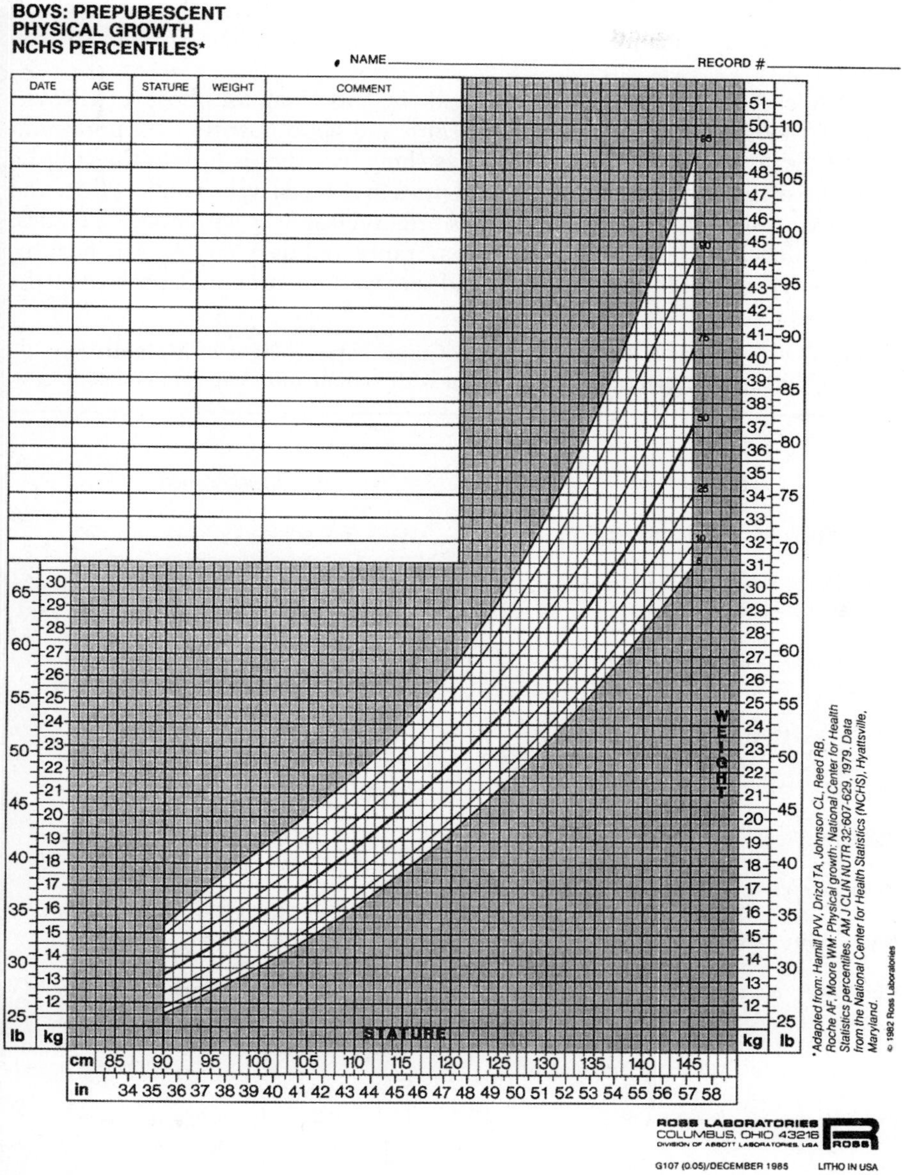
BOYS: PREPUBESCENT
PHYSICAL GROWTH
NCHS PERCENTILES*
NAME
RECORD #
DATE
AGE
STATURE
WEIGHT
COMMENT
STATURE
WEIGHT
kg
lb
cm
in
*Adapted from: Hamill PVV, Drizd TA, Johnson CL, Reed RB, Roche AF, Moore WM: Physical growth: National Center for Health Statistics percentiles. AM J CLIN NUTR 32:607-629, 1979. Data from the National Center for Health Statistics (NCHS), Hyattsville, Maryland.
© 1982 Ross Laboratories
ROSS LABORATORIES
COLUMBUS, OHIO 43216
DIVISION OF ABBOTT LABORATORIES, USA
ROSS
G107 (0.05)/DECEMBER 1985
LITHO IN USA

Source: Used with permission of Ross Laboratories, Columbus, OH 43216, from NCHS Growth Charts, ©1982 Ross Laboratories.

Physical examination The effects of malnutrition during childhood are not limited to growth impairment; they are diverse and numerous, and include behavioral as well as physical effects. If a child looks unhealthy and acts abnormally, consider that the cause *may* be malnourishment. This may sound obvious, but surprisingly, parents and medical practitioners often overlook the possibility that malnutrition may account for abnormalities of appearance and behavior. Any departure from normal, healthy appearance and behavior is a possible sign of poor nutrition. Figure 7–4 shows the physical signs to watch for in assessing nutrition status; Table 7–1 lists the signs specific to iron deficiency.

While the signs of malnutrition can be mistaken for those of disease, neglect, or other causes, the signs of health and good nutrition are unmistakable. A healthy, well-nourished child has shiny hair that is firm in the scalp; a malnourished child, especially one with protein deficiency, has dull, brittle, and loose hair. The well-nourished child has bright, clear eyes with no dark circles; the iron-deficient child has pale eye membranes and dark circles. The healthy, well-nourished child has bright eyes and sees well; in vitamin A deficiency, the

Figure 7–4 Nutrition Assessment of the Child: Physical Signs of Malnutrition
The physical signs shown here are consistent with malnutrition but not diagnostic of it.

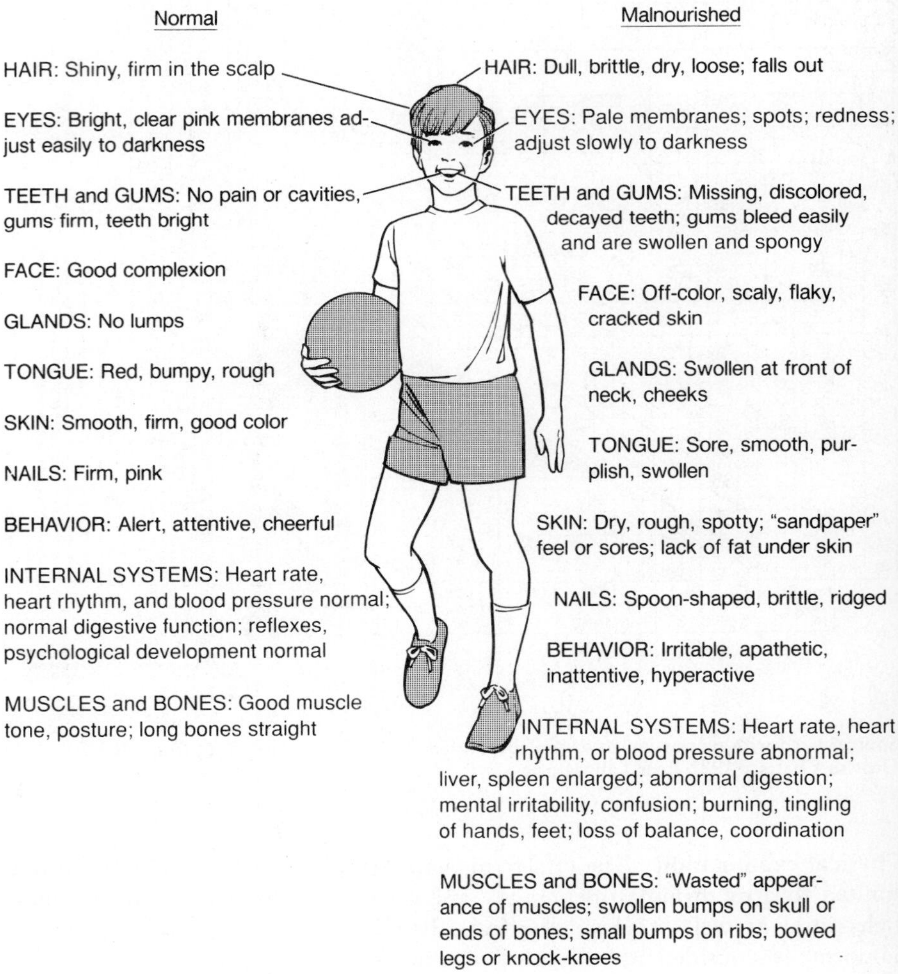

Table 7–1 Nutrition Assessment of the Child: Signs of Iron Deficiency

GI Tract

- Lactose intolerance, and possibly intolerance to other sugars.
- Increased risk of lead and cadmium poisoning.

Immune System

- Reduced resistance to infection (lowered immunity).

Nervous/Muscular Systems

- Reduced work productivity, tolerance to work, and voluntary work.
- Reduced physical fitness.
- Weakness.
- Fatigue.
- Impaired cognitive function.
- Reduced learning ability.
- Increased distractibility (inability to pay attention).
- Impaired visual discrimination. Impaired reactivity and coordination (also in infants).

Skin

- Itching.
- Pale nailbeds, eye membranes, and palm creases.
- Concave nails.
- Impaired wound healing.

General

- Reduced resistance to cold, inability to regulate body temperature.
- Pica (clay eating, ice eating).

Source: Adapted from L. Hallberg, Iron absorption and iron deficiency, *Human Nutrition: Clinical Nutrition* 36C (1982): 259–278; N. S. Scrimshaw, Functional consequences of iron deficiency in human populations, *Journal of Nutrition Science and Vitaminology* 30 (1984): 47–63.

eyes adjust slowly to dark, and the skin is dry. The teeth of the well-nourished child are bright and healthy, and the gums are firm; if the gums bleed easily, this may be a symptom of vitamin C deficiency. Strong, straight bones are indicative of adequate nutrition; bent bones, of vitamin D malnutrition. Strong muscles reflect adequate energy and protein intake; a wasted appearance of muscles occurs with energy deprivation, as the body breaks down its own protein for energy. Poor growth, apathy, weakness, and poor skin color also suggest protein deficiency; edema in the belly and legs indicates that fluid balance is disturbed. Frequent infections also suggest malnutrition with respect to protein and other nutrients. Poor appetite and impaired taste acuity may indicate zinc deficiency.[6] None of these symptoms is diagnostic of a particular deficiency, but they do suggest the need for further tests.

Biochemical analyses The biochemical tests used for adults are suitable for children also, although in some cases the standards are different. To assess PEM, some of the same measures as for adults are useful, chiefly serum total protein, serum albumin, total lymphocyte count, and urinary creatinine. Standards for these tests for children are shown in Tables 7–2 and 7–3.

The cautions presented in Chapter 1 regarding assessment of nutrition status are especially applicable to protein-energy status. The PEM indicators selected for presentation here, like all such indicators, are sensitive to PEM but not specific to it; they can also reflect altered hormonal states, drug effects, disease states, and others. References on assessment of nutrition status in children provide important qualifications without which the application of these standards could easily lead to wrong results.[7] However, Tables 7–2 and 7–3 provide a reminder important to people working with children: PEM does occur and can be diagnosed.

Standards for vitamin A deficiency are shown in Table 7–4. As for iron assessment, a complete discussion is in Appendix A, together with tables of standards. Recommended screening ages to detect iron deficiency in children are one, between two and three, five, and adolescence.[8]

Determinations of serum cholesterol are recommended, especially for those children with a family history of heart disease (family members who have had heart attacks before age 60 or a history of obesity, hypertension, or hyperlipidemia). Elevated plasma cholesterol concentrations correlate with an increased risk of developing heart disease. Children with plasma cholesterol concentrations above the 50th percentile are at increased risk of having high cholesterol concentrations as adults.[9] Table 7–5 shows plasma cholesterol percentile ratings for children.

Malnutrition in Children

Malnutrition is not common in U.S. or Canadian children in general. A Canadian survey typifies the situation: it examined the nutrient intakes of close to 200 preschool children 3½ to 4 years of age.[10] Energy intakes of the

Table 7–2 Nutrition Assessment from 1 to 17 Years: Standards for Selected PEM Indicators

Biochemical Test	Age	Deficient	Acceptable
Serum total protein (g/dL)[a]	1 to 5 yr	[b]	≥5.5
	6 to 17 yr	[b]	≥6.0
Serum albumin (g/dL)[a]	1 to 5 yr	<2.8	≥3.0
	6 to 17 yr	<2.8	≥3.5
Total lymphocyte count (mm^3)	All ages	<1500	2500
Creatinine-height index[c]	3 mo to 17 yr	<0.5	≥0.9

[a]To convert g/dL to standard international units (g/L), multiply by 10.
[b]The reference provided no line to demarcate deficiency, but the lower the value, the more likely deficiency is.
[c]For creatinine standards for children, see Table 7–3.

Source: Adapted with permission from A. Grant and S. DeHoog, *Nutritional Assessment and Support,* 3rd ed. (Seattle, Wash.: Anne Grant and Susan DeHoog, 1985); H. E. Sauberlich, J. H. Skale, and R. P. Dowdy, *Laboratory Tests for the Assessment of Nutritional Status* (Boca Raton, Fla.: CRC Press, 1979).

Table 7–3 Nutrition Assessment of the Child and Teenager: Standards for Creatinine Excretion (milligrams per 24 hours)

Height (cm)	Both Sexes	Males	Females
55	50.0		
60	65.2		
65	80.5		
70	97.5		
75	118.0		
80	139.6		
85	167.6		
90	199.9		
95	239.8		
100	278.7		
105	305.4		
110	349.8		
115	394.5		
120	456.0		
125	535.1		
130		448.1	525.2
135		480.1	589.2
140		556.3	653.1
145		684.3	717.2
150		812.3	780.9
155		940.3	844.8
160		1068.3	908.8
165		1196.3	
170		1324.3	
175		1452.3	
180		1580.3	

Source: Adapted with permission from D. B. Cheek and coauthors, in *Pediatric Research* 4 (1970): 135–144, and F. E. Viteri and J. Alvarado, in *Pediatrics* 46 (1970): 696–706, as adapted by R. J. Merritt and G. L. Blackburn, Nutritional assessment and metabolic response to illness of the hospitalized child, in *Textbook of Pediatric Nutrition,* ed. R. M. Suskind (New York: Raven Press, 1981), pp. 285–307.

children were sufficient to support normal growth. Average nutrient intakes were above current Canadian recommendations. With the exception of iron, nutrient intakes were above U.S. recommendations as well. Preschoolers' vitamin C intakes in this survey far exceeded recommendations. Two-thirds of these children were taking vitamin supplements, which partially explains the

Table 7–4 Nutrition Assessment from 6 months to 17 years: Standards for Serum Vitamin A

Nutrient	Deficient (μg/dL)[a]	Acceptable (μg/dL)[a]
Vitamin A	<20	≥30

[a]To convert μg/dL to standard international units (μmol/L), multiply by 0.03491.

Source: Adapted with permission from A. Grant and S. DeHoog, *Nutritional Assessment and Support*, 3rd ed. (Seattle, Wash.: Anne Grant and Susan DeHoog, 1985).

Table 7–5 Nutrition Assessment from 0 to 19 Years: Percentile Classifications of Plasma Cholesterol

Age	Cholesterol Mean (mg/100 ml)	Range[a] (mg/100 ml)	Mean (mmol/L)[b]	Range[a] (mmol/L)[b]
0 to 4 yr:				
Males	155	114 to 203	4.00	2.94 to 5.24
Females	156	112 to 200	4.03	2.89 to 5.17
5 to 9 yr:				
Males	160	121 to 203	4.13	3.12 to 5.24
Females	164	126 to 205	4.24	3.25 to 5.30
10 to 14 yr:				
Males	158	119 to 202	4.08	3.07 to 5.22
Females	160	124 to 204	4.13	3.20 to 5.27
15 to 19 yr:				
Males	150	113 to 197	3.87	2.92 to 5.09
Females	158	120 to 203	4.08	3.10 to 5.24

[a]The range represents the values from the 5th percentile to the 95th percentile.
[b]To convert cholesterol (mg/100 ml) to standard international units (mmol/L), multiply by 0.02586.

Source: Adapted from the *The Lipid Research Clinics: Population Studies Data Book*, DHHS (NIH) publication no. 80–1527 (Washington, D.C.: Government Printing Office, 1980), pp. 28–29.

excess. Even without the supplements, however, the children's vitamin intakes were sufficient according to Canadian recommendations.

All of the children in the survey were eating three meals a day and at least one snack. Dinner was the most important meal, providing the most energy and protein. Milk and milk products provided the main source of energy, protein, fat, and calcium. Breakfast supplied the most iron, with cereals and cereal products representing the primary iron food sources. Snacks were less high in kcalories than meals, but still made a substantial contribution to daily energy intake. Without them, energy intake would have been insufficient to support normal growth.

In contrast to average-income children, children of low-income groups experience more malnutrition than most citizens would like to believe. This problem receives attention here because it provides some background for appreciating the importance of meeting nutrient needs—the subject of a later section.

Different malnutrition problems characterize different stages of the life span. To review for a moment, the previous chapter focused on those nutrition disorders that are particularly severe or have particular impact in infancy. Foremost among the diseases that are most severe for infants are PEM and iron deficiency. PEM and iron deficiencies continue to take a toll throughout childhood, and iron deficiency becomes aggravated by lead toxicity. Other deficiency diseases become severe as well—notably vitamins A and D and zinc deficiencies. The distinction between malnutrition problems of infancy and childhood is somewhat artificial; they grade into each other just as the periods of life do, but the emphasis here reflects the diseases that remain or first become severe in childhood.

Protein-Energy Malnutrition

Protein-energy malnutrition, which takes a toll both domestically and worldwide, continues to afflict children after the first year of life, even in the United States. Worldwide, PEM takes a devastating toll on children. It affects over 100 million children in Africa, Latin America, and Asia, and causes or contributes to nearly half of the deaths of children under four worldwide.[11] It is less common in most parts of developed countries but is still prevalent among sick children. In the United States, PEM affects a third to a half of all hospitalized children.[12] Disease and infection are its most common precursors there, while outside the hospital, poverty is its most common antecedent. In all cases, other nutrient deficiencies accompany PEM.

Vitamin A Deficiency

Vitamin A deficiency is seldom seen in developed countries, but it is a vast problem worldwide, affecting more than 5 million children. Hundreds of thousands of children in the developing countries of the world go blind each year from vitamin A deficiency. They may also experience stunted growth, decreased appetite, and increased infections and illness. Approximately 20 to 35 percent of all childhood diseases in developing countries are related to vitamin A deficiency.[13] Vitamin A deficiency is an enormous, yet preventable, problem.

When children who are vitamin A deficient are given a vitamin A supplement, they gain weight and grow taller.[14] Vitamin A–supplemented children also benefit from a stronger immune system. In a study in Indonesia of 25,000 preschool children, the distribution of two large-dose vitamin A capsules per year resulted in a 34 percent reduction in childhood deaths.[15]

Vitamin D Deficiency

Worldwide, the vitamin D–deficiency disease rickets afflicts large numbers of children. The clinical signs of rickets include growth failure, bone deformity, listlessness, and delayed motor development. In the United States and Canada, rickets was at one time a common disease, but it has become much less prevalent since the addition of vitamin D to cow's milk. Most milk available in the United States and Canada, including nonfat and low-fat milk, is fortified with vitamin D. In addition, vitamin D forms in response to the action of sunlight on the skin, although the amount formed depends on skin color, duration of exposure, and atmospheric pollution. Fresh air and sunshine provide not only a pleasant outing, but needed vitamin D as well.

Rickets was virtually nonexistent domestically during the 1950s, 1960s, and early 1970s, but medical workers reported several cases around 1980. In Connecticut, physicians diagnosed four cases of rickets in children between 18 months and three years of age within one year.[16] In Philadelphia, physicians reported 24 cases of vitamin D–deficiency rickets in the late 1970s; in Chicago, they noted another ten cases of children with the disease.[17] Based on observation of these children, researchers have identified several risk factors for rickets.[18] These include premature birth; pigmented skin; lack of exposure to sunlight; prolonged, unsupplemented breastfeeding; and vegetarian diets.

Iron Deficiency

Iron-deficiency anemia is another major worldwide problem, and is the most prevalent nutrient deficiency among children in the United States and Canada.[19] In all national surveys, children's iron intakes are below recommended levels. The effects of iron deficiency are evident in the physical, mental, behavioral, and biochemical signs described in the assessment sections of this chapter and Appendix A. Almost one out of every ten children between the ages of one and two is iron deficient.[20] The prevalence of iron deficiency is slightly lower (approximately 1 out of every 15) in children between the ages of three and ten. The high iron needs of growth combined with typically low iron intakes leave many children with marginal iron status.

Of all of the nutrition problems accused of causing abnormal behavior, iron deficiency is probably most often guilty as charged. See "Nutrition and Behavior," later in this chapter, for further details of its effects.

Zinc Deficiency

Zinc deficiency is as widespread as protein-energy and vitamin A deficiencies, and the three typically occur together. Mild cases of zinc deficiency are identifiable in less severe conditions and have been studied as a separate entity. Poor appetite and depressed growth have been observed in children with chronic, mild zinc deficiency.[21] Pronounced growth retardation and impaired sexual development result from severe zinc deficiency.[22]

Surveys show that zinc intakes of young children in the United States are low.[23] Researchers gave zinc supplements to preschool children who had been eating diets low in zinc and who were short for their age. Compared with controls, the supplemented children grew at a significantly greater rate in one year.[24] Although the energy and protein intakes of both groups of children did not differ significantly at the beginning of the study, the zinc-supplemented children's intakes were greater by the end of the study. This suggests that the accelerated growth of the supplemented children may, in part, be due to improved appetite. More research is needed to confirm that mild zinc deficiency contributes to poor growth in children, but such research has implications for children worldwide.

Lead Toxicity

Lead toxicity occurs in much the same population as iron deficiency and often in the same individuals, and each of the two conditions makes the other more likely. Lead toxicity is widespread—one out of every six children between the ages of six months and five years are exposed to threatening levels of lead.[25] Mild lead toxicity has nonspecific effects, including diarrhea, irritability, reduced ability of the blood to carry oxygen, intestinal cramps, fatigue, and kidney abnormalities; the symptoms may be reversible if exposure stops soon enough. With higher levels of lead, the signs become more pronounced, yet still difficult to pinpoint to a cause. Children lose their general cognitive, verbal, and perceptual abilities, developing learning disabilities and behavior problems. Still more severe lead toxicity can cause irreversible nerve damage, paralysis, mental retardation in children, and death.

Lead absorption is greatest during times of rapid growth.[26] Therefore, infants and young children are more susceptible to lead poisoning and absorb five to ten times as much lead as adults do. Indeed, the incidence of lead toxicity is highest in children less than six years old.

High blood lead concentrations are associated with iron, calcium, or zinc deficiencies—nutrient deficiencies that are common in young children. Iron deficiency cannot be held totally accountable for the high blood lead concentrations, but it does impair the body's defenses against absorption of lead. A child with iron-deficiency anemia is three times as likely as a child with adequate iron stores to have elevated blood lead concentrations.[27] Not only does iron deficiency contribute to lead toxicity, but lead toxicity contributes to iron deficiency. One of lead's actions is to interfere with iron's incorporation into heme, resulting in symptoms of anemia. Thus the combination of iron deficiency and lead toxicity acts synergistically, having a more severe impact than one would predict from adding the effects of the two considered separately.

Deficiencies of other essential minerals are associated with a high lead burden. A significant, inverse relationship between dietary calcium and blood lead levels has been observed.[28] It is possible that inadequacy of calcium intake renders the body susceptible to lead absorption and retention. Zinc deficiency also appears to open the way to lead toxicity.[29] Serum zinc concentrations are frequently low in children with elevated blood lead concentrations.[30] Recent experiments have shown that lead's effects occur with lower doses than has been thought in the past. Even children who have only moderately elevated blood lead levels and who have never had high exposures show deficits in school performance—in speed, dexterity, verbal memory, language functions, concentration, and reasoning. Based on these experiments, 1 million children in the United States may now be at risk of showing permanent damage caused by lead.[31] Three trends are occurring simultaneously. Scientists are discovering that lead poisoning has more *subtle* effects than had heretofore been appreciated; these effects are more *permanent* than had been known earlier; and they are being found at *lower levels of exposure* than before.

Lead appears in all foods, and is also the nation's most significant contaminant in drinking water.[32] Whether some of it is naturally present is not known, but much of it is known to come from industrial pollution. People use lead in gasoline, paint, batteries, and pesticides, as well as in industrial processes that release it into the air and water. Exposures are highest in urban and industrial areas, but elevated blood lead concentrations are also found in the children of families who engage in hobbies such as stained glasswork or remodeling old houses where lead-based paints were used.[33] Lead poisoning is especially high near highways and in slums, where children may accidentally ingest leaded paint by teething on old furniture, toys, and the railings of old buildings. Old plumbing is made of lead, and it dissolves into water—especially soft water. Food in the fields is also contaminated by the air pollution from leaded gasoline, by way of rainfall and soil. Scientists have recommended a standard for weekly acceptable intakes of lead, but there is no monitoring system to keep track of the amounts to which people are actually exposed.[34]

Reductions in the use of leaded gasoline for automobiles mandated by federal law in recent years have helped to limit the amounts of lead in the environment and, thereby, in the blood. The decline in blood lead concentrations in children during the 1970s paralleled exactly the decline in the nation's use of leaded gasoline.[35] Even so, preschool children's blood lead concentrations are still unacceptably high in 4 to 6 percent of cases. More leaded gas is still being sold than anticipated, so considerable lead is still being deposited in the soil. Consumers would be wise to take ultraconservative measures to protect themselves, and especially their children, from lead poisoning. A first step is to check with the local public health department about the lead in the water supply.

Adverse Reactions to Foods

Adverse reactions to foods can threaten children's nutritional health to varying extents, depending on their severity and duration and on the classes of foods they involve. Diagnosis is often elusive. Many food aversions are labeled allergies when they are not, and some real allergies go undetected. Adverse reactions that are only temporary may lead to permanent avoidance of foods, to the detriment of a child's health; conversely, permanent adverse reactions that go undetected can cause chronic illness.

The term *food allergy* is used loosely, even by many physicians, as a catchall term for any unexplained adverse reaction to foods. Thus a parent whose child has any kind of discomfort after eating—stomachache, headache, pain, rapid pulse rate, nausea, wheezing, hives, bronchial irritation, cough, or any other—may conclude that an allergy is responsible, when in fact something else is the cause. Only careful, skilled testing can distinguish among the many possibilities.

Possibilities other than allergy, some of which were already mentioned in Chapter 6, include reactions to bacterial toxins; reactions to the chemicals in foods, such as monosodium glutamate or the natural laxative in prunes; digestive tract disorders such as obstructions or injuries; enzyme deficiencies such as inborn errors of metabolism or lactose intolerance; and even psychological aversions. These are not allergies; they are food intolerances.

food allergy: a term generally synonymous with food hypersensitivity, although often used to denote any unusual response to food; properly defined, an adverse reaction to food that involves an immune response. Allergies may be **symptomatic** or **asymptomatic**.

antigen: defined on p. 243 Food antigens are usually glycoproteins (large proteins with glucose molecules attached).

antibody: defined on p. 243.

histamine: a substance produced by cells of the immune system as part of a local immune reaction to an antigen; participates in causing inflammation.

food intolerance: an adverse reaction to foods that does not involve an immune response.

Food allergies The prevalence of food allergies is highest in the first several years of life and tends to decline with age.[36] A true food allergy—appropriately called a food-hypersensitivity reaction—occurs when a large molecule, most commonly a protein, enters the system and elicits an immunological response. (Recall that large molecules of food are normally dismantled in the digestive tract to smaller ones that are absorbed without problem.) The body's immune system reacts to a food protein or other large molecule as it does to other antigens—by producing antibodies, histamines, or other defensive agents.

Allergies may have one or two components. They always involve antibodies; they may or may not produce symptoms. A person may produce antibodies *without* having any symptoms (known as asymptomatic allergy) or may produce antibodies *and* have symptoms (known as symptomatic allergy). Symptoms without antibody production are food *intolerances;* they are *not* due to allergy. This means that allergies cannot be diagnosed from symptoms alone; they have to be diagnosed by testing for antibodies. Even if a child's symptoms seem exactly like those of an allergy, they may not be caused by one.

Depending on the location of the allergic reaction in the body, a symptomatic allergy will produce different symptoms. In the digestive tract, it may cause nausea or vomiting; in the skin, it may cause rashes; and in the nasal passages and lungs, it can cause inflammation or asthma. A generalized, all-systems reaction is anaphylactic shock.

anaphylactic (an-AFF-ill-LAC-tic) **shock:** a whole-body allergic reaction to an offending substance. Symptoms: abdominal pain, nausea, vomiting, diarrhea, inflamed nasal membranes, chest pain, hives, swelling, low blood pressure.

Allergic reactions to food can occur with different timings, simply classified as immediate and delayed. In both, the interaction of the antigen with the immune system is immediate, but the appearance of symptoms may come within minutes or after several (up to 24) hours.[37] Identifying the food that causes an immediate allergic reaction is easy, because symptoms correlate closely with the time of eating the food. Identifying the food that may cause a delayed reaction is more difficult, because the symptoms may not appear until a day after the offending food is eaten; by this time, many other foods will have been eaten, too, complicating the picture.

The foods that most often cause immediate allergic reactions are listed in Table 7–6. According to one investigator, 91 percent of adverse reactions in children are caused by only four major foods: nuts (43 percent), eggs (21 percent), milk (18 percent), and soy (9 percent).[38] Allergic reactions to single foods are common. Reactions to multiple foods are the exception, not the rule.

A number of tests and food challenges are required to identify a true food allergy. A simple elimination diet that enables a person to avoid the offending food is the preferred test diet.[39] This requires starting with a diet that omits the suspected food altogether for a week or two. If the symptoms do not disappear, the suspected food is not guilty. If symptoms do resolve, the food is reintroduced into the diet in small quantities. If there is a reaction, the food is eliminated for a month or two, and then reintroduced. Unless the reaction is severe, food challenges are performed at regular intervals until the food is either tolerated or clearly never will be. A large majority of allergic reactions resolve themselves.[40]

Allergies are not always diagnosed by these time-consuming, laborious methods, however. A number of unreliable tests, such as cytotoxic testing, are available that offer people quick, unfounded results.[41] In cytotoxic testing, one test tube of blood is taken from an individual; the white blood cells are mixed with plasma and sterile water and placed on microscope slides. Each slide is coated with a dried extract of a particular food to see how the cells react.

cytotoxic testing: an unreliable test touted as a means of tracking down and curing ills by on-the-spot blood testing for food allergies.

Table 7–6 Foods That Most Often Cause Allergies

Nuts	Peanuts
Eggs	Chicken
Milk	Fish
Soybeans	Shellfish
Wheat	Mollusks

Source: Adapted from R. H. Buckley and D. Metcalfe, Food allergy, *Journal of the American Medical Association* 248 (1982): 2627–2631; D. D. Metcalfe, Diseases of food hypersensitivity, *New England Journal of Medicine* 321 (1989): 255–257; C. D. May, Food allergy: Perspective, principles, practical management, *Nutrition Today,* November–December 1980, pp. 28–31.

Usually about 200 slides representing 200 foods are examined. If the cells collapse, disintegrate, or change shape, the individual is supposedly allergic to that food. However, there are no controlled experiments showing that white cells shrink or change shape when exposed to food substances to which a person is allergic.

celiac disease: a sensitivity to gliadin, a fraction of the wheat protein gluten, that causes flattening of the intestinal villi and generalized malabsorption; also called **gluten-sensitive enteropathy** or **celiac sprue.**

The grains that must be restricted in celiac disease:

- Barley.
- Rye.
- Oats.
- Wheat.

Celiac disease About 1 in 3000 U.S. children is afflicted with a type of food hypersensitivity known as celiac disease, or gluten-sensitive enteropathy.[42] A child with celiac disease is sensitive to gluten, a protein in barley, rye, oats, and wheat that injures the GI tract, causing malabsorption, diarrhea, and malnutrition. Treatment consists of feeding the child a gluten-free diet from which these foods are omitted. Such treatment is easier described than administered, since barley, rye, oats, and wheat are common in many foods. Rice, corn, and potatoes are well tolerated by children on a gluten-free diet. Rice cereal is frequently the first food introduced to infants because it contains almost no gluten. Children can outgrow celiac disease, but it may reappear in adulthood.

Adverse reactions to milk In infancy, cow's milk can cause GI bleeding, as already described in Chapter 6. After infancy, adverse reactions to milk fall into two major categories: milk allergy (technically, a type of food hypersensitivity), an immune reaction to the protein in milk; and lactose intolerance (technically, a type of food intolerance), caused by an inability to digest the milk sugar, lactose. No other adverse reactions to milk are clearly defined, although people often describe themselves vaguely as "milk intolerant," without specifying the causes of their "intolerance." In some cases, simple dislike may lead to milk avoidance. It is possible that substances in milk other than its protein and lactose may cause adverse reactions in some people, but until research confirms that they do, the interest of clarity is best served by avoiding the term *milk intolerance.*

lactose intolerance: inability to digest the milk sugar, lactose, due to inactivity or insufficiency of the enzyme lactase. Symptoms are gas, abdominal cramping, nausea, and/or watery stools after ingestion of lactose (either in milk, in other dairy foods, or as purified sugar).

lactase nonpersistence: the most common inherited cause of lactose intolerance, characterized by genetically preset low lactase activity by about five to six years of age; formerly called *primary lactase deficiency.*

congenital lactase deficiency: a rare, life-threatening form of lactose intolerance caused by a genetic enzyme defect that manifests itself at birth.

secondary lactase deficiency: a transient absence of lactase or temporary lactase inactivity in previously lactase-persistent individuals following injury to the small intestinal mucosa.

Lactose intolerance Lactose intolerance may arise in several different ways. One is genetic: a predetermined, age-related decline in the activity of the enzyme lactase in the GI villi. Called primary lactase deficiency, this condition results in low lactase activity usually by about five or six years of age. Because low intestinal lactase is not an abnormal state, nor actually a deficiency, the World Health Organization has suggested that the term *primary lactase deficiency* be replaced with *lactase nonpersistence.*[43] About 70 percent of the world's population is lactase nonpersistent, the incidence ranging from 1 or 2 percent among Scandinavians to about 100 percent among Asians.[44]

Lactose intolerance may have two other causes. One is congenital lactase deficiency, a rare, life-threatening condition caused by an enzyme defect present at birth. The other is secondary lactase deficiency, a temporary state of low lactase activity following injury to the small intestine—for example, from protein malnutrition or GI infection.

Adjusting the diet Parents are advised to watch for signs of food dislikes and take them seriously: "Dislike of a food may be only a whim or fancy, but it should be regarded as significant until proven otherwise."[45] For a parent to demand that a child eat a specific food is to create unnecessary, often harmful conflict. Although many cases of suspected allergies and intolerances turn out to be minor and temporary, real adverse reactions do occur. Do not prejudge, in any case. Test, and then apply nutrition knowledge conscientiously in deciding how to alter the diet.

When parents stop serving a suspected food to their child, they risk feeding the child an unbalanced diet that could lead to nutrient deficiencies. Whenever a food is excluded from the diet, care must be taken to include other foods that offer the same nutrients as the omitted food contains. Milk is often perceived as an offending food and all too casually omitted from a child's diet. Continued restriction of milk in the child's diet without appropriate substitutions can result in energy, protein, calcium, vitamin A, vitamin D, and riboflavin deficiencies. It is critical that nutritionally comparable milk substitutes be introduced.

Sometimes milk allergy can be treated by elimination of the offending milk protein from the diet for a period of time, followed by gradual reintroduction. During the elimination time, possible substitutes for milk are boiled milk (milk protein is altered with cooking), goat's milk (but it should be state certified or pasteurized), soy milk (although soy is also a common allergen), and cheeses. Liquid or powdered milk can be cooked into foods such as custards, soups, and casseroles.

The child who is lactose intolerant can usually handle small amounts of milk periodically throughout the day—up to half a glass each time, especially if offered with other foods. Fermented milk products such as yogurt offer the same nutrients as milk but with a lower lactose content.

As a last resort, the milk's nutrients may be offered in supplement form. Chapter 2 discusses calcium supplements for adults, but the factors bearing on the choice apply equally to children. Figure 7–5 offers a decision tree for choosing a milk substitute.

Nutrition and Behavior

Many different relationships between nutrition and behavior have been suggested in recent years. Among the nutrition-related problems proposed to influence behavior are nutrient deficiencies, allergies to foods, reactions to food additives, reactions to sugar, stimulation by caffeine, and more. While reading the information presented here, remember that although poor nutrition does affect behavior, abnormal behavior can also be the result of a variety of other factors.

All children at times become excitable, rambunctious, and unruly. The most common cause of such "hyper" behavior is not nutrition but the tension-fatigue syndrome, which arises from a combination of factors: lack of exercise, a craving for attention, lack of sleep, overstimulation, and too much television. Caretakers can counter these effects by insisting on regular hours of sleep, regular mealtimes, and regular outdoor exercise.

tension-fatigue syndrome: apparent hyperactivity produced in a child by the combination of lack of sleep and overstimulation with anxiety.

A nutrition-related condition that does *not* cause behavioral abnormalities—except for the general misery associated with any sickness—is food allergy.

Figure 7–5 Choosing a Milk Substitute

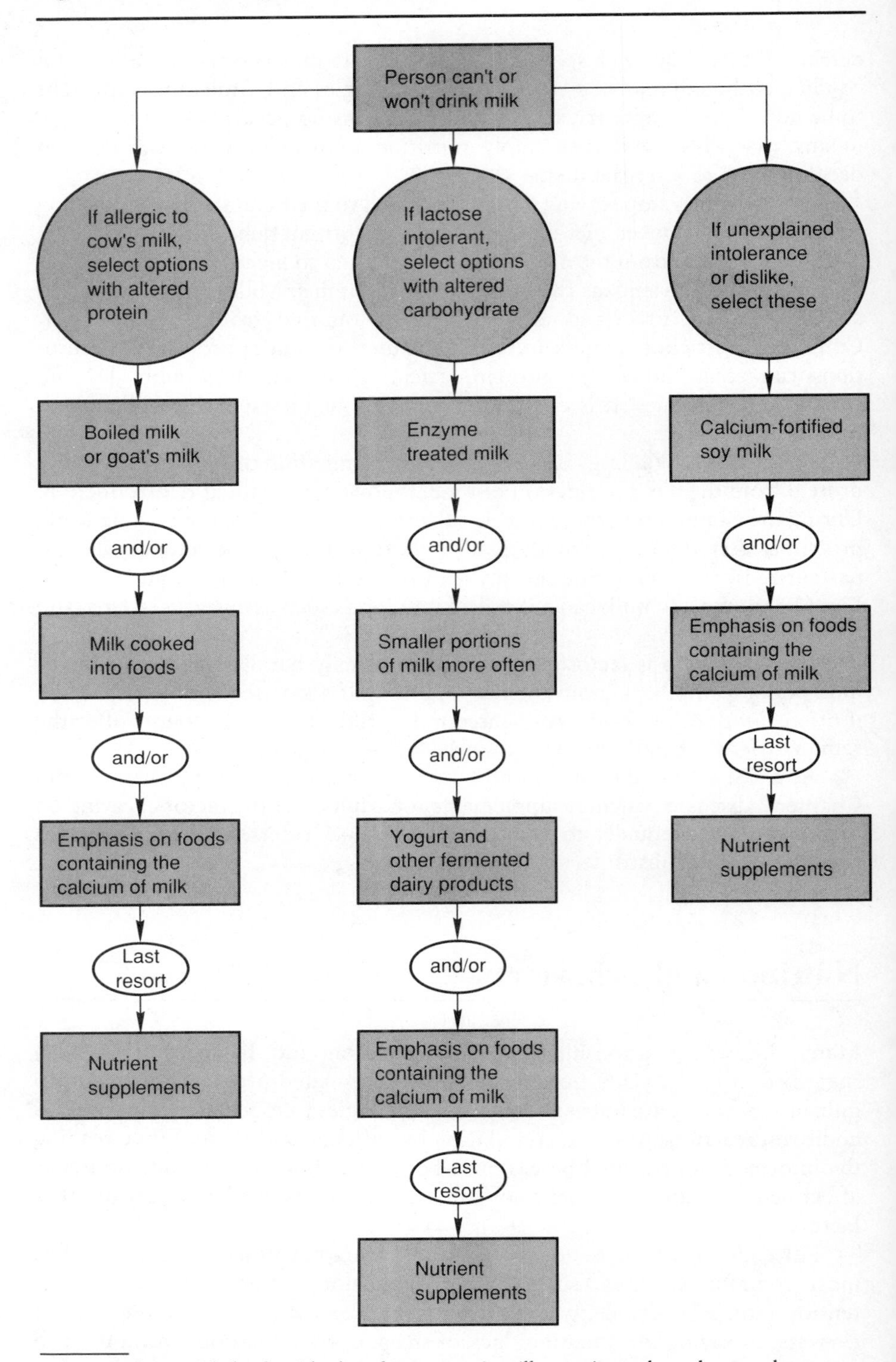

[a]Choose calcium-rich foods with altered or no cow's-milk protein, such as cheese, salmon, or broccoli.
[b]Buy milk already treated, or add the enzyme (LactAid) yourself. Enzyme treatment may not reduce lactose content sufficiently to relieve symptoms, and you may have to try the other alternatives.
[c]Choose calcium-rich foods with less or no lactose, such as cheese, yogurt, salmon, or broccoli.
Source: Adapted from E. N. Whitney, C. B. Cataldo, and S. R. Rolfes, *Understanding Normal and Clinical Nutrition,* 2nd ed. (St. Paul, Minn.: West, 1987), p. 397.

Another is the learning disability known as attention deficit disorder with hyperactivity. Both of these disorders have been wrongly represented as diet related, and for that reason, both are treated here. Food allergies are real, need diagnosis, and demand dietary treatment, but they do not cause hyperactive behavior; allergies had an earlier section of their own. Learning disabilities are also real, need diagnosis, and may cause hyperactive behavior, but they do not require dietary treatment. (This is not to deny that the general health of any child may be improved by attention to diet.)

Sugar does not cause behavioral abnormalities in children except as it contributes to nutrient deficiencies by displacing nutrients in the diet. The following section describes how nutrient-poor diets, especially iron-deficient diets, influence children's behavior.

Nutrient Deficiency Effects

Iron deficiency presents the best-known and most widespread effects on behavior. Most people are familiar with the role of iron in hemoglobin, which carries oxygen in the blood, but iron also acts as part of many intracellular proteins (cytochromes), which use oxygen to help produce energy. A lack of iron not only causes an energy crisis but also directly affects behavior, mood, attention span, and learning ability. Oxygen use, facilitated by iron, is imperative for normal functioning of the brain and nervous system. Deficiencies of iron produced experimentally in animals have caused abnormal synthesis and degradation of neurotransmitters, most notably those that regulate the ability to pay attention, which is crucial to learning.[46] Lead toxicity, which can accompany iron deficiency, causes similar symptoms: children lose their cognitive, verbal, and perceptual abilities, developing learning disabilities and behavior problems.

Iron deficiency is usually diagnosed by use of iron indicators in the *blood* when it has progressed all the way to overt anemia. A child's *brain,* however, is sensitive to slightly lowered iron levels long before the blood effects appear. Iron's effects are hard to distinguish from the effects of other factors in children's lives, but it is likely that iron deficiency manifests itself in a lowering of the "motivation to persist in intellectually challenging tasks," a shortening of the attention span, and a reduction of overall intellectual performance. Research shows that anemic children perform less well on tests and have more conduct disturbances than their classmates.[47] The results of studies of iron-deficiency effects have been challenged from time to time on the basis of weak study design. More recent, well-controlled double-blind studies confirm that iron deficiency in children can cause behavioral abnormalities such as reduction of learning and cognitive processes.[48] In both preschool and school-aged iron-deficient children, iron supplementation improves performance on tests of cognition.[49]

Iron is only one of several dozen nutrients that can be displaced in a diet high in empty-kcalorie foods. Any of the others may be lacking as well; and the deficiencies of those nutrients may also cause behavioral, as well as physical, symptoms, as Table 7–7 shows.

A child with the behavioral symptoms of nutrient deficiencies might be irritable, aggressive, disagreeable, or sad and withdrawn. One might label such

Table 7–7 Behavioral Symptoms of Nutrient Deficiencies

- *Protein-energy deficiency*—apathy, fretfulness, lack of energy, and lack of interest in food.
- *Thiamin deficiency*—confusion, uncoordinated movements, depressed appetite, irritability, insomnia, fatigue, and general misery.[a]
- *Riboflavin deficiency*—depression, hysteria, psychopathic behavior, lethargy, and hypochondria evident before deficiency can be detected by clinical symptoms.[b]
- *Niacin deficiency*—irritability, agitated depression, headaches, sleeplessness, memory loss, emotional instability (early signs of pellagra onset), and mental confusion progressing to psychosis or delirium.
- *Vitamin B_6 deficiency*—irritability, insomnia, weakness, mental depression, abnormal brainwave pattern, convulsions, the mental symptoms of anemia,[c] fatigue, and headache.[a]
- *Folate deficiency*—the mental symptoms of anemia,[c] tiredness, weakness, forgetfulness, mild depression, abnormal nerve function, headache, disorientation, confusion, and inability to perform simple calculations.[d]
- *Vitamin B_{12} deficiency*—degeneration of peripheral nervous system and anemia.[c]
- *Vitamin C deficiency*—hysteria, depression, listlessness, lassitude, weakness, aversion to work, hypochondria, social introversion, possible iron-deficiency anemia,[c] and fatigue.
- *Vitamin A deficiency*—anemia.[c]
- *Iron deficiency*—fatigue, weakness, headaches, pallor, listlessness, and the mental symptoms of anemia.[c]
- *Magnesium deficiency*—apathy, personality changes, and hyperirritability.
- *Copper deficiency*—iron-deficiency anemia.[c]
- *Zinc deficiency*—poor appetite, failure to grow, iron-deficiency anemia,[c] irritability, emotional disorders, and mental lethargy.[e]

---

[a]Symptoms of thiamin and vitamin B_6 deficiency are from Marginal vitamin deficiency, *Nutrition and the M.D.*, July 1983, p. 3.
[b]Symptoms of riboflavin deficiency are from R. Sterner and W. Price, Restricted riboflavin: With [error in title as published] subject behavioral effects in humans, *American Journal of Clinical Nutrition* 26 (1973): 150–160.
[c]The mental symptoms of anemia can include any or all of the following: lack of appetite, apathy, listlessness, clumsiness, conduct disturbances, shortened attention span, hyperactivity, irritability, learning disorders (vocabulary, perception), lowered IQ, low scores on latency and associative reactions, reduced physical work capacity, and repetitive hand and foot movements. These symptoms are not caused by anemia itself but by iron deficiency in the brain. Children with much more severe anemias from other causes, such as sickle-cell anemia and thalassemia, show no reduction in IQ when compared with children without anemia.
[d]Symptoms of folate deficiency are from J. H. Pincus and coauthors, Subacute combined system degeneration with folate deficiency, *Journal of the American Medical Association* 221 (1972): 496–497. (Possibly, because folate is required for many steps in amino acid metabolism, its lack causes altered amino acid levels in the blood, and therefore altered neurotransmitter levels in the brain; Neurological disease in folic acid deficiency, *Nutrition Reviews* 39 [1981]: 337–338.)
[e]Symptoms of zinc deficiency are from A. S. Prasad, Clinical, biochemical and nutritional spectrum of zinc deficiency in human subjects: An update, *Nutrition Reviews* 7 (1983): 197–206.

Source: Unless otherwise cited, all of the listed symptoms appeared in *Modern Nutrition in Health and Disease,* 6th ed., eds. R. S. Goodhart and M. E. Shils, (Philadelphia: Lea and Febiger, 1980).

a child "hyperactive," "depressed," or "unlikeable," when in fact, the cause for these behaviors may be simple, albeit marginal, malnutrition. Should suspicion of dietary inadequacies be raised, *no matter what other causes may be implicated,* the people responsible for feeding the child should take steps to correct those inadequacies promptly.

General Hunger Effects

Not only the content, but also the timing, of meals makes a difference in children's behavior. Children who eat no breakfast perform less well in tasks of concentration, their attention spans are shorter, and they even show lower IQ test results than their well-fed peers.[50] Common sense dictates that it is unreasonable to expect anyone to learn and perform work when no fuel has been provided. By the late morning, discomfort from hunger may become distracting even if a child has eaten breakfast.

The problem that arises for children who attempt morning schoolwork on an empty stomach appears to be at least partly due to hypoglycemia. The average child up to the age of ten or so needs to eat at least every four to six hours to maintain a blood glucose concentration high enough to support the activity of the brain and nervous system.[51] A child's brain is as big as an adult's—and the brain is the body's chief glucose consumer. A child's liver is considerably smaller—and the liver is the organ responsible for storing glucose (as glycogen) and releasing it into the blood as needed. A child's modest liver stores of glycogen are soon depleted; hence the need to eat fairly often. Teachers aware of the late-morning slump in their classrooms wisely request that a midmorning snack be provided; it improves classroom performance all the way to lunchtime.[52]

Nonnutritional Hyperactivity

Unlike the apathy, inattention, and poor learning ability caused by nutrient deficiencies and hunger, true hyperactivity, or attention deficit disorder with hyperactivity (ADDH), is not caused by diet. It is a type of learning disability that occurs in 5 to 10 percent of young school-age children—that is, in 1 or 2 in every classroom of 20 children. It can lead to academic failure and major behavior problems. Parents and teachers need to deal effectively with it wherever it appears in order to avert the grief that can otherwise result.

The idea that hyperactivity might be caused by diet became popular in 1973, when Dr. Benjamin Feingold proposed that hyperactive children suffer adverse reactions to natural salicylates (compounds with a chemical structure similar to aspirin) and the artificial flavors and colors in foods. The Feingold diet eliminates foods with high levels of natural salicylates (apples, berries, tomatoes, and peaches, for example) and foods with artificial flavors and colors. However, when controlled, double-blind experiments with additive-containing and additive-free foods are conducted, the benefits of the Feingold diet do not materialize as they do in some individual family situations. In family situations, diet is not the only thing that changes. Children receive special attention along with the special diet, and this may well have beneficial effects on their behavior. Both the children and their parents (and teachers) are likely to be influenced by the hope that the experiment will work—and by

hyperactivity syndrome in children: a cluster of symptoms in which "the essential features are signs of developmentally inappropriate inattention, impulsivity, and hyperactivity." Other important features are onset before age seven, duration of six months or more, and proven absence of mental illness or mental retardation.[53] Other names associated with hyperactivity: **attention deficit disorder, hyperkinesis, minimal brain damage, minimal brain dysfunction, minor cerebral dysfunction.**

learning disability: a disorder in one or more cognitive processes involved in understanding or using language, which manifests itself in an imperfect ability to listen, think, speak, read, write, spell, or do mathematical calculations.

suggestibility, a factor that is difficult to rule out.[54] The dietary approach to hyperactivity, when it appears to work, seems to do so by suggestion.[55] It appears not to be the additive-free nature of the diet that works but the changed lifestyle that the diet demands.[56]

Physicians often diagnose hyperactivity by conducting a trial with stimulant drugs. Stimulants normally speed up people's activity, but they have a paradoxical effect with hyperactivity: they normalize it. (Perhaps they stimulate control centers in the brain.) If a child responds to stimulant drugs by calming down, that indicates that the drugs may be correcting a biochemical imbalance in the nervous system, and can be used to help control the behavior. *In children who are responsive,* prescription medication should at least be considered as the treatment of choice.

Caffeine

Caffeine is often overlooked as a source of "hyper" behavior in children, but it is a matter of some concern to pediatricians. A survey of over 1000 children between the ages of 1 and 17 years found that 77 percent of them were caffeine consumers.[57] Children not accustomed to caffeine who are given doses equivalent to about two cola beverages a day become noticeably inattentive and restless.[58] Children who are troubled by sleeplessness, restlessness, and irregular heartbeats may need to control their caffeine consumption. A 12-ounce cola beverage may contain as much as 50 milligrams of caffeine; two or more such beverages are equivalent in the body of a 60-pound child to the caffeine in eight cups of coffee for a 175-pound man. Chocolate bars also contribute caffeine. As long as undeniably attractive temptations such as cola beverages and candy bars surround children, barriers against their abuse have to be provided by concerned adults until the children learn to control consumption themselves.

Energy and Nutrient Needs of Children

Growth during childhood necessitates increasing intakes of all nutrients except vitamin D and iron. (The RDA for iron and vitamin D increase during infancy and remain there throughout childhood.) The RDA table and the RNI for Canadians list nutrient averages for each span of three years. Recommended nutrient allowances for children are, for the most part, extrapolated from studies conducted on infants and adults. These allowances consider body size and growth needs.

Beyond their roles of maintaining a healthy body, most of the nutrients actively contribute to growth. Consider, for example, the growth of a bone. Vitamin A is required for the resorption process; vitamin C helps form the collagen matrix on which the bone is formed; and vitamin D, calcium, phosphorus, magnesium, and fluoride are required to mineralize the bone. Many nutrients perform roles during growth in a variety of body systems. For example, in addition to its contributions to bone growth, vitamin A helps manufacture red blood cells.

Energy Intake and Activity

A one-year-old child needs perhaps 1000 kcalories a day; a three-year-old needs only 300 kcalories more. At age ten, a child needs about 2000 kcalories a day. Total energy needs increase slightly with age, but in proportion to the child's size, the energy needs are actually declining gradually.

Energy RDA for children:
1300 kcal/day (1 to 3 yr).
1800 kcal/day (4 to 6 yr).
2000 kcal/day (7 to 10 yr).

The wide variation in the physical activity of children creates big differences in their energy needs. Inactive children can become obese even when consuming diets that are below average in food energy. Childhood is a time for children to develop the habit of being physically active in order to prevent obesity; this chapter's Focal Point discusses obesity prevention in detail.

Protein

Like energy needs, total protein needs increase slightly with age, but when the child's body weight is considered, the protein requirement actually declines gradually. The estimation of protein needs considers the requirements for maintaining nitrogen balance, the quality of protein consumed, and the added needs of growth.[59]

Protein RDA for children:
16 g/day (1 to 3 yr).
24 g/day (4 to 6 yr).
28 g/day (7 to 10 yr).

Vitamins and Minerals

The RDA tables show the increments by which vitamin and mineral needs of children increase with their ages. A balanced intake of whole foods can meet children's needs for these nutrients well, with the notable exception of iron.

Iron RDA for children:
10 mg/day (1 to 10 yr).
Other vitamin and mineral RDA: see inside front cover.

The food intakes of children between the ages of one and two provide about three-fourths of the RDA for iron. Children between the ages of three and ten meet the iron RDA.[60] Some researchers suggest that iron recommendations for children, rather than being expressed on the basis of age, should be based on adequate body weight for age.[61]

Critical to the consideration of dietary iron intake and iron status is the amount of iron available for absorption. The RDA assume that 10 to 15 percent of dietary iron is available. If children eat only small amounts of meat, fish, or poultry, less than 10 percent of the iron in their diets may be available.[62]

The body's iron status reflects its ability to conserve and recycle iron once it has been absorbed. Thus iron status reflects more than dietary intake alone, but diet is more critical to iron balance in children than it is in adults. In adult males, about 95 percent of the required iron is recycled, and only 5 percent need come from the diet. In children, the percentages are 70 and 30, respectively.[63]

Supplements for Children

Many parents provide their children with vitamin and mineral supplements, relying on them as insurance against possible dietary insufficiencies. Supplements are reasonably inexpensive and available without a prescription. Most children's vitamin supplements are of two kinds: the first contains vitamins A,

C, and D, and the second contains vitamins A, B_6, B_{12}, C, D, E, thiamin, folate, riboflavin, and niacin. Both supplements are available with or without iron and sometimes, zinc. A few children's supplements include other minerals as well.

Routine supplement use for most children, when growth is slowing down, is unnecessary. Diets of normal, well-fed children generally supply sufficient amounts of vitamins; supplements do not improve their nutrition status.[64] However, particular groups of children may benefit from supplementation.[65] These groups include:

- Children who suffer from malnutrition.
- Children with anorexia or poor eating habits.
- Children adhering to restricted diets, such as weight control or vegetarian diets.

For a discussion of vegetarianism for growing children, read Focal Point 4.

Unfortunately, disadvantaged children, who most need more food or supplements to provide the essential nutrients they lack, are the ones most likely not to have access to them. For children who do take supplements, the kind chosen should be any multivitamin-mineral product that provides a complete array of vitamins and minerals at approximately the RDA levels.

Appendix E compares several multivitamin-mineral supplements for children.

To prevent possible deficiencies of iron, fortification of foods seems effective—certainly for children over five, in any case. For most cases of iron deficiency, the treatment of choice is oral ferrous sulfate administered between meals for maximum absorption. Blood measures should indicate improvement within two months, and iron stores should be replete within five months.

Little effort is required to convince a child to take a vitamin supplement. Fruit-flavored, chewable vitamins shaped like cartoon characters entice young children to accept them. These cute, flavorful tablets also have the potential to cause poisoning in children. Of most concern are the supplements that contain iron. Iron-containing supplements should be packaged in childproof containers and labeled with a warning to parents to keep out of reach of children. A mild overdose of iron-containing vitamins causes GI distress, nausea, and black diarrhea (indicating bleeding in the upper GI tract). More severe overdoses result in bloody diarrhea, shock, liver damage, coma, and in some cases, death.[66] The ingestion of 30 milligrams of iron per kilogram of body weight is toxic, and if it occurs, vomiting should be induced immediately.

In regions without water fluoridation, fluoride supplements are recommended for all children. The recommended fluoride dosage varies from 0.25 to 1.0 milligrams per day, depending on the child's age and the fluoride concentration of the water.[67] (See Table FP5–2 in Focal Point 5 for recommended dosages.)

Feeding Children

Feeding children requires not only providing a variety of nutritious foods but also nurturing the children's self-esteem and well-being. Parents face a number of challenges in preparing meals that appeal to their children's tastes as well as providing needed nutrients. The interactions between parents and children regarding food intake can set the stage for lifelong attitudes and habits.

Eating Patterns

To provide all the needed nutrients, a variety of foods from each of the food groups is recommended. Table 7–8 shows children's daily food patterns; portion sizes increase with the children's ages. For fruits and vegetables, the pattern loosely defines a portion as 1 tablespoon for each year of the child's life. Thus, at four years of age, a portion of fruit or vegetable would be 4 tablespoons (¼ cup). Because the portion size adjusts as the child grows older, this rule of thumb is appropriate from age two to the teen years.

Parents and other caretakers must balance food choices skillfully to ensure adequate nutrition. Some food intakes that seem fairly nutritious are not, as the following comparison demonstrates. Consider the two sample days' menus, A and B, for two fairly typical five-year-old children. Note that the planners of both sets of menus were conscientious: both sets consist of three meals and snacks that deliver nutrients from all four food groups.

Sample Menu A

Breakfast:
½ c cereal with ½ c 2% milk
½ banana
¾ c orange juice

Snack:
2 oz raisin/sunflower seed mixture

Lunch:
Peanut butter and jelly sandwich with 1 tbsp peanut butter and 1 tbsp jelly on 2 slices of whole-wheat bread
1 c 2% milk
3 small carrot sticks
1 small apple

Snack:
6 oz fruit yogurt

Table 7–8 Children's Daily Food Patterns for Good Nutrition

		Average Size of Serving		
Food Group	Servings per Day	*1 to 3 Yr*	*4 to 6 Yr*	*7 to 12 Yr*
Milk and milk products[a]	4	½ to ¾ c	¾ c	¾ to 1 c
Meat and meat alternates[b]	2 or more	1 to 2 oz	1 to 2 oz	2 to 3 oz
Fruits and vegetables[c]	4 or more	2 to 4 tbsp or ½ c juice	¼ to ½ c or ½ c juice	½ to ¾ c or ½ c juice
Bread and cereals (whole grain or enriched)[d]	4 or more	½ slice	1 slice	1 to 2 slices

[a]½ c milk = ½ c cottage cheese, pudding, or yogurt; ¾ oz cheese; or 2 tbsp dried milk.
[b]1 oz meat, fish, or poultry = 1 egg, 1 frankfurter, 2 tbsp peanut butter, or ½ c legumes.
[c]Vitamin C source (citrus fruits, berries, tomatoes, broccoli, cabbage, or cantaloupe) daily; Vitamin A source (spinach, carrots, squash, tomatoes, or cantaloupe) three to four times weekly.
[d]1 slice bread = ¾ c dry cereal; ½ c cooked cereal; or ½ c potato, rice, or noodles.

Source: Adapted from P. M. Queen and R. R. Henry, Growth and nutrient requirements of children, in *Pediatric Nutrition,* eds. R. J. Grand, Jr., L. Sutphen, and W. H. Dietz, Jr. (Boston: Butterworth's, 1987), p. 347.

Dinner:
¼ c macaroni and cheese
¾ c apple juice
¼ c broccoli with 1 tbsp butter
Snack:
2 peanut butter cookies

Sample Menu B

Breakfast:
1 glazed doughnut and 1 cup 2% milk
Lunch:
Peanut butter and jelly sandwich with 1 tbsp peanut butter and 1 tbsp jelly on 2 slices of white bread
1 c lemonade
1 small apple
10 potato chips
Snack:
6 oz fruit yogurt
Dinner:
1 baked chicken drumstick
10 large french fries with 1 tbsp catsup
½ c applesauce
6 oz grape soda
Snack:
1 oz jelly beans

Figure 7–6 shows that sample menu A provided adequate amounts of all the recommended nutrients, while sample menu B fell short of supplying some important nutrients.

Inspection of the two sample menus as compared with the recommended food pattern in Table 7–8 shows some foods missing in menu B, which accounts for its nutrient deficits. The child's intake of foods in the milk and milk products group fell short of the recommended four (¾ cup) servings, so the child's calcium intake that day was low. The cereal food group was inadequately represented as well; as a consequence, iron and zinc were lacking for the day. An additional small serving of whole-grain cereal or bread, and an additional serving of legumes or meat, would have boosted the intakes of these nutrients considerably. Fruits and vegetables were minimal on this day, consisting of an apple, applesauce, and french fries. This child needed at least one more serving of fruits and vegetables, according to Table 7–8, and the right choices would have improved the child's vitamin C and vitamin A intakes. The foods in menu A provided more nutrients for a similar energy intake and offered a greater variety of foods; they also did a lot more than the foods in menu B to promote healthful eating habits for the long-term future. A little extra effort and planning can make a big difference in a child's day-to-day nutrient intake.

The child represented by menu B did have three meals and a nourishing snack on this day. Imagine how much worse the picture is when a child skips breakfast, or when more sugary foods take the place of some of the nourishing

Figure 7–6 Typical Children's Intakes
Sample menu A meets a child's nutrient needs much better than sample menu B.

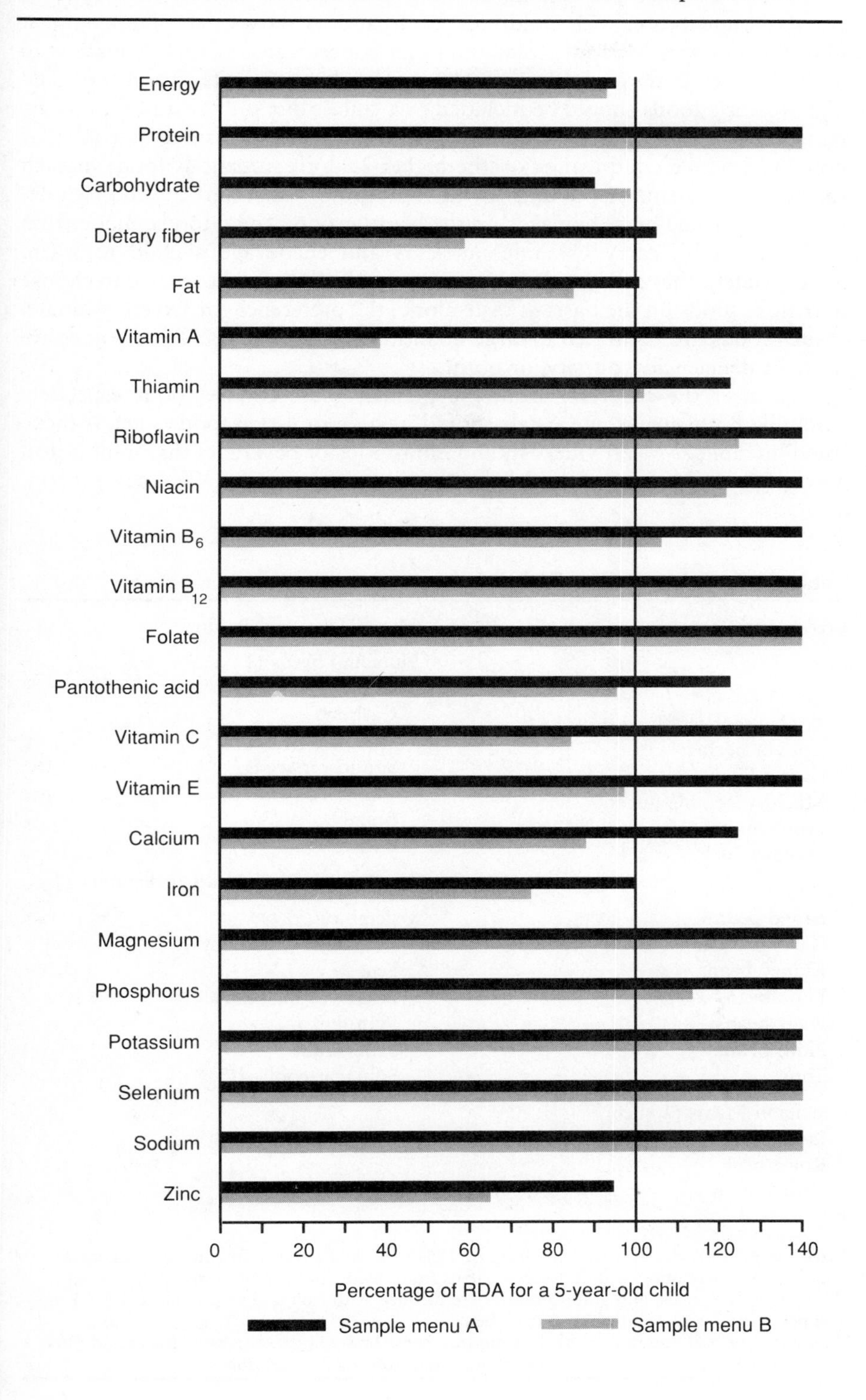

ones. Chances are, the nutrients missed from a skipped breakfast will not be "made up" at lunch and dinner, but will be completely left out that day.

Experimentation with children's food patterns shows that planners must limit the amounts of candy, cola, and other concentrated sweets they allow in a child's diet if they wish to supply the needed nutrients. These optional high-kcalorie foods should be included in a child's diet only in addition to the required servings of nutritious foods, and then only in small portions. A nonobese, active child can enjoy the higher-kcalorie nutritious foods in each category: ice cream or pudding in the milk group, cake and cookies (whole-grain or enriched only, however) in the bread group. These foods, made from milk and grain, carry valuable nutrients and encourage a child to learn, appropriately, that eating is fun. However, children cannot be trusted to choose nutritious foods on the basis of taste alone; the preference for sweets is innate. If such foods are permitted in large quantities, the only possible outcomes are nutrient deficiencies, obesity, or both.

To meet the RDA for iron, the planner must choose foods especially carefully. Both snacks and meals should include iron-rich foods such as those listed in Table 7–9. Providers should limit foods or beverages that inhibit iron absorption and use liberally those that enhance absorption. Tea, soy protein,

Table 7–9 Selected Iron-Rich Foods[a]

Breads and Cereals[b]

Eggs

Juices (1 c)
- Apple-cranberry juice
- Apricot nectar
- Carrot juice
- Mixed vegetable juice
- Prune juice
- Tomato juice

Legumes (½ c)
- Baked beans
- Garbanzo beans
- Kidney beans
- Lima beans
- Navy beans
- Pinto beans
- Tofu

Luncheon Meats (1 slice)
- Liverwurst
- Roast beef
- Turkey ham

Meats, Fish, and Poultry[c]

Nuts and Seeds (1 oz)
- Pine
- Cashews
- Almonds
- Sunflower seeds
- Mixed assortment
- Brazil
- Filberts

Soups made with meats or legumes (1 c)

Vegetables (½ c)
- Artichoke (1 medium)
- Peas
- Potato (1 medium)
- Pumpkin
- Sauerkraut
- Spinach (cooked)

[a]Each serving provides at least 1 mg iron, or 10% of a child's RDA. To enhance iron absorption, include a vitamin C source (such as citrus fruits) with snacks and meals.
[b]Many breads, cereals, and grain products are fortified with iron and so provide at least 1 mg iron per slice or ½ c serving. Read the label.
[c]Many cuts of beef and pork provide 1 mg iron per 1 oz serving; other cuts of beef and pork, and most poultry and seafood provide about 1 mg iron per 3 oz serving.

and nuts are potent inhibitors of iron absorption.[68] Foods and beverages rich in vitamin C enhance iron absorption from a meal by twofold to fivefold. Meat, fish, and poultry also enhance the absorption of nonheme iron. Dairy products are important sources of energy, protein, calcium, and riboflavin in children's diets, but they are poor iron sources. Excessive consumption of milk and milk products can displace other foods that make equally important nutrient contributions.

Children above the age of one or so are no longer passive recipients of food from spoons held by parents, but are active in feeding themselves. The Practical Point: Mealtimes with Children, suggests ways to support the learning of desired behaviors.

▸▸ PRACTICAL POINT

Mealtimes with Children

Children's appetites often diminish around the age of one year, in line with the reduction in their growth rates. A wise parent will accept the one-year-old's slowing down, and not try to force food on a child who does not want or need to eat. After one, children's appetites fluctuate; at times they seem to be insatiable, and at other times they seem to live on air and water. Parents need not worry about varying appetites; children need and demand more food during periods of rapid growth than during slow periods.

In children of normal weight, appetite regulates the energy intake so that it is appropriate for each stage of growth. Unfortunately, many people mistakenly try to overcontrol the amounts of foods children eat. An innovative study showed this to be unnecessary.[a] Researchers wanted to determine how much lunch children would eat if given snacks of different energy value before lunch. The children ate puddings that contained either 40 kcalories or 150 kcalories but were otherwise identical. The children were then given lunch. They compensated for the kcaloric difference: those who had eaten the high-kcalorie pudding ate less lunch than those who had eaten the low-kcalorie pudding. (Overweight children, however, may eat in response to external cues, disregarding appetite regulation signals.)

Not surprisingly, conflicts over food often arise during the second or third year of life, when children are asserting their independence and parents are attempting to make them eat what they think is best for them. Such conflicts can disrupt children's abilities to regulate their own food intakes and to determine their own likes and dislikes.

Research with children indicates that allowed to govern their own development, they often do better than when pushed. For example, when they are urged to try new foods, even by way of rewards, they are less likely to try those foods again than children who are left to decide for themselves.[b] In describing the attitude parents might best adopt, one authority on children's eating behaviors suggests that the parent should be responsible for what the child is offered to eat, and the child should be responsible for *how much* and even *whether* to eat.[c]

▸▸ **PRACTICAL POINT** *continued*

Parents are encouraged to allow toddlers to explore and satisfy their curiosity about the world around them. This is as true with respect to food as it is with other aspects of their lives. Young children learn about food with all their senses; this includes feeling it as well as tasting and smelling it. Table manners can be taught with time. Young children make messes while eating. They are learning new skills and should be allowed to feed themselves and drink from a cup when the interest arises, even if much of the food or beverage ends up on the floor. The parents should gently guide them in learning these skills and protect them while doing so by ensuring that they consume adequate amounts of nutritious food.

It is desirable for children to learn to like nutritious foods in all the food groups. With one exception, this liking usually develops naturally. The exception is vegetables, which young children frequently dislike and refuse.[d] Even a tiny serving of spinach, cooked carrots, or squash may elicit an expression that registers the utmost in negative feelings (as well as great pride in the ability to make an ugly face). Since most youngsters need to eat more vegetables, the next few paragraphs are addressed to this problem.[e]

Try to remember how you felt when first offered a cup of vegetable soup, a serving of runny spinach, or a pile of peas and carrots. If the soup burned your tongue, it may have been years before you were willing to try it again. As for the spinach, it was suspiciously murky looking. (Who could tell what might be lurking in that dark, stringy stuff?) The peas and carrots troubled your sense of order. Before you could eat them, you felt compelled to sort the peas onto one side of the plate and the carrots onto the other. Then you had to separate, into a reject pile, all those that got mashed in the process or contaminated with gravy from the mashed potatoes. Only then might you be willing to eat the intact, clean peas and carrots one by one—perhaps with your fingers, since the peas, especially, kept rolling off the fork.

Why children respond in this way to foods that look uninviting or messy to them is a matter for conjecture. Parents need only be aware that this is how many children feel and then honor those feelings. Researchers attempting to explain children's food preferences are met with contradictions. Children describe liking colorful foods, yet vegetables are often rejected, and brown peanut butter and white potatoes, apple wedges, and breads are among their favorites. Raw vegetables are better accepted than cooked ones, so it is wise to offer vegetables that are raw or slightly undercooked and crunchy, bright in color, served separately, and easy to eat.[f] They should be warm, not hot, because a child's mouth is much more sensitive than an adult's. The flavor should be mild. A child has more taste buds, and they are more sensitive than those of an adult. Strong flavors offend them. Smooth foods such as mashed potatoes or pea soup should have no lumps in them (a child wonders, with some disgust, what the lumps might be). Irrational as the fear of strangeness may seem, the parent must realize that it is practically universal among children. Children prefer foods that are familiar to them.[g] One study showed that the more times two-year-old children are exposed to a new food, even if they do not taste the food each time, the higher the preference for the food becomes.[h]

When feeding children, parents must always be alert to the dangers of choking. A choking child is a silent child. An adult knowledgeable about what to do if choking does occur should be present whenever a child is eating.

Encouraging the child to sit down to eat is also a good practice. The possibility of choking is more likely when a child is running or falling. Round foods such as grapes, nuts, hard candies, and hot dog pieces are dangerous. They are difficult to control in a mouth with few teeth and can easily become lodged in the small opening of a child's trachea. If served, grapes should be crushed and hot dogs, diced. Other potentially dangerous foods include tough meat, hard pieces of fruit or vegetables, popcorn, chips, and peanut butter. Parents of toddlers need to remember that topical anesthetics used to numb gums when a child is teething interfere with the ability to chew and swallow.

Wise parents allow children to help make some of the family's choices, including sometimes giving two-year-olds the right to say their favorite word (No!)—at the same time, of course, sensibly preventing them from dominating the family. Allowing children to help plan and prepare the family's meals provides enjoyable learning experiences that encourage children to eat the foods they have prepared. Vegetables are pretty, especially when fresh, and provide opportunities to learn about color, about growing things and their seeds, about shapes and textures—all of which are fascinating to young children. Measuring, stirring, decorating, cutting, and arranging vegetables are skills even a young child can practice with enjoyment and pride.

Before sitting down to eat, small children should be helped to wash their hands and faces thoroughly. Ideally, outdoor playtime will have preceded the meal. If fun and games *follow* the meal, children are likely to hurry out to play, leaving food on their plates that they were hungry for and otherwise would have eaten.

Little children like to eat at little tables and to be served little portions of food. Teaching children how to serve themselves the quantity they will eat minimizes waste. Parents need to remember that food is wasted not only when it is dumped in the garbage but also when it is stored as excess fat on the child's body. Never force children to clean their plates. This practice can lead to behaviors that encourage obesity. Instead, allow children to stop eating when they are full; encourage them to listen to their bodies. The remaining food can be recycled as leftovers for another meal. One word of caution is in order: children who are too full to eat their dinner must also be too full to eat dessert.

Children like to eat with other children and have been observed to stay at the table longer and eat much more when in the company of their peers. Children are also more likely to eat nonpreferred foods when their peers are eating those foods.[i] Eating is fun. It is healthy to look forward to and enjoy meals.

When introducing new foods at the table, parents are advised to offer them one at a time—and only a small amount at first. As mentioned earlier, the more often a food is presented to a young child, the more likely the child will like that food, so foods should continue to be offered regardless of acceptance. Whenever possible, the new food should be presented before the other foods are revealed at the beginning of the meal, when the child is hungry. Offer the new food, and allow the child to make the decision. Whether the child accepts or rejects the new food is not an issue; do not make it one, even to reward acceptance. Parents often mistakenly use rewards or bribes to train their children to eat specific foods. "When you finish eating your spinach, you will be allowed to watch television." Children learn that they must work to earn the reward. In this case, the "work" is spinach, and if it is work, it must not be

▸▸ PRACTICAL POINT *continued*

▸▸PRACTICAL POINT *continued*

desirable. The end result is a decreasing preference for the food—in this case, spinach.[j] Sometimes parents use foods as rewards to train their children to perform specific tasks. "When you finish putting your toys away, you will be allowed to eat ice cream." In this case, the child learns to give ice cream an enhanced preference.[k]

Parents may find that their children often snack so much that they are not hungry at mealtimes. Some parents find they can live with the philosophy not of teaching children *not* to snack, but of teaching them *how* to snack. Provide snacks that are as nutritious as the foods served at mealtime. Milk and water are appropriate beverages at snack time. Sweet drinks contribute to problems of dental caries and obesity. Snacks can even be mealtime foods that are served individually over time, instead of all at once on one plate. When providing snacks to children, a smart parent thinks of the four food groups and offers pieces of cheese, tangerine slices, carrot sticks, and peanut butter on whole-wheat crackers. Snacks need to be easy to prepare and readily available to children. This is particularly important to children who return home after school without parental supervision.

A bright, unhurried atmosphere free of conflict is conducive to good appetite. Parents who serve meals in a relaxed and casual manner, without anxiety, provide the climate in which a child's negative emotions will be minimized. Conflicts can be promoted by unaware parents, even those with the best of intentions. Parents who beg, cajole, and demand that their child eat deny the child an opportunity to develop self-control. Instead, the child enters a battle that takes on more importance than hunger. The power struggle almost invariably results in a confirmed pattern of resistance and a permanently closed mind on the child's part. Mealtimes can be nightmarish for the child who is struggling with personal and parental problems. If, as a child sits down to the table, a barrage of accusations are shouted—"Your hands are filthy . . . your report card . . . and clean your plate! Your mother cooked that food!"—mealtimes may be unbearable. The stomach may recoil, because the body, as well as the mind, reacts to stress of this kind.[l]

In an effort to practice these many tips, parents may overlook perhaps the single most important influence on their child's food habits—their own example. Parents who do not eat carrots should not be surprised when their children refuse to eat carrots. Likewise, parents who comment on the odor of brussels sprouts do not convince children that these vegetables are delicious. Much of the learning a child accomplishes is through imitation. By setting an example, parents can show children how to enjoy nutritious foods.

At each age, food can be served and enjoyed in the context of encouraging emotional, as well as physical, growth. If the beginnings are right, children will grow without the confusion over food that can lead to nutrition problems. In the interest of promoting both a positive self-concept and a positive attitude toward good food, it is important for parents to help their children remember that they are good kids. What they *do* may sometimes be unacceptable; but what they *are*, on the inside, are normal, healthy, growing, fine human beings.

---

[a]L. L. Birch and M. Deysher, Caloric compensation and sensory specific satiety: Evidence for self-regulation of food intake by young children, *Appetite* 7 (1986): 323–331.

[b]L. L. Birch, D. W. Marlin, and J. Rotter, Eating as the "means" activity in a contingency: Effects on young children's food preference, *Child Development* 55 (1984): 431–439.
[c]E. M. Satter, *Child of Mine: Feeding with Love and Good Sense* (Palo Alto, Calif.: Bull Publishing, 1983).
[d]N. R. Beyer and P. M. Morris, Food attitudes and snacking patterns of young children, *Journal of Nutrition Education* 6 (1974): 100–103.
[e]Adapted with permission from E. N. Whitney, E. M. N. Hamilton, and S.R. Rolfes, *Understanding Nutrition,* 5th ed. (St. Paul, Minn.: West, 1990).
[f]M. E. Breckenridge, Food attitudes of five- to twelve-year-old children, *Journal of the American Dietetic Association* 35 (1959): 704–709.
[g]B. K. Phillips and K. K. Kolasa, Vegetable preferences of preschoolers in day care, *Journal of Nutrition Education* 12 (1980): 192–195.
[h]L. L. Birch and D. W. Marlin, I don't like it; I never tried it: Effects of exposure on two-year-old children's food preferences, *Appetite* 3 (1982): 353–360, as cited in Manipulation of children's eating preferences, *Nutrition Reviews* 44 (1986): 327–328.
[i]L. L. Birch, Effects of peer models' food choices and eating behaviors on preschoolers' food preferences, *Child Development* 51 (1980): 489–496, as cited in Manipulation of children's eating preferences, 1986.
[j]Birch, Marlin, and Rotter, 1984.
[k]L. L. Birch, S. Zimmerman, and H. Hind, The influence of social effective context on preschool children's food preferences, *Child Development* 51 (1980): 856–861.
[l]S. L. King and E. S. Parham, The diet-stress connection, *Journal of Home Economics,* Fall 1981, pp. 25–28.

▸▸ PRACTICAL POINT *continued*

Nutrition at School

While parents are doing what they can to establish favorable eating behaviors during the transition from infancy to childhood, other factors are entering the picture. During preschool or grade school, children encounter foods prepared and served by outsiders. The U.S. government funds several programs to provide nutritious meals for children at school. (School lunches in Canada are administered locally and therefore vary from area to area.) School lunches are required to meet certain criteria. They must include specified servings of milk, protein-rich foods (meat, poultry, fish, cheese, eggs, legumes, or peanut butter), vegetables, fruits, and breads or other grain foods. The design is intended to provide at least a third of the RDA for each of the nutrients. The U.S. school lunch patterns are adjusted to age, as shown in Table 7–10.

Parents rely on the school lunch program to meet a significant part of their children's nutrient needs on school days. Indeed, students who participate in the school lunch program have higher intakes of food energy and nutrients than students who do not. Because children do not always like what they are served, school lunch programs attempt to offer children both what they want and what will nourish them. In response to children's differing needs and tastes, the best programs:

- Present a variety of offerings and allow children to choose what they are served.
- Vary portion sizes so that little children can take little servings.
- Schedule lunches so that children can eat when they are hungry and can have enough time to eat well.

Table 7–10 School Lunch Patterns for Different Ages

Food Group	Preschool		Grades		
	Ages 1 to 2	*Ages 3 to 4*	*K to 3*	*4 to 6*	*7 to 12*
Meat or meat alternate (1 serving per day)					
Lean meat, poultry, or fish	1 oz	1½ oz	1½ oz	2 oz	3 oz
Cheese	1 oz	1½ oz	1½ oz	2 oz	3 oz
Large egg(s)	1	1½	1½	2	3
Cooked dry beans or peas	½ c	¾ c	¾ c	1 c	1½ c
Peanut butter	2 tbsp	3 tbsp	3 tbsp	4 tbsp	6 tbsp
Vegetable and/or fruit (2 or more servings per day, both to total)	½ c	½ c	½ c	¾ c	¾ c
Bread or bread alternate (servings)[a]	5 per week	8 per week	8 per week	8 per week	10 per week
Milk (1 serving)					
Fluid milk	¾ c	¾ c	1 c	1 c	1 c

[a]A serving is 1 slice of whole-grain or enriched bread; a whole-grain or enriched biscuit, roll, muffin, and so on; or ½ c cooked pasta or other cereal grain such as bulgur or grits.

Source: School lunch patterns: Ready, set, go! *School Food Service Journal* 34 (August 1980): 31.

Children can learn from:

- Bakeries.
- Mills.
- Dairy farms.
- Milk-bottling plants.
- Farmers' markets.
- Vegetable farms and fields.
- Food-processing plants.
- Fast-food places.
- Institutional kitchens.
- Supermarkets.
- Convenience stores.
- Neighborhood gardens.
- Natural-food stores.
- Food salvage banks.

Coincident with the school lunch program is a program of nutrition education and training (NET program) in all the public schools. This program is minimally funded, but program administrators, if highly motivated, can be ingenious and creative in accomplishing the program's highest-priority objectives.[69] Creative methods for teaching nutrition to children can center on the use of stories, puppets, and games. For classes with computers, a number of nutrition education programs are available beginning at the preschool level. One way in which children can learn about nutrition, sometimes with only small expense to the school, is to take field trips to nearby food operations facilities. Among the possible places to visit are those listed in the margin. At the very least, children can go to the depots where the school food comes in and to the kitchens where it is prepared. Knowledgeable teachers can then use the questions that arise as opportunities to teach nutrition. Children need not only to be fed well, but to learn enough about nutrition to become able to make healthy food choices when the choices become theirs to make.

On the average, children in the United States spend as much time watching television as they do attending school.[70] It is little wonder that television viewing has become associated with a variety of child and teen behaviors. Television's influence is evident in dental health and childhood obesity, and is discussed in Focal Points 5 and 7.

Preparing Children to Be Healthy Adults

The task of parents and health care professionals is not simply to keep children healthy. They need to look ahead and put forth efforts to create future healthy adults. Many adult diseases develop out of personal habits and styles of living that take root in childhood. (A whole chapter, Chapter 9, is devoted to lifestyle diseases and the efforts adults can make to prevent or forestall them.) Among faulty behavior patterns children can pick up are lack of exercise and dietary habits that lead to obesity, elevated cholesterol, and hypertension. Faulty behaviors that develop in childhood and persist into adulthood contribute the major risk factors of today's major killers of adults, coronary heart disease and other diseases.

The childhood years may be a parent's last chance to influence the development of children's habits. In addition to providing nutritious foods, parents need to foster children's preferences for them. The task is to promote the development of healthful behaviors.

Prevention of obesity To prevent obesity, train preschool children to eat slowly, to pause and enjoy their table companions, and to stop eating when they are full. Stamp out the "clean your plate" dictum for all time, and in its place learn to serve small portions that can be followed by additional servings, if needed. Encourage physical activity on a daily basis to promote strong skeletal and muscular development and to establish exercise habits that will support good health throughout life.

The child who is already obese needs careful management. Weight loss is not ordinarily recommended, because it may easily impair growth in children.[71] Instead, aim to maintain a constant weight while the obese child grows taller. The object is to support normal lean body development, while letting children "grow out" of their obesity (see Focal Point 7).

Prevention of cardiovascular disease Primary prevention of cardiovascular disease begins in childhood. The symptoms of atherosclerosis present themselves in adulthood, but fatty streaks and fibrous plaques are evident early in life. The gradual accumulation of lipids along the inner walls of the arteries is a lifelong process influenced by several factors. Two major predisposing factors are obesity and elevated plasma cholesterol, and these often go together. Early treatment of obesity, maintenance of ideal weight, and regular exercise may well be sufficient to normalize cholesterol levels and benefit cardiac health. The AAP advises that before instituting special dietary measures to correct plasma cholesterol, it is sensible to deal with obesity.[72]

Elevated plasma cholesterol may exist in children independently of obesity. A follow-up study of children who were reexamined as adults showed that elevated cholesterol concentrations during childhood pose a high risk for elevated cholesterol in adult life.[73] In populations where the incidence of adult coronary artery disease is high, the children have average to high plasma cholesterol. In contrast, where the disease incidence is low, the children's cholesterol is low to average.[74] In the United States, children have higher cholesterol than children in other populations where atherosclerosis is less widespread.

Many experts seem to agree that the risk of adult heart disease will be lessened if a preventive diet is implemented early in life. The American Heart Association recommends "prudent" modification of diet in healthy children over the age of two.[75] The dietary recommendations are:

1. The diet should be nutritionally adequate, consisting of a variety of foods.
2. The energy intake should be based on growth rate, activity level, and subcutaneous fat so as to maintain desirable body weight.
3. The total fat intake should be approximately 30 percent of kcalories, with 10 percent or less from saturated fat, about 10 percent from monounsaturated fat, and less than 10 percent from polyunsaturated fat. The emphasis should be on reduction of total fat and saturated fat rather than on increasing polyunsaturated fat.
4. The daily cholesterol intake should be approximately 100 milligrams cholesterol per 1000 kcalories, and should not exceed 300 milligrams.
5. Protein intake should be about 15 percent of kcalories, derived from varied sources.
6. Carbohydrate intake should be about 55 percent of kcalories, derived primarily from complex carbohydrate sources to provide necessary vitamins and minerals.
7. High-salt processed foods, sodium-containing condiments, and added salt at the table should be limited.

The proposed diet, designed for a healthy child, is adequate in all essential nutrients. It provides lean meat, poultry, fish, and vegetable sources of protein; fruits and vegetables; whole grains; and low-fat dairy products. This dietary effort to moderate the consumption of fat, cholesterol, sodium, and sugar in children appears to be safe; it does not impair growth and development.[76] The AAP warns parents to avoid extremes; follow a prudent diet with moderation.

Not all children are healthy, nor do they all receive adequate nourishment regularly. For these reasons, some members of the pediatric health profession oppose the adoption of a prudent-type diet for children. They caution that while the intentions are good, such dietary restrictions may compound nutrient deficiencies in deprived children. One physician warns, "If parents get too zealous, such diets might stunt a child's growth."[77] Findings confirm that dietary restrictions of fat and cholesterol can impair normal growth and development in children.[78] Perhaps dietary restrictions are not needed for all children.

Routine screening of all children can determine what conditions are likely to develop, but family histories are the primary method of identifying "high-risk" children, or children who might need additional dietary restriction. Children of parents with elevated cholesterol levels are almost three times as likely to have cholesterol levels in the 95th percentile as children in the general population. The blood-lipid profiles of children whose fathers were diagnosed with coronary artery disease closely resembled the blood-lipid profiles of the fathers.[79] Once a child is thus identified, at least two cholesterol measurements should be taken to confirm the diagnosis; then appropriate dietary and medical intervention can be implemented.

Based on family histories, different dietary measures should be adopted for different children. For example, it is advisable to feed the child of a parent with high blood pressure a diet relatively low in salt, the child of a parent with diabetes a diet low in sugar and high in complex carbohydrates, and the child of a parent with coronary artery disease a diet low in fat—especially saturated fat—and possibly low in cholesterol. In all these situations, the greatest success is likely to be achieved if the whole family, and not just the child, follows the diet rules.

Obesity and cardiovascular disease have received emphasis here not only because they are widespread in adulthood but also because efforts aimed at preventing them will inevitably help to prevent many other common adult diseases as well. A diet that maintains appropriate weight, and that is low in fat and high in nutrients and fiber, is preventive against cancer, diabetes, diverticulosis, and other ills. Chapter 9 describes in detail nutrition's role in helping prevent disease; suffice it to say here that the earlier in life children's health-promoting habits become established, the better they will stick.

The eventful period of infancy, followed by the steady progress of childhood, brings the child to the brink of a time of major transitions, adolescence. The next chapter describes the nutrition implications of those transitions.

Chapter 7 Notes

1. D. Sinclair, *Human Growth after Birth*, 4th ed. (New York: Oxford University Press, 1985), pp. 23–50.
2. Sinclair, 1985, pp. 102–122.
3. S. E. Morris, *The Normal Acquisition of Oral Feeding Skills: Implications for Assessment and Treatment* (New York: Therapeutic Media, 1982).
4. E. E. Eid, Follow-up study of physical growth of children who had excessive weight gain in first six months of life, *British Medical Journal* 2 (1970): 74–76.
5. W. C. MacLean, Jr., Protein-energy malnutrition, in *Pediatric Nutrition: Theory and Practice*, eds. R. J. Grand, J. L. Sutphen, and W. H. Dietz, Jr. (Boston: Butterworth's, 1987), pp. 421–431.
6. K. M. Hambidge and coauthors, Low levels of zinc in hair, anorexia, poor growth, and hypogeusia in children, *Pediatric Research* 6 (1972): 868–874; C. Xue-Cun and coauthors, Low levels of zinc in hair and blood, pica, anorexia, and poor growth in Chinese preschool children, *American Journal of Clinical Nutrition* 42 (1985): 694–700.
7. N. S. LeLeiko and coauthors, The nutritional assessment of the pediatric patient, in *Pediatric Nutrition: Theory and Practice*, eds. R. J. Grand, J. L. Sutphen, and W. H. Dietz, Jr. (Boston: Butterworth's, 1987), pp. 395–420; MacLean, 1987; F. Viteri, Primary protein-calorie malnutrition—Clinical, biochemical, and metabolic changes, in *Textbook of Pediatric Nutrition*, ed. R. M. Suskind (New York: Raven Press, 1981), pp. 189–216; and D. M. Picou, Evaluation and treatment of the malnourished child, in *Textbook of Pediatric Nutrition*, ed. R. M. Suskind (New York: Raven Press, 1981), pp. 217–228.
8. P. R. Dallman, M. A. Siimes, and A. Stekel, Iron deficiency in infancy and childhood, *American Journal of Clinical Nutrition* 33 (1980): 86–118.
9. R. M. Lauer, J. Lee, and W. R. Clarke, Factors affecting the relationship between childhood and adult cholesterol levels: The Muscatine Study, *Pediatrics* 82 (1988): 309–318.
10. M. Leung and coauthors, Dietary intakes of preschoolers, *Journal of the American Dietetic Association* 84 (1984): 551–554.
11. MacLean, 1987.
12. MacLean, 1987.
13. U.S. House of Representatives, Select Committee on Hunger, *Vitamin A: An Urgent Nutritional Need for the World's Children* (Washington, D.C.; Government Printing Office, 1985), pp. 1–20.
14. K. P. West and coauthors, Vitamin A supplementation and growth: A randomized community trial, *American Journal of Clinical Nutrition* 48 (1988): 1257–1264; Muhilal and coauthors, Vitamin A–fortified monosodium glutamate and health, growth, and survival of children: A controlled field study, *American Journal of Clinical Nutrition* 48 (1988): 1271–1276.
15. A. Sommer and coauthors, Impact of vitamin A supplementation on childhood mortality: A randomized controlled community trial, *Lancet* 1 (1986): 1169–1173.
16. M. Rudolf, K. Arulanantham, and R. M. Greenstein, Unsuspected nutritional rickets, *Pediatrics* 66 (1980): 72–76.

17. S. Bachrach, J. Fisher, and J. S. Parks, An outbreak of vitamin D deficiency rickets in a susceptible population, *Pediatrics* 64 (1979): 871–877; D. V. Edidin and coauthors, Resurgence of nutritional rickets associated with breast-feeding and special dietary practices, *Pediatrics* 65 (1980): 232–235.
18. Rudolf, Arulanantham, and Greenstein, 1980.
19. *Iron Nutrition Revisited—Infancy, Childhood, Adolescence,* report of the 82nd Ross Conference on Pediatric Research (Columbus, Ohio: Ross Laboratories, 1981), p. 1.
20. Summary of a report on assessment of the iron nutritional status of the United States population, *American Journal of Clinical Nutrition* 42 (1985): 1318–1330.
21. K. M. Hambidge and coauthors, Zinc nutrition of preschool children in the Denver Head Start program, *American Journal of Clinical Nutrition* 29 (1976): 734–738; Xue-Cun and coauthors, 1985; R. Gibson and coauthors, A growth-limiting, mild zinc deficiency syndrome in some Southern Ontario boys with low height percentiles, *American Journal of Clinical Nutrition* 49 (1989): 1266–1273.
22. H. H. Sandstead and coauthors, Human zinc deficiency endocrine manifestations, and response to treatment, *American Journal of Clinical Nutrition* 20 (1967): 422–442, as cited by Xue-Cun and coauthors, 1985.
23. J. A. T. Pennington, B. E. Young, and D. B. Wilson, Nutritional elements in U.S. diets: Results from the Total Diet Study, 1982 to 1986, *Journal of the American Dietetic Association* 89 (1989): 659–664.
24. P. A. Walravens and coauthors, Linear growth of low income preschool children receiving a zinc supplement, *American Journal of Clinical Nutrition* 38 (1983): 195–201.
25. R. W. Miller, The metal in our mettle, *FDA Consumer,* December 1988–January 1989, pp. 24–27.
26. Y. Neggers and K. R. Stitt, Effects of high lead intake in children, *Journal of the American Dietetic Association* 86 (1986): 938–940.
27. M. Clark, J. Royal, and R. Seeler, Interaction of iron deficiency and lead and the hematologic findings in children with lead poisoning, *Pediatrics* 81 (1988): 247–254; W. S. Watson and coauthors, Food iron and lead absorption in humans, *American Journal of Clinical Nutrition* 44 (1986): 248–256.
28. K. Mahaffey and coauthors, Blood lead levels and dietary calcium intake in 1- to 11-year-old children: The Second National Health and Nutrition Examination Survey, 1976 to 1980, *Pediatrics* 78 (1986): 257–262.
29. F. L. Cerklewski and R. M. Forbes, Influence of dietary zinc and lead toxicity in the rat, *Journal of Nutrition* 106 (1976): 689–696, as cited by Neggers and Stitt, 1986.
30. M. E. Markowitz and J. F. Rosen, Zinc and copper metabolism in $CaNa_2EDTA$–treated children with plumbism (abstract), *Pediatric Research* 15 (1981): 635.
31. D. Faust and J. Brown, Moderately elevated blood lead levels: Effect on neuropsychological functioning in children, *Pediatrics* 80 (1987): 623–629.
32. Getting the lead out, *Science News,* 24 October 1987, p. 269.
33. M. B. Rabinowitz, Lead and nutrition, *Nutrition and the M.D.,* March 1986.
34. The World Health Organization suggests not more than 3 milligrams per individual for adults; Evaluation of mercury, lead, cadmium and the food additives amaranth, diethylpyrocarbonate and octyl gallate, *WHO Food Additives Series No. 4* (Geneva: World Health Organization), as cited by D. G. Lindsay and J. C. Sherlock, Environmental contaminants, in *Adverse Effects of Foods,* eds. E. F. P. Jelliffe and D. B. Jelliffe (New York: Plenum Press, 1982), pp. 85–110.
35. E. Yetley, Nutritional applications of the Health and Nutrition Examination Surveys (HANES), *Annual Review of Nutrition* 7 (1987): 441–463.
36. S. A. Bock, The natural history of adverse reactions to food in young children. Address presented at the 70th Annual Meeting of the American Dietetic Association, Atlanta, Georgia, 19 October 1987.
37. S. L. Taylor, Food allergy—The enigma and some potential solutions, *Journal of Food Protection* 43 (1980): 300–306.
38. C. D. May, Food allergy: Perspective, principles, practical management, *Nutrition Today,* November–December 1980, pp. 28–31.
39. Bock, 1987.
40. Bock, 1987.
41. R. C. Thompson, The flaw in cytotoxic testing: There's no proof it works, *FDA Consumer,* October 1984, pp. 34–36.
42. American Council on Science and Health, *Baby Foods* (Summit, N.J.: American Council on Science and Health, 1987).
43. World Health Organization–sponsored workshop on "Lactose Malabsorption," Moscow, 10–11 June 1985, as cited in National Dairy Council, Nutritional implications of lactose and lactase activity, *Dairy Council Digest* 56 (1985): 25–30.
44. A. D. Newcomer, Lactase deficiency, *Contemporary Nutrition* 4 (1979): 1–2.
45. V. J. Fontana and F. Moreno-Pagan, Allergy and diet, in *Modern Nutrition in Health and Disease,* 6th ed., eds. R. S. Goodhart and M. E. Shils (Philadelphia: Lea and Febiger, 1980), pp. 1071–1081.
46. D. M. Tucker and H. H. Sandstead, Body iron stores and cortical arousal, in *Iron Deficiency: Brain Biochemistry and Behavior,* eds. E. Pollitt and R. L. Leibel (New York: Raven Press, 1982), pp. 161–182.
47. T. E. Webb and F. A. Oski, Iron deficiency anemia and scholastic achievement in young adolescents, *Journal of Pediatrics* 82 (1973): 827–830; T. E. Webb and F. A. Oski, Behavioral status of young adolescents with iron deficiency anemia, *Journal of Special Education* 8 (1974): 153–156.
48. B. Lozoff, Behavioral alteration in iron deficiency, *Advances in Pediatrics* 35 (1988): 331–360.
49. S. Soesmalijah, M. Husaini, and E. Pollitt, Effects of iron deficiency on attention and learning processes in preschool children: Bantung, Indonesia, *American Journal of Clinical Nutrition* 50 (1989): 667–674; S. Seshadri and T. Gopaldas, Impact of iron supplementation on cognitive functions in preschool and school-aged children: The Indian experience, *American Journal of Clinical Nutrition* 50 (1989): 675–686.
50. E. Pollitt, R. L. Leibel, and D. Greenfield, Brief fasting, stress and cognition in children, *American Journal of Clinical Nutrition* 34 (1981): 1526–1533.
51. Pollitt, Leibel, and Greenfield, 1981.
52. M. Kiester, Relation of mid-morning feeding to behavior of nursery school children, *Journal of the American Dietetic Association* 26 (1950): 25–29, as cited by E. Pollitt, M. Gersovitz, and M. Gargiulo, Educational benefits of the United States School Feeding Program: A critical review of the literature, *American Journal of Public Health* 68 (1978): 477–481.
53. *Diagnostic and Statistical Manual (DSM III),* 3rd ed. (Washington, D.C.: American Psychiatric Association, 1980), p. 41.
54. National Advisory Committee on Hyperkinesis and Food Additives, *Final Report to the Nutrition Foundation* (New York: Nutrition Foundation, 1980).
55. National Advisory Committee on Hyperkinesis and Food Additives, 1980; National Institutes of Health Consensus Development Panel, Consensus development con-

ference statement: Defined diets and childhood hyperactivity, *American Journal of Clinical Nutrition* 37 (1983): 161–165.

56. National Institutes of Health Consensus Development Panel, Defined diets and childhood hyperactivity, *Journal of the American Medical Association* 248 (1982): 290–292.
57. M. L. Arbeit and coauthors, Caffeine intakes of children from a biracial population: The Bogalusa Heart Study, *Journal of the American Dietetic Association* 88 (1988): 466–471.
58. J. L. Rapoport and coauthors, Behavioral effects of caffeine in children, *Archives of General Psychiatry* 41 (1984): 1073-1079.
59. Food and Nutrition Board, *Recommended Dietary Allowances*, 10th ed. (Washington, D.C.: National Academy of Sciences, 1989), pp. 53–77.
60. N. R. Raper, J. C. Rosenthal, and C. E. Woteki, Estimates of available iron in diets of individuals 1 year old and older in the Nationwide Food Consumption Survey, *Journal of the American Dietetic Association* 84 (1984): 783–787. Original data reviewed and restated in terms of 1989 RDA.
61. P. G. Taylor and coauthors, Daily physiological iron requirements in children, *Journal of the American Dietetic Association* 88 (1988): 454–458.
62. Raper, Rosenthal, and Woteki, 1984.
63. P. R. Dallman, M. A. Siimes, and A. Stekel, Iron deficiency in infancy and childhood, *American Journal of Clinical Nutrition* 33 (1980): 86–118.
64. M. W. Breskin and coauthors, Supplement use: Vitamin intakes and biochemical indexes in 40- to 108-month-old children, *Journal of the American Dietetic Association* 85 (1985): 49–56.
65. American Academy of Pediatrics, Committee on Nutrition, Vitamin and mineral supplement needs in normal children in the United States, *Pediatrics* 66 (1980): 1015–1021.
66. W. B. Deichmann and H. W. Gerarde, *Toxicology of Drugs and Chemicals* (New York: Academic Press, 1969), pp. 333–334.
67. American Academy of Pediatrics, Committee on Nutrition, *Pediatric Nutrition Handbook*, 2nd ed., eds. G. B. Forbes and C. W. Woodruff (Elk Grove Village, Ill.: American Academy of Pediatrics, 1985), pp. 170–173.
68. T. A. Morck, S. R. Lynch, and J. D. Cook, Inhibition of food iron absorption by coffee, *American Journal of Clinical Nutrition* 37 (1983): 416–420; S. R. Lynch and A. M. Covell, Iron in soybean flour is bound to phytoferritin (abstract), *American Journal of Clinical Nutrition* 45 (1987): 866; B. J. Macfarlane and coauthors, Inhibitory effect of nuts on iron absorption, *American Journal of Clinical Nutrition* 47 (1988): 270–274.
69. H. R. Armstrong and D. B. Root, Managing a lean NET program, *Community Nutritionist* 2 (1983): 8–10.
70. W. H. Dietz and S. L. Gortmaker, Do we fatten our children at the television set? Obesity and television viewing in children and adolescents, *Pediatrics* 75 (1985): 807–812.
71. W. H. Dietz and R. Hartung, Changes in height velocity of obese preadolescents during weight reduction, *American Journal of Diseases of Children* 139 (1985): 705–707.
72. American Academy of Pediatrics, Committee on Nutrition, Toward a prudent diet for children, *Pediatrics* 71 (1983): 78–80.
73. R. M. Lauer, J. Lee, and W. R. Clarke, Factors affecting the relationship between childhood and adult cholesterol levels: The Muscatine Study, *Pediatrics* 82 (1988): 309–318.
74. Task Force Committee of the Nutrition Committee and the Cardiovascular Disease in the Young Council of the American Heart Association, Diet in the healthy child, *Circulation/American Heart Association Report* 67 (1983): 1411A-1414A.
75. Task Force Committee of the Nutrition Committee and the Cardiovascular Disease in the Young Council of the American Heart Association, 1983.
76. C. J. Glueck, Pediatric primary prevention of atherosclerosis, *New England Journal of Medicine* 314 (1986): 175-177.
77. R. E. Olson, M.D., professor of pharmacology and medicine, as quoted in Low-cholesterol diet "not for everyone," *Modern Medicine,* August 1986, p. 13.
78. M. T. Pugliese and coauthors, Parental health beliefs as a cause of nonorganic failure to thrive, *Pediatrics* 80 (1987): 175–182.
79. J. L. Lee, R. M. Lauer, and W. R. Clarke, Lipoproteins in the progeny of young men with coronary artery disease: Children with increased risk, *Pediatrics* 78 (1986): 330–337.

▶ Focal Point 7

Childhood Obesity

What causes obesity? To put it quite simply, obesity results from an excess of food energy eaten over energy expended. And yet, obesity is not a simple problem to solve. Its underlying causes are complex, its effects far-reaching, and its treatment difficult.

Obesity is the most prevalent and serious nutrition problem in the United States. The best solution to this national health problem is prevention.

Prevention depends on our understanding of how and when obesity arises. Simple questions to ask are: At what age can obesity be predicted? When does it truly set in and persist? It seems that the critical period of development with ramifications on adult obesity occurs during the first five years of life.[1] It can be *predicted* as early as age two for girls and three for boys.[2] Actual obesity, however, usually begins to *set in* between the ages of six and nine. Estimates of its incidence among our nation's youth range from 5 to 25 percent. A large majority (80 percent) of overweight children remain overweight into adulthood. Additional obesity sets in among adults, so two-thirds of obese adults were actually normal-weight children.[3]

Researchers studying the growth patterns of children from one month to 16 years identified trends that could predict obesity.[4] An infant's level of obesity increases during the first year and then decreases. At about age six, another increase in obesity, termed adiposity rebound, occurs and continues into adolescence. A relationship is seen between the age of rebound and later obesity. Obese one-year-olds with an early rebound (before 5½ years of age) tend to remain obese; obese one-year-olds with a normal (six years of age) or delayed (seven years of age) rebound eventually join the average group; nonobese one-year-olds with an early rebound reach the higher percentiles by adolescence; and nonobese one-year-olds with a normal or delayed rebound remain in the low-to-average weight range.

adiposity rebound: the onset of the second period of rapid growth in body fat.

obesity: excessive body fatness.

Overweight infants most often grow into children who are no longer heavy in relation to their height; they are not destined to become obese adults. Almost 80 percent of obese infants lose their obesity in childhood.[5] The early high-risk period for the onset of obesity, then, is not infancy, but the early childhood years. Most obese infants do not become obese children; many obese children do become obese adults. The earlier the onset of obesity, the greater the likelihood of spontaneous resolution with age; obesity of later onset has a higher risk of persistence.[6] One study found that children who were overweight as toddlers were even more overweight as teenagers.[7] Wherever obesity has shown signs of truly setting in, in children, prevention efforts need to be instituted promptly. Assessment is important in identifying those in need of intervention.

The simplest way to define obesity is as excessive body fatness. But then we must ask, excessive compared with what standard? Beyond the point at which a person can live a healthy life? Beyond the population's average? Beyond what is considered beautiful by fashion designers and movie directors?

How is obesity identified? Most commonly, it is defined by body weight. Researchers study the distribution of weight in a population and set the definition of obesity at a standard weight for height and age. If obesity is defined as above the 95th percentile, then we label 5 percent of the population "obese." Likewise, if we defined the cutoff at 50 percent, then half of us would be obese. The definition arbitrarily determines obesity. We have no well-defined biological basis to establish the cutoff point between "normal" and "obese." Researchers generally agree that the cutoff percentile defining obesity in children is 95. The criterion for "overweight," then, lies between "normal" and "obese." While arbitrary, these points do mark a level that is associated with increased morbidity.

Appendix A provides a discussion of assessment, including forms, indexes, tables, and graphs.

percentile: one of 100 equal divisions of data. For example, if a value, such as a person's test score, is higher than that of 75% of the rest of the population, then it is at the 75th percentile in the range of test scores. A person whose weight is at the 75th percentile weighs more than 75% of the population being used for comparison.

Causes of Obesity

As stated in the introductory remarks, obesity is the result of an energy intake in excess of energy expenditure. Therefore, the causes of obesity are those that increase kcaloric intake, decrease kcaloric expenditure, or otherwise disturb the balance of this energy budget. Many possible causes have been explored, and most not proven. Quite likely, more than one cause operates in most instances, and different sets of causes may operate in different individuals. The following sections describe the many areas in which research is currently proceeding in the attempt to understand the root causes of obesity.

Fat-Cell Hypothesis

The fat-cell hypothesis has two major premises. First, overfeeding in early life causes the production of an excess number of fat cells. Second, this excess alters the mechanisms of fat storage and release in adipose tissue in such a way as to predispose the individual to obesity.[8]

The hypothesis contends that obesity begins to develop during critical periods of cellular growth. Fat cells first grow by increasing their numbers (hyperplasia). If excess energy is consumed during this time, the number of cells increases, and a number is reached at some time before adulthood that remains fairly constant throughout life. After this number is reached, cells then grow in size (hypertrophy), but no longer increase in number: subsequent weight gains and losses result in changes in cell size only. Thus the onset of obesity in childhood is considered to be primarily hyperplastic, reflecting an increased number of fat cells, whereas adult-onset obesity is considered to be hypertrophic, a result of existing fat cells becoming overfilled.

hyperplasia: see definition on p. 82.

hypertrophy: see definition on p. 82.

Findings support this distinction between juvenile- and adult-onset obesity, but there is always overlap in both directions. Studies have confirmed that obese children do have a greater number of fat cells than nonobese children. The fat cells are also larger. The fat-cell size measurements of obese children become similar to those of adults even before the children enter their teen years.[9] It is most likely that an obese child whose number of fat cells equals or exceeds that of the average adult will remain obese. Obese children with the greatest number of fat cells are least likely to lose weight successfully. However, a chubby child whose number of fat cells is within the normal range (whose obesity, in other words, is hypertrophic) will more likely outgrow the "baby fat."

Thus the time of cell number determination is critical, and the number arrived at is considered irreversible. Recent studies, however, indicate that cell growth and development do not necessarily follow the simple pattern implied by this theory.

It is important to take notice of whether studies were conducted on normal-weight infants or obese infants.

Adipocyte development One study approached the question of how obesity arises by first studying the normal development of adipose tissues in normal-weight infants.[10] The researchers took samples of fat from beneath the skin (subcutaneous fat biopsies) and determined both the cell size and number of cells. They also determined total body fat, based on whole-body counting of potassium. They repeated these procedures and measures over time in order to determine the ways in which both cell size and cell number change in normal-weight infants. The results from this study show that normal expansion of fat depots in the first year of life is due almost exclusively to fat cell size enlargement. Fat-cell number, however, remains unchanged. During the next six months (the first half of the second year), no further increase in fat-cell size takes place, but a significant increase is noted in cell number. The size of fat cells, in fact, does not change significantly from the age of 18 months all the way to adulthood: in normal 18-month-old infants it is no different from that of normal 8-year-old children or normal 22-year-old females. On the other hand, fat-cell number at 18 months is much lower than that of a normal-weight 8-year-old child (one-half) and the still higher number of a nonobese 22-year-old adult (one-fourth). Cell multiplication begins when adipocytes are filled to a size similar to that of normal adults. That is, when cells reach a "peak" size, more fat cells are produced.

Another study found it important to determine how and at what age obese children deviate from normal fat-cell development and at what age they exceed nonobese adult values. Adipocyte size and number were measured (using methods similar to those described in the previous study) several times in nonobese and obese infants, children, and adolescents.[11] Before age two, all infants were nonobese. From age six months to one year, cell size increased until it reached adult levels. Then it decreased slightly until age two. By age two, children were identified as obese based on weight-for-height measures. In the comparison between nonobese and obese children at age two, the researchers found that the adipocytes of the obese children had become significantly larger; they remained at the size normal for adults instead of shrinking, as those of the nonobese children had done. This difference persisted throughout all ages studied; that is, obese children's fat cells were larger and continued increasing in number from age two onwards. Cell size did not change in obese children from ages 2 through 16, but increased again after age 17. In nonobese children, cells did not change in size from age two until early adolescence; then they reached the adult size. No increases in cell number occurred until after age ten. (In contrast, as mentioned, obese children's fat cells showed significant increases in number throughout the growing years.)

In summary, the normal course of events is as follows. From six months to one year, increments in fat depots are primarily due to increases in cell size. During the next year, cell size decreases, and cell number slightly increases. Fat depots remain fairly inactive until age ten, when both number and size again increase. In contrast, obese children's fat cells attain adult sizes by age two and then show increases in cell number continuously thereafter.

The researchers only counted and measured cells that had already begun to fill with fat and thus were clearly identifiable as adipocytes. Investigators differ on the question of whether and how to count preadipocytes; it is hard to tell whether a cell with no fat in it is destined to become an adipocyte or some other kind of cell. The method, with this limitation, seems to make a real distinction between nonobese and obese children's cell differentiation: whether a cell is called preadipocyte is less important than whether it fills with fat. Obese children's cells do; nonobese children's do not.[12]

This study reveals two time intervals that appear critical in adipose cell development. One is before age two, and the other is during the adolescent growth spurt. These periods may prove to have important consequences for the onset and persistence of obesity.

Critical periods The theory that the number of fat cells becomes fixed at a critical point in time, and thereafter sets a person's tendency to be normal weight or obese, has been considered and debated without confirmation over the past 20 years. It has theoretical appeal because it agrees with other findings about critical periods. However, it has drawbacks because it does not seem to concur with some research findings.

A parallel between malnutrition and obesity becomes evident when considering critical periods of cell development. When a child is malnourished during cell multiplication, cell number is permanently reduced, and growth is stunted forever. The child cannot "catch up" when adequately refed. However, if the child is malnourished during cell size enlargement, growth of cells and the child's overall growth will catch up when the child is properly fed; no number increase is needed. Likewise, if factors encourage cell multiplication, obesity will result without hope of treatment. However, if it is the cell size enlargement that is altered, obesity becomes a treatable disease. By manipulating nutrition during early childhood, the parent (or other caretaker) can control cell number. Even if the range of that cell number is genetically determined, at least it can be kept to the lower end. Such dietary manipulation at critical times during the development of rats has altered the expression of their obesity within genetic limits.

While we are unable to determine exactly when critical periods begin and end, times of exceptionally rapid cell growth have been identified. Three peak periods for the development of obesity in children are late infancy, early childhood (age six), and adolescence.[13]

Adipose enzymes If we accept that fat-cell number and size are the determining factors in weight regulation, then we must ask what mechanisms in turn determine fat-cell number and size. One obvious place to look is to the lipoprotein lipase enzyme system. Lipoprotein lipase is a membrane-bound enzyme, characteristic of fat cells, that breaks down circulating triglycerides for storage in the adipocytes. When lipoprotein lipase activity is high, more fat is stored.[14] Its activity is altered by dietary changes. For example, in fasting, lipoprotein lipase activity decreases in adipose tissue; fatty acids are not stored, but instead are used to meet energy demands. High-carbohydrate diets, on the other hand, increase adipose lipoprotein lipase activity. Obese people who have lost weight tend to have higher lipoprotein lipase activity than when they were maintaining their weight.[15] Unfortunately, their fat cells seem to be more efficient at saving energy than is normal.

Set Point

Many internal physiological variables, such as blood glucose, blood pH, and body temperature, remain fairly stable under a variety of environmental conditions. Constant monitoring of the body's internal status and delicate changes maintain these variables within certain limits. The stability of such complex systems as the human body may depend on set-point regulators. These regulators maintain variables at specified values.

set point: the biological weight above which the body tends to lose weight and below which it tends to gain weight.

Research on the regulation of body weight has been influenced by this set-point concept. Researchers propose that each individual body has a set biological weight determined by genetic and environmental factors, a "set point." However, unlike body temperature, the range of body weight in human beings is large. For example, a reasonable weight for an adult woman, 5 feet 4 inches tall, is about 120 pounds. Yet it is easy to find women of that same height who weigh less than 100 pounds and others who weigh more than 200 pounds. Such large variation does not seem consistent with a tightly regulated set-point system. Such is the picture when we look at the population. A look at individuals reveals another pattern. The range of one individual's body weight remains fairly small over periods of time.

It is thought that the body sends out signals to establish, regulate, and maintain the set point.[16] Yet these set-point regulators do more than maintain a constant body weight; they *defend* that body weight when it is challenged. People who have lost 25 percent of their body weight by restricting their dietary intake return to their normal weights when allowed to eat as they please. Similarly, people who increase their energy intake and gain 15 to 25 percent of their body weight return to their normal weight when allowed to eat as usual. This tenacious defense of body weight is unfortunate for obese people. It deters them from losing weight and encourages regaining of any weight that is successfully lost.

In using set-point regulation to explain obesity, researchers have asked several questions. How is set point regulated? Regulation occurs in two ways. Weight loss triggers signals to increase food intake and reduce kcaloric expenditure. Weight gain triggers the opposite—reduced food intake and increased kcaloric expenditure. These changes in kcaloric expenditure may be due in part to activity level. However, the real difference is seen in an altered metabolism that becomes either more efficient with, or more wasteful of, its energy.

Another question asked is what mechanisms are involved in determining and regulating the set point. No one mechanism holds the answer. A body's set point could operate in many different ways. One of the ways is by fat cells;

another is by inherited enzyme deficiencies; another way is by central nervous system sensitivities; another might be by the thermogenesis of brown fat.

If this theory is valid, it explains why many obese persons find it so hard to lose weight; perhaps their obesity should be accepted as normal. One proponent of the set point states, "Thus, if we view obese individuals as differing not in how they regulate body weight but rather in terms of the set point each is prepared to defend, we might better understand why [some people] remain at essentially the same body weight, without so much as trying, while others remain obese no matter how hard they try to change."[17] The theory is still controversial; even if it is valid, it is not yet possible to determine a person's set point independent of body weight—it can only be estimated from a person's weight over time. For example, a person who has weighed within a few pounds of 150 over the past several years is considered to have a set point of 150.

Genetics

We know very little about the heritability of human obesity. Genetic influences may help determine fat-cell number and size, regulate the efficiency of metabolic processes, and establish the sensitivity of the central nervous system to nutrient deprivation and repletion. Research on these possibilities is progressing.

Genetic studies involve observations of familial resemblances, twins, and adoptees. Familial resemblance studies strongly support the impression that obesity runs in families, but they cannot distinguish between genetic and environmental explanations for this tendency. Twin studies use estimates of heritability based on differences between the intrapair similarities of monozygotic and dizygotic twins. Adoption studies seem ideal in resolving the genetic-versus-environmental question, yet results of studies are frequently inconclusive.

monozygotic twins: twins originating from a single fertilized ovum; identical twins.

dizygotic twins: twins originating from two fertilized ova; fraternal twins.

Familial resemblance The observation that obesity tends to run in families has stimulated debate over genetic versus environmental factors. No doubt, both genes and environment play a role, and interesting "family" factors emerge. When one parent is obese, the chances of infantile obesity's persisting are greater (40 percent) than when neither parent is obese (7 percent). If both parents are obese, the chances become quite likely (80 percent).[18] Similarities of "fatness" between people can be seen the longer they live together. Even the family dog shares the family's fatness profile.

A study of preschool children found strong relationships between parent and child body measurements. Both mothers' and fathers' weight-for-height measurements correlated with the children's weight for height, although mothers' measurements correlated more closely.[19]

Such family relationships are useful in the screening of children at risk for obesity. In addition, the degree of parental fatness can help in predicting the eventual severity of the child's obesity. Most important, in obese families, treatment should involve all family members.

Twin studies Results from twin studies do suggest a genetic potential for obesity.[20] In fact, the genetic impact on fatness appears to be at least as strong as that seen in schizophrenia, alcoholism, and coronary heart disease. The concordance rate for weight of monozygotic twins is approximately twice as high as that for dizygotic twins. The influence of heredity on obesity appears to increase late in childhood, remaining high and fairly stable throughout adulthood.

Adoption studies Adoption studies provide a natural experiment for contrasting the effects of heredity and environment. They assume that similarities between adopted children and their biological parents are genetically determined, while resemblances with adoptive parents must be environmental. Results from adoption studies usually demonstrate a significant similarity in obesity incidence between biological parents and their natural children, but not between adoptive parents and their adopted children.[21] The time of adoption is variable and may be an important factor, considering the effects of early nutrition on the later development of obesity.

It is probably safe to say that fatness is not primarily inherited—at least not in the inflexible sense that blue eyes are inherited. Genetic influence appears to play an important role in determining obesity, but its expression is also influenced by environmental factors. Truly powerful genetic determinants of obesity governing metabolic pathways, such as the Prader-Willi syndrome, account for only a small percentage of obesity in human beings.

Prader-Willi syndrome: a debilitating hereditary disorder marked by childhood-onset obesity, mental retardation, small stature, small extremities, and a propensity toward diabetes.

Energy Balance and Metabolism

Several studies have reported that the energy intake of obese children is actually no greater than their normal-weight peers. Rather than answering any questions, such findings stimulate the search for explanations.

One explanation relies on the wide fluctuation in appetite that children experience. Children's appetites change daily, as do their energy intakes. The maximum daily intake for a child may be two to three times the minimum intake for that same child at any given age. Furthermore, the quantity of food eaten may not be reflected in the child's growth. So, if a study unknowingly compared an obese child's minimum kcaloric intake with a lean child's maximum kcaloric intake, it might reveal no difference. Or it might even seem to show that lean children ate more than obese ones.

Results that seem to imply that lean children eat more than obese ones may reflect the method of assessing kcaloric intake more than the reality of how much of what is eaten. For example, data based on recall interviews are of questionable reliability. Obese people will defensively minimize their food intake on self-report.

One study directly observed the eating and exercise habits of four families, each with an obese and nonobese brother within two years of each other in age. It made measurements weekly, for four to five months.[22] While findings from such a small sample are limited, the strengths of the method used make the results noteworthy. The obese boys consumed significantly more kcalories than did their nonobese brothers. Observers noted that at home, the mothers tended to serve their obese sons larger amounts of food and that the nonobese boys left more food on their plates. At school lunches, the obese boys either purchased more food or bartered, begged, and bullied food from other children.

These results are in contrast to other studies indicating that obese children eat no more than lean children. Even the most basic question about childhood obesity—Do obese children eat more than nonobese children?—has yet to be satisfactorily answered.

In one study of children, no correlation between individual energy intake and degree of fatness was apparent.[23] A relationship between energy intake and social class, however, was apparent, with the privileged-class diet providing less energy than that of the other social classes. The difference in kcaloric density of the diets was evident in the children's fatness. The lower social class had four times as many obese children as the upper social class. However, the energy intake of the children within a social class did not differ; that is, obese children had the same energy intake as their normal-weight and lean peers.

Therefore, it seems appropriate to examine the diet of the social and cultural groups in which a child lives. A diet higher in fat contributes more to obesity than a diet low in fat, even when total energy intake is equal.

Many obese people would like to believe they are overweight due to a "slow metabolism." Actually, metabolic disturbances account for a small percentage of childhood obesity. In fact, the basal metabolic rate (BMR) in obese people is often equal to or greater than the BMR in nonobese people. Because lean body mass is a primary determinant of the metabolic rate, and lean body mass, as well as fat, is increased in childhood obesity, the increased rate is easily explained.

On the other side of the energy balance scale is physical activity. Obese boys have been found to be far less active than their normal-weight brothers inside the home, slightly less active outside the home, and equally active on the school playground.[24] The biggest differences noted have been that obese boys spent more time sitting and standing and less time running. This picture changed dramatically when these activity measurements were converted from time spent into actual kcalorie expenditure. The kcalorie expenditure of obese boys at rest was greater than that of their nonobese brothers; being bigger, they required more kcalories just to maintain their weight. They also needed more kcalories to support their activity, of course: during activity, their kcalorie expenditure tripled as compared with their siblings' twofold increase.

Societal and Psychological Factors

Physical inactivity may be the most important environmental factor contributing to obesity. In the case of children, it may in turn be television that contributes most to physical inactivity. On the average, children in the United States spend as much time watching television as they do attending school. Several effects of watching television could contribute to obesity. First, television viewing requires no energy beyond the resting metabolic rate. Second, it replaces time spent in more vigorous activities. Third, watching television correlates with between-meal snacking, eating the kcalorically dense foods most heavily advertised on children's programs, and influencing family food purchases. The foods advertised on television tend not to be natural, whole, nutrient-dense foods, but rather the more processed, packaged foods that tend to be higher in fat, sugar, and kcalories and lower in nutrients. Nonnutritious foods appear not only in commercials, but also within the television programs themselves. Children may miss the message that eating and drinking high-kcalorie foods will effect weight gain when they see television stars indulging in such behavior and remaining thin.

One study examined the data from the National Health and Nutrition Examination Survey (HANES) to determine whether obesity was associated with increased television watching.[25] Two cross-sectional samples and one longitudinal sample of children and adolescents in the United States were used. The findings reflect that children who watched more television had a greater prevalence of obesity (triceps fatfold at or above the 85th percentile) and superobesity (triceps fatfold at or above the 95th percentile) than children watching less television. In addition, evidence supported a dose-response relationship between obesity and time spent watching television. The prevalence of obesity increased by 2 percent for each additional hour of television viewed. This relationship between television and obesity held strong when control variables such as prior obesity and socioeconomic class were considered.

A **dose-response** relationship shows a correlation of increasing doses of one variable (in this case, television watching) with responses (in this case, obesity) such that the conclusion can be drawn that the more children watch television, the more obese they are. It is stronger than a simple yes-no response (one group watches television and is obese; the other does not and is thin), because the correlation appears at all points along a continuum.

Food satisfies physical hunger and the biological need for energy and nutrients. Yet this simple function of food is often overshadowed by the role food plays in comforting people psychologically or connecting them with others socially. Food satisfies emotional hunger for some people.

The comfort function of food manifests itself in a particular pattern of obesity that reveals no family history of obesity. Described as a reactive obesity, it is an emotional overeating. It begins when a normal-weight child grows fatter in response to a specific stressful event. This may be a divorce, a cross-country move, or rejection by a friend. Oftentimes a look through the family photo album will alert a counselor to a major development in the young person's life.

Eating Habits

One of the ways we learn is by observing the behavior of family members. Young children look to their parents and older siblings as role models in acquiring values, beliefs, and behavior patterns. Much of what we learn as youngsters persists into adulthood. This socialization process can be seen at the family dining table. Mealtime behavior is quite different for fatter and thinner children and their mothers.[26] For thinner children, mealtime provides social interaction as well as food. Fatter children and their mothers frequently have less social interactions during a meal. Obese parents also tend to overfeed their children and lack interest in physical activities. Passing this combination of overeating and underexercising on to another generation perpetuates familial obesity.

Patterns can be seen in the relationship between a child's degree of obesity and mother-child interactions.[27] Mothers of thinner children respond to their children more frequently with encouragement and approval. Mothers of fatter children provide fewer suggestions for performing a task, even when the children request assistance. In general, they offer less guidance and feedback. These observations, noted during playtime and mealtime, are consistent with opinions that obese children receive inappropriate or inconsistent responses to their expressions of need.[28] These are not limited to food-related situations, but involve all areas of daily life. This inconsistency in a family situation appears to have the power to precipitate eating disorders and weaken the child's ability to deal with problems.

This study did not report the degree of fatness of the mothers.[29] If the mother was obese, then perhaps, like the child, she too has an inability to deal with problems. She would be less able to help the child and to teach coping skills if they were lacking in herself.

Obesity seems to be a uniquely human phenomenon. Most animals in the wild or those fed standard rations do not overeat or become obese even when food is plentiful. This suggests that obesity is the result of psychological, cultural, and social influences. Animals do not face the many culinary challenges people encounter regularly. If they did, would they get fat? The possibility has been tested using laboratory rats.

Rats that had maintained normal weight eating lab chow were offered cookies, chocolate, salami, cheese, marshmallows, bananas, and peanut butter. Even though they still had the option of eating lab chow, the animals preferred the supermarket diet and gained more than 2½ times as much weight as the controls.[30] Apparently, the animals found the supermarket foods more palatable than their accustomed meals, and so increased their food intake. The supermarket diet combined the three characteristics thought to be most effective in producing obesity—sweetness, greasiness, and variety—and it worked. The question was answered: special diets can make free-eating animals obese.

This experiment and others like it have demonstrated that the natural preference of both animals and human beings is for energy-dense foods. They are programmed to store energy. This was a useful mechanism during the evolutionary history of humankind when food was scarce and the food supply was unreliable. The body had to be prepared to survive times of famine. It relied on fat stores to get through those times. This same survival mechanism is detrimental in today's developed world, because food scarcities rarely occur. In fact, this society surrounds us with abundant food and limits our physical exercise. Even those of us with only a mild susceptibility to obesity are bound to gain excess weight if we do not actively work to prevent it. Animals, equipped to store energy for the same reasons, become obese for the same reasons when they are placed in lab situations.

Another important consideration may be not what or how much children eat, but rather *how* they eat. Indeed, obese children do eat differently than their normal-weight peers. They take less time to complete a meal, and even more strikingly, they differ in their rates of kcalorie consumption. Obese boys eat twice as many kcalories per minute as their normal-weight siblings.[31] In general, obese children take more bites per minute and chew fewer times per bite than nonobese children.[32]

Whether an infant is breastfed and the age of introduction to solid foods are two other feeding habits that have been examined as possible links to obesity.[33] Most often, breastfed infants receive solid foods later than formula-fed infants. In fact, the longer a child is breastfed, the later solids are introduced (another dose-response relationship). The assumption that an early introduction of solids will lead to obesity has a physiological basis. Infants regulate their intake by volume, not by kcaloric density. A full stomach is a fine mechanism for monitoring breast milk or formula intake, but less adequate in regulating solid foods, because they pack more kcalories per unit volume.

Many solid foods offered to infants are more kcalorically dense than breast milk or formula.[34] In contrast to these theories and assumptions, studies do not support the hypothesis that early introduction of solids and duration of breastfeeding are related to obesity.[35]

Effects of Obesity

The single most important problem for obese children is the potential of becoming obese adults with all the social, economic, and medical ramifications. They have other problems, though, arising from differences in their growth, physical health, and psychological development.

On Growth

Obese children develop a characteristic set of physical traits. They begin puberty earlier and are taller than their peers, although they stop growing at a shorter height.[36] They develop greater bone and muscle mass, possibly because their skeletons respond to the demand of having to carry more weight—not only fat, but also fat-free weight. This causes them to appear "stocky" even when they lose any excess fat. They have a faster BMR, apparently due to their abundant lean body mass.

Factors seem to affect both fatness and maturity so that a relationship between obesity and maturation is evident, although not clearly understood. Early-maturing females are not only shorter than their peers, but fatter as well.[37] By age 30, they have accumulated 30 percent more fat. Such evidence of the inverse relationship between age of menarche and fatness in women is apparent throughout their lives.

On Physical Health

Obese infants do not have a higher mortality rate than normal-weight infants, but they are more likely to have respiratory infections and other illnesses. Like obese adults, obese children display an atherogenic profile—high total serum cholesterol, high serum triglycerides, high low-density lipoprotein (LDL) cholesterol and very-low-density lipoprotein (VLDL) cholesterol. These signs indicate that atherosclerosis is beginning to develop. Obese children also tend to have high blood pressure; in fact, obesity is the leading cause of pediatric hypertension.[38]

For all these reasons, prevention of excessive weight gain in childhood is important. It serves as an effective intervention in cardiovascular disease; atherosclerosis may be reversible with proper diet and exercise. Obese adolescents at high risk for the development of coronary heart disease can reduce that risk with exercise and moderate dietary restriction.[39] Weight reduction prior to or—at the latest—during adolescence is critical to forestalling this lifelong disease.

On Psychological Health

Obesity often causes psychological problems. People frequently react to others' body shapes and send signals that collectively create a "body concept." This body concept becomes incorporated into the person's total self-concept and behavior patterns.

Obese children have many of the same characteristics as minorities; they, too, are victims of prejudice. Many suffer discrimination by adults and rejection by their peers. They are teased and called names. They often have poor self-images, a sense of failure, and a passive approach to life. Television shows, which are a major influence in children's lives, frequently portray the fat person as the bumbling misfit.

The meaning body shape has for children is illustrated by some research in which children were asked to assign adjectives ascribing various behavior and personality traits to silhouettes representing extreme endomorph, mesomorph, and ectomorph body types.[40] Their responses revealed that they associated a common stereotype with each body image. They described the mesomorph favorably; they saw the endomorph in a socially unfavorable light; and they gave personally unfavorable adjectives to the ectomorph. These stereotypes were not related to the body type of the child assigning the description; overweight children themselves share the negative view of fat children. The general consensus is that obese children were children that they "did not like so well." No doubt, childhood obesity creates heartaches. Many children in this study were reasonably accurate in their perception of their own body shape. Children clearly preferred to look like the mesomorph image. In general, children believe thin children have more friends and are better looking and smarter than fat children. In fact, a common belief among obese people is that thin people have no problems.

W. H. Sheldon's classification of body types:

- Mesomorphic: having a husky, muscular body type.
- Endomorphic: having a heavy, rounded body build, often with a marked tendency to become fat.
- Ectomorphic: characterized by a light body build with slight muscular development.

Parents of obese children sometimes are overprotective. The children may not learn to trust their own impulses, becoming passive, demanding, frustrated, and dependent. Parental relationships that do not provide appropriate responses may give the child a feeling of being extra special, of needing to be bigger and better. Unfortunately, with such unrealistic goals, frustration is often encountered and is relieved through overeating—and so, a cycle begins.

Personality problems may also arise from the other physical changes that accompany obesity. Puberty comes earlier. For some young teens, staying fat is a way to avoid sexuality. They may overeat to postpone dealing with sexual feelings that accompany puberty. What effect a shortened childhood, or being one of the first in the class to begin the transition, has on a child is difficult to determine. The teen years are a time of finding a place in the social system. With that comes the overwhelming importance of what you look like, who your friends are, and what activities you are involved in. They are not easy years for most teens; obese teens have an added obstacle in their path.

Psychological reasons may or may not explain obesity, but obesity can cause psychological distress. Some part of treatment must focus on this distress; the counselor should at least be sensitive to it.

Treatment of Obesity

With such a wide array of possible contributing causes, it is little wonder that a successful treatment for obesity has yet to be found. Medical science has worked wonders in preventing or curing many of even the most serious childhood diseases. Yet obesity remains without a surefire remedy. Once excess fat has been stored, it is stubbornly difficult to remove. As one authority on obesity noted, "If 'cure' from obesity is defined as reduction to ideal weight and maintenance of that weight for five years, then a person is more likely to recover from many forms of cancer than from obesity."[41]

Professionals need to recognize that obesity is indeed an obstinate disorder, and that each person's circumstances are unique. Not everyone is motivated to lose weight, and instilling motivation is difficult, if not impossible. Not all persons will lose weight even with strict adherence to a program, and those who do may not keep the weight off. Professionals and clients alike meet with frustration in treating obesity.

Parents and professionals need to realize that pressures on children to conform to the image presented by our society—"fat is ugly" and "thin is in"—create problems. Although not always easy to accomplish, a relaxed, nonjudgmental attitude toward the child, whatever the child's weight, will assure the best emotional growth potential. The child's individual needs must be considered and incorporated into treatment. Treatment also needs to consider the many aspects of the problem and possible solutions. An integrated approach involving diet, exercise, behavioral changes, and psychological support is recommended.

When evaluating obesity, fatness needs to be seen in light of the total needs of the child. Excess weight may only be a side effect of a psychosocial problem. In many cases, parents are part of the problem and the solution. Their ability to understand and follow through with a treatment is as critical to success as the involvement of the child. Because many treatment programs have a high rate of failure, it is important to ask if the consequences of failing at a weight loss attempt are worse than being overweight. Other questions to consider include: When is intervention most appropriate? Will the problem resolve itself without intervention? What are the child's needs and wants?

Before implementing a treatment program for a child, a caretaker must assess the medical, social, and economic risks of obesity. These must be balanced against the potential benefit of success and the possible harm of failure.

Diet

The only way to lose body fat is to establish a negative energy balance. Almost all of the causes of obesity respond to some degree to a combination of diet and exercise.

The initial goal for obese children is not to lose weight but rather to stop gaining weight. Continued growth in height will then accomplish the desired change in weight for height. Treatment should begin with this conservative approach before other, more drastic measures are taken.

The diet plan for weight control is easy to formulate; carrying it through to accomplish a permanent weight loss, however, is more difficult. Whether the goal is to treat or prevent obesity, the concept to teach is controlled eating. Children need to learn to eat foods that supply essential nutrients without consuming excess kcalories. Family meals should reflect kcalorie control in the foods selected and in the method of preparation.

Some physicians recommend a prudent diet to prevent and treat obesity. A prudent diet encourages normal weight gain without inhibiting growth or increasing the incidence of disease. By adopting the taste preferences and eating habits associated with a prudent diet, the child learns to select foods that prevent the deleterious effects on blood-lipid profile and blood pressure of a high-kcalorie, high-fat, high-salt diet.

Children are oftentimes unsuccessful in maintaining new eating habits and weight control. The best we can hope for when children follow a specified diet program is that they will either lose a small amount of weight (less than 10 pounds) or maintain their weight without a gain.

Diet therapy involves dietary counseling at intervals throughout various amounts of time. If the counseling is intense and effective, then short-term success may be seen. For the most part, dietary counseling alone is ineffective in the long run.

A prudent diet was developed by the American Heart Association for use with adults, but has been tested successfully and is recommended by some physicians for children over the age of two. The text describes its use for children on p. 296.

Prudent diet:

- 15% protein.
- 30% fat.
- 55% carbohydrate.

Eat more:

- Nonfat milk.
- Fish and poultry.
- Fruit.
- Vegetables.
- Whole grains.

Eat less:

- Whole milk.
- Red meat and eggs.
- Sugar.
- Salt.
- Fat.

Benefits of exercise:

- Increases energy expenditure.
- Enhances the thermogenesis produced by eating.
- Increases metabolic rate.
- Lowers blood pressure.
- Changes blood lipid levels.
- Increases self-esteem.
- Improves coronary efficiency.
- Suppresses appetite.

Physical Activity

Whether physical inactivity is a cause or a consequence of obesity, obese people are usually less physically active than their thinner peers.[42] They do not eagerly respond to the suggestion that they get more exercise. Their reluctance to exercise stems from a variety of experiences. The added weight and bulk of their bodies increases the effort required to perform activities, making some exercises uncomfortable or even painful. They may be self-conscious about the size of their bodies or have had embarrassing moments that discourage them from exercising where others can watch. Needless to say, after years of avoiding exercise, many do not perform well at athletic activities. The many benefits of exercise are well known, but often are not incentive enough to motivate obese persons, especially children.[43] A large majority will quit exercise programs.

One key to a successful program is to begin at a level that is attainable and enjoyable, gradually increasing the intensity, duration, and frequency of exercise. Even though a short, slow-paced walk may have limited physical benefits, the psychological effects will be positive. The person is actively doing something constructive toward attacking obesity.

Social Support

Social factors play a key role in influencing people. Support or lack of support from family and friends can help or hinder a person's recovery process. The long-term nature of a social system offers the potential for a lasting effect. Two social-support systems have been studied in obesity treatment for children: parental involvement and school programs.

Parental involvement Just as parents may have a critical role in the development of their child's obesity, so too can their involvement in a treatment program be significant. Programs that involve parents in treatment report greater weight losses. When parents are not involved, obesity treatment programs are less successful. Because obesity in parents and children tends to be positively correlated, both benefit from a weight loss program.

Parental attitudes about food greatly influence their children's eating behavior. Unaware that they are teaching their children, parents pass on lessons at the dinner table, on television trays, and in drive-through restaurants. Parents who have battled against being overweight themselves and failed may model for their children the behaviors that have led them to failure—eating too much, dieting inappropriately, exercising too little. Those who have fears about their offspring's being overweight may teach their children fears that are unproductive.

Many childhood obesity treatment programs do include parents. Some provide specific instructions, while others allow passive observation. In general, the parent's role is to:

- Provide a nutritionally adequate diet.
- Serve food under pleasant conditions.
- Allow the child to eat according to need.
- Allow the child to eat without being nagged or cajoled.
- Encourage the child to enjoy nutritious foods.
- Encourage the child to exercise.

One program was designed to assess three methods of parental involvement in the treatment of obese adolescents (12 to 16 years old).[44] The groups were mother-child separately (children and mothers attended separate groups), mother-child together (children and mothers met together in the same group), and child alone (children met in groups, and mothers were not involved). All children received the same treatment; the only difference was parental involvement. The program involved behavior modification, nutrition education, exercise instruction, and social support. The mother-child separately group lost more weight during treatment than did the other two groups. This group also maintained the loss for a year, compared with gains in the other two groups. While parental involvement can have an effect on weight loss, the kind of involvement is as critical as its presence or absence.

Parental involvement had a powerful effect on the children's weight loss, yet there was no relation between changes in children's weight and changes in mothers' weight. Perhaps this was because not all mothers needed to lose weight, or no effort was made to encourage mothers to lose weight. A trend was seen in obese mothers losing more weight than nonobese mothers, but this was not significant. Nor did it seem to matter which group the mother was in.

School programs The school can play a three-part role in treating childhood obesity, by way of the foods served, the exercise program, and the classroom teaching. The school cafeteria can offer lunches that not only meet the nutrient needs of children but also are low in kcalories. Nutrient-dense snacks can replace empty-kcalorie ones.

In physical education class, the emphasis should be on lifelong activities. Physical activities that can be done alone and at any age with minimal equipment are most valuable in adulthood. An instructor who instills motivation and enthusiasm for athletics serves students well.

In the classroom, nutrition education can affect eating behaviors. Teachers may want to include lessons on:

- Body fatness.
- Relative kcalorie value of foods.
- Relative kcalorie value of activities.
- Nutrient density.
- Energy balance.
- Benefits of exercise.
- Changing needs throughout the life span.
- Food sources and variety.
- Cooking.
- Learning to enjoy tastes of new foods.

The public school system offers many opportunities to attack the problem of obesity.[45] It can reach a large number of children, the treatment can be continuous, and it can be offered at minimal cost. The setting is also conducive to nutrition education, an important facet of treatment. Perhaps most important, a broad base of social support can be established. School personnel, family, and peers can offer an obese child the reinforcement needed to continue efforts in a weight loss program.

One school-based program using behavior modification, nutrition education, and physical activity involved such a social network.[46] A positive social environment was created in which the child received support from parents, a nurse's aide, teachers, peers, lunchroom personnel, and school administrators. These people received information on the child's treatment program and were instructed in ways of providing emotional support. The results showed a significant weight loss in contrast to the steady gains noted in the three years prior to treatment. Children not included in the program continued to gain weight relative to their heights, clearly headed for obesity as adults.

Behavioral Therapy

Behavior modification techniques have shown modest success when used to control obesity in children. In contrast to traditional weight loss programs that focus on *what* to eat, behavioral programs focus also on *how* to eat. These techniques involve changing learned habits that lead a child to eat excessively. Several principles are involved.

Self-monitoring The child keeps a careful record of food eaten and physical activity. This procedure provides baseline information and heightens awareness of dietary patterns such as portion sizes and snack times. In some cases, the procedure itself favorably alters the behavior.

Stimulus control Stimulus control limits problems by keeping high-kcalorie foods out of the house. By eliminating the cues that trigger the desire to eat, children can learn skills in modifying eating behavior. They are encouraged to eat at scheduled times and to avoid other activities (such as watching television) while eating.

Family involvement Family involvement is crucial to the successs of treatment programs for childhood obesity. Family members can provide emotional support and reinforcement for adhering to a diet and exercise program.

Reinforcement Items are given to reward attaining goals. It is most effective to use small, tangible items and to reward frequently at the attainment of small goals.

Cognitive restructuring Cognitive restructuring encourages positive, realistic thinking in place of illogical and self-defeating thoughts. Cognitive restructuring has three major components. The first, success rehearsal, mentally prepares a child for handling new or difficult situations. By role-playing and rehearsing high-risk situations, a child can learn to anticipate possible setbacks and develop appropriate coping strategies. The second component minimizes self-disappointment by recognizing failings and renewing plans to do better. Rather than overindulging after having "blown it," a person can simply return to the program without feelings of guilt. The third component teaches children how to handle teasing and criticism about their weight and how to identify their positive characteristics. Learning to cope with size and appearance and to increase self-esteem is a part of growing up for children of all shapes and sizes. Like other children, the overweight child needs to grow and mature in a positive environment.

Harmful Treatments

While drugs are occasionally used to treat adult obesity, they should not be used to manage obesity in children. The possible side effects and potential for abuse make drug therapy a dangerous route for treating growing children.

Severe energy restriction (less than 1000 to 1200 kcalories per day) must also be avoided, especially in the first few years of life. The growth of the child can be compromised with inadequate nutrients and energy. The aim of a restrictive diet is to have the child maintain weight while "outgrowing" obesity.

Weight loss without exercise can also have a negative effect on body composition, especially if the weight is regained. A person who diets without exercising loses both lean and fat tissue. If the person then gains weight without exercising, more fat than lean is gained. Fat tissue is less active metabolically than lean tissue, and so the person's daily energy expenditure is less. Each time a person loses weight and regains it without exercising, that person's metabolism requires fewer kcalories. If the person eats the same amount as before the last diet, the person will not maintain, but will gain, weight. This is one explanation for the ratchet effect of dieting and underlies the importance of exercise as part of a weight loss plan.

Obesity is prevalent in our society. Its far-reaching effects make the need to remedy it urgent. Yet the wide and varied spectrum of interacting factors causing obesity makes solutions difficult. The prevention and treatment of obesity is frequently unsuccessful. Perhaps as research continues and our understanding of obesity becomes clearer, the answers will become more evident and the treatment, more successful.

ratchet effect: the effect of repeated rounds of dieting; the person rebounds to a higher weight (and higher body fat content) at the end of each round.

Focal Point 7 Notes

1. J. L. Knittle, Obesity in childhood: A problem in adipose tissue cellular development, *Journal of Pediatrics* 81 (1972): 1048–1059.
2. K. J. Morgan, M. E. Zabik, and G. A. Leveille, Food consumption and weight/height ratios of children, *Agricultural Experiment Station Research Report* 459 (1984): 1–12.
3. W. H. Dietz, Jr., Childhood obesity: Susceptibility, cause, and management, *Journal of Pediatrics* 103 (1983): 676–686.
4. M. F. Rolland-Cachera and coauthors, Adiposity rebound in children: A simple indicator for predicting obesity, *American Journal of Clinical Nutrition* 39 (1984): 129–135.
5. Morgan, Zabik, and Leveille, 1984.
6. Dietz, 1983.
7. C. L. Shear and coauthors, Secular trends of obesity in early life: The Bogalusa Heart Study, *American Journal of Public Health* 78 (1988): 75–77.
8. J. Kirtland and M. I. Gurr, Adipose cellularity: A review, *International Journal of Obesity* 3 (1979): 15–55.
9. Knittle, 1972.
10. A. Hager and coauthors, Body fat and adipose tissue cellularity in infants: A longitudinal study, *Metabolism* 26 (1977): 607–613.
11. J. L. Knittle and coauthors, The growth of adipose tissue in children and adolescents, *Journal of Clinical Investigation* 63 (1979): 239–246.
12. Knittle and coauthors, 1979.
13. M. Winick, Childhood obesity, *Nutrition Today,* May–June 1974, pp. 6–12.
14. M. R. McMinn, Mechanisms of energy balance in obesity, *Behavioral Neuroscience* 98 (1984): 375–393.
15. McMinn, 1984.
16. R. E. Keesey, A set-point analysis of the regulation of body weight, in *Obesity,* ed. A. J. Stunkard (Philadelphia: Saunders, 1980), pp. 144–165
17. Keesey, 1980.
18. Winick, 1974.
19. R. E. Patterson and coauthors, Factors related to obesity in preschool children, *Journal of the American Dietetic Association* 86 (1986): 1376–1381.
20. A. J. Stunkard, T. T. Foch, and Z. Hrubec, A twin study of human obesity, *Journal of the American Medical Association* 256 (1986): 51–54.
21. A. J. Stunkard and coauthors, An adoption study of human obesity, *New England Journal of Medicine* 314 (1986): 193–198.
22. M. Waxman and A. J. Stunkard, Caloric intake and expenditure of obese boys, *Journal of Pediatrics* 96 (1980): 187–193.
23. M. F. Rolland-Cachera and F. Bellisle, No correlation between adiposity and food intake: Why are working class children fatter? *American Journal of Clinical Nutrition* 44 (1986): 779–787.
24. Waxman and Stunkard, 1980.
25. W. H. Dietz, Jr., and S. L. Gortmaker, Do we fatten our children at the television set? Obesity and television viewing in children and adolescents, *Pediatrics* 75 (1985): 807–812.
26. L. L. Birch and coauthors, Mother-child interaction patterns and the degree of fatness in children, *Journal of Nutrition Education* 13 (1981): 17–21.
27. Birch and coauthors, 1981.
28. H. Bruch, Family transactions in eating disorders, *Comprehensive Psychiatry* 12 (1971): 238–248.
29. Birch and coauthors, 1981.
30. A. Sclafani and D. Springer, Dietary obesity in adult rats: Similarities to hypothalamic and human obesity syndromes, *Physiology and Behavior* 17 (1976): 461–471.
31. Waxman and Stunkard, 1980.
32. R. S. Drabman, D. Hammer, and G. J. Jarvie, Eating rates of elementary school children, *Journal of Nutrition Education* 9 (1977): 80–82.
33. J. H. Himes, Infant feeding practices and obesity, *Journal of the American Dietetic Association* 75 (1979): 122–125.
34. Himes, 1979.
35. P. G. Wolman, Feeding practices in infancy and prevalence of obesity in preschool children, *Journal of the American Dietetic Association* 84 (1984): 436–438; S. Dubois, D. E. Hill, and G. H. Beaton, An examination of factors believed to be associated with infantile obesity, *American Journal of Clinical Nutrition* 32 (1979): 1997–2004; D. L. Yeung and coauthors, Infant fatness and feeding practices: A longitudinal assessment, *Journal of the American Dietetic Association* 79 (1981): 531–535; R. E. Patterson and coauthors, Factors related to obesity in preschool children, *Journal of the American Dietetic Association* 86 (1986): 1376–1381.
36. J. L. Knittle and coauthors, Childhood obesity, in *Textbook of Pediatric Nutrition,* ed. R. M. Suskind (New York: Raven Press, 1981), pp. 415–434.
37. S. M. Garn and coauthors, Maturational timing as a factor in female fatness and obesity, *American Journal of Clinical Nutrition* 43 (1986): 879–883.
38. L. K. Rames and coauthors, Normal blood pressures and the evaluation of sustained blood pressure elevation in children: The Muscatine Study, *Pediatrics* 61 (1978): 245–251.
39. M. D. Becque and coauthors, Coronary risk incidence of obese adolescents: Reduc-

tion by exercise plus diet intervention, *Pediatrics* 81 (1988) 605–612.
40. J. R. Staffieri, A study of social stereotype of body image in children, *Journal of Personality and Social Psychology* 7 (1967): 101–104.
41. K. D. Brownell, Ph.D.
42. Shear and coauthors, 1988.
43. K. D. Brownell and A. J. Stunkard, Physical activity in the development and control of obesity, in *Obesity*, ed. A. J. Stunkard (Philadelphia: Saunders, 1980).
44. K. D. Brownell, J. H. Kelman, and A. J. Stunkard, Treatment of obese children with and without their mothers: Changes in weight and blood pressure, *Pediatrics* 71 (1983): 515–523.
45. C. C. Seltzer and J. Mayer, An effective weight control program in a public school system, *American Journal of Public Health* 60 (1970): 679–689.
46. K. D. Brownell and F. S. Kaye, A school-based behavior modification, nutrition education, and physical activity program for obese children, *American Journal of Clinical Nutrition* 35 (1982): 277–282.

Nutrition during Adolescence

8

Girl on a Swing by Richard Fleishner.

The teen years are a time of change—the child is becoming an adult. Changes are evident in the physical body, emotional maturity, and intellectual achievements. Nutrient needs are high during this time of growth, and the challenge to make sure that nutrient needs are met continues.

A transition of importance is that teens make many more choices for themselves than they did as children. Teens are not fed; they eat. They are not sent out to play; they choose whether or not to invest their energy in athletics. The person concerned with nutrition of teenagers cannot simply deliver food, but must instead deliver motivation. That means becoming knowledgeable about the subjects teens themselves are interested in, and showing the relationship of nutrition to those subjects.

Many teenagers become interested in physical fitness, and some develop self-images as athletes. Many begin to care about weight control and about their complexions. In all of their interests, they are susceptible to pressures from the media, from their peers, and from adults they look up to. Thus a whole new set of topics appears in this chapter—topics related to the interests of adolescents themselves.

Growth, Development, and Assessment

Perhaps the most notable feature of the growth and development of teenagers is its unpredictability. No two teens grow and develop in the same way, and this makes growth standards almost useless, as the "Assessment of Nutrition Status" section will show.

Growth and Development

The rate of growth, which has been fairly steady throughout childhood, rises abruptly and dramatically with the onset of adolescence. The adolescent growth spurt is genetically controlled, with its intensity and duration mediated by hormones. For females, the spurt begins between the ages of 10½ and 11 years and peaks around age 12; for males, the spurt begins between the ages of 12½ and 13 years and peaks around age 14.[1] The duration is between 2 and 2½ years, but of course, wide variations are seen for individuals.

The early stage of adolescent growth is linear; the child "shoots up." Males add approximately 8 inches in height during the growth spurt; females, 6 inches. Skeletal growth ceases with the closure of the epiphyses. The later stage of the spurt is lateral; the child "fills out." Males add approximately 45 pounds to their weight; females, about 35 pounds.[2]

epiphyses: the end segments of long bones that contain a thin area of active bone growth; this growth eventually stops, after which no further significant growth can occur.

Many changes in body shape and posture become evident during adolescence. The first areas to accelerate growth are the feet and hands, then the calves and forearms, followed by the hips and chest, and then the shoulders.[3] The trunk of the body is the last to go through a growth spurt. This sequence of development takes teenagers through an awkward phase of having large and ungainly limbs compared with the rest of their bodies.

Before puberty, the differences between male and female body composition are minimal. Sex differences in the skeletal system, lean body mass, and fat stores become apparent during the adolescent spurt.[4] In males, the shoulders grow substantially. In females, the pelvis becomes wide and spacious in preparation for childbearing. During the growth spurt, males increase their lean body mass more than females.

Just prior to the adolescent growth spurt, body fat begins to increase.[5] During the male growth spurt, total body fat decreases. Females have no such decrease and, in fact, lay down additional fat. Additional fat deposition dramatically alters body shape. Why females deposit fat in specific locations is unknown, although the fat in the breasts does serve to protect the mammary glands and provide energy for future gestation and lactation needs.

Hormones control the secondary sex changes experienced by both males and females. The earliest sign of puberty in males is the growth of the testicles and penis.[6] Changes in the larynx, skin, and hair distribution follow. The earliest internal sign of puberty for females is growth of the ovaries.[7] Externally, the breasts enlarge, and pubic hair appears. Menarche occurs after the peak of growth in height and coincides with maximum growth deceleration.[8] A female can expect to grow only about three more inches in height after menarche.

menarche: defined on p. 42.

Growth is reflected not only in the changing size of the body but also in the relative proportions of the body's water, fat, and lean tissues. The greatest change in body composition in females during the adolescent growth spurt is seen in the ratio of lean body weight to fat.

Researchers have attempted to define the relationship between growth and menarche, suggesting that a minimum amount of body fat is required for the onset of menses.[9] The total amount of body fat differs in females of different sizes, but the ratio of lean body weight to fat at menarche appears always to fall within a specific range (about 3:1), and the percentage of body weight as fat at that time approximates 22 percent.

Not all studies support the hypothesis that menarche depends on a minimal amount of body fat. One study found no relationship between body size or composition and the onset of sexual maturity in female adolescents who had been undernourished as young children.[10] These adolescents were much smaller than their well-nourished peers even at the same stage of sexual development.

Throughout childhood, hemoglobin and red blood cells steadily increase. During the adolescent growth spurt, hemoglobin concentration and red blood cell number rise considerably; the red cell count in males rises above that for females and remains higher for life. No corresponding spurt in the number of white blood cells takes place.[11] The heart muscle grows markedly, resulting in a rise in blood pressure and a fall in heart rate.

Development during the adolescent period depends not only on the individual's present nutrition status but also on previous nutrient intake. Several studies have examined the impact of undernutrition in early childhood on adolescent growth. A follow-up study of adolescent females who had suffered varying degrees of undernutrition during early childhood revealed that undernutrition in early life diminished eventual body size.[12] Malnourished children are shorter and lighter than their well-nourished peers at the end of the adolescent growth spurt. But given adequate food, undernourished children add more to their height during their growth spurt than do previously well-nourished children; weight gains are comparable. In other words, adolescence can serve as a catch-up period to regain some growth losses due to undernutrition in the early years.

A possible explanation for the greater increment in height gain seen in previously undernourished females could be delayed menarche. Menarche is delayed by two years in females undernourished as children.[13] With delayed menarche, epiphyseal fusion is delayed, and this allows for a longer period of rapid growth. Delayed sexual maturation is also reported in males undernourished as children. Their attainment of sexual maturity is postponed by up to three years.

Assessment of Nutrition Status

Growth in adolescence is so variable and individual that norms are useless. The best standard against which to evaluate an adolescent's growth in height is a set of previous measures of the same adolescent. Wide variations in weight are seen as growth spurts take place, and can be expected to smooth out as the steadier time of adulthood approaches. Obesity or underweight both can set in, and adult norms are useful as rough indicators; however, deviations are not unusual, and great significance need not be attached to them unless they persist.

The standards for biochemical tests shift from those of children to those of adults during adolescence (many of the nutrition assessment tables for the child presented in Chapter 7 included values for the adolescent; see also Appendix A). When assessing adolescent growth and nutrient status, the health care provider must consider many changes and individual variations that occur during this time. For example, the stage of sexual maturation affects assessment of adolescent serum zinc concentrations. Males and females who have reached adult sexual maturity have higher serum zinc concentrations than adolescents at less mature stages.[14]

Energy and Nutrient Needs of Adolescents

The rapid growth that occurs during adolescence is reflected in the high nutrient needs of this period. In general, *total* nutrient needs are greater during adolescence than at any other time of life, with the exception of pregnancy and lactation. Nutrient and food energy deficiencies during this time can retard growth and delay sexual maturation.

Chronological age provides an inappropriate timetable for evaluation of energy and nutrient requirements. Growth would be preferable. Yet age is most often used in establishing guidelines, because it is easy to keep track of and because it works in a general way. The RDA split the teen years into four-year-long blocks, on the assumption that the physical changes that occur during those periods are similar for most adolescents. The amount of change in each nutrient need during adolescent growth varies with each individual. Needs depend on the age of puberty onset, the velocity of growth, individual activity level, and the length of time required to complete the maturation process.

Energy-Yielding Nutrients

The body requires energy and protein for metabolic processes, physical activity, and growth. When compared on a per-kilogram-body-weight basis, the energy and protein allowances for adolescents are lower than for younger age groups. However, when total values for the average individual are compared, energy and protein allowances for adolescents are greater than for any other age group. Caution is offered that the energy recommendations for the adolescent years may not be appropriate for all individuals. Activity, growth, and tendencies toward obesity or underweight must be considered in determining the appropriate energy allowance for each individual.

Energy RDA for adolescents:
2500 kcal (11 to 14 yr, males).
2200 kcal (11 to 14 yr, females).
3000 kcal (15 to 18 yr, males).
2200 kcal (15 to 18 yr, females).

Protein RDA for adolescents:
45 g/day (11 to 14 yr, males).
46 g/day (11 to 14 yr, females).
59 g/day (15 to 18 yr, males).
44 g/day (15 to 18 yr, females).

Vitamins

The RDA for most vitamins increase during the teen years (see RDA table on the inside front cover). Several of the nutrient recommendations for adolescents are the same as those for adults. One exception is vitamin D. The special role of vitamin D in skeletal growth is reflected in a higher RDA during the adolescent growing years (10 micrograms) than in adulthood (5 micrograms).

Minerals

A look at the RDA table reveals that the RDA for zinc and iodine rise from low values for children to high values at adolescence and remain there for adults. Iron RDA for females follow that same pattern until the later years when they return to preadolescent values. For calcium and phosphorus, and for iron in males, the RDA reach a peak during the adolescent years and return to preadolescent values for adults. Magnesium and selenium RDA are based on body weight and change accordingly.

Calcium needs are high during adolescence, because the bones are growing in length and in density during the adolescent growth spurt. Absorption of calcium is high during this time of growth, and food sources such as milk best provide the calcium.

Calcium RDA for adolescents:
1200 mg/day.

Iron needs are also high during adolescence. Iron status is most affected during four times of life: in infancy, during growth spurts, during the female reproductive years, and in pregnancy.[15] All adolescents are in growth spurts, and half of them are females in their reproductive years. During periods of growth, blood volume and muscle mass increase, thus increasing the iron need for hemoglobin and myoglobin synthesis. The RDA for iron during adolescence differs for males and females. Even though males are building and maintaining a larger blood volume and muscle mass, females have greater iron losses from menstruation.[16] For females, the values at adolescence remain high into adulthood. For males, they return to preadolescent values in adulthood.

Iron RDA for adolescents:
12 mg/day (males).
15 mg/day (females).

Supplements for Adolescents

No across-the-board recommendations are made for supplements for teenagers. Like adults, they can get the nutrients they need from foods. However, approximately 10 percent of adolescents report taking vitamin and mineral supplements, and do obtain a few nutrients from them.[17] These adolescents are more likely to take single-nutrient supplements than multivitamin and mineral preparations, and their choices are not likely to match their specific deficits. If a teenager takes a supplement, the guidelines should be the same as for adults; it should provide RDA-level amounts of the whole spectrum of needed nutrients (see discussion in Chapter 2). All teenagers' iron status should be monitored by health care providers, and supplementation or dietary modification should be instituted where needed.

Eating Patterns

Perhaps the safest generalization that can be made about adolescent eating patterns is that no one pattern exists. At any given time on any given day, a teenager may be skipping a meal, eating a snack, preparing a meal, or consuming food prepared by a parent or restaurant.

Using 24-hour food records, one study examined the meals and snacks of adolescent students and offered a glimpse of what might be considered typical of teenage eating patterns.[18] Approximately one-third of the adolescents skipped breakfast. An extensive survey of adolescent eating habits confirms this finding.[19] Not too surprisingly, breads and cereals were the most popular breakfast foods and milk the most common beverage consumed by those students who did eat breakfast. Students who omitted breakfast had lower total nutrient and food energy intakes for the whole day than those who did eat breakfast. These differences were even greater than the intake quantities of the breakfast meals, indicating that the breakfast eaters made better food choices and consumed more energy throughout the day. Breakfast skippers did not compensate for lost nutrients and energy at other eating times.

Almost 75 percent of the students ate lunches scored as "good"—that is, meals that provided adequate protein, vitamins, and minerals. Most of the lunch-eating students ate food brought from home or purchased from the school cafeteria. The most popular lunch food was a sandwich.

Almost all of the adolescents ate an evening meal at home or at the home of a friend or relative. These evening meals contributed significantly to daily intakes of nutrients and provided more than one-third of the total food energy. Most evening meals included foods such as cheese, meat, eggs, potatoes, bread, and pasta. The beverage of choice was a soft drink.

Most of the adolescents ate one or more snacks, providing approximately one-third of the daily food energy intake. On the average, the snacks were lower in nutrients than the meals. However, the snacks were not always "empty kcalories," but did contribute to daily nutrient intakes. Typical snack foods such as carbonated beverages, candies, desserts, and chips were popular, but many teenagers selected breads, cereals, meats, and milk products for snacks.

Teenagers especially value their independence, and one way they express it is through food selection. Half of the breakfast eaters prepared their own meals, and one-third of the adolescents either prepared their evening meals or selected from restaurant menus. However, independence has its price: adolescents who prepared their own meals consumed less of several nutrients than those who ate parent-prepared meals.

In this study, males met the RDA for all nutrients except iron; females met the RDA for all nutrients except iron, calcium, and vitamin A. These inadequacies are frequently reported in nutrient intake studies of teenagers.[20] In addition, various studies have reported dietary inadequacies of vitamin B_6, zinc, folate, iodine, vitamin D, and magnesium prevalent among adolescent girls.[21] The gender difference in diet adequacy primarily reflects the males' consuming larger quantities of food. Females' lower food energy intakes make more evident the less-than-perfect nutrient density of their diets.

Fast foods make an important contribution to many teenagers' intakes. The high energy contents of these foods in relation to nutrient contents are not equally affordable to all teenagers; they can compromise the quality of the diet for those with lower energy allowances. But fast foods do not have to mean total abandonment of nutrition. They can have an acceptable place in a teenager's diet, provided they do not dominate the diet and that their shortcomings are compensated for at other meals. A fast-food meal once a week or so has little impact on a teenager's overall nutrition status. Teenagers who consume fast-food meals frequently, however, need to vary their menu selections and pay close attention to their other meal selections.

Two major shortcomings of fast-food meals are their high energy and high fat contents. A single large fast-food sandwich (with double beef and cheese) can provide up to 950 kcalories, over half of them from fat. Over half of the kcalories in french fries also are from fat. In general, the items that are high in fat, like these, are items that either contain meat or are fried; others are made with whole milk. Many fast-food establishments publish the kcalorie, fat, and cholesterol contents of their products so that customers can theoretically choose meals compatible with their needs, but teenagers might more easily simply learn some rules of thumb. If you want to limit your energy intake, select lower-kcalorie menu items, eliminate high-kcalorie toppings such as mayonnaise and salad dressing, choose fewer items, or limit energy intake at other meals.

Active teenagers may be able to afford kcalorically expensive fast-food meals, provided they also deliver substantial percentages of their recommended intakes for several nutrients. A moderate fast-food meal of a hamburger, french fries, and milk shake, which offers 820 kcalories (35 percent from fat) contributes its share of protein, thiamin, riboflavin, niacin, vitamin B_{12}, calcium, and iron to a day's intake. As is true of most fast-food meals, however, this one falls short in its contribution to a day's intake of fiber, vitamin A, folate, and vitamin C. Teenagers wanting to eat fast-food meals can compensate for their lack of these vitamins and fiber by eating a large salad, several fruits, and a generous serving of dark green vegetables at other meals and snacks during the day.

Fast foods also contribute large amounts of salt, as do other snacks that teenagers typically eat. Adolescents' sodium intakes typically exceed the upper limit of the NRC's recommended daily intake for sodium.[22] As discussed in Chapter 9, epidemiological evidence links high salt intakes with an increased incidence of hypertension.[23]

Snacks are a part of the teenage lifestyle. They fit in nicely when socializing, studying, working, playing, and relaxing. Because they are a part of the daily food intake, snacks need to provide nutrients. Snacks rich in the nutrients that are missing from meals are a bonus. To get extra iron into the day's intake, a teenager might snack on bran muffins, hard-boiled eggs, or crackers with peanut butter. Calcium is available from yogurt, cheese, and ice cream (if the kcalories can be afforded). Good vitamin A snack ideas include carrot and broccoli pieces, apricots, peaches, and cheddar cheese.

Teenage beverage choices are another subject of interest. A study focusing on adolescent beverage consumption patterns used 24-hour recall and two-day diet records.[24] Similar to the findings from the previous study, this study found that milk is more likely to be consumed with a meal (especially breakfast) than as a snack. Most of the adolescents drink milk, but males drink larger quantities than females. Fruit juice is consumed primarily at breakfast. Soft drinks are likely to be consumed with lunch, supper, or snacks.

The study then asked what impact soft drinks had on the nutrient intakes of adolescents. The greatest impact found was on calcium intake, which varied inversely with soft drink consumption. For males, high soft drink users met almost 90 percent of the calcium RDA, whereas female high soft drink users met only approximately two-thirds of the RDA.

It has been suggested that soft drinks may threaten bone integrity both by displacing milk (and therefore calcium intake), and also by inhibiting calcium absorption with their high phosphorus content.[25] Manufacturers add phosphoric acid to colas and root beers and phosphate salts to powdered beverage mixes. That a high phosphorus-to-calcium ratio in the diet impairs calcium absorption was of concern in the past.[26] An expert reviewer on the subject of calcium absorption cites convincing evidence to the contrary.[27] When adult males consumed as much as 2000 milligrams of phosphorus per day, at several levels of calcium intake, calcium absorption was unaffected.[28] From another study, high phosphate loads may markedly increase serum phosphorus, but serum calcium and the hormone that regulates calcium absorption are still unaffected.[29] Thus the phosphate contents of soft drinks present no problem, and for teenagers who can afford the kcalories and are meeting their calcium needs, they are an acceptable part of the diet.

Soft drinks may present a problem when caffeine intake becomes excessive. Caffeine is a true stimulant drug found in many soft drinks. It increases the respiration rate, heart rate, blood pressure, and secretion of stress and other hormones. However, caffeine seems to be relatively harmless when used in moderate doses (the equivalent of fewer than, say, four 12-ounce cola beverages a day). In amounts greater than this, it can cause the symptoms associated with anxiety—sweating, tenseness, and the inability to concentrate. It is also suspected of increasing the risk for cardiovascular disease and heart attack, as well as the possibly painful, but benign, fibrocystic breast disease.

Problems Adolescents Face

A major element, new in the life of teens, is that they are engaged in the effort to define themselves. They are all wrapped up in their own body images and self-images. Some are becoming sexually active; a few teenage girls face pregnancies, while their own nutrition needs for growth are still high and difficult to meet. Some become intensely interested in weight control, even to the point of obsession. Some become concerned about their complexions, especially if they have acne (or even one pimple). The following sections deal with these problems.

Teenage Pregnancy

The teen years are a time of sexual awareness. Young people are experiencing new changes in their bodies and in relationships with others. The number of females having sexual intercourse increases from less than 20 percent to 55 percent between the ages of 15 and 19.[30] Percentages for males are slightly higher. With this sexual activity comes the responsibility of contraception. Teenagers who do practice contraception most often use oral contraceptives.[31]

The impact of oral contraceptive use on nutrient status is discussed in Chapter 2.

Health and nutrition implications Nine out of ten sexually active teenagers who do not use contraception become pregnant within a year. The birthrate among teenagers in the United States is higher than that in most developed countries (see Figure 8–1).[32] Even though it has declined in all countries in the last decade, teenage pregnancy continues to be a major problem in the United States.[33] One out of every ten teenage girls becomes pregnant each year. Of these approximately 1 million teens, more than half give birth.[34] Putting it another way, one out of five U.S. women bears a child before reaching her 20th birthday.[35]

The timing of adolescent growth and sexual maturation varies greatly among teenagers, but in general, females attain physiological maturity about four years after menarche.[36] Pregnancy before the end of these four years, while a teenage girl is still growing, presents serious threats to health. The demands of pregnancy compete with those of growth, placing the young female at high risk for pregnancy complications.

Perhaps the greatest risk of a teenage pregnancy is death of the infant. The infant mortality rate for mothers under age 20 is high, with mothers under 15 having the highest rate of all age groups.[37] A closer look at infant mortality reveals that the risks to infants of young mothers are greatest during the first month of life. Problems that arise shortly after birth originated during gestation.

When reviewing infant mortality rates, researchers must consider the role of premature, low-birthweight infants, for these infants have the highest risk of mortality. The percentage of low-birthweight infants is greater for teenage mothers than for any other age group.[38] In fact, premature birth is the most critical aspect of teenage pregnancy because of its associated risk of mortality.

In contrast to the above statistics, maternal mortality is *lowest* for mothers under age 20.[39] However, maternal illness is especially common in teenage pregnancies, with the rates for pregnancy-induced hypertension 50 percent higher than for older women. Other complications common to teenage pregnancy are iron-deficiency anemia, which may reflect poor diet and inadequate prenatal care, and prolonged labor, which reflects physical immaturity of the mother. Only about half of all teenage mothers receive prenatal care in the first trimester of pregnancy.[40] A note from research: women giving birth before the age of 18 have a low risk of breast cancer.[41] The risk of breast cancer increases as age at first birth increases.

Figure 8–1 Birthrates among Adolescents
This figure shows the national birthrates per 1000 women aged 15 to 19 as reported in 1980 in rank order. Pregnancy rates were only available for one-third of the nations listed here but, in general, seemed to follow a similar ranking. As you can see, the United States has a higher pregnancy rate than all but a few developed countries.

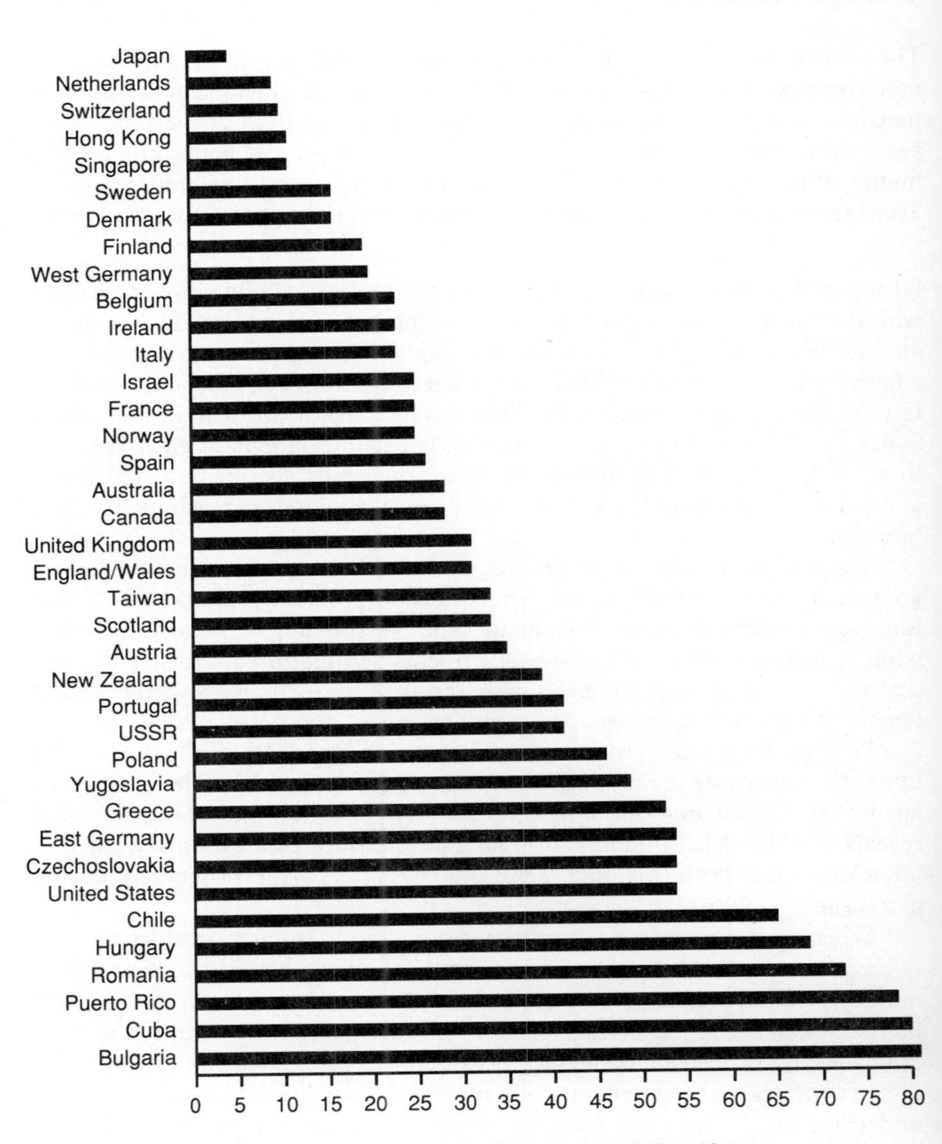

Source: Adapted from E. F. Jones and coauthors, *Teenage Pregnancy in Industrialized Countries: A Study Sponsored by the Alan Guttmacher Institute* (New Haven, Conn., Yale University Press, 1986), pp. 251–255.

Energy needs, nutrient needs, and weight gain Not much specific information is available on the energy and nutrient needs of the pregnant adolescent. For some nutrients, estimates are made by adding the increment for the pregnant adult woman to the RDA for the nonpregnant teenager 15 to 18 years of age; for others, an allowance is provided for pregnant women of all ages.[42] In some cases, estimates may overstate needs due to adjustments the body makes during pregnancy. Reduced physical activity of the pregnant teenager may reduce her energy needs compared with those of her nonpregnant peers.

Figure 8–2 shows that a teenage pregnant girl's needs for many nutrients increase above prepregnant levels much more than her energy allowance does. This raises the question of how to determine the proper energy intake for a pregnant teenage girl. The best way is to monitor her weight gain over time. Younger mothers (13 to 16 years of age) have smaller newborns than older

Figure 8–2 Nutrient Needs of Pregnant and Nonpregnant Teenage Girls Compared Most of the values derive from adding the increment of recommended nutrients for the pregnant adult woman to the RDA for females 15 to 18 years of age; the values for vitamin D, folate, calcium, phosphorus, and zinc are allowances for pregnant women of all ages. The nonpregnant teenager's nutrient recommendations are set at 100% RDA for females 15 to 18 years old.

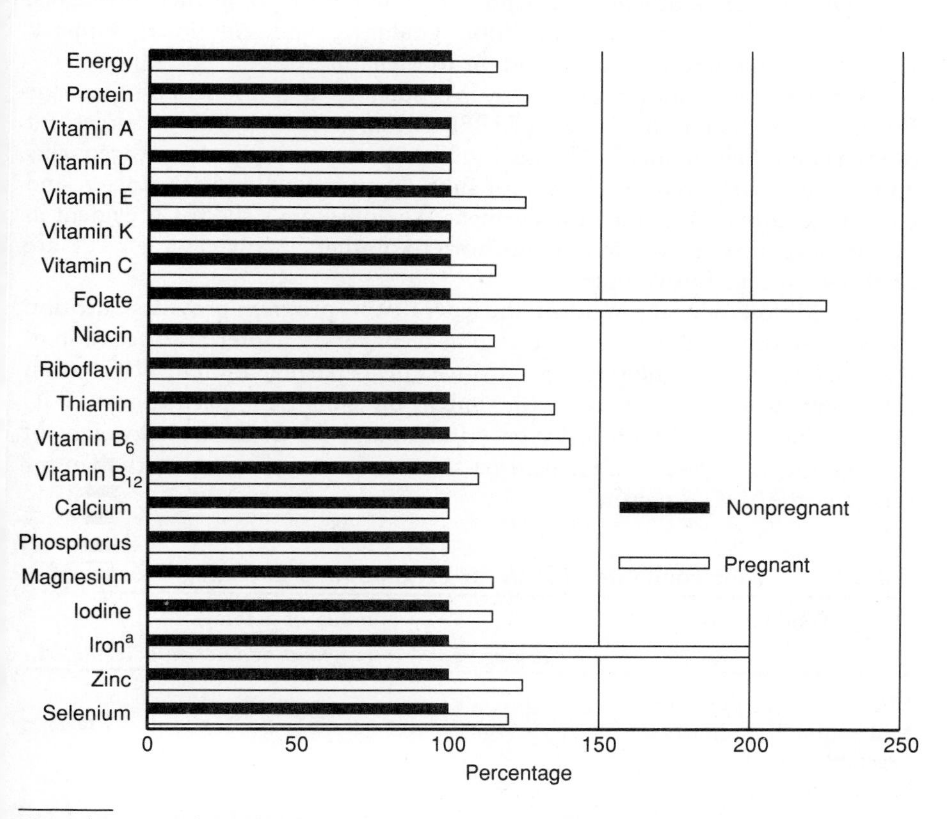

[a]Since the increased iron requirement cannot be met by typical diets or by iron stores, supplements are recommended.

adolescent and adult mothers (17 to 25 years of age) with the same prepregnant weights and pregnancy weight gains (24 pounds).[43] A young teenage mother (aged 13 to 16) needs to gain approximately 30 to 35 pounds in order to deliver an infant of optimum birthweight. The total weight gain needs to equal the weight gain expected for growth plus the weight gain required for pregnancy, if a young adolescent is to give birth to an infant of the same size as an older woman's infant.

Like the nutrient needs of pregnant teens, those of lactating teens deserve mention. When lactating adolescents consume their usual diets containing 900 milligrams of calcium, they lose 10 percent of their bone minerals by 16 weeks postpartum.[44] In contrast, lactating adolescents consuming diets that meet their calcium RDA lose no bone minerals. Adequate calcium intake is obviously important to preventing bone loss during lactation in adolescents. Table 8–1 shows a daily eating pattern for pregnant and lactating teens based on the Four Food Group Plan.

Support needs The plight of a pregnant teenager is often compounded by a multitude of social and economic problems. Most are from low-income families. Oftentimes, they have several children during their teen years, and this further increases their pregnancy risks.

Pregnant teenagers often find themselves without resources or access to prenatal care. They urgently need programs addressed to all their problems, including medical attention, nutrition guidance, and emotional support. Continued schooling is most important to their future.

Many communities offer programs, such as San Francisco's Teenage Pregnancy and Parenting Project (TAPP), to improve the health of adolescent mothers and their infants.[45] The goals of TAPP are to keep pregnant teenagers enrolled in school, reduce the rate of subsequent unwanted pregnancies, and encourage father and family involvement. The program's success is evident in that it meets these goals, and infants born to mothers involved with TAPP are of above-average birthweight.

In addition to local programs, the federal WIC program provides nutrition supplementation and education to eligible teenagers. Chapter 3's discussion of maternal nutrition highlighted the positive effects participation in WIC has on infant birthweight. Compared with similar, but non-WIC, participants, WIC mothers have fewer low-birthweight infants, and the average birthweight of their infants is higher.[46] These benefits are even greater for teenagers than for others in the WIC program.

Table 8–1 Four Food Group Plan for Pregnant and Lactating Teenagers[a]

Food Group	Number of Servings	
	Teenagers	*Pregnant or Lactating Teenagers*
Meat and meat alternates	2	3
Milk and milk products	4	5
Vegetables and fruits	4	4
Breads and cereals	4	4

[a]See Appendix C for a more detailed summary of serving sizes and food sources.

Counseling the pregnant teenager is a challenge. It is extremely important to establish rapport. Teenagers typically turn a deaf ear to lectures. Empathy and support are critical. Open-ended questions will encourage the young woman to talk about herself. Once she has revealed something of herself, the counselor can use this information to employ nutrition counseling. For example, if she says she eats meals away from home, the counselor can provide her with tips on how to get the most nutrients for the money she spends. If she enjoys snacking, then she will welcome ideas for nutritious snacks. She needs respect and support for her good judgment whenever possible. With emotional support and adequate prenatal care, she will look and feel better throughout the pregnancy, and her infant will have a better chance of being born healthy and staying that way.

Eating Disorders

To teenagers, growing means getting taller and heavier. For many female teenagers, this is not a welcome occurrence. They do not want to grow bigger; they become dissatisfied with their appearances. Adolescence is a time of increased concern for body image, anyway; and peer pressure and media pressure make it worse. Fear of obesity is prevalent even among those who are at or below normal weight. The result of being "terrified about being overweight" and "preoccupied with the thought of having fat on their body" incites teenagers to frequent dieting. In fact, one survey of females (ages 14 to 18) reported that over 70 percent had dieted to lose weight; almost 40 percent were dieting at the time of the survey.[47]

The development of obesity is discussed in Focal Point 7.

Most women see themselves as one-fourth larger than they actually are.[48] The more inaccurately they perceive their body size, the worse they feel about themselves. Female adolescents often diet to lose weight even when they are within or below the average range for body weight.[49] Such a goal is neither healthy nor obtainable for most teenagers. Females 15 to 17 years of age consume an average of 450 kcalories less than the RDA. Distorted attitudes and self-perceptions can lead to such eating disorders as compulsive overeating, anorexia nervosa, or bulimia.

Anorexia nervosa or bulimia affects an estimated 1 million teenagers.[50] Specific causes of these eating disorders still baffle clinicians. Some speculate that society's excessive pressure to be thin is to blame, others point to neurological links with depression and impulsive behaviors or other biological malfunctions, and still others believe the disorders to be the result of an inability to cope. All agree that the conditions are multifactorial—sociocultural, neurochemical, and psychological—and that treatment requires a multidisciplinary approach. The nutrition component of treatment involves both intervention and education.[51]

Anorexia nervosa Anorexia nervosa is characterized by extreme weight loss, distorted body image, a preoccupation with food, and an intense fear of becoming obese.[52] This pathological fear of gaining weight leads to unusual eating patterns, malnutrition, and excessive weight loss.[53] Most persons with anorexia nervosa are white females from middle- or upper-class families. Males account for less than 10 percent of the cases.[54] Most often the saga

anorexia nervosa: a disorder involving compulsive self-starvation and extreme weight loss, not explainable by disease, most common in adolescent females.
an = not
orexis = appetite
nervos = of nervous origin

begins when a teenager who either is overweight, or perceives herself to be overweight, begins a weight loss regimen. Oftentimes she receives encouragement and praise from friends and family members. Her efforts to lose weight are heroic and include a severe kcalorie-restricted diet and excessive physical activity. A person with anorexia nervosa would consider it a normal day's workout to swim 96 laps at the pool, run 5 miles, attend an aerobic dance class, and do body-building exercises.[55]

The physical consequences of anorexia nervosa are similar to those of starvation. They include diminished norepinephrine synthesis, elevated growth hormone concentrations, defective temperature regulation, and lowered basal metabolic rate.[56] In addition, estrogen concentrations are depressed, and this contributes to the amenorrhea and osteoporosis common in women with anorexia. In males, serum testosterone concentrations are low, depressing sexual desire.[57] One study found medical problems such as cardiovascular abnormalities, hypothermia, renal dysfunction, and electrolyte imbalance present in over half the adolescents with anorexia and in almost one-fourth of those with bulimia.[58]

A typical psychological profile of a female with anorexia nervosa includes depression, early developmental failure, and family dysfunction.[59] For these reasons, treatment usually involves parents, to help develop appropriate family interactions.[60] A teenager with anorexia strives for perfection and control over her life, finds fat disgusting and thin admirable, and believes gaining weight means being out of control. For a teenager in a growth phase, gaining weight is normal. Thus part of recovery for the teenager with anorexia is an understanding of growth and its relationship to food.[61] For some teenagers with mild anorexia nervosa, this is all that is necessary to elicit a positive response. Those with strong resistance to treatment and severe weight loss require hospitalization.

The specific causes of anorexia nervosa are difficult to define, because they are often interwoven, and the symptoms manifest themselves in a variety of physical and psychological ways. Treatment must therefore artfully combine medical and dietary intervention (to initiate and sustain weight gain) with psychological techniques (to resolve personal and family problems). Teams of physicians, nurses, psychiatrists, family psychologists, and dietitians work together to treat clients with anorexia nervosa.

Part of treatment involves recognition of the power struggle between therapists and clients.[62] When treatment programs try to manage their clients' eating regimens by counting kcalories and checking food trays, they may be heading for failure. Instead, wherever possible, they should encourage clients to maintain control over their own food intakes, but require that they design healthy diet plans.[63]

bulimia: a disorder involving recurring binge eating combined with a morbid fear of becoming fat, sometimes followed by self-induced vomiting or purging, most common in women; other terms used to describe this eating behavior include **bulimarexia** and **bulimia nervosa.**

binge eating: rapid consumption of a large quantity of food.

Bulimia Bulimia is a distinct eating disorder that shares some characteristics with anorexia nervosa. Like the person with anorexia nervosa, the teenager with bulimia spends much time thinking about body weight and food. The preoccupation with food manifests itself in secretive binge-eating episodes followed by self-induced vomiting, fasting, or the use of laxatives and diuretics.[64] Such behaviors typically begin in late adolescence after a long series of various unsuccessful weight reduction diets. People with bulimia commonly follow a pattern of restrictive dieting interspersed with binge behaviors and experience weight fluctuations of 10 pounds gained and lost—up and down—over short periods of time.

People with bulimia are difficult to identify; they hide their secret well. To the world outside, they may appear bright, successful, near normal weight, and without obvious psychological problems.[65] Internally, though, a person with bulimia has a "fat" self-image, fears of being unable to stop eating and of getting fatter, and the ultimate ambition of being slim.

In an attempt to attain the thin ideal, the person with bulimia severely restricts food intake, especially "forbidden" sweet treats. When unable to resist the temptation any longer, the person loses control and begins to binge. The average binge takes about an hour, and the person consumes about 3400 kcalories.[66] Binge episodes are followed by guilt, depression, and self-condemnation. With the guilt comes a renewed compulsion to lose weight—by either starving or purging.

Purging involves inducing vomiting (either manually or with emetic drugs) or using enemas, laxatives, or diuretics. Not all people with bulimia purge. Those who do believe they have found the ideal solution to their weight problem—that purging will help to eliminate the fattening effect of the excess food eaten, as well as the guilt associated with the overindulgence in food. For them, it serves to cleanse both the body and the soul. But in "cleansing" the body, a person loses water and minerals. This incurs dehydration and electrolyte imbalances, which can lead to fatigue, seizures, muscle cramps, and irregular heartbeats, and over the long term, to decreased bone density and osteoporosis. Table 8–2 lists other physical ramifications of purging.

Even more severe than the physical consequences of bulimia are the associated behavioral and psychological problems.[67] The rates of alcohol, marijuana, and cigarette use and of depression among bulimics are high.[68] For

Table 8–2 Consequences of Purging

In general:
- Electrolyte imbalances

Specifically:
- Diuretic use causes:
  - Hypokalemia
- Emetic use causes:
  - Poisoning
  - Cardiac arrest
- Laxative use causes:
  - Rectal bleeding
  - Hypokalemia
- Vomiting causes:
  - Damage to the esophagus and stomach
  - Swelling of the salivary glands
  - Recession of the gums
  - Erosion of tooth enamel
  - Skin rashes
  - Broken blood vessels on the face
  - Aspiration pneumonia

Source: Adapted from D. B. Herzog and P. M. Copeland, Eating disorders, *New England Journal of Medicine* 313 (1985): 295–303; D. Farley, Eating disorders: When thinness becomes an obsession, *FDA Consumer*, May 1986, pp. 20–23.

these reasons, a mental health professional should be one of the members on the multidisciplinary treatment team.

The goal of a dietary plan to treat bulimia is to help clients gain control and establish regular eating patterns.[69] One strategy is obvious: since most binges begin after a round of strict dieting, the person needs to learn to eat a quantity of nutritious food sufficient to nourish her body and leave her satisfied without bringing on the anathema of weight gain. Such an approach has been used successfully in the treatment of many people with severe bulimia; it requires that they eat no less than 1500 kcalories a day.[70] Initial dietary management requires a structured eating plan with little flexibility. Such a plan prevents the client from making decisions and reduces anxiety about eating. Binge foods are avoided at first and gradually reintroduced in moderate amounts.

These eating disorders illustrate how a desired end (in this case, weight loss) can alter a person's eating habits and food choices, and therefore nutrition status. Teenagers are willing to go to considerable trouble and expense, at the risk of jeopardizing their health, to achieve their perception of physical beauty. A similar effort (although usually to a lesser degree and with fewer ill consequences) is made by the teenager in search of a remedy for acne.

Acne

acne: a chronic inflammation of the skin's follicles and oil-producing glands that involves the accumulation of sebum inside the ducts that surround hairs, usually associated with the maturation of young adults.

No one knows why some people get acne while others do not, but heredity is one factor. In addition, the hormones of adolescence increase the activity of the sebaceous glands in the skin (see Figure 8–3).[71]

Many a teenager would pay dearly for the remedy for acne. Some teenagers hopefully believe that if they stop eating certain foods and drinking certain beverages, they can prevent acne. Among foods charged with aggravating acne are chocolate, cola beverages, fatty or greasy foods, milk, nuts, sugar, and foods or salt containing iodine. None of these foods has proven to worsen acne.[72] Even though it is true that the skin of a person with acne is oily, the diet's fat and oil content is not at fault. Of course, teenagers could omit these foods, with the exception of milk and foods or salt containing iodine, without harming their health, and to do so might actually benefit their nutrition status.

Miniglossary of Acne Terms

blackhead: an open lesion with an accumulation of the natural dark pigments of the skin (not dirt) in the opening.

cyst: an enlarged, deep pimple.

sebaceous gland: the oil-secreting gland of the skin.

sebum: the skin's natural mixture of oils and waxes that helps keep skin and hair moist.

whitehead: a pimple caused by the plugging of oil-gland ducts with shed material from the duct lining.

Figure 8–3 Acne

The skin's natural oil, sebum, is made in deep sebaceous glands and is supposed to flow out through the tiny ducts around the hairs to the skin surface. In acne, the oily secretions exceed the skin's clearance capacity.

Inside each of the ducts is a skinlike lining that regularly sheds cells. These cells mix with the oil and then are pushed to the surface of the skin. When acne develops, they stick together, forming a plug that blocks the duct. The duct enlarges, allowing oil and the skin-surface bacteria to leak into the surrounding skin. The oil and bacterial enzymes are irritating and cause redness, swelling, pus formation—and the beginning of a whitehead, or pimple. A cyst may form, or the skin may open above the plug, revealing an accumulation of dark skin pigments just below the surface—a blackhead.

Note that acne is not caused by the skin bacteria, although once the process has begun, they make it worse. Also note that the color of a blackhead is caused by skin pigments, not by dirt. Squeezing or picking at the lesions of acne in an attempt to remove their contents can cause more scars than the acne.

Among the over-the-counter acne treatments, preparations that contain benzoyl peroxide are safe and effective. Careful washing helps remove skin-surface bacteria and oil and keeps the oil ducts open; surface treatment with antibiotics also helps control the bacteria. A cream or gel containing retinoic acid can help: retinoic acid loosens the plugs that form in the ducts, allowing the oil to flow again so that the ducts will not burst. Care is necessary, because the acid may burn the skin and even cause pimples to form, making the acne look worse rather than better at first.

Stress, with its accompanying hormonal secretions, clearly worsens acne. Vacations from school pressures help to bring acne relief. The sun, the beach, and swimming also help, perhaps because they are relaxing, and also because the sun's rays kill bacteria and water cleanses the skin.

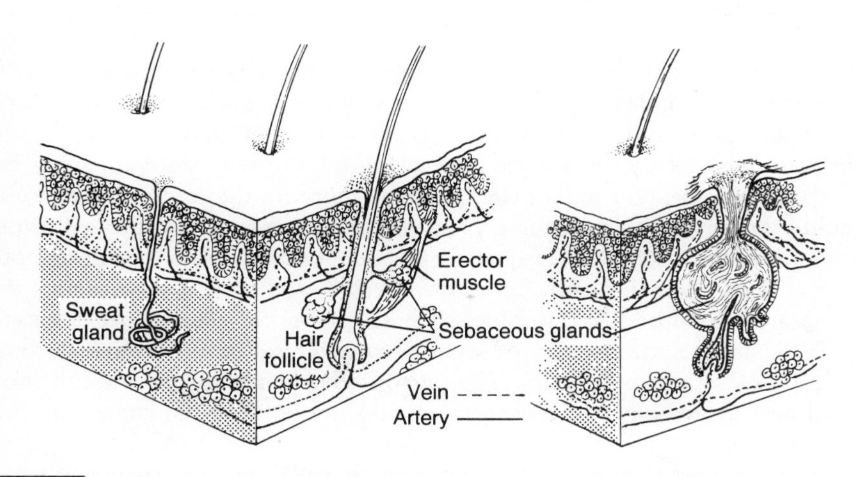

Source: Adapted from *Acne,* a pamphlet (May 1980) available from the National Institute of Allergy and Infectious Diseases, Bethesda, MD 20205, NIH publication no. 80–188; *Stubborn and Vexing, That's Acne,* a pamphlet (May 1980) available from the Food and Drug Administration, 5600 Fishers Lane, Rockville, MD 20857, HHS publication no. (FDA) 80–3107.

Not all dietary practices attempting to treat acne are innocuous. Misinformed teenagers, taking vitamin A supplements, will obtain no relief from acne, but may induce vitamin A toxicity. Vitamin A is needed for the health of the skin, as is zinc; but as with other nutrients, excesses are toxic.

Because stress worsens acne, and because adolescents easily develop guilt feelings over what they eat, perhaps the best nutrition advice is to dispense with food-related guilt. The guidelines for the acne-plagued teenager, as for anyone, are to eat nutritious foods in abundance and to enjoy sweet treats in moderation as a harmless pleasure.

Dermatologists sometimes treat cystic acne with medicines that are related to vitamin A but chemically different from over-the-counter supplements. Excellent results have been obtained with the topical use of a salve containing the vitamin A relative tretinoin (retinoic acid); the trade name is Retin-A. Retin-A reddens the skin and causes tenderness and peeling, but the subsequent healing produces healthier skin.

Isotretinoin (13-*cis*-retinoic acid) is available in soft gelatin capsules for oral administration under the trade name Accutane. Its mechanism of action is not completely understood, although it seems to inhibit sebaceous gland function and keratinization. Accutane is responsible for a number of adverse effects, including serious birth defects in infants of women using it during pregnancy. As mentioned in Chapter 2, an effective form of contraception is advised beginning one month before, and continuing until one month after, Accutane's use.

Fitness for Teens and Adults

Fitness is important throughout life. Each chapter has extolled its virtues, recommending regular exercise prior to pregnancy, during pregnancy, and during lactation, and recommending regular outdoor play in infancy and childhood. The teen years are the time when outdoor play shifts to sports for many, and to sedentary indoor life for many more. In the teen years, exercise should become a habit as regular as eating, practiced in a way that will bring optimal benefits. Regular, frequent exercise does much to promote both physical and mental health, prevent obesity, and retard or reverse the degenerative conditions of later life, such as osteoporosis and heart disease.

Teens choose for themselves whether to exercise or not. Some become seriously interested in athletics. The person who wants to provide nutrition and fitness education for teenagers should speak their language on this important topic.

Fitness expert Dr. George Sheehan says that "We are all athletes, it's just that some of us are in training and others are not." He means that the human body—any human body—can adapt to intense, prolonged exercise by developing a greater and greater capacity for it. All that is required is the appropriate training, supported by appropriate nutrition. Teens who admire the image of the athlete should be encouraged to see themselves this way—as people with bodies that will respond to athletic training. Accordingly, the following sections are devoted to the special nutrition needs of all bodies during exercise, not just the bodies of those already excelling in athletics.

An athlete is defined here as anyone who participates in any kind of competitive sport; exercises on a regular, frequent basis; or is serious about improving fitness and health through exercise. Athletes include world-class Olympians; college basketball players; high school football players; members of the Little League team; and all individuals who spend at least 30 minutes a day, three days a week or more, running, walking, swimming, cycling, rowing boats, dancing, or engaging in any other activity that makes the heart beat fast and the muscles work hard for most of that time. Athletes all along this spectrum reap the benefits that come with sound nutrition.

The two major contributors to athletic achievement are heredity and training, while nutrition and psychological preparation play lesser, but important, roles. Heredity sets the ultimate potential; training permits realization of the potential; nutrition supports training by supplying the fuels, building materials, and metabolic regulators the muscles demand; and psychological preparation sparks the whole process with motivation.

Many sports nutrition experts agree that the diet most athletes need is simply an ordinary balanced diet, consisting of a variety of nutrient-dense foods and ample fluids. This is a simple nutrition prescription and is easy to follow, provided that athletes and their coaches are informed as to what it means. The discussion that follows goes into depth to show the relationships between nutrients and fitness, and then revisits the question of what diet will best deliver those nutrients. Along the way, it explores some misconceptions relevant to the topics at hand.

Energy and Fuels for Exercise

The body receives usable energy from food in the form of three classes of compounds—carbohydrates, lipids, and proteins. The first two of these give rise to glucose and fatty acids, respectively, and these are normally the body's preferred energy sources. Protein contributes a certain minimum share of the body's energy at all times, and can contribute more if necessary. In the absence of food, the body draws glucose and fatty acids from its own stores of carbohydrate (glycogen), lipids (body fat), and tissue protein.

The energy from these fuels is available to muscle cells in several forms. Glucose is available both from muscle glycogen and from circulating blood glucose. Fatty acids are available both from fat stored in muscle and from circulating fatty acids. Amino acids are available within the muscle and also from the blood. These fuels are converted within the muscle to the compounds ATP (adenosine triphosphate) and PC (phosphocreatine), which are held in direct proximity to their contractile fibers for instant use. ATP and PC are interconvertible: between contractions, ATP can generate PC; during contractions, PC regenerates ATP (see Figure 8–4). This conversion is virtually instantaneous; it is not dependent on a person's eating food and does not require oxygen (it makes energy available anaerobically). Whenever a muscle first contracts, whether after a prolonged period of rest or after a few-second interval of relaxation, it uses the energy available from ATP-PC. But the muscle's ready supply of these compounds lasts only for about 30 seconds, and thereafter, it must draw on the other energy fuels mentioned to replenish its ATP-PC to fuel further contractions.

Figure 8–4 ATP and PC in Muscle
ATP is available instantaneously and is the muscles' immediate energy source—that is, the one they actually use when they are working. A temporary nearby storage form of ATP's energy is PC (phosphocreatine); limited amounts are made from ATP and held in readiness near working muscle fibers. PC's energy can be transferred back into ATP at a moment's notice when the ATP supply runs short.

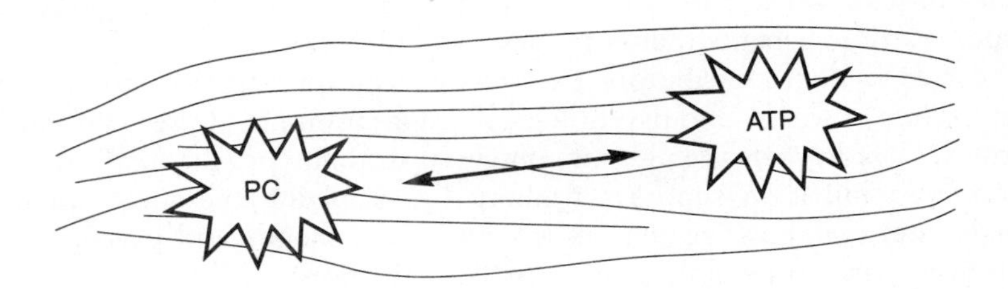

Figure 8–5 shows the pathways of metabolism by which glucose, fatty acids, and protein yield ATP (and thereby, PC) for the muscles' continued use. Which are used as fuel, and to what extent, depends partly on whether oxygen is available for their breakdown. The figure shows that anaerobic metabolism yields a little ATP, and that aerobic metabolism yields much more ATP. Normally, exercise metabolism of any duration beyond about 2 minutes requires that the muscles receive oxygen and the two main muscle fuels—glucose and fatty acids. The oxygen comes from the lungs, which pass it to the blood, which carries it to the muscles. The glucose originates mainly from the glycogen stores within the muscle itself; some comes from the general circulation, contributed by diet or by liver glycogen. The fatty acids come partly from fat stores within the muscles, but mostly from the blood, contributed either by diet or by the body's fat tissues.

Let us briefly get an overview of each of these sources of ATP in turn, before presenting details about each compound in the following sections. First, consider glycogen. If glycogen can be broken down aerobically all the way to carbon dioxide and water, it can yield abundant ATP. If it cannot be (if insufficient oxygen is available to permit breakdown below the dotted line in Figure 8–5), then it must stop at pyruvate and take a temporary side route to lactic acid. The pathway to lactic acid indirectly generates an important, quickly available, but small amount of ATP. (Later, the lactic acid will be reconverted to pyruvate and completely broken down—but it has to be released from the muscles and returned to the liver for this to take place.) Much more ATP is generated when glucose breakdown occurs in the presence of oxygen and can proceed all the way to carbon dioxide and water.

Fatty acids also generate abundant ATP when they can be used, because they proceed aerobically all the way to carbon dioxide and water. As long as oxygen is available, fatty acids can serve as an inexhaustible source of abundant ATP. But the other ATP sources, PC and the glycogen-to-lactic acid pathway, have to stand behind the fatty acids to permit exercise to continue during bursts of intensive activity, when oxygen supply momentarily fails to meet demand.

Figure 8–5 Food, Fuels, and Exercise
This is the central metabolic pathway that generates energy (ATP). The two main fuels for exercise, carbohydrate (glucose, from food or body glycogen stores) and fat (fatty acids, from food or body fat stores), are broken down by separate pathways to a common intermediate, acetyl CoA, and then by a common pathway to carbon dioxide and water. In the process, energy is released, and some is used to make the compound ATP, which serves in all cells as a source of readily available energy. Some of the earlier steps in glucose breakdown yield ATP without oxygen's involvement (they are anaerobic), but the later ones, including all of the steps by which fat is broken down require oxygen (they are aerobic). These later, aerobic steps yield by far the most ATP and so are the most important for endurance exercise.

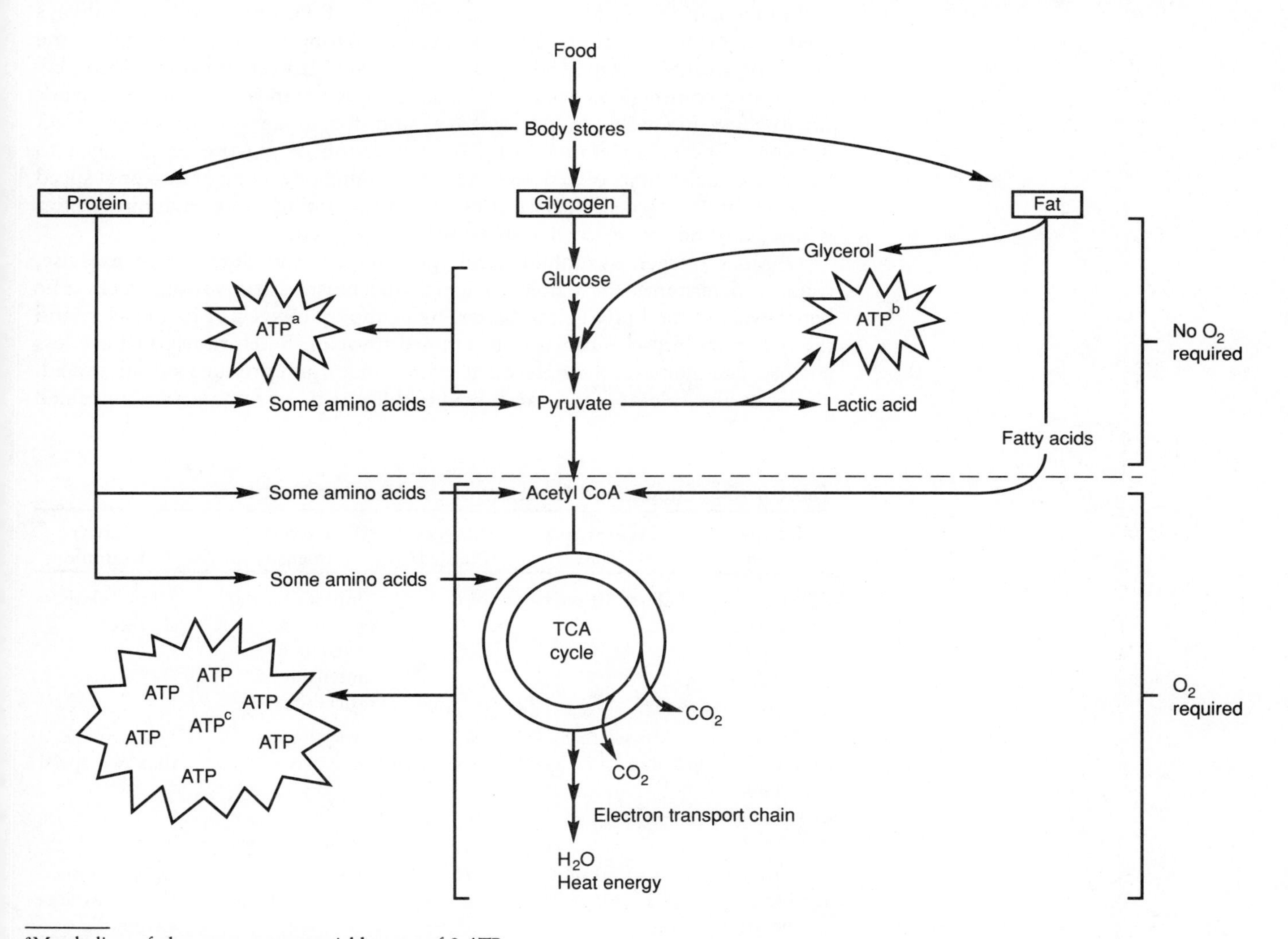

[a]Metabolism of glucose to pyruvate yields a net of 2 ATP.
[b]Conversion of glucose to lactic acid can yield 2 more ATP, but when lactic acid is reconstructed to pyruvate, these must be paid back.
[c]Conversion of the two acetyl CoA (derived from glucose) to carbon dioxide and water yields 36 to 38 more ATP, for a total (from glucose) of 38 to 40 ATP. Two acetyl CoA from fat or protein also yield 36 to 38 ATP.

Protein stores can also be used for energy. Some amino acids can be broken down to pyruvate and hence be used anaerobically like glucose, if need be. Other amino acids can be broken down to acetyl CoA and require oxygen, as fatty acids do, to be used as fuel. Still others can enter the TCA cycle directly to generate ATP.

Normally, during exercise, all fuels are in use to varying degrees from moment to moment, depending on their availability, on the availability of oxygen, on the duration and intensity of the exercise, and on local conditions within the muscle. Within a given muscle, the ATP-PC energy system is always used at the start of a contraction, such as the lifting of a heavy weight or the taking of a running step. The pyruvate-lactic acid conversion comes into play as repeated contractions occur in exercise that demands that the muscles work slightly less intensely on the average, and for longer durations (½ to 3 minutes). Energy generated from the conversion of glucose or glycogen to lactic acid fuels bursts of activities such as running or cycling at maximal speed for a half mile or so. Table 8–3 summarizes the timing and exercise intensities with which the various fuel systems tend to be used.

Besides oxygen availability and the intensity and duration of exercise, training determines the extent to which the muscles use various fuels. The better trained the body is, the better it can supply oxygen to its muscles, and the better the muscles can use it. Trained muscles therefore need to use less glucose, can maintain their glycogen stores longer, and produce less lactic acid. They can also tolerate more lactic acid buildup without fatigue than untrained

Table 8–3 Timing and Intensity of the Muscles' Fuel Use

Energy System	Performance Time	Oxygen Needed?	Exercise Intensity	Activity Example
ATP-PC (immediate availability)	First 30 sec	No	Initiation of all activities; extreme intensities thereafter	100-yd dash, shot put
ATP from anaerobic conversion of carbohydrate to lactic acid	30 sec to 3 min	No	Very high	¼-mi run at maximal speed
ATP from aerobic/ carbohydrate	3 to 20 min	Yes	High	Cross-county skiing, distance swimming or running
ATP from aerobic/fat	More than 20 min	Yes	Moderate	Distance running or jogging

Source: Adpated in part from M.H. Williams, Human energy, in *Nutritional Aspects of Human Physical and Athletic Performance,* 2nd ed. (Springfield, Ill.: Charles C Thomas, 1985), pp. 21–57; E.L. Fox, Sports activities and the energy continuum, in *Sports Physiology,* 2nd ed. (New York: Saunders, 1984), pp. 26–39.

muscles. For all these reasons, trained muscles can exercise at high intensities longer than untrained muscles.[73]

In summary, the working muscles always use both glucose and fatty acids to replenish their ATP-PC and keep working after they have depleted their immediate pool of ATP-PC. Depending on the availability of oxygen, on the intensity and duration of the exercise, and on the muscles' own capacity, they may use more or less of one or the other fuel. They also use body protein stores for energy to some extent, and use them to a greater extent if the other fuels are in short supply. The following sections describe the energy fuels in greater detail.

Carbohydrate The body stores carbohydrate as glycogen, most of it in the muscle. The liver contains a small amount (less than a pound) of glycogen, which it breaks apart into glucose and releases into the bloodstream when the body needs it. During exercise, the muscles pick up and use the glucose donated by the liver, along with glucose from their own private glycogen stores. The more glycogen stored, the longer the stores will last during exercise. Compared with fat, the carbohydrate stores of the body are limited.

Body glycogen is closely related to the carbohydrate content of the diet. How much carbohydrate the athlete eats both determines how much glycogen is stored and influences the rate at which glycogen is used in any given exercise.[74] The rate at which an athlete uses glucose also depends partly on the duration of the exercise and partly on its intensity.

Exercise intensity is expressed as a proportion of an individual's maximum aerobic capacity, or maximum oxygen consumption (percent VO_2 max). At rest or at intensities below 50 percent of maximum aerobic capacity, the primary fuel is fat, while muscle glycogen use is minimal. As exercise *intensity* rises above 70 percent of maximum oxygen consumption (the average intensity of most athletic events), *glycogen* becomes the primary fuel.[75] As exercise *duration* continues to increase and intensity lessens, *fat* becomes the preferred fuel, but only to a point. The endurance athlete continues to use glycogen and will eventually run out of it.

VO_2 max: the maximum volume of oxygen that a person can consume during a minute of heavy work; a measure of cardiovascular and muscular fitness. Also called **MOC (maximum oxygen consumption).**

Fatigue is the inevitable consequence of glycogen depletion, and it sets in earlier when the muscle glycogen stores are small. Glycogen depletion does not limit short-term, intense exercise, but it has a profound effect on performance during endurance exercise. Glycogen use is rapid at the beginning of intense exercise.[76] As exercise continues, glycogen use slows down. The body begins to rely more on fat for fuel, conserving the remaining glycogen supply. If exercise continues long enough and at a high enough intensity, glycogen will run out almost completely. Muscle glycogen depletion occurs within about 90 to 120 minutes when exercise intensity averages 70 percent of aerobic capacity.

It has long been known that the composition of the diet affects fuel availability during exercise. A classic study in 1939 compared fuel usage during exercise among runners who had consumed three different diets.[77] Researchers determined fuel utilization by observing the respiratory exchange ratio (R value). The R value rises as carbohydrate utilization during exercise increases.[78] The study participants consumed either a high-fat diet (94 percent fat), a normal diet (55 percent carbohydrate), or a high-carbohydrate diet (83 percent carbohydrate) for several days before exercising. The study was one of the first to demonstrate the importance of carbohydrate in supporting long-duration exercise (see Figure 8–6).

respiratory exchange ratio (R value): the ratio of carbon dioxide production to oxygen consumption and an indicator of the type of fuel being utilized. At rest, the R value is 0.85, which indicates that 50 to 60% of the energy for metabolism is fat derived, and the remainder is carbohydrate derived. During exercise, as intensity increases, the R value rises, indicating greater carbohydrate utilization.

Figure 8–6 The Effect of Diet on Physical Endurance
A high-carbohydrate diet can triple an athlete's endurance.

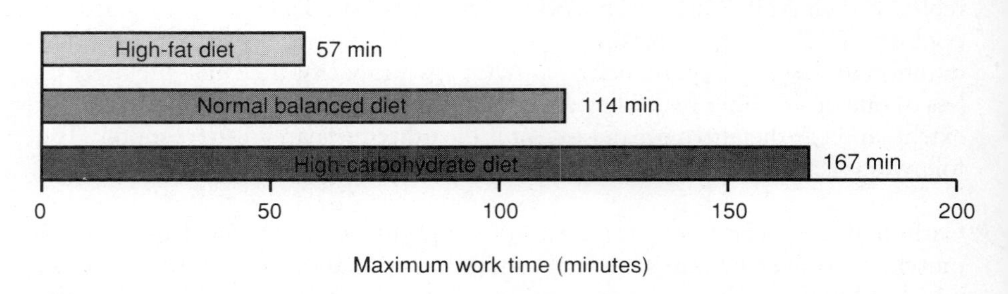

Source: Data from P. Astrand, Something old and something new . . . very new, *Nutrition Today*, June 1968, pp. 9–11.

Thus ample stores of glycogen enhance an athlete's endurance.[79] The greater the muscle glycogen stores, the longer the athlete is able to perform at a sustained high intensity. Athletes who want to perform well in long-duration events, then, are wise to maintain their glycogen stores with a high-carbohydrate diet.

"If some is good, more may be better" is an adage that may be true for glycogen, at least up to a point. Muscle glycogen supercompensation has been demonstrated by studying the thigh muscles of individuals who had exercised one leg heavily to deplete glycogen stores. When this regimen was followed by a high-carbohydrate diet for several days, glycogen renewal was rapid and exceeded normal levels. This supercompensation of glycogen occurred only in the exercised muscles. The highest concentrations of muscle glycogen were obtained when depletion was followed by a low-carbohydrate diet for several days and then by a high-carbohydrate diet. This procedure was, until recently, the strategy followed by endurance athletes wishing to prolong performance at a given intensity. Such extreme manipulation of an individual's diet can be dangerous, however, and is not recommended.

glycogen supercompensation: a technique of exercising, followed by eating a high-carbohydrate diet, that enables muscles to store glycogen beyond their normal capacity; also called **glycogen loading**.

Later research indicates that the same glycogen supercompensation effect can be obtained in a safer way than from a low-carbohydrate diet and then a high-carbohydrate diet; it can be obtained from exhaustive exercise followed by a high-carbohydrate diet.[80] Researchers speculate that muscle glycogen supercompensation occurs partially because exhaustive exercise increases insulin sensitivity, which in turn enhances glucose transport from the high-carbohydrate diet into the glycogen-depleted muscle cells.[81]

The type as well as the timing of carbohydrate consumption affects the amount stored. Researchers compared the effect of simple versus complex carbohydrates on muscle glycogen synthesis after strenuous exercise.[82] During the first 24 hours after exercise, the type of carbohydrate an individual eats makes no difference in muscle glycogen concentration. After 48 hours, however, muscle glycogen concentrations are significantly higher in runners who ingest complex carbohydrates. Such is the basis for recommendations that the athlete or active person consume a large proportion of kcalories from complex carbohydrates (50 to 55 percent) and additional kcalories from simple carbohydrates (to bring the total to 60 percent or more).[83] Athletes who wish to pack their muscles with extra glycogen can try for a diet that is even higher in carbohydrate, up to 80 percent, for a few days before competition. A later section discusses the athlete's diet.

Fat Body fat is an important metabolic fuel for exercise and is a virtually unlimited source of energy for the human body. Fat is available as fuel to the muscles in two forms: as free fatty acids (FFA) transported from fat stores, and as triglyceride stores within the muscles themselves. As with carbohydrate, the availability and utilization of fat as a fuel for the muscles is somewhat conditional. As already discussed, diet influences fuel preference, but even more influential is exercise intensity. Unlike carbohydrate, which can be metabolized with oxygen via the TCA cycle or without oxygen via the lactic acid system, fat metabolism absolutely requires oxygen. Accordingly, at low exercise intensity (less than 50 percent of maximal aerobic capacity) or moderate exercise intensity (50 to 70 percent), when oxygen is plentiful, fat is a major fuel source. As exercise intensity increases, the energy contribution of fat diminishes, and carbohydrate-derived energy dominates.[84]

Researchers have investigated the question whether greater FFA availability might slow down glycogen depletion during exercise.[85] Athletes consumed a high-fat, low-carbohydrate diet in the effort to increase FFA utilization, but could not sustain intensity because the glycogen stores were low and were rapidly depleted.[86] In short, the high-fat diet does enhance FFA utilization, but not athletic performance. Furthermore, such a diet is a risk factor for heart disease and other conditions, and so is best avoided for health's sake.

An alternative to using a high-fat diet to increase FFA availability is to ingest caffeine prior to exercise. Moderate caffeine consumption (2 milligrams per pound of body weight, or two to three cups of coffee) one hour before exercise improves endurance and makes the work seem easier.[87] Caffeine facilitates the utilization of the body's fat stores, thus sparing muscle glycogen during exercise. Caffeine ingestion is not without risk, however. Caffeine is a stimulant and a diuretic. Its use may pose hazards to some individuals who are particularly sensitive to it, or who perform endurance exercise in a hot environment. Caffeine-containing beverages should be used in moderation and in *addition* to other fluids, not as a substitute for them.[88] Caffeine is banned as an illegal drug for international and Olympic competition, and the competitors' urine is tested for it.

The use of FFA by muscles depends partly on how much is available in the blood. Eating a high-fat diet and ingesting caffeine are dietary means of enhancing fat utilization—a desirable goal, because this will improve endurance by sparing glycogen. However, as mentioned, a high-fat diet is incompatible with optimal health, and caffeine ingestion is inappropriate at times. Fortunately, the athlete has another advantage to draw on: conditioning. With endurance training, muscles develop a greater capacity for FFA oxidation.[89] The physically conditioned athlete can therefore rely on fat for energy at higher exercise intensities, and in doing so, conserve limited glycogen stores. Enhanced fat metabolism is critical to the endurance athlete's performance.

Protein As with carbohydrate and fat, changes in protein metabolism during exercise appear to be related to diet, glycogen stores, physical conditioning, and exercise intensity and duration. During exercise of moderate intensity, protein metabolism is greater in glycogen-depleted individuals than in those who are not glycogen depleted.[90] Thus carbohydrate appears to exert a protein-sparing effect during exercise.

Protein always contributes about 15 percent of the fuel consumed, whether a person is resting or exercising.[91] However, because total energy needs are higher in exercise, total protein use is also higher in exercise.[92] Nevertheless, the protein content of diets in developed countries is ample. If athletes meet their increased energy needs by selecting and consuming nutrient-dense foods, they will easily meet their protein needs as well.

In summary, carbohydrate and fat are the major fuel sources of exercise, and the degree of energy contribution by each is dependent on several factors. Training state, diet, and intensity and duration of exercise influence fuel use in physical activity. Although many questions about energy sources and exercise remain from the evidence to date, the following conclusions emerge:

- During activity of low intensity (less than 50 percent of maximal aerobic capacity), fat is the primary fuel.
- As exercise intensity increases, so does the contribution of carbohydrate as an energy source.
- Above 70 percent of maximal aerobic capacity, carbohydrate is the predominant fuel.[93]
- Muscle glycogen depletion by way of exercise, in addition to a high-carbohydrate diet, results in greater-than-normal muscle glycogen concentrations.
- Physical conditioning enhances the capacity for fat metabolism at greater intensities, which in turn spares limited carbohydrate stores (glycogen).

Vitamins

Do physically active people need more vitamins than sedentary people? The answer to this question lies in the metabolic workings of the muscles in relation to each vitamin. Table 8–4 shows some exercise-related functions of vitamins and minerals. The table illustrates the point that all vitamins and minerals play roles in muscles, as they do in all tissues at all times. Of course they are necessary for physically active people, but research does not support the idea that in general, physical activity increases the need for them. A few possible exceptions have been investigated—among them the B vitamins, which are involved in energy release from fuels.

The B vitamins Vitamins are not oxidized for energy, but without the B vitamins, the production of energy for the body's use would be impossible. Thus it seems logical to assume that a person who uses more energy in a day might need more of the B vitamins. Accordingly, the RDA for the B vitamins required in glucose breakdown are expressed in amounts per 1000 kcalories of food energy consumed, because the more food the body processes, the more of these vitamins it needs to extract energy from the food.

Table 8–4 Exercise-Related Functions of Vitamins and Minerals

Vitamin or Mineral	Function
Thiamin, riboflavin, niacin, magnesium	Energy-releasing reactions
Vitamin B_6, zinc	Building of muscle protein
Folate, vitamin B_{12}	Building of red blood cells to carry oxygen
Vitamin C	Collagen formation for joint and other tissue integrity; hormone synthesis
Vitamin E	Protection of red blood cells and others from oxidation
Iron	Transport of oxygen in blood and in muscle tissue; energy transformation reactions
Calcium, vitamin D, vitamin A, phosphorus	Building of bone structure; muscle contractions; nerve transmissions
Sodium, potassium, chloride	Maintenance of fluid balance; transmission of nerve impulses for muscle contraction
Chromium	Assistance in insulin's energy-storage function
Magnesium	Cardiac and other muscle contraction

Note: This is just a sampling. Other vitamins and minerals play equally indispensable roles in exercise.

Source: E. M. N. Hamilton, E. N. Whitney, and F. S. Sizer, *Nutrition: Concepts and Controversies,* 4th ed. (St. Paul, Minn.: West, 1988).

To answer the question of whether athletes in training need thiamin in amounts greater than the RDA, scientists supplemented athletes with the vitamin and compared their physical performance with that of athletes eating only a regular, adequate diet.[94] They found that extra thiamin conferred no benefits on performance; the thiamin in an *adequate* diet is all a person needs, even for heavy work. This is because almost any kind of whole, unprocessed food supplies thiamin, and when an athlete meets energy demands with nourishing food, thiamin is sure to be plentiful. For thiamin, then, the answer is that athletes do not benefit from more than the RDA amount; they can get what they need from food.

The link between riboflavin and physical performance arises from its role as part of the coenzyme flavin adenine dinucleotide (FAD), central to the mitochondrial oxidative reactions of electron transport. To try to answer the question of whether riboflavin at levels beyond the RDA assists in athletic performance, researchers studied groups of overweight, sedentary women who began an exercise regimen in addition to weight loss dieting.[95] One group of women consumed 1.2 milligrams riboflavin per day, while the other group consumed slightly more (1.4 milligrams). Blood tests to evaluate riboflavin activity seemed to indicate a deficiency in the group that consumed the lesser amount. However, aerobic capacity of both groups increased similarly. Higher riboflavin intakes did not improve physical performance, as would be expected if a true riboflavin deficiency had existed. For both groups of women, however, exercise resulted in decreased urinary excretion of riboflavin, suggesting a greater need for riboflavin during exercise. Questions about riboflavin requirements during exercise remain to be answered by more research.

Vitamin B_6 is a part of over 60 enzyme systems involved in the metabolism of protein and the other energy nutrients. It functions in the breakdown of glycogen and is involved in the formation of hemoglobin. Such roles are the basis for claims that vitamin B_6 in amounts greater than the RDA promotes aerobic endurance, but thus far, research does not support these claims.[96]

Nor does research support the notion that vitamin B_{12} supplementation will enhance performance (see "Supplements for Athletes," later in this chapter). Such beliefs about vitamin B_{12} stem from its role in the production of red blood cells. In anemia, a diminished number of circulating red blood cells robs the blood of its oxygen-carrying capacity, starving the cells and restricting aerobic energy metabolism. Vitamin B_{12} deficiency is only one cause of anemia. Iron and folate deficiencies are equally destructive to performance, as are medical anemias.

Vitamin E Vitamin E has enjoyed many claims to fame. For the most part, these claims are exactly that, unsubstantiated claims. The relevance of vitamin E to physical performance has to do with its role as an antioxidant of polyunsaturated fatty acids. Vitamin E deficiency in animals causes anemia and limits oxidative phosphorylation.[97] Oxidation of phospholipids in the red blood cell membrane leads to hemolysis and thereby causes the anemia. This would no doubt hinder athletic performance, should it occur, but that is unlikely. Vitamin E is so widespread in foods that scientists who wish to study its deficiency effects in animals must go to some trouble to concoct a vitamin E-free diet to induce the condition.

Minerals

Like the vitamins, minerals play roles in exercise metabolism. For example, the major minerals calcium and phosphorus are structural components of bone; phosphorus is part of the high-energy compounds ATP and PC; and iron in hemoglobin carries the body's oxygen to the exercising muscles. Phosphorus is widely distributed in food, so a deficiency is unlikely, but calcium and iron deficiencies are common. Therefore, calcium and iron need particular attention in planning diets for athletes.

Calcium Moderate exercise is thought to be protective against bone loss, but intense exercise may be detrimental to the bone health of some young women.[98] A side effect of endurance training for some women is athletic amenorrhea, characterized by low estrogen concentrations and possibly increased calcium requirements. Because the food energy intakes of some amenorrheic athletes are abnormally low, and the calcium intakes of women in the United States may also be low, amenorrheic athletes are at greater risk for osteoporosis than other women.[99]

athletic amenorrhea: cessation of menstruation associated with strenuous athletic training.

A survey of almost 200 female athletes indicated that one-third of them practiced pathological eating behaviors such as using laxatives and diet pills, inducing vomiting, and binge eating—bulimia.[100] The damage was greatest for the athletes who practiced these behaviors in conjunction with an abnormally low food intake—anorexia nervosa. These behaviors are harmful to bone health and general health, and they impair physical performance. As mentioned earlier, treatment of anorexia nervosa and bulimia requires a team of medical personnel, dietitians, psychologists, and family counselors for greatest efficacy, and it is well beyond the scope of nutrition alone. Treatment of the eating disorder should receive high priority and the dietary component of that treatment should encourage an adequate intake of calcium to help protect the bones.[101] Some coaches encourage and teach bulimic behaviors to gymnasts, wrestlers, and ballet dancers to keep their weight down; they should discontinue this practice.

Iron As repeatedly emphasized, iron deficiency is the most common nutrient deficiency in both developing and developed countries.[102] It is most prevalent among infants, young children, teenagers, and menstruating young women.[103] In fact, the only group with reliably adequate iron nutriture is adult men, and even they can experience anemia if they run long distances in training. Endurance athletes, especially women, are prone to iron deficiency; the condition has been named runner's anemia.[104]

runners' anemia: a true iron-deficiency anemia that develops in many high-mileage runners.

In a recent nutrition study, 35 percent of women runners had diminished iron stores, as indicated by serum ferritin concentrations.[105] In another study, some female high school runners were given supplements, some containing iron and vitamin C (to both deliver added iron and enhance the absorption of the iron from their food) and some containing vitamin C alone.[106] All but one of the women taking the combination supplement maintained sufficient iron stores throughout the study, while 40 percent of those taking only vitamin C did not. This indicates that in high school–age female athletes, iron supplements should be useful for maintaining adequate iron status. Habitually low intakes of iron-rich foods and increased iron losses usually cause iron deficiency in young women athletes. In addition, blood losses through the GI tract correlate with strenuous exercise, and at least some iron is lost in sweat.[107]

Iron-deficiency anemia dramatically impairs physical performance by reducing the oxygen-carrying capacity of the blood and inhibiting mitochondrial enzyme function.[108] Even marginal iron deficiency without frank anemia may impair physical performance to some extent, although this effect is still under study. On the other hand, decreased iron indicators in the blood do not always accompany decreases in performance. In a 20-day study of 12 male marathon runners, hematocrit and hemoglobin measures decreased significantly, indicating marginal anemia, even though running speeds remained unchanged.[109] The below-normal hematocrit and hemoglobin measures observed in these distance runners are not unusual.

sports anemia: a transient condition of low hemoglobin in the blood, associated with the early stages of sports training or other strenuous activity.

Sports anemia is a condition distinct from the true iron-deficiency condition, runner's anemia. Sports anemia is characterized by a temporary decrease in hemoglobin concentration after a sudden increase in aerobic exercise.[110]

The exact cause of the condition remains unclear, but it is thought that marginal iron intakes may contribute to it.[111] Strenuous aerobic exercise promotes destruction of fragile, older red blood cells and increases the plasma volume of the blood.[112] Sports anemia appears to be an adaptive, temporary response to endurance training. Unlike iron-deficiency anemia, which requires iron-supplementation therapy, sports anemia usually corrects itself after a few weeks of training.

The best advice about iron and iron supplements may be to individualize the recommendation. Many young menstruating women probably border on iron deficiency even without the additional iron losses incurred through exercise, and vigorous exercise worsens iron status. To prevent depletion of iron stores for women and teens, low hematocrit and hemoglobin measures may warrant the use of supplements, along with attention to the diet.

Supplements for Athletes

Years ago, protein was thought to be a major fuel source for muscular work. This thought gave rise to the misconception that athletes required protein supplementation. Despite later proof that protein is not a major fuel source during exercise, many athletes continue to place protein on a performance pedestal by using expensive, unnecessary protein supplements. Even in the face of new evidence that exercise induces more changes in protein metabolism than previously thought, protein supplementation still appears to be unnecessary for the healthy athlete.[113] For a variety of reasons, protein supplementation may actually impair performance in endurance events. The waste products of an excess of protein are excreted in the urine, placing a burden on the kidneys and possibly leading to dehydration. Also, a diet that emphasizes protein is likely to be higher in fat and lower in carbohydrate than is ideal.

Athletes may also take protein supplements in the false hope that since muscles are made of protein, eating extra protein will build bigger muscles. Of course, muscle work builds muscle; protein supplements do not, and athletes do not need them. Protein and carbohydrate spare body protein equally well, and carbohydrate is safer. A balanced diet supplies enough protein, so protein or amino acid supplements are not needed by normal, healthy people, even vegetarian athletes.

Furthermore, protein supplements are expensive and less well digested than protein-rich food; when used as a total replacement for food, they are often dangerous. The "liquid protein" diet, once advocated for weight loss, caused deaths in some users.

That athletes use protein supplements is not a reflection on them. They are the targets of misguided coaches and trainers as well as quacks who try to convince them that they need supplements of all kinds, and that the established scientific community, and especially the nutrition community, is not up on the latest research. These tactics work; athletes spend millions of dollars on supplements each year. They do this because they have one correct piece of knowledge: nutrition makes a difference. In a world where body condition and skill are hard won, promises of performance gains in pills or potions are enticing.

The lure of supplements encourages sales not only of protein potions but also of vitamins and minerals. According to one survey, 84 percent of world-class athletes use vitamin supplements.[114] No research validates their use. Needs for extra thiamin and riboflavin are unsupported, as mentioned, and vitamin B_6 supplements are known to have toxic effects in amounts not far above the RDA.

Some athletes take vitamin B_{12} injections or pills prior to competition because they believe that they can enhance endurance and oxygen delivery in this way. The limited research thus far does not lend support to the concept that vitamin B_{12} injections or pills enhance the performance of the well-nourished athlete. In fact, taking *any* vitamin directly before a competition runs contrary to science. Vitamins usually function only as small parts of larger working units, usually involving proteins. A molecule of a vitamin, floating around in the blood, is simply waiting for the tissues to combine it with its appropriate enzymes or other parts so that it can do its work. This takes time—hours or days. Vitamins taken right before an event are still waiting in the blood during that event and are useless for improving performance, even if the person happens to be deficient in the vitamins. Vitamin B_{12} supplements are the appropriate treatment for a vitamin B_{12} deficiency. For a well-nourished athlete, they have not been shown to improve performance.[115]

Preliminary research reporting that excretion of vitamin C increased after exercise seemed to indicate that vitamin C in amounts two or three times the RDA might best serve the needs of the athlete. Since then, the great bulk of research designed to explore this theory has disproved it—athletes perform no better when taking vitamin C supplements than when they receive the RDA amount from food. Even so, athletes are often told by "advisers" in health-food stores to ingest huge quantities of vitamin C, measured in multiples of a gram. These amounts are clearly beyond those indicated as potentially useful, and could be harmful.

In general, vitamin and mineral supplementation for the healthy athlete appears unwarranted. Excessive amounts of vitamins and minerals can be toxic and should be avoided. The athlete's nutrient needs can be met by a balanced diet consisting of a variety of nutrient-dense foods from the four food groups, as described in the "Diets for Exercisers" section.

Fluids and Electrolytes

The active person's need for water far surpasses the need for any other nutrient. Water's role in temperature regulation is of critical importance in exercise. Working muscles release much of the energy they spend as heat. Body water absorbs the heat generated during exercise and transports it to the skin for release as sweat, thus cooling the body by evaporation. Sweating accounts for most of the water lost during exercise; some water vapor is also lost through breathing.

In hot environments, evaporation of sweat is the body's cooling system, but high humidity renders this system inefficient. In humid air, sweat evaporates less readily, and body heat rises. This signals greater sweating, and dehydration can result, just as it does in dry weather. Water loss beyond a 2 percent loss of body weight impairs temperature control and aerobic endurance. Water loss equal to 5 percent of the body weight reduces muscular work capacity by 20 to 30 percent.[116] A rise in body heat accompanied by fluid loss through sweating is dangerous: it sets the stage for life-threatening heat stroke.

heat stroke: an acute and dangerous reaction to heat buildup in the body, characterized by high body temperature, loss of consciousness, low blood pressure, and possible death. **Heat exhaustion** precedes it, with warning signs of fatigue, nausea, dizziness, and stomach cramps.

Water loss during strenuous exercise can amount to as much as 2 to 4 liters per hour. During vigorous exercise in a hot, humid environment, thirst is an inadequate indicator of fluid need, because it occurs only after significant fluid depletion.[117] Fluid replacement under these conditions requires a planned schedule. A person can defend against heat stroke and promote optimal performance and endurance by drinking cold water before, during, and after exercise, as shown in Table 8–5. The human body cannot be conditioned to use less water. Coaches who restrict water during practice are liable for sanction by the American College of Sports Medicine. Cold water (40 to 50 degrees Fahrenheit) promotes rapid gastric emptying and prevents bloating discomfort.

Plain, cold water is the optimal fluid for people exercising under most conditions. Sugar- and electrolyte-containing beverages (sports drinks) have a higher concentration of dissolved solids than the body fluids do (that is, they are hypertonic) and demand dilution in the digestive tract. This dilution steals fluid from the tissues and takes time, and so the beverage is held in the stomach, away from the thirsty tissues; plain, cold water passes through the stomach unimpeded and rushes to the tissues that need it.[118]

A glucose-containing beverage may be desirable in one extreme case—that of the endurance athlete who competes in events lasting more than two hours. In this instance, once the event is well under way, a dilute solution containing 2 tablespoons of sugar or 1 cup of fruit juice in a quart of water can provide a glucose alternative to glycogen, and so forestall exhaustion. For the typical weekend exerciser, however, plain water holds the fluid advantage.

glucose polymers: compounds that supply glucose not as single molecules but linked in chains somewhat like starch. The object is to attract less water from the body into the digestive tract (osmotic attraction depends on the number, not the size, of particles).

In an effort to provide the endurance athlete with a carbohydrate-containing drink that does not slow gastric emptying, researchers have developed glucose polymer beverages. Glucose polymer beverages supply glucose in chains of three or four linked glucose units rather than singly. Research results on these beverages are mixed: some show the polymers to demand less dilution in the stomach, and others show them to slow down gastric emptying just as much as the glucose drinks do.[119]

Losses of water exceed losses of electrolytes such as sodium, potassium, and chloride during exercise; most athletes' diets supply ample sodium, potassium, chloride, and other electrolytes to replace losses in sweat. Even if losses are high, replacement of these minerals need not occur right away during

Table 8–5 Schedule of Hydration before, during, and after Exercise

When to Drink	Amount of Fluid
2 hr before exercise	About 3 c
10 to 15 min before exercise	About 2 c
Every 10 to 20 min during exercise	About ½ c or more
After exercise	Replace each pound of body wieght lost with 2 c fluid

Source: Adapted from J. B. Marcus, ed., *Sports Nutrition* (Chicago: American Dietetic Association, 1986), p. 57.

exercise; later meals can deliver them soon enough. Also, the body adapts to conserve electrolytes if training and sweating occur repeatedly.

A caution is in order regarding fluid and electrolyte replacement for the athlete. Elite, ultraendurance athletes, such as triathletes who compete and train for many grueling hours on consecutive days in extreme high heat and humidity, can lose 5 to 10 pounds of body weight in fluid a day. If these athletes were to drink large amounts of water throughout competition, they would risk water intoxication—excessive dilution of the blood with water, called hyponatremia. This can be dangerous; diarrhea, fatigue, and central nervous system disturbances signal the onset, and death can occur soon if treatment is delayed.[120] Water intoxication is rare, but for the few athletes at risk for it, electrolyte replacement during the activity is necessary.[121] In these instances, commercial glucose-electrolyte solutions are appropriate; salt tablets are never advisable, as they cause stomach distress and worsen dehydration.

Diets for Exercisers

Most exercisers, and especially high school and college athletes, do not choose foods based on their nutrient needs. In a study of men athletes, only half knew even the basic nutrition facts.[122] In a study of women athletes, all of them knew the importance of calcium, but only 12 percent chose diets that would provide the RDA amount.[123] This section is therefore dedicated to diets for exercisers.

Notice that the word *diets* is plural—there is no one best diet for performance, and many variations on the advice given here will support the athlete's performance superbly. However, choices must be made within a framework of two absolutely unbreakable rules. The person who obeys these rules can be confident that the diet is adequate.

The first absolutely unbreakable rule is to eat a nutrient-dense diet composed mostly of whole foods. The word *whole* means that foods are as close to the farm-fresh state as possible, for it is in this state that foods provide maximum vitamins and minerals for the kcalories they contain. When people rely heavily on processed foods that have lost nutrients and gained sugar and fat, nutrient status suffers. Even if these foods are fortified or enriched, manufacturers cannot replace the full array of nutrients and nonnutrients lost in processing. For example, manufacturers mill and process vitamin B_6 out of foods but do not replace it—and vitamin B_6 is essential to optimal performance. This does not mean that active people can *never* choose a white bread, bologna, and mayonnaise sandwich for lunch, but only that they should later eat a large fresh salad or big portions of vegetables and drink a glass of milk. The whole foods provide the needed nutrients; the bologna sandwich was extra.

The following advice to the active person may seem timeworn, but it is the other absolutely unbreakable rule for dietary adequacy—build the diet according to a food group plan to obtain the complete array of vitamins and minerals. Their metabolic roles in exercise necessitate that the diet supply each one amply if a person is to perform and compete successfully. While any food group plan will do, the "basic four" plan is most familiar and therefore easiest to use (see Appendix C).

Beyond adequacy, energy needs may be immense, and the endurance performer may want full glycogen stores as well. Simply stated, a diet that is high in carbohydrate, not too high in fat, and adequate in protein while meeting the athlete's energy needs works best. Even if the person does not engage in glycogen-depleting events, such a diet is also recommended to control weight, to provide adequate fiber, and to reduce the risk of diabetes while supplying abundant nutrients. A day's worth of food according to such a plan might look like the example in Table 8–6.

The foods listed represent realistic choices for the adolescent, based on eating habits of individuals in this age group. Figure 8–7 shows that most of the nutrients in these meals meet the RDA for an active adolescent.

Many diet variations are suitable, as long as nutritious foods are the basis. This day's meals provide about 2600 kcalories, 57 percent of them from carbohydrate, 28 percent from fat, and 15 percent from protein. Table 8–7 shows some sample diet plans for people who wish to increase their energy and carbohydrate intakes by using whole foods. These plans are effective only if the user chooses whole foods to provide nutrients as well as energy—extra milk for calcium and riboflavin, many vegetables for B vitamins, meat or alternates for iron and other vitamins and minerals, and whole grains for magnesium, zinc, and chromium. These foods provide plenty of sodium, potassium, and chloride.

Adding more carbohydrate-rich foods is, up to a point, a sound and reasonable option for increasing kcalories. It becomes unreasonable when the

Table 8–6 A Day's Meals for a Physically Active Teenager

Breakfast:
1 c oatmeal
¼ c raisins
½ c 2% low-fat milk
2 pieces whole-wheat toast
1 tbsp preserves
1 c orange juice

Lunch:
Cheese, lettuce, tomato, and mayonnaise sandwich on whole-wheat bread
Carrot sticks (1 carrot)
1 large apple
4 chocolate chip cookies
1 c 2% low-fat milk
1 Popsicle

Dinner:
2 pieces fried chicken
⅓ c baked beans
⅓ c coleslaw
1 biscuit
1 12-oz cola

Snack:
1 banana
1 brownie
1 c 2% low-fat milk

Figure 8–7 Dietary Analysis of a Typical Day's Meals and the Adolescent RDA Compared
The meals listed in Table 8–6 meet more than 100% of a teenager's recommended intakes for almost all nutrients, as shown. The RDA used in this comparison is for a moderately active 16-year-old female of average height (5 ft 4 in) and weight (125 lb), and for a moderately active 16-year-old male of average height (5 ft 8 in) and weight (150 lb).

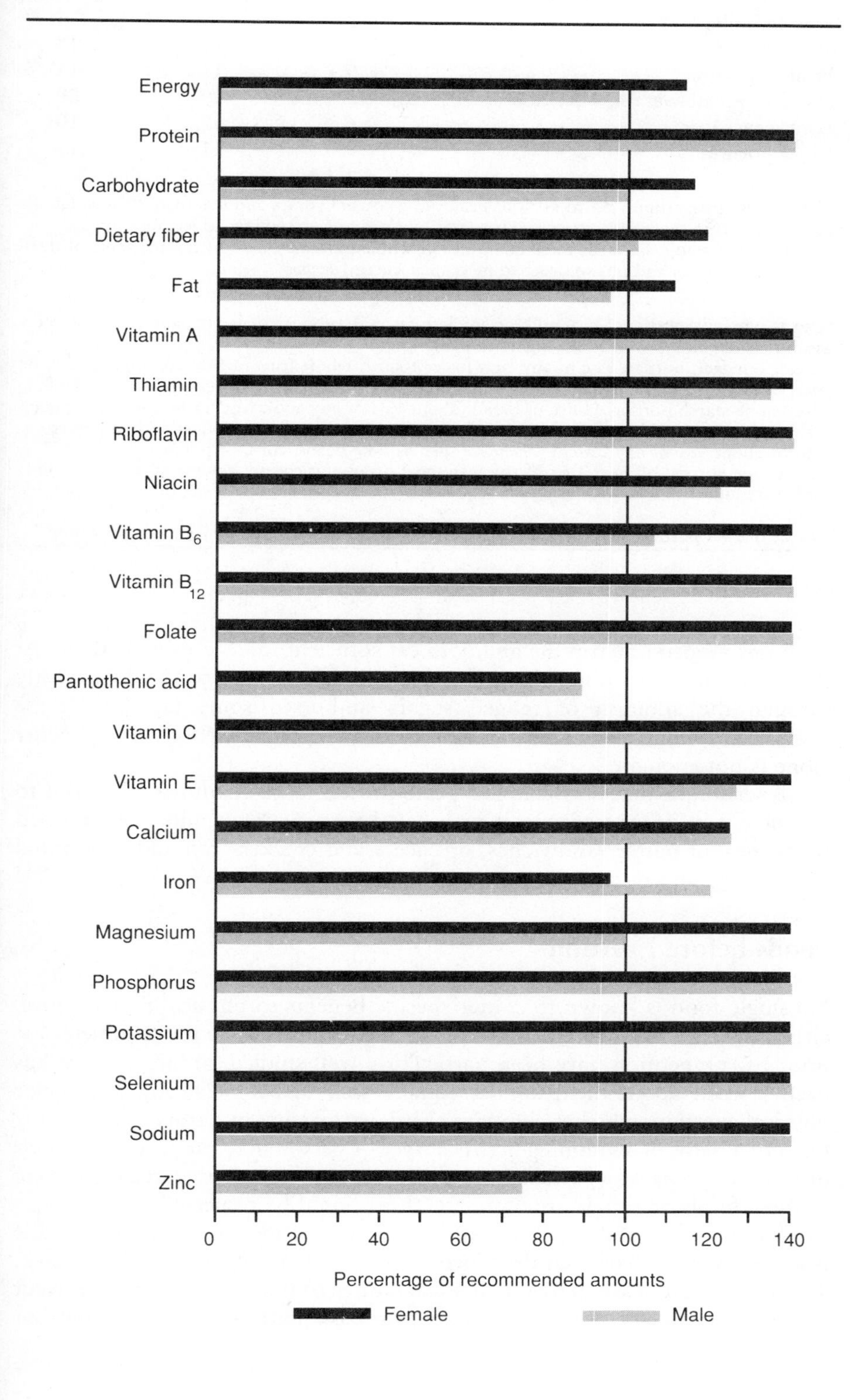

Table 8–7 Diet Plans for High-kCalorie, High-Carbohydrate Intakes

	Energy Level (kcal)		
Food, Portion	**3100**	**3900**	**4600**
Low-fat milk, 1 c[a]	3	5	6
Vegetable, ½ c	9	11	13
Fruit, 1 portion[b]	9	11	15
Starchy vegetable/grain, 1 portion[c]	15	19	20
Lean meat, 1 oz[d]	6	7	10
Fat, 1 portion[e]	9	11	11

Note: These plans supply 55 to 60% of kcalories as carbohydrate and less than 30% as fat. To increase the carbohydrate content to over 60%, substitute ⅓-c servings of legumes for the meats. People who cannot eat these quantities of whole food may have to replace some of them with refined sugars and fats in order to meet their energy needs.

[a]Use as one milk portion 1 c of milk or any low-fat milk product such as yogurt or buttermilk; assume whole milk products are higher in kcalories.
[b]Use as one fruit portion ½ c of any fruit juice; 1 small whole fruit such as an apple, peach, or pear; or about ½ to 1 c of any bite-sized fruit such as berries, melon pieces, or fruit cocktail.
[c]Use as one starch portion 1 slice of bread; about ½ c of any cooked grain product or legume such as rice, hot cereal, or baked beans; about 1 oz of any dry grain product such as dry cereal; or about ½ c of any starchy vegetable such as lima beans, corn, or potato.
[d]Use as one portion of meat 1 oz of any lean meat or low-fat cheese; assume fattier meats and cheese are higher in kcalories. Alternatively, substitute ⅓ c legumes or 1 egg for 1 meat portion.
[e]Use as one fat portion 1 tsp or 1 pat of butter or margarine, or 1 tbsp oil or salad dressing.

kcalories needed outstrip the ability to eat sufficient food to provide them. At that point, the person must find ways of adding kcalories to the diet, mostly through the addition of refined sugars and even some fat. Still, these kcalorie-rich foods must be superimposed on nutrient-rich choices. Energy alone is not enough.

A technique that may enable a person to eat large quantities of food is to consume it in six or eight meals each day. Large snacks of milk shakes, dried fruits, peanut butter sandwiches, or cheese and crackers can add substantial food energy and nutrients.

Foods before Exertion

No single food is known to confer specific benefits to physical performance, although some kinds of foods are preferable to others. The meals athletes eat prior to competition have been particularly well studied, although what has been learned about them applies equally well to people facing any major physical exertion, whether climbing a cliff, performing in a trapeze act, giving birth to a baby, or swimming across a river. A person may eat particular foods or practice preparatory rituals that convey psychological advantages. As long as these foods or rituals are harmless, they should be respected.

From studies of competitive athletes, a type of meal particularly suitable prior to exertion has been described—the so-called pregame meal or snack. The meal is light, easy to digest, and eaten three to four hours before the event to allow time for the stomach to empty before the start of activity. The meal or snack should contain between 300 and 1000 kcalories, although the lighter the better.[124] Table 8–8 shows some sample pregame meals. Breads, potatoes, pasta, and fruit juices—carbohydrate-rich foods low in fat, protein, and fiber—are the basis of the pregame meal. Fiber-rich carbohydrate foods such as raw vegetables or whole grains, while usually desirable, are best avoided just prior to heavy exercise. Fiber in the digestive tract attracts water out of the blood, and can cause stomach discomfort during performance. Some people prefer liquid meals that are commercially available and easily digested.

Rapidly absorbed high-sugar snacks are counterproductive for some exercisers; insulin production is stimulated, which lowers blood glucose and deprives the working muscles of the fuel they need. Other research contradicts this finding.[125] It seems that people differ in their response to sugar consumption prior to heavy exertion. When cyclists consumed sugar immediately (five minutes) prior to a one-hour cycling session, exercise performance improved

Table 8–8 Sample Pregame Meals

These foods, to be eaten three to four hours before competition, do *not* have many vitamins or minerals for the most part, and should not be overemphasized in the daily diet. The hours before competition are too late for vitamins and minerals, but the special needs for energy and fluid can be met by eating the following meals.

Food	Serving Size	Energy (kcal)	Energy Donated by Carbohydrate (% of kcal)
Sample Meal #1			
White bread	2 slices	140	74
Jam or jelly	2 tbsp	130	100
Gelatin dessert	1 c	70	97
Grape juice	1 c	165	100
Total		505	94
Sample Meal #2			
Spaghetti and tomato sauce	1 c	190	80
Roll (white flour)	1	85	71
Popsicle (3 oz)	2	140	100
Limeade	1 c	100	100
Total		515	90
Sample Meal #3			
Banana	1	100	100
Sweetened dry cereal	¾ c	115	90
Nonfat milk	½ c	45	53
Cranberry juice cocktail	⅔ c	109	100
Gumdrops	1 oz	100	100
Total		469	95

Note: Substitutions can be made for cereal (1 c), spaghetti (1 c), or bread (2 slices):
- 1 c white or flavored rice.
- 1 3-inch-diameter muffin (plain).
- 1 piece angel food cake.
- 3 small pancakes and syrup.

To substitute liquids (gelatin, juices, limeade), use 1 c for each: 1 c any sugar-sweetened beverage.

Source: E. M. N. Hamilton, E. N. Whitney, and F. S. Sizer, *Nutrition: Concepts and Controversies*, 4th ed. (St. Paul, Minn.: West, 1988).

compared with that of cyclists receiving placebo.[126] For those who may be susceptible to a lowering of blood glucose, sugar consumption just a few minutes before exercise inhibits the insulin response because of the exercise-induced rise in epinephrine and norepinephrine.[127]

People who want to excel in endeavors will apply in their daily routines the most accurate possible nutrition knowledge, along with dedication to rigorous training. A diet that provides ample fluid and consists of a variety of nutrient-dense, whole foods in quantities to meet energy needs will enhance not only physical performance but overall health as well. Training and genetics being equal, it is easy to guess who would turn in the more outstanding performance—the athlete who habitually consumes half or less of the needed nutrients, or the one who arrives at the event with a long history of full nutrient stores and well-met metabolic needs.

The teen years are exciting years in preparation for adulthood. The lifestyle choices people make as teenagers can facilitate or undermine the transition into adulthood. Habits established during the teen years lay a foundation for health throughout the rest of life.

Chapter 8 Notes

1. D. Sinclair, *Human Growth after Birth,* 4th ed. (New York: Oxford University Press, 1985), pp. 23–50.
2. Sinclair, 1985, pp. 23–50.
3. Sinclair, 1985, pp. 123–147.
4. Sinclair, 1985, pp. 123–147.
5. Sinclair, 1985, pp. 51–72.
6. Sinclair, 1985, pp. 102–122.
7. Sinclair, 1985, pp. 102–122.
8. Sinclair, 1985, pp. 102–122.
9. R. E. Frisch, Fatness, menarche, and female fertility, *Perspectives in Biology and Medicine* 28 (1985): 611–633.
10. H. E. Kulin and coauthors, The effect of chronic childhood malnutrition on pubertal growth and development, *American Journal of Clinical Nutrition* 36 (1982): 527–536.
11. Sinclair, 1985, pp. 73–101.
12. K. Satyanarayana and coauthors, Effect of nutritional deprivation in early childhood on later growth—A community study without intervention, *American Journal of Clinical Nutrition* 34 (1981): 1636–1637.
13. Kulin and coauthors, 1982.
14. P. A. Wagner and coauthors, Serum zinc concentrations in adolescents as related to sexual maturation, *Human Nutrition: Clinical Nutrition* 39C (1985): 459–462.
15. Food and Nutrition Board, *Recommended Dietary Allowances,* 10th ed. (Washington, D.C.: National Academy of Sciences, 1989), p. 197.
16. U.S. Department of Health, Education, and Welfare, *Iron Nutriture in Adolescence,* HHS publication no. (HSA) 77–5100 (Washington, D.C.: Government Printing Office, 1977).
17. J. D. Skinner and coauthors, Appalachian adolescents' eating patterns and nutrient intakes, *Journal of the American Dietetic Association* 85 (1985): 1093–1099; J. Bowering and K. L. Clancy, Nutritional status of children and teenagers in relation to vitamin and mineral use, *Journal of the American Dietetic Association* 86 (1986): 1033–1038.
18. Skinner and coauthors, 1985.
19. Results from the National Adolescent Student Health Survey, *Journal of the American Medical Association* 261 (1989): 2025, 2031.
20. Canadian Paediatric Society, Nutrition Committee, Adolescent nutrition: II. Normal nutritional requirements, *Canadian Medical Association Journal* 129 (1983): 420–422.
21. J. A. Driskell, A. J. Clark, and S. W. Moak, Longitudinal assessment of vitamin B_6 status in Southern adolescent girls, *Journal of the American Dietetic Association* 87 (1987): 307–310; P. Thompson and coauthors, Zinc status and sexual development in adolescent girls, *Journal of the American Dietetic Association* 86 (1986): 892–897; H. McCoy and coauthors, Nutrient intakes of female adolescents from eight southern states, *Journal of the American Dietetic Association* 84 (1984): 1453–1460; A. J. Clark, S. Mossholder, and R. Gates, Folacin status in adolescent females, *American Journal of Clinical Nutrition* 46 (1987): 302-306.
22. National Research Council, *Diet and Health: Implications for Reducing Chronic Disease Risk,* (Washington, D.C.: National Academy Press, 1989).
23. F. C. Luft, Dietary sodium, potassium and chloride intake and arterial hypertension, *Nutrition Today,* May-June, 1989, pp. 9–14.
24. P. M. Guenther, Beverages in the diets of American teenagers, *Journal of the American Dietetic Association* 86 (1986): 493–499.
25. L. K. Massey, Soft drink consumption, phosphorus intake, and osteoporosis, *Journal of the American Dietetic Association* 80 (1982): 581–583.
26. Food and Nutrition Board, 1989, pp. 178–179.
27. L. H. Allen, Calcium bioavailability and absorption: A review, *American Journal of Clinical Nutrition* 35 (1982): 783–808.
28. H. Spencer and coauthors, Effect of phosphorus on the absorption of calcium and on the calcium balance in man, *Journal of Nutrition* 108 (1978): 447–457, as cited by Allen, 1982.
29. M. S. Calvo and H. Heath, Acute effects of oral phosphate-salt ingestion on serum

phosphorus, serum ionized calcium, and parathyroid hormone in young adults, *American Journal of Clinical Nutrition* 47 (1988): 1025–1029.

30. M. Zelnik and J. F. Kantner, Sexual and contraceptive experience of young unmarried women in the United States, 1976 and 1971, in *Adolescent Pregnancy and Childbearing: Findings from Research,* NIH publication no. 81–2077 (Washington, D.C.: Government Printing Office, 1980), pp. 43–81.
31. C. A. Bachrach, Contraceptive practice among American women, 1973–1982, *Family Planning Perspectives* 16 (1984): 253–259.
32. 1985 natality report available from USPHS, *American Journal of Public Health* 77 (1987): 1473; K. Davis, A theory of teenage pregnancy in the United States, in *Adolescent Pregnancy and Childbearing,* ed. C. S. Chilman, NIH publication no. 81–2077 (Washington, D.C.: Government Printing Office, 1981), pp. 309-339.
33. V. Ktsanes, The teenager and the family planning experience, in *Adolescent Pregnancy and Childbearing,* ed. C. S. Chilman, NIH publication no. 81–2077 (Washington, D.C.: Government Printing Office, 1981), pp. 83–100.
34. U.S. Department of Agriculture, Food and Dietary Service, and U.S. Department of Health and Human Services, March of Dimes Birth Defects Foundation, *Working with the Pregnant Teenager: A Guide for Nutrition Educators* (Washington, D.C.: Government Printing Office, 1981), p. 1.
35. A. A. Campbell, Trends in teenage childbearing in the United States, in *Adolescent Pregnancy and Childbearing,* ed. C. S. Chilman, NIH publication no. 81–2077 (Washington, D.C.: Government Printing Office, 1981), pp. 3–13.
36. Sinclair, 1985, pp. 102–122.
37. J. Menken, The health and demographic consequences of adolescent pregnancy and childbearing, in *Adolescent Pregnancy and Childbearing,* ed. C. S. Chilman, NIH publication no. 81–2077 (Washington, D.C.: Government Printing Office, 1981), pp. 177–205.
38. Menken, 1981.
39. Menken, 1981.
40. Menken, 1981.
41. Menken, 1981.
42. Adolescent pregnancy—Counseling considerations, *Nutrition and the M.D.,* January 1986, p. 4.
43. A. R. Frisancho, J. Matos, and L. A. Bollettino, Influence of growth status and placental function on birth weight of infants born to young still-growing teenagers, *American Journal of Clinical Nutrition* 40 (1984): 801–807.
44. G. M. Chan and coauthors, Effects of increased dietary calcium intake upon the calcium and bone mineral status of lactating adolescent and adult women, *American Journal of Clinical Nutrition* 46 (1987): 319–323.
45. Evaluation of the Teenage Pregnancy and Parenting (TAPP) Project, Family Service Agency of San Francisco, October 1, 1982–September 30, 1983, a report submitted to the Office of Adolescent Pregnancy Programs, Department of Health and Human Services (San Francisco: Center for Population and Reproductive Health, Institute for Health Policy Studies, University of California, 1984).
46. E. T. Kennedy and M. Kotelchuk, The effect of WIC supplemental feeding on birth weight: A case-control study, *American Journal of Clinical Nutrition* 40 (1984): 579–585.
47. N. S. Moses, M. Banilivy, and F. Lifshitz, Fear of obesity among adolescent females (abstract), cited in *American Journal of Clinical Nutrition* 43 (1986): 664.
48. J. K. Thompson, Larger than life, *Psychology Today,* April 1986, pp. 38–44.
49. N. S. Storz and W. H. Greene, Body weight, body image, and perception of fad diets in adolescent girls, *Journal of Nutrition Education,* March 1983, pp. 15–18.
50. D. Farley, Eating disorders: When thinness becomes an obsession, *FDA Consumer,* May 1986, pp. 20–23.
51. Position of the American Dietetic Association: Nutrition intervention in the treatment of anorexia nervosa and bulimia, *Journal of the American Dietetic Association* 88 (1988): 68.
52. D. B. Herzog and P. M. Copeland, Eating disorders, *New England Journal of Medicine* 313 (1985): 295–303.
53. American College of Physicians, Health and Public Policy Committee, Eating disorders: Anorexia nervosa and bulimia, *Annals of Internal Medicine* 105 (1986): 790–794.
54. Farley, 1986.
55. J. Chalmers and coauthors, Anorexia nervosa presenting as morbid exercising (letter), *Lancet* 1 (1985): 286, as cited in Anorexia nervosa presenting as morbid obesity (abstract), *Journal of the American Dietetic Association* 85 (1985): 762.
56. Herzog and Copeland, 1985; N. Vaisman and coauthors, Effect of refeeding on the basal energy metabolism and substrate utilization of adolescents with anorexia nervosa (abstract), *American Journal of Clinical Nutrition* 43 (1986): 670.
57. Herzog and Copeland, 1985.
58. B. Palla and I. F. Litt, Medical complications of eating disorders in adolescents, *Pediatrics* 81 (1988): 613–623.
59. Herzog and Copeland, 1985.
60. Y. Danziger and coauthors, Parental involvement in treatment of patients with anorexia nervosa in a pediatric day-care unit, *Pediatrics* 81 (1988): 159–162.
61. D. M. Huse and A. R. Lucas, Dietary treatment of anorexia nervosa, *Journal of the American Dietetic Association* 83 (1983): 687–690.
62. E. Sanger and T. Cassino, Eating disorders: Avoiding the power struggle, *American Journal of Nursing* 84 (1984): 31–33.
63. Huse and Lucas, 1983.
64. Herzog and Copeland, 1985.
65. Farley, 1986.
66. Farley, 1986.
67. B. G. Kirkley, Bulimia: Clinical characteristics, development, and etiology, *Journal of the American Dietetic Association* 86 (1986): 468–472.
68. J. D. Killen and coauthors, Depressive symptoms and substance use among adolescent binge eaters and purgers: A defined population study, *American Journal of Public Health* 77 (1987): 1539–1541; American College of Physicians, 1986.
69. M. Story, Nutrition management and dietary treatment of bulimia, *Journal of the American Dietetic Association* 86 (1986): 517–519.
70. S. Dalvit-McPhillips, A dietary approach to bulimia treatment, *Physiology and Behavior* 33 (1984): 769–775.
71. *Acne,* a pamphlet (May 1980) available from the National Institute of Allergy and Infectious Diseases, Bethesda, MD 20205, NIH publication no. 80–188; *Stubborn and Vexing, That's Acne,* a pamphlet (May 1980) available from the Food and Drug Administration, 5600 Fishers Lane, Rockville, MD 20857, HHS publication no. (FDA) 80–3107.
72. R. M. Reisner, Acne vulgaris, *Pediatric Clinics of North America* 20 (1973): 851–864.
73. J. O. Holloszy and E. F. Coyle, Adaptations of skeletal muscle to endurance exercise and their metabolic consequences, *Journal of Applied Physiology:*

Respiratory, Environmental and Exercise Physiology 56 (1984): 831–838.

74. J. P. Flatt, Dietary fat, carbohydrate balance, weight maintenance: Effects of exercise, *American Journal of Clinical Nutrition* 45 (1987): 296–306.
75. B. Essen, Intramuscular substrate utilization during prolonged exercise, *Annals of the New York Academy of Science* 301 (1977): 30–44.
76. J. Bergstrom and E. Hultman, Nutrition for maximal sports performance, *Journal of the American Medical Association* 28 (1972): 999–1006.
77. E. H. Christensen and O. Hansen, Arbeitsfahigkeit und ehrnahrung, *Skandinavisches Archiv fuer Physiologie* 8 (1939): 160–175, as cited by E. L. Fox, *Sports Physiology,* 2nd ed. (New York: Saunders, 1984), pp. 40–57.
78. M. H. Williams, *Nutritional Aspects of Human Physical and Athletic Performance,* 2nd ed. (Springfield, Ill.: Charles C Thomas, 1985), pp. 20–57 and Figure 2.
79. Bergstrom and Hultman, 1972.
80. W. M. Sherman and coauthors, Effect of exercise-diet manipulation on muscle glycogen and its subsequent utilization during performance, *International Journal of Sports Medicine* 2 (1981): 114–118.
81. L. P. Garetto and coauthors, Enhanced insulin sensitivity of skeletal muscle following exercise, Proceedings of the 5th International Symposium on Biochemistry of Exercise, in *Biochemistry of Exercise,* ed. H. G. Knuttgen (Champaign, Ill.: Human Kinetics, 1983), pp. 681–687.
82. D. L. Costill and coauthors, The role of dietary carbohydrates in muscle glycogen resynthesis after strenuous running, *American Journal of Clinical Nutrition* 34 (1981): 1831–1836.
83. American Dietetic Association, Position paper: Nutrition for physical fitness and athletic performance for adults, *Journal of the American Dietetic Association* 87 (1987): 933–939.
84. Bergstrom and Hultman, 1972.
85. D. L. Costill and coauthors, Effects of elevated plasma FFA and insulin on muscle glycogen usage during exercise, *Journal of Applied Physiology: Respiratory, Environmental, and Exercise Physiology* 43 (1977): 695–699.
86. W. J. Evans and V. A. Hughes, Dietary carbohydrates and endurance exercise, *American Journal of Clinical Nutrition* 41: (1985): 1146–1154.
87. D. L. Costill, G. P. Dalsky, and W. J. Fink, Effects of caffeine ingestion on metabolism and exercise performance, *Medicine and Science in Sports* 10 (1978): 155–158.
88. F. T. O'Neil, M. T. Hynak-Hankinson, and J. Gorman, Research and application of current topics in sports nutrition, *Journal of the American Dietetic Association* 86 (1986): 1007–1015.
89. B. Essen, Intramuscular substrate utilization during prolonged exercise, *Annals of the New York Academy of Science* 301 (1977): 30–44.
90. P. W. R. Lemon and J. P. Mullin, Effect of initial muscle glycogen levels on protein catabolism during exercise, *Journal of Applied Physiology: Respiratory Environmental and Exercise Physiology* 48 (1980): 624–629, as cited by E. R. Buskirk, Some nutritional considerations in the conditioning of athletes, *Annual Review of Nutrition* 1 (1981): 319–350.
91. M. N. Goodman and N. B. Ruderman, Influence of muscle use on amino acid metabolism, *Exercise and Sport Sciences Reviews* 10 (1982): 1–26.
92. Williams, 1985, pp. 120–146.
93. Essen, 1977.
94. Williams, 1985, pp. 147–185.
95. A. Belko and coauthors, Effects of exercise on riboflavin requirements: Biological validation in weight reducing women, *American Journal of Clinical Nutrition* 41 (1985): 270–277.
96. Williams, 1985.
97. Williams, 1985.
98. M. E. Nelson and coauthors, Diet and bone status in amenorrheic runners, *American Journal of Clinical Nutrition* 43 (1986): 910–916.
99. B. B. Peterkin, Women's diets: 1977 and 1985, *Journal of Nutrition Education* 18 (1986): 251–257.
100. L. W. Rosen, Pathogenic weight-control behavior in female athletes, *Physician and Sportsmedicine* 14 (1986): 79–86.
101. Nelson and coauthors, 1986.
102. P. R. Dallman, M. A. Siimes, and A. Stekel, Iron deficiency in infancy and childhood, *American Journal of Clinical Nutrition* 33 (1980): 86–118.
103. Expert Scientific Working Group of the Federation of American Societies for Experimental Biology, Summary of a report on assessment of the iron nutritional status of the United States population, *American Journal of Clinical Nutrition* 42 (1985): 1318–1330.
104. R. B. Parr, L. A. Bachman, and R. A. Moss, Iron deficiency in female athletes, *Physician and Sportsmedicine* 12 (1984): 81-86.
105. P. A. Deuster and coauthors, Nutritional survey of highly trained women runners, *American Journal of Clinical Nutrition* 44 (1986): 954–962.
106. H. J. Nickerson and coauthors, Decreased iron stores in high school female runners, *American Journal of Diseases of Children* 139 (1985): 1115–1119.
107. J. G. Stewart and coauthors, Gastrointestinal blood loss and anemia in runners, *Annals of Internal Medicine* 100 (1984): 843–845; M. Brune and coauthors, Iron losses in sweat, *American Journal of Clinical Nutrition* 43 (1986): 438–443.
108. G. W. Gardner and coauthors, Physical work capacity and metabolic stress in subjects with iron-deficiency anemia, *American Journal of Clinical Nutrition* 30 (1977): 910–917.
109. R. H. Dressendorfer, C. E. Wade, and E. A. Amsterdam, Development of pseudoanemia in marathon runners during a 20-day road race, *Journal of the American Medical Association* 246 (1981): 1215–1218.
110. Parr, Bachman, and Moss, 1984.
111. American Dietetic Association, Position Paper: Nutrition for physical fitness and athletic performance for adults, *Journal of the American Dietetic Association* 87 (1987): 933–939.
112. Buskirk, 1981; Dressendorfer, Wade, and Amsterdam, 1981.
113. Williams, 1985, pp. 120–146.
114. J. R. Brotherhood, Nutrition and sports performance, *Sports Medicine* 1 (1984): 350–389.
115. Williams, 1985, pp. 147–186.
116. Bergstrom and Hultman, 1972.
117. J. E. Greenleaf and coauthors, Drinking and water balance during exercise and heat acclimation, *Journal of Applied Physiology: Respiratory, Environmental and Exercise Physiology* 54 (1983): 414–419.
118. D. L. Costill and B. Saltin, Factors limiting gastric emptying during rest and exercise, *Journal of Applied Physiology* 37 (1974): 679–683.
119. C. Foster, Gastric-emptying characteristics of glucose polymers, in *Ross Symposium on Nutrient Utilization during Exercise,* ed. E. L. Fox (Columbus, Ohio: Ross Laboratories, 1983): 80–84.
120. R. T. Frizzel and coauthors, Hyponatremia and ultramarathon running, *Jour-*

nal of the American Medical Association 255 (1986): 772–775.

121. F. T. O'Neil, M. T. Hynak-Hankinson, and J. Gorman, Research and application of current topics in sports nutrition, *Journal of the American Dietetic Association* 86 (1986): 1007–1015.

122. L. R. Shoaf, P. D. McClellan, and K. A. Birskovich, Nutrition knowledge, interests, and information sources of male athletes, *Journal of Nutrition Education* 18 (1986): 243–245.

123. M. Perron and J. Endres, Knowledge, attitudes, and dietary practices of female athletes, *Journal of the American Dietetic Association* 85 (1985): 573–576.

124. American Dietetic Association, 1987.

125. M. Hargreaves and coauthors, Effect of pre-existing carbohydrate feedings on endurance cycling after meals, *Medicine and Science in Sports and Exercise* 196 (1987): 33–36.

126. P. D. Neufer and coauthors, Improvements in exercise performance: Effects of carbohydrate feedings and diet, *Journal of Applied Physiology* 62 (1987): 983–988.

127. E. Coleman, Sugar feedings in the pre-exercise period, *Sports Medicine Digest,* July 1989, p. 5.

▶ Focal Point 8

Drugs, Alcohol, and Tobacco

The physical maturity and growing independence of the teen years present adolescents with a new set of responsibilities and decisions to handle. The choices they make and the consequences of their actions will affect their lives, for better or worse. Some of these behaviors may influence their lives only for today; others can have lifelong effects. The connections between some of these behaviors and nutrition are explored in this focal point.

Drugs

The teen years are a critical time in the development of problem behaviors such as drug use. With the exception of cocaine use, illicit drug use among adolescents in this country has declined since its peak in the late 1970s.[1] Still, three of every five high school seniors report that they have at least tried an illicit drug, most commonly marijuana, amphetamines, and cocaine. This discussion focuses on the nutrition-related effects of these drugs.

Marijuana

Half of all high school seniors surveyed report having tried marijuana, with half of them smoking marijuana within the past 30 days.[2] Like all substances entering the body, marijuana must be processed. The active ingredients are rapidly and almost completely (90 percent) absorbed from the lungs.[3] Then, being fat soluble, these substances are packaged (most likely in lipoproteins) before being transported by the blood to the various body tissues.[4] They are processed by many tissues (not just by the liver), and they persist for several days in the body, being excreted over a period of a week or more after the smoking of a single marijuana cigarette.[5] With repeated exposure, these substances accumulate and become concentrated in body fat, the lungs, liver, reproductive organs, and the brain.[6]

Smoking a marijuana cigarette has several characteristic effects on the body, altering, among other things, the sense of taste. Among the apparent taste changes induced by marijuana is an enhanced enjoyment of eating, especially of sweets, commonly known as "the munchies." Why or how this effect occurs is not known.[7] The drug does not change blood glucose concentrations.[8] Some investigators speculate that the hunger induced by marijuana is actually a social effect caused by the suggestibility of the group in which it is smoked.[9] The heightened appetite and food consumption effects of

THC were tested in clients with anorexia nervosa without success.[10] Prolonged use of the drug does not seem to bring about a weight gain; in one small sample (30 smokers), regular users weighed less than comparable nonsmokers by about 7 pounds.[11]

THC: delta-9-tetrahydrocannabinol, the active ingredient of marijuana primarily responsible for its intoxicating effects.

Amphetamines

Statistics on amphetamine use are difficult to determine. Incidence reports vary in their inclusion of over-the-counter stimulants, medically prescribed stimulants, and illicit amphetamine use. Whatever the actual numbers, the prevalence of amphetamine use appears to be declining.[12] This is due, in part, to physicians' reducing their prescriptions of amphetamines to adolescents. When amphetamine users were surveyed, almost 30 percent indicated that their first use was via a medical prescription, whereas approximately 20 percent had never received a prescription.[13]

amphetamine: a central nervous system stimulant.

Physicians prescribe amphetamines to treat hyperkinesis, narcolepsy, and obesity. Amphetamines raise the pulse rate and blood pressure. Their effects include increased alertness, excitation, and euphoria; insomnia; and loss of appetite.

Cocaine

Cocaine use has not followed the declining trend of other illicit drugs. Of the seniors surveyed, 17 percent reported having tried cocaine, with approximately one-third of those students using it within the past month.[14] Cocaine's properties are like those of both amphetamines and anesthetics. The drug elicits such effects as intense euphoria, restlessness, heightened self-confidence, irritability, insomnia, and loss of appetite. Weight loss is a common side effect, and cocaine abusers often meet the criteria for eating disorders.[15] Repeated use can cause rapid heart rate, irregular heartbeats, heart attacks, and even death.

Drug abusers face multiple nutrition problems:

- They spend money for drugs that could be spent on food.
- They lose interest in food during "high" times.
- Some drugs induce at least a temporary depression of appetite.
- Their lifestyle often lacks the regularity and routine that promote good eating habits.
- They may contract hepatitis, a liver disease common in drug abusers, which causes taste changes and loss of appetite.
- Their nutrient status may be altered by treatments and medicines.
- They often become ill with infectious diseases, which increase their need for nutrients.

During withdrawal from drugs, an important aspect of treatment is the identification and correction of nutrition problems.

Alcohol

drink: a dose of any alcoholic beverage that delivers ½ oz of pure ethanol:

- 3 to 4 oz wine.
- 8 to 12 oz beer.
- 1 oz hard liquor (whiskey, scotch, rum, or vodka).

At some point during adolescence, teenagers face a critical choice: to drink or not to drink alcohol. Even though the law forbids sale of alcohol to people under a specific age, alcohol is still available to many teenagers who seek it. Over 90 percent of high school seniors decide to try a drink; 1 in 20 reports daily consumption of alcohol.[16] The motivating factors are numerous. A person observing adolescents might glean the following reasons for their drinking alcohol:

- To be popular.
- To be grown-up.
- To defy parents.
- To drown feelings of inadequacy.

For some adults, drinking an alcoholic beverage is a custom that accompanies social relations. It provides them pleasure without problems. Many teenagers find that alcohol and marijuana serve similar purposes, and the pattern of substance use indicates parallel consumption, not a displacement of one by the other.[17] Some teenagers use alcohol as an escape or for support—an ineffective way to cope with problems that leads to greater problems. Dependency on alcohol or any drug has major adverse effects on the growth and development of adolescents and deserves attention, but is beyond the scope of this text.

People use alcohol to help them relax or to relieve anxiety. They think that alcohol is a stimulant because it seems to make them lively and uninhibited at first. Actually, though, the way it does this is by sedating inhibitory nerves, which are more numerous than excitatory nerves. Ultimately, alcohol acts as a depressant, and sedates all the nerve cells.

Alcohol is an empty-kcalorie beverage and can displace needed nutrients from the diet while simultaneously altering metabolism so that even good nutrition cannot normalize it. A discussion of alcohol absorption and metabolism provides a basis for understanding the effects alcohol has on a person's body and nutrition status.[18]

ethanol: the alcohol in beer, wine, and hard liquor.

To the chemist, *alcohol* refers to a class of compounds containing reactive hydroxyl (OH) groups. To most other people, *alcohol* refers to the intoxicating ingredient in beer, wine, and hard liquor (distilled spirits). The chemist's name for this particular alcohol is *ethyl alcohol*, or *ethanol*.

The alcohols affect living things profoundly, partly because they act as lipid solvents. They can dissolve the lipids out of cell membranes, destroying the cell structure and thereby killing the cells. For this reason, most alcohols are toxic.

euphoria (you-FORE-ee-uh): a feeling of great well-being, which people often seek through the use of drugs such as alcohol.
eu = good
phoria = bearing

Like the other alcohols, ethanol is toxic—but less so than some. Sufficiently diluted and taken in small enough doses, it produces euphoria—not without risk, but with a risk that some find tolerable (if the doses are low enough). Used to achieve these effects, alcohol is a drug—that is, a substance that can modify one or more of the body's functions.

From the moment ethanol enters the body in a beverage, it is treated as if it has special privileges. Unlike foods, the tiny ethanol molecules need no digestion; they can quickly diffuse through the walls of an empty stomach and reach the brain within a minute, creating a feeling of euphoria almost immediately. When the stomach is full of food, the ethanol molecules have less chance of touching the walls and diffusing through, so a person does not feel the effects so quickly. Therefore, to slow ethanol absorption, a person should eat something, preferably carbohydrate snacks. High-fat snacks help, too, because they slow peristalsis.[19] The presence of food is less influential when stomach contents empty into the duodenum; intestinal absorption of alcohol is rapid, "as if it were a V.I.P. (Very Important Person)."[20]

The capillaries that surround the digestive tract merge into the veins that carry the alcohol-laden blood to the liver. Here the veins branch and rebranch into capillaries that touch every liver cell. Liver cells are the only cells in the body that can make enough alcohol dehydrogenase to oxidize ethanol at an appreciable rate. The rate of clearance is determined by the number of alcohol dehydrogenase enzyme molecules that reside in the liver. If more molecules of ethanol arrive at the liver cells than the enzymes can handle, the extra ethanol must wait. It enters the general circulation and is carried past the liver to all parts of the body, circulating again and again through the liver until enzymes are available to convert it to acetaldehyde. Prudent drinkers drink slowly, with food in their stomachs, to allow the ethanol molecules to move to the liver cells gradually enough for the enzymes to handle the load. Spacing of drinks is important, too. It takes about an hour and a half to metabolize one drink, depending on the person's body size, previous drinking experience, recent food consumption, and general health. Figure FP8–1 diagrams the primary metabolic pathway of ethanol.

alcohol dehydrogenase: a liver enzyme that converts ethanol to **acetaldehyde** (ass-et-AL-duh-hide).

That each person has a particular concentration of alcohol dehydrogenase that limits the rate of ethanol clearance explains why only time will restore sobriety. Walking will not; it makes the muscles work, but since they cannot metabolize ethanol, they do not help clear it from the blood. Drinking a cup of coffee is of no use either; caffeine is a stimulant, but it does not speed up the metabolism of ethanol.

Careful study of Figure FP8–1 reveals that ethanol metabolism uses the niacin coenzyme NAD^+, creating an accumulation of $NADH + H^+$. This consequence alters the body's "redox state," because NAD^+ can oxidize, and $NADH + H^+$ can reduce, many other body compounds. During ethanol metabolism, NAD^+ becomes unavailable for the multitude of reactions for which it is required. Consider that the metabolism of glucose, fatty acids, and amino acids all require NAD^+. For these nutrients to be completely metabolized to energy, the TCA cycle must be operating, and this also requires NAD^+. Without NAD^+, the metabolic pathways are blocked, causing metabolites to accumulate or take alternate routes. Such changes in the normal metabolism of nutrients produce altered biochemistry.

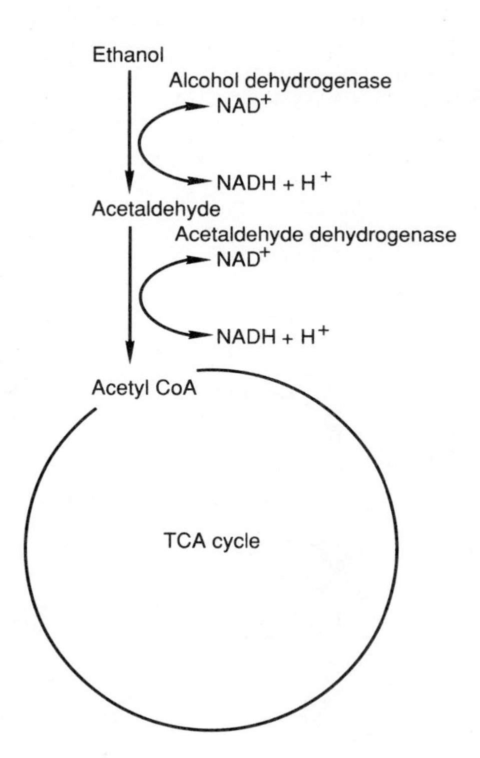

Figure FP8–1 Metabolism of Ethanol to Acetyl CoA

Alcohol dehydrogenase oxidizes ethanol to acetaldehyde within the liver. Simultaneously, this reaction reduces a molecule of the niacin coenzyme NAD^+ to $NADH + H^+$. A related enzyme, acetaldehyde dehydrogenase, reduces another NAD^+ to $NADH + H^+$, while it oxidizes acetaldehyde to acetyl CoA, the compound that enters the TCA cycle to generate energy. (All cells possess acetaldehyde dehydrogenase, so this step can take place outside the liver.)

Wherever NAD^+ is converted to $NADH + H^+$ in ethanol metabolism, hydrogen ions accumulate, resulting in a dangerous shift of the acid-base balance toward acid. The accumulation of $NADH + H^+$ depresses the TCA cycle so that pyruvate and acetyl CoA build up. The excess acetyl CoA then takes the route to the synthesis of fatty acids, and fat clogs the liver so it cannot function.[21]

The synthesis of fatty acids also accelerates as a result of the liver's exposure to ethanol. Fat accumulation can be seen in the liver after a single night of heavy drinking. Fatty liver, the first stage of liver deterioration seen in

fatty liver: an early stage of liver deterioration seen in several diseases, including kwashiorkor and alcoholic liver disease. Fatty liver is characterized by accumulation of fat in the liver cells.

fibrosis: an intermediate stage of liver deterioration seen in several diseases, including viral hepatitis and alcoholic liver disease. In fibrosis, the liver cells lose their function and assume the characteristics of connective tissue cells (fibers).

cirrhosis (seer-OH-sis): advanced liver disease, in which liver cells have died, hardened, and turned orange; often associated with alcoholism.
cirrhos = an orange

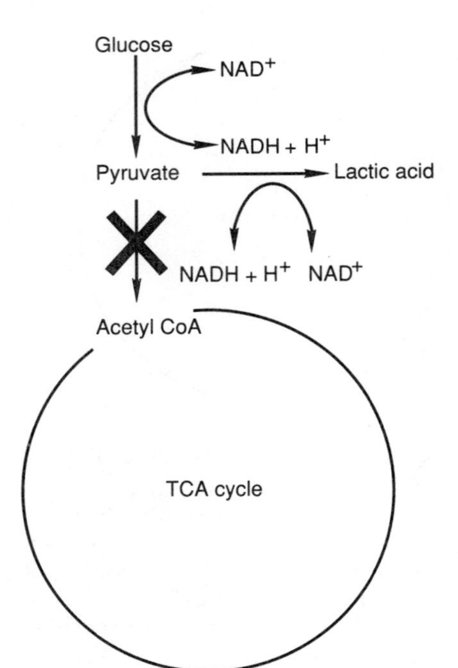

Figure FP8–2 Blocked Metabolism of Pyruvate to Acetyl CoA
Pyruvate is converted to lactic acid if the pathway to acetyl CoA is blocked.

heavy drinkers, interferes with the distribution of nutrients and oxygen to the liver cells. If the condition lasts long enough, the liver cells die, and fibrous scar tissue invades the area—the second stage of liver deterioration, called fibrosis. Fibrosis is reversible with good nutrition and abstinence from alcohol, but the next (last) stage—cirrhosis—is not.

The body's altered redox state inhibits gluconeogenesis and can lead to hypoglycemia if glycogen stores are not repleted.[22] Limited glucose combined with the overabundance of acetyl CoA sets the stage for a shift into ketosis. The making of ketone bodies consumes acetyl CoA, but some ketone bodies are acids, so they push the acid-base balance further toward acid.

Figure FP8–2 illustrates the conversion of pyruvate to lactic acid when the path to acetyl CoA is blocked. The surplus of NADH + H^+ also favors the conversion of pyruvate to lactic acid, which serves as a temporary storage place for hydrogens from NADH + H^+. The conversion of pyruvate to lactic acid restores some NAD^+, but a lactic acid buildup adds still further to the body's acid burden.

Liver metabolism clears most of the ethanol from the blood. However, about 10 percent is excreted through the breath and in the urine. This is the basis for the breath-analyzing test for drunkenness administered by the police. The amount of alcohol in the breath is in proportion to that in the bloodstream. In most states, legal drunkenness is set at 0.10 percent or lower. Table FP8–1 shows the blood-alcohol levels that correspond with progressively greater intoxication.

Figure FP8–3 illustrates alcohol's effects on the brain. Brain cells are particularly sensitive to excessive exposure to alcohol. Like liver cells, they die; however, unlike liver cells, brain cells cannot regenerate. This is one reason for the permanent brain damage observed in some heavy drinkers.

It is lucky that the brain centers respond to ethanol in the order described in Figure FP8–3, because an individual passes out before drinking enough to reach a lethal dose. It is possible, though, to drink fast enough that the effects continue to accelerate after one has gone to sleep. The occasional death that takes place during a drinking contest is attributed to this effect. The drinker drinks fast enough, before passing out, to receive a lethal dose.

Ethanol interferes with a multitude of chemical and hormonal reactions in the body. It depresses production of antidiuretic hormone (ADH) by the pituitary gland in the brain. Loss of body water leads to thirst, and thirst leads

Table FP8–1 Alcohol Doses and Brain Responses

Number of Drinks	Blood Alcohol (%)	Effect on Brain
2	0.05	Impaired judgment
4	0.10	Impaired control
6	0.15	Impaired muscle coordination and reflexes
8	0.20	Impaired vision
12	0.30	Drunk, out of control
14 or more	0.50 to 0.60	Amnesia, finally death

to more drinking. The only fluid that will relieve dehydration is water, but the thirsty drinker may choose another alcoholic beverage instead. A person who tries to use alcoholic beverages to quench thirst, however, only worsens the problem. The smart drinker, then, either drinks beer (which contains plenty of water), or drinks mixers or chasers with wine or hard liquor. Better still, the drinker alternates alcoholic beverages with glasses of water and limits the total amount consumed.

antidiuretic hormone (ADH): a hormone produced by the pituitary gland in response to dehydration (or a high sodium concentration in the blood); stimulates the kidneys to reabsorb more water and so excrete less. This ADH should not be confused with the enzyme alcohol dehydrogenase, which is also abbreviated ADH.

The water loss caused by ADH depression involves loss of more than just water. The water takes with it minerals such as magnesium, potassium, calcium, and zinc, depleting the body's reserves.

Ethanol also depresses appetite by the euphoria it produces, as well as by its attack on the mucosa of the stomach, so that heavy drinkers usually eat poorly, if at all. With a large portion of their energy fuel coming from the empty kcalories of alcohol, they find it difficult to obtain the essential nutrients. Thus some of their malnutrition is due to lack of food—but even if they eat well, the direct effects of alcohol will take their toll, because ethanol affects every tissue's metabolism of nutrients. Stomach cells oversecrete histamine and acid, becoming vulnerable to ulcer formation. Intestinal cells fail to absorb thiamin, folate, and vitamin B_{12}. Liver cells lose efficiency in activating vitamin D, and alter their production and excretion of bile. Rod cells in the retina, which normally process retinol (the vitamin A alcohol) to retinal (its aldehyde form) needed in vision, process ethanol to acetaldehyde instead. The kidneys excrete increased quantities of magnesium, calcium, potassium, and zinc.

Acetaldehyde interferes with nutrient metabolism, too. For example, it dislodges vitamin B_6 from its protective binding protein so that it is destroyed, creating a secondary vitamin B_6 deficiency and, thereby, lowered production of red blood cells.

In summary, ethanol hinders the absorption, alters the metabolism, and increases the excretion of many nutrients, so that malnutrition can occur even in the well-fed drinker. Ethanol disturbs many normal body processes—many more than have been enumerated here. Since the liver is a crossroads for all nutrients in the body, its domination by ethanol or injury results in many side effects. Table FP8–2 lists some long-term effects of alcohol abuse.

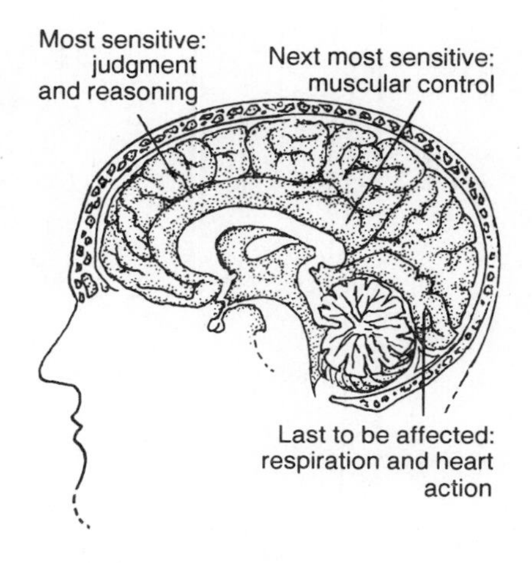

Figure FP8–3 Alcohol's Effects on the Brain

When ethanol flows to the brain, it first sedates the frontal lobe, the reasoning part. As the ethanol molecules diffuse into the cells of this lobe, they interfere with reasoning and judgment. If the drinker drinks faster than the rate at which the liver can oxidize the ethanol, then the speech and vision centers of the brain become sedated, and the area that governs reasoning becomes more incapacitated. Later, the cells of the brain responsible for large-muscle control are affected; at this point, people "under the influence" stagger or weave when they try to walk. Finally, the conscious brain is completely subdued, and the person "passes out." Now, luckily, the person can drink no more; this is fortunate, because a higher dose's anesthetic effect could reach the deepest brain centers, which control breathing and heartbeat, and the person could die.

Tobacco

Young people searching for role models may be taken in by the highly persuasive advertisements distributed by the tobacco industry. In their efforts to achieve sophistication, young people imitate these attractive advertising models, adopting the habit of smoking tobacco. A person whose mind is open to using tobacco begins by trying it once. That one time is followed by another, and in a short while, because nicotine is a powerfully addictive drug, the person becomes a user for life.[23]

Statistics on smoking are impressive. Each day, 5000 children light up for the first time—some of them only seven or eight years old, in a hurry to grow up. In 1968, 3 million teenagers were smoking; by 1978, that number had more than doubled.[24] Cigarette smoking among high school seniors peaked in

Table FP8–2 Long-Term Effects of Alcohol Abuse

Hepatitis and cirrhosis.
Vitamin and trace mineral deficiencies.
Brain damage.
Psychological depression.
Loss of testicular function and damage to the adrenal glands, leading to feminization and sexual impotence in men.
Failure of the ovaries and early menopause in women.
Hypertension and an increased risk of stroke.
Increased risk of cancer of the tongue, mouth, esophagus, and liver.
Intestinal inflammation; ulcers.
Deterioration of muscles, including the heart muscle.
Sedation of the bone marrow, with consequent blood abnormalities.
Reduced capacity for exercise; heart pain sooner with exercise.
Suppression of the immune system; reduced resistance to disease.
Kidney damage; bladder damage; prostate gland damage.
Failure to maintain the skin's health; rashes and sores.
Increased susceptibility to lung infections.
Adverse drug reactions.

Source: Adapted from M. J. Eckardt and coauthors, Health hazards associated with alcohol consumption, *Journal of the American Medical Association* 246 (1981): 648–666.

the mid-1970s, declined until 1980, and seems to have leveled out. Surveys of high school seniors report that seven of ten have tried cigarettes, and one in five smokes cigarettes regularly.[25] This shift in smoking behavior reflects society's reassessment of a previously accepted normative behavior. A folder written by young people for young people describes how smokers rationalize their choice to smoke:

- I'm young now. Why not smoke? I can quit later.
- I don't inhale. Smoking can't hurt me.
- Smoking makes me look grown-up and mature.
- I smoke filter cigarettes; that will protect me.
- My parents smoke. Why shouldn't I?
- If I don't spend the money on cigarettes, I'll spend it on something else.
- All my friends smoke. Why shouldn't I?
- It keeps me from biting my nails.
- It keeps me from eating. Smoking is better than putting on weight.
- It gives me something to do when I'm mad or bored or hurt or unhappy or restless.[26]

Each of these reasons may be invalid, but the new smoker believes them.

Cigarette smoking is a pervasive health problem causing thousands of people to suffer from cancer and diseases of the cardiovascular, digestive, and respiratory systems. These effects are beyond the scope of this text, but there are a few nutrition connections to be explored. Smoking cigarettes does influence hunger, body weight, and nutrient status. There are also links between nutrients and lung cancer.

Smoking a cigarette eases feelings of hunger. Nicotine inhibits the hunger contractions of the stomach and causes a temporary, but rapid, rise in blood glucose concentrations.[27] So when smokers receive a hunger signal, they can quiet it with a cigarette instead of food. Such behavior ignores body signals and deters energy and nutrient intake.

Indeed, smokers tend to weigh less than nonsmokers and to gain weight upon cessation of smoking.[28] This phenomenon is not easily explained. Common belief held that smokers weighed less because they ate less—that cigarette smoking affected their eating behaviors. When they quit smoking, they began to eat more, and therefore gained weight. However, studies have indicated that smokers actually consume at least as many kcalories per day as nonsmokers.[29] Some smokers show only a slight increase in their energy intakes after cessation, while others increase their energy intakes by more than 200 kcalories per day.[30] However, increased energy intake does not fully account for the weight gains seen upon cessation.[31]

The lower body weight and subsequent weight gain upon cessation of smoking appear to be due to effects of cigarette smoking beyond those on food intake. One study suggests that smoking and nicotine lower the efficiency of energy storage or increase the metabolic rate.[32] Researchers have found that smoking increases daily energy expenditure by approximately 10 percent, even though changes in physical activity or basal metabolic rate were not noted.[33] Any increase noted in metabolic rate does not appear to be dose related. If it were, heavy smokers would be the lightest in weight of all smokers. Such is not the case; moderate smokers (15 to 24 cigarettes per day) are the lightest.[34] The effects of cigarette smoking on metabolic rate are not clear, nor are they always evident.[35]

Another possible metabolic explanation for the body weight effects of smoking involves the enzyme lipoprotein lipase. Lipoprotein lipase is the enzyme on the fat cells that is responsible for hydrolyzing triglycerides from blood-borne lipoproteins, making fatty acids available for fat storage. Smokers have higher lipoprotein lipase activity than nonsmokers. When researchers measured the activity of this enzyme in smokers before and after they quit smoking, they found a striking correlation.[36] The higher the enzyme activity when smoking, the greater the weight gain upon cessation.

Weight gain is often a concern for people contemplating giving up cigarettes. The decision to quit weighs unhealthy smoking against unattractive (and potentially unhealthy) weight gain. The message to smokers wanting to quit is to adjust diet and exercise habits in order to maintain weight during and after cessation.

Smokeless or chewing tobacco is gaining popularity in this country, especially among young people. In addition to the cancer problems such a practice presents, these tobacco products contain large quantities of sodium. Sodium is added to the tobacco for flavor and may be found at levels comparable to that found in dill pickles or cured bacon.

Nutrient intakes of smokers and nonsmokers differ. Smokers have been found to have lower intakes of dietary fiber, vitamins, and minerals, even when their energy intakes are quite similar.[37] The association between smoking and low vitamin intake may be noteworthy, considering the altered metabolism of vitamin C in smokers and the protective effect of beta-carotene against lung cancer.

The vitamin C requirement of smokers exceeds that of nonsmokers. The plasma concentration of vitamin C in smokers is commonly low, and the metabolic turnover of vitamin C in smokers is higher than in nonsmokers; that is, smokers break down vitamin C faster, thus requiring more vitamin C to achieve steady body pools comparable to those of nonsmokers.[38] The RDA for regular smokers is 100 milligrams of vitamin C per day, 40 milligrams per day higher than for nonsmokers.

Recent research findings suggest that beta-carotene, a precursor to vitamin A found in vegetables, has anticancer activity.[39] Specifically noted is an inverse correlation between dietary carotene and the incidence of lung cancer. That is, the risk of lung cancer is greatest for smokers who have the lowest intake of carotene. Of course, conclusions from such evidence cannot be made in haste. Teenagers cannot be led to believe that as long as they eat their carrots they can safely smoke their cigarettes. However, it is important to encourage teenagers to eat foods rich in carotene, a nutrient many of them lack in their diets.

The teen years are similar to an obstacle course, both requiring careful consideration of the pathways and the consequences that follow each decision. To make the best choice, adolescents need to learn of the associated risks and benefits of various behaviors. Perhaps the most effective way to educate teens is by example. The teen years are a time of identity formation, a time of seeking and emulating models. Adults who enthusiastically maintain their own health can have a positive impact on teenagers. Teenagers who realize that a life of health and happiness begins today are on their way. The rapid changes of the teen years offer a flexibility and opportunity to make positive choices that will improve the rest of life's journey.

Focal Point 8 Notes

1. Data presented in this discussion are from the national surveys of roughly 17,000 high school seniors entitled Monitoring the Future: A Continuing Study of the Lifestyles and Values of Youth, funded by the National Institute on Drug Abuse, conducted every spring since 1975, as cited by L. D. Johnston, P. M. O'Malley, and J. G. Bachman, Psychotherapeutic, licit, and illicit use of drugs among adolescents, *Journal of Adolescent Health Care* 8 (1987): 36–51.
2. Johnston, O'Malley, and Bachman, 1987.
3. L. J. King, J. D. Teale, and V. Marks, Biochemical aspects of cannabis, in *Cannabis and Health*, ed. J. D. Graham (New York: Academic Press, 1976).
4. L. E. Hollister, Marihuana in man: Three years later, *Science* 172 (1971): 21–29.
5. King, Teale, and Marks, 1976; Hollister, 1971.
6. N. C. Doyle, *Marihuana and the Lungs,* a bulletin (November 1979) distributed by the American Lung Association.
7. E. L. Abel, Effects of marihuana on the solution of anagrams, memory and appetite, *Nature* 231 (1971): 260–261; C. T. Tart, Marihuana intoxication: Common experiences, *Nature* 226 (1970): 701–704.
8. L. E. Hollister, Hunger and appetite after single doses of marihuana, alcohol, and dextroamphetamine, *Clinical Pharmacology and Therapeutics* 12 (1971): 44–49; J. D. P. Graham and D. M. F. Li, The pharmacology of cannabis and cannabinoids, in *Cannabis and Health*, ed. J. D. Graham (New York: Academic Press, 1976).
9. Hollister, Hunger and appetite, 1971.
10. L. E. Hollister, Health aspects of cannabis, *Pharmacological Reviews* 38 (1986): 1–20.
11. Marihuana: Truth on health problems, *Science News* 22 (1975): 117.
12. Johnston, O'Malley, and Bachman, 1987.
13. Johnston, O'Malley, and Bachman, 1987.
14. Johnston, O'Malley, and Bachman, 1987.
15. J. M. Jonas and M. S. Gold, Cocaine abuse and eating disorders, *Lancet* 1 (1986): 390–391.
16. *Alcohol and the Adolescent,* a pamphlet available from National Council on Alcoholism, 733 Third Ave., New York, NY 10017; B. Bower, Teen drug use: Ups and downs, *Science News* 128 (1985): 310; Johnston, O'Malley, and Bachman, 1987.
17. Johnston, O'Malley, and Bachman, 1987.
18. Parts of this discussion are adapted from E. N. Whitney, E. M. N. Hamilton, and S. R. Rolfes, *Understanding Nutrition,* 5th ed. (St. Paul, Minn.: West, 1990), pp. 186–192.
19. A. B. Eisenstein, Nutritional and metabolic effects of alcohol, *Journal of the American Dietetic Association* 81 (1982): 247–251.
20. F. Iber, In alcoholism, the liver sets the pace, *Nutrition Today,* January–February 1971, pp. 2–9.
21. C. S. Lieber, Liver adaptation and injury in alcoholism, *New England Journal of Medicine* 228 (1973): 356–361.
22. H. L. Bleich and E. S. Boro, Metabolic and hepatic effects of alcohol, *New England Journal of Medicine* 296 (1977): 612–616.
23. Parts of this discussion are adapted from F. S. Sizer and E. N. Whitney, *Life Choices: Health Concepts and Strategies* (St. Paul, Minn.: West, 1988).

24. R. Keeshan, Children and smoking, in *Smoking and Health,* proceedings of a conference commemorating the 20th anniversary of the first Surgeon General's Report on Smoking and Health, 11 January 1984, available from the American Council on Science and Health, 47 Maple St., Summit, NJ 07901.
25. Johnston, O'Malley, and Bachman, 1987.
26. The bulleted items are adapted from *8 Reasons Young People Smoke,* HHS publication no. 200-75-0516, (Washington, D.C.: Government Printing Office, 1975).
27. L. Willian-Olsson, Smoking and platelet stickiness (letter), *Lancet* 2 (1965): 908–909.
28. R. M. Carney and A. P. Goldberg, Weight gain after cessation of cigarette smoking: A possible role for adipose-tissue lipoprotein lipase, *New England Journal of Medicine* 310 (1984): 614–616.
29. D. R. Jacobs and S. Gottenborg, Smoking and weight: The Minnesota lipid research clinic, *American Journal of Public Health* 71 (1981): 391–396.
30. A. M. Fehily, K. M. Phillips, and J. W. G. Yarnell, Diet, smoking, social class, and body mass index in the Caerphilly Heart Disease Study, *American Journal of Clinical Nutrition* 40 (1984): 827–833; B. A. Stamford and coauthors, Effects of smoking cessation on weight gain, metabolic rate, caloric consumption, and blood lipids, *American Journal of Clinical Nutrition* 43 (1986): 486–494.
31. Stamford and coauthors, 1986.
32. J. T. Wack and J. Rodin, Smoking and its effect on body weight and the systems of caloric regulation, *American Journal of Clinical Nutrition* 35 (1982): 366–380.
33. A. Hofstetter and coauthors, Increased 24-hour energy expenditure in cigarette smokers, *New England Journal of Medicine* 314 (1986): 79–82.
34. Fehily, Phillips, and Yarnell, 1984.
35. Stamford and coauthors, 1986.
36. Carney and Goldberg, 1984.
37. Fehily, Phillips, and Yarnell, 1984.
38. A. B. Kallner, D. Hartmann, and D. H. Hornig, On the requirements of ascorbic acid in man: Steady-state turnover and body pool in smokers, *American Journal of Clinical Nutrition* 34 (1981): 1347–1355.
39. Dietary carotene and the risk of lung cancer, *Nutrition Reviews* 40 (1982): 265–268.

Adulthood: Health Promotion and Disease Prevention

9

Singing Man by Ernst Barlach.

Beyond the childbearing years, a person's primary nutrition-related responsibility is to self. The time has passed when one's nutrition choices will physically affect the next generation. The quality of the adult's own life now is paramount, and there is much that he or she can do to enhance it. The right choices, made throughout adulthood, can support the person's ability to meet physical, emotional, and mental challenges, and to achieve freedom from disease. Three goals inspire adults to take responsibility for their nutritional health: prevention of disease, promotion of overall wellness, and slowing of aging. This chapter is dedicated to the prevention of disease and promotion of wellness; the next chapter focuses on the aging process.

The Concept of Prevention

The idea that nutrition can help promote health in adult human beings is of recent origin. Nutrition itself is a young science—the first vitamin was not discovered until around 1900—and its primary focus at first was on the roles of vitamins and minerals, and on the identification and prevention of deficiency diseases. The first food program was designed during World War I to prevent nutrient deficiencies among the members of the armed forces. By 1960, it was recognized that malnutrition could impair children's growth and health, and the School Lunch Program was in force. In the 1970s, the health of ordinary, middle-class adults became of interest.

Prevention in the Limelight

During the 1970s, the focus of the nation's interest in nutrition changed from those who were undernourished to those who were overnourished. It was becoming apparent that the primary killers of adults were not the infectious diseases of an earlier time (tuberculosis, smallpox, and the like) but degenerative diseases—particularly cardiovascular disease and cancer. Overnutrition was perceived to have something to do with those diseases, and preventive nutrition measures to defend against them came under consideration. Evidence on these subjects was marshaled and presented before several Senate hearings during the 1970s, and the result was a spate of documents beginning with the *Dietary Goals for the United States* in 1977.

During the 1980s, another focus developed with respect to nutrition. The immune system had begun to reveal its secrets, and its importance to disease resistance was becoming apparent. The interest in immune-system research was sparked by concerns about allergy, the need for protection against infectious diseases, and the relations of the immune system to human health in general. Immune resistance is affected by stress, by substance abuse, by radiation, by environmental contamination—so it both affects, and is affected by, lifestyle factors. A strong immune system can help protect people not only against infectious disease but also against cancer, a so-called degenerative disease. Nutrition plays a major role in supporting the immune system, and research into nutrition and immunity has yielded much information on both the prevention and the treatment of disease. A number of unreliable books have popularized the subject of diet's effects on immunity, but their unreliability does not invalidate the general premise that immunity and nutrition work hand in hand to protect people's health.

Then in the 1980s, AIDS struck, providing further impetus to research into immunity. So deadly is AIDS that the Centers for Disease Control predict that it will soon be one of the leading causes of death in the United States, surpassing all other causes of death for those between the ages of 25 and 44. Severe malnutrition and wasting are common findings in people with AIDS, due largely to poor food intake, increased nutrient requirements, and GI malabsorption.[1] Malnutrition caused by AIDS further impairs immune function and sets in motion a vicious cycle of recurrent infections and, eventually, death.[2] For people with AIDS, adequate nutrition can enhance strength, provide comfort, and generally improve functioning.[3] Research into the links between nutrition and immunity thus can benefit not only those who wish to prevent diseases but also those who have them. But to return to prevention: nutrition-aware people focus on prevention of both infectious and degenerative diseases.

The *U.S. Dietary Goals,* published in 1977, was the first major government document advancing the point of view that it would be prudent to reduce our diets' fat and food energy contents in hopes of defending against the diseases of overnutrition, even though the evidence in favor of such a move was not conclusive. In presenting the goals, D. M. Hegsted, a member of the Senate Select Committee on Nutrition and Human Needs at the time, gave a speech that has been much-quoted since then:[4]

> This diet which affluent people generally consume is everywhere associated with a similar disease pattern—high rates of heart disease, certain forms of cancer, diabetes, and obesity. These are the major causes of death and disability in the United States. . . . It is not correct, strictly speaking, to say that they are caused by malnutrition but rather that an inappropriate diet contributes to their causation. . . . Not all people are equally susceptible. Yet those who are genetically susceptible, most of us, are those who would profit most from an appropriate diet. . . .
>
> The diet we eat today was not planned or developed for any particular purpose. It is a happenstance related to our affluence, the productivity of our farmers and the activities of our food industry. The risks associated with eating this diet are demonstrably large. The question to be asked, therefore, is not why should we change our diet but why not? What are the risks . . . ? There are none that can be identified and important benefits can be expected.

Since then, many different health agencies have approached the importance to health of adult nutrition, and many additional sets of recommendations have been made. Table 1–3 in Chapter 1 compared several of them.

In 1979, the surgeon general issued his first *Healthy People: Report on Health Promotion and Disease Prevention.* The report reviewed a history of dramatic improvements in the health of the nation's people, reviewed present preventable threats to health, and identified priorities for further efforts in five areas—health of infants, children, adolescents and young adults, adults, and older adults. The overall objective under the heading of "Adults" was "by 1990, to reduce deaths among people ages 25 to 64 by at least 25 percent, to fewer than 400 per 100,000."[5] In 1980, the Department of Health and Human Services developed a set of specific objectives under 15 headings ranging from "High Blood Pressure Control" to "Behavior," in a document entitled *Promot-*

ing Health, Preventing Disease: Objectives for the Nation. The Public Health Service has been working on promoting these goals ever since. A distillation of the nation's most important nutrition goals for disease prevention appeared in *Promoting Health,* and that distillation serves as the starting point for this chapter, which addresses the nutrition efforts adults can make to promote and prolong their health into the later years.

The Dietary Recommendations of the NRC Report, presented in Table 1–4 of Chapter 1, spell out specific dietary measures for promoting health and preventing cardiovascular disease. The authors of the NRC report are nutrition experts. The authors of *Promoting Health* are medical experts, and while the two reports are similar in some respects, they differ in perspective.

In *Promoting Health,* the U.S. Public Health Service not only identified the major health concerns of our citizens but also singled out strategies to make significant headway against them by 1990. Of the health concerns identified by the U.S. Public Health Service, these have known relationships to human nutrition:

- Infectious diseases, because nutrition supports the immune system, and nutritional deficiencies impair it.
- Atherosclerosis, because nutrition supports a healthy cardiovascular system, and many dietary factors affect the disease process.
- Hypertension, for the same reasons.
- Diabetes, because nutritional factors (including obesity) seem to predispose people to it, because diet is important in its management, and because diabetes is itself a risk factor for cardiovascular disease.
- Ulcer and stress-related disease, because adequate nutrition helps protect against the effects of stress and support the body during times of stress.
- Cancer, because many links with nutrition are known, for both its prevention and treatment.
- Osteoporosis, because nutrition contributes to both its prevention and its management.
- Substance abuse, because it causes nutrient deficiencies that impair health.
- Dental caries, because the frequent consumption of cariogenic foods nullifies preventive efforts.

The following sections of this chapter will discuss all of these health concerns except for dental caries (which was covered in Focal Point 5) and stress and substance abuse (which are in the next chapter and Focal Point 8, respectively). Diet recommendations are more important for some people than for others, for some people are genetically predisposed to certain diseases. People who examine their family histories and note which tests are out of line on their physical examinations can begin to decide which diet recommendations are most important. Table 9–1 presents a summary of the signs to watch for in both categories.

Obesity a Central Issue

The concern of obesity pervades many of the others. Obesity increases risk of adult-onset diabetes and high blood pressure, as well as gallbladder disease,

Table 9–1 Priorities for People with Personal, Genetic, and Lifestyle Risk Factors

Lifestyle Changes Recommended for All People	Especially Important If Your Family History Indicates:	And/Or If Your Medical History Indicates:
Fats and cholesterol: reduce consumption of fat (especially saturated fat) and cholesterol.	Diabetes, obesity, cancer, or any form of cardiovascular disease (atherosclerosis, hypertension, heart attacks, strokes)	Glucose intolerance, high blood cholesterol or triglycerides, or hypertension
Energy and weight control: achieve and maintain a desirable body weight.	Diabetes, obesity, cancer, or any form of cardiovascular disease (atherosclerosis, hypertension, heart attacks, strokes)	
Carbohydrate: increase consumption of complex carbohydrates and fiber.	Diabetes, obesity, cancer, or any form of cardiovascular disease (atherosclerosis, hypertension, heart attacks, strokes)	Glucose intolerance, high blood cholesterol or triglycerides, or hypertension
Salt/sodium: reduce intake of salt/sodium.	Hypertension, diabetes, or any form of cardiovascular disease (atherosclerosis, hypertension, heart attacks, strokes)	Hypertension
Alcohol: to reduce the risk for chronic disease, take alcohol only in moderation (no more than two drinks a day), if at all.	Liver disease (cirrhosis), cancer, any form of cardiovascular disease (atherosclerosis, hypertension, heart attacks, strokes), or osteoporosis	Gluscose intolerance, high blood cholesterol or triglycerides, hypertension, or any sign of adult bone loss
Other Issues for Some People	**Especially Important If Your Family History Indicates:**	**Recommended Diet Especially For:**
Fluoride: community water systems should contain fluoride at optimal levels for prevention of tooth decay; if such water is not available, use other appropriate sources of fluoride.	Osteoporosis or dental problems	Children and women
Sugars: limit the consumption and frequency of use of foods high in sugars.	Susceptibility to dental caries	Children
Calcium: increase consumption of foods high in calcium, including low-fat dairy products.	Osteoporosis or cancer	Children and women
Iron: be sure to consume good sources of iron.	Iron-deficiency anemia	Low-income families, and children, teens, girls, and women

Source: *Surgeon General's Report on Nutrition and Health: Summary and Recommendations,* DHHS (PHS) publication no. 88-50211 (Washington, D. C.: Government Printing Office, 1988), Table 1, p. 3.

joint diseases, and some types of cancer—and the extent of obesity in this nation is growing steadily greater. (Some subsets of the population are more prone to obesity than others—women, especially black women, for example). One of the foremost objectives of the Public Health Service was to take whatever measures would help to reduce its incidence. In terms of education, a goal was to "increase weight consciousness." In terms of service, a goal was to offer "support groups for weight control and maintenance." In terms of technology, a goal was to support the development of foods conducive to weight control, such as lean meats and products low in fat, and to label them informatively. Another goal was to provide incentives to corporations offering health promotion with a nutrition emphasis—for example, adjusting insurance premiums in favor of people who control their weight successfully. The U.S. Public Health Service stated that compared with 1980, by 1990, significantly more people should know what foods are low in fat, low in sodium, low in kcalories, and good fiber sources, and should know how to lose weight by cutting kcalories, increasing activity, or both.

This book's Focal Point 7 is on the prevention of obesity in children, and Chapter 8 offers information on fitness for teens and adults—the other half of weight control. But weight control remains important throughout the adult years, and its relevance will be apparent throughout the sections that follow.

Nutrition, the Immune System, and Infectious Disease

Focal Point 6 summarized the action of the immune system and defined related terms. It showed how malnutrition weakens the system and makes children susceptible to infection, and it emphasized the importance of fever in the defense against it. It also presented a table of nutrients important to immunity, and made clear that water was the most important nutrient of all.

In adulthood, nutrition continues to provide major support to immunity. People who suffer from malnutrition become more ill from infections than people who are well nourished, and they may become more prone to infections as well. A vicious cycle develops in which malnutrition aggravates infection, and infection further aggravates malnutrition; it has been shown that nutrition intervention can improve resistance to infection.[6] Researchers continue to uncover bits and pieces of information that are helping to form a more complete picture of the nutrition-immunology interaction. Table 9–2 summarizes some of this information.

The first barriers a foreign material encounters when trying to enter the body are the skin and the mucous membranes. Malnutrition can cause changes in these tissues that compromise their function. The skin becomes thinner, with

Table 9–2 Effects of Malnutrition on the Immune System

Immune System Component	Effects of Malnutrition
Skin	Thinner, with less connective tissue
Mucous membranes	Microvilli flattened; antibody secretions reduced
Lymph tissues	Thymus gland, lymph nodes, and spleen smaller; T-cell areas depleted of lymphocytes
Phagocytosis	Kill time delayed
Cell-mediated immunity	Circulating T-cells reduced
Humoral immunity	Circulating immunoglobulin levels normal; antibody response may be impaired

less connective tissue. The microvilli of the mucous membranes become flattened. The mucosa of the GI tract allow entry of antigens through the intestine that would normally be barred. One type of antibody present in mucous membrane secretions (including those of the lungs and GI tract) is depressed in malnutrition. Phagocytosis is altered by malnutrition.[7] For these and other reasons, malnourished people have repeated lung and GI tract infections.

Cell-mediated immunity, which depends on the white cells known as T-lymphocytes (T-cells), appears to be markedly affected by malnutrition. People with severe malnutrition have reduced numbers of these lymphocytes in circulation, and also experience atrophy of the organs that produce, maintain, and house them—the spleen, lymph nodes, and lymph-associated areas of the intestinal tract.[8]

Malnutrition seems to affect humoral immunity less than cell-mediated immunity. Antibodies secreted from B-cells belong to a family of proteins commonly called immunoglobulins. In malnourished people, the blood levels of immunoglobulins are not reduced, although as already noted, one type of immunoglobulin found in mucous secretions is depressed. Data support the concept that immunoglobulin production is generally not affected during malnutrition.

immunoglobulin: a protein that is capable of acting as an antibody.

Studies of malnutrition and immunity are difficult to conduct. Part of the reason is that malnourished people or animals often have infections, which greatly influence metabolism and the use of nutrients. It is hard to sort out the variables and determine which effects result from infection and which from malnutrition.

Changes in immune function in human beings have been associated with deficiencies of folate, iron, and zinc, and with excesses of vitamin E and essential fatty acids. Clinical reports suggest that changes in immune function accompany deficiencies of vitamins A, B_6, B_{12}, and C; deficiencies of pantothenic acid; and elevated blood cholesterol levels. Studies in animals link the levels of many vitamins, minerals, trace elements, amino acids, fatty acids, and cholesterol with immunological changes. For maximal support of a strong defense against infection, then, people should attempt to make their diets adequate, and not excessive, in all nutrients.

Cardiovascular Disease: Atherosclerosis

Diseases of the heart and blood vessels (cardiovascular diseases, or CVD) account for more than half of the nation's deaths each year, mostly by way of heart attacks and strokes.[9] Efforts to fight CVD have led to valuable discoveries and public education. We now know that smoking, high blood pressure, and high blood cholesterol levels are the three major risk factors for CVD, and many people have changed their lifestyles accordingly. Many have quit smoking, or refused to begin. Many who have hypertension have learned to control it. Many have been willing to change their diets, eating fewer kcalories and less fat, and are exercising more. The rate of CVD has fallen steadily since 1950, for some or all of these reasons; still, it is high. The question of how diet relates to CVD is an important one.

CVD (cardiovascular disease): a general term for all diseases of the heart and blood vessels. Atherosclerosis is the main form of CVD. Related terms to describe atherosclerosis in the arteries feeding the heart muscle are **CAD (coronary artery disease)**, and **CHD (coronary heart disease)**.

atherosclerosis (ath-er-oh-scler-OH-sis): the most common form of artery disease, characterized by plaques along the inner walls of the arteries (The related term **arteriosclerosis** means *all* forms of hardening of the arteries and includes some rare diseases.)
athero = porridge or soft
scleros = hard
osis = too much

hypertension: chronic elevated blood pressure.

plaques (PLACKS): mounds of lipid material, mixed with smooth muscle cells and calcium, which develop in the artery walls in atherosclerosis. The same word is also used to describe an entirely different kind of accumulation of material on teeth, which promotes dental caries.
placken = patch

The two major forms of most CVD are atherosclerosis and hypertension. Atherosclerosis, the subject of this section, is the common form of hardening of the arteries; hypertension (the subject of the next section) is high blood pressure; and each makes the other worse.

How Atherosclerosis Develops

No one is free of atherosclerosis.[10] The question is not whether a person has it but how far advanced it is and what the person can do to retard or reverse it. It usually begins with the accumulation of soft mounds of lipid, known as plaques, along the inner walls of the arteries, especially at branch points (see Figure 9–1). These plaques gradually enlarge, making the artery walls lose their elasticity and narrowing the passage through them. Most people have well-developed plaques by the time they are 30.

Normally, the arteries expand with each heartbeat to accommodate the pulses of blood that flow through them. Arteries hardened and narrowed by plaques cannot expand, and so the blood pressure rises. The increased pressure puts a strain on the heart and damages the artery walls further. At damaged points, plaques are especially likely to form; thus the development of atherosclerosis is a self-accelerating process.

Figure 9–1 The Formation of Plaques in Atherosclerosis

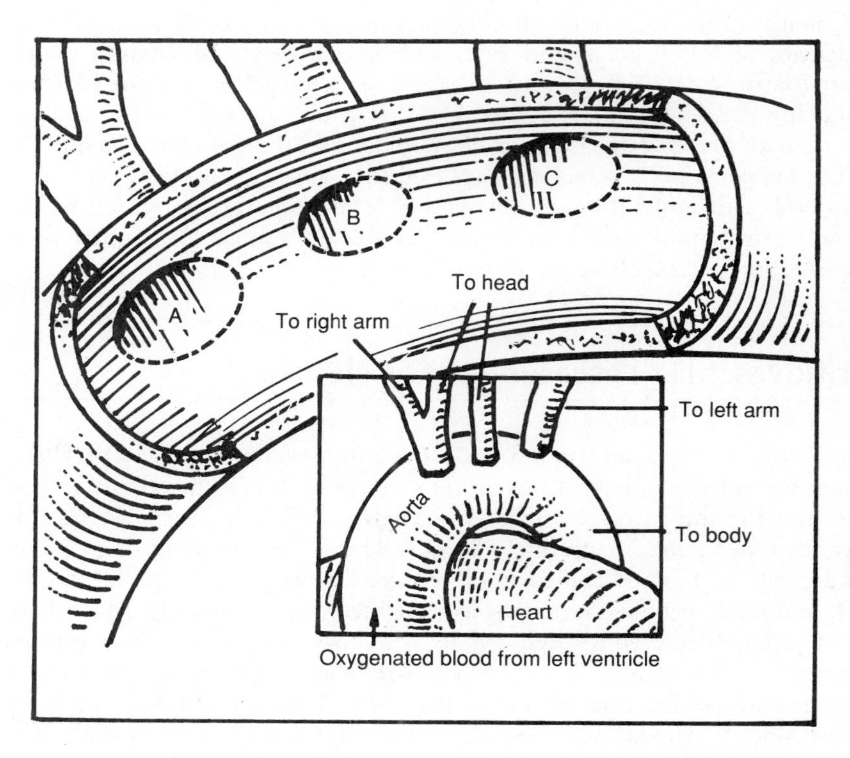

The aorta carries oxygenated blood from the heart's left ventricle to the body. A, B, and C are normal-sized openings.

As pressure builds up in an artery, the arterial wall may become weakened and balloon out, forming an aneurysm. An aneurysm can burst, and when this happens in a major artery such as the aorta, it leads to massive bleeding and death.

Abnormal blood clotting also contributes to life-threatening events. Clots form and dissolve in the blood all the time, and the balance between these processes ensures that they do no harm. That balance is disturbed in atherosclerosis. Small, cell-like bodies in the blood, known as platelets, are supposed to cause clots to form whenever they encounter injuries in blood vessels, but they respond in the same way to plaques and form clots when none are needed. Eicosanoids control the action of the platelets, and an imbalance among these compounds may contribute to the formation of clots. Substances released by platelets also may aggravate the growth of plaques.

A clot, once formed, may remain attached to a plaque in an artery and gradually grow until it shuts off the blood supply to that portion of the tissue supplied by the artery. That tissue may die slowly and be replaced by scar tissue. The slow death of heart tissue caused by reduced blood flow is ischemia, and the stationary clot is called a thrombus. When a thrombus has grown large enough to close off a blood vessel, it becomes a thrombosis. A coronary thrombosis is the closing off of a vessel that feeds the heart muscle. A cerebral thrombosis is the closing off of a vessel that feeds the brain.

aneurysm (AN-you-rism): the ballooning out of an artery wall at a point where it has been weakened by deterioration.

platelets: tiny, disc-shaped bodies in the blood, important in blood clot formation.
platelet = little plate

eicosanoids (eye-COSS-uh-noyds): derivatives of 20-carbon polyunsaturated fatty acids, which are, in turn, derived from essential fatty acids; hormonelike compounds that regulate blood pressure, clotting, and other body functions. Among eicosanoids are prostaglandins and thromboxanes.

prostaglandins (PROS-tah-glandins): eicosanoid compounds with a multitude of diverse effects on the body, including blood vessel contractions, nerve impulses, and hormone responses; synthesized from the 20-carbon polyunsaturated fatty acids.

thromboxanes: eicosanoid compounds with effects on the blood-clotting system; synthesized from the 20-carbon polyunsaturated fatty acids.

ischemia (iss-KEY-me-uh): the deterioration and death of tissue (for example, of heart muscle), often caused by atherosclerosis.
is = to restrain
heme = blood

thrombus: a stationary clot. When it has grown enough to close off a blood vessel, it becomes a **thrombosis.** A **coronary thrombosis** is the closing off of a vessel that feeds the heart muscle. A **cerebral thrombosis** is the closing off of a vessel that feeds the brain.
coronary = crowning (the heart)
thrombo = clot
cerebrum = part of the brain

Figure 9–1 Continued

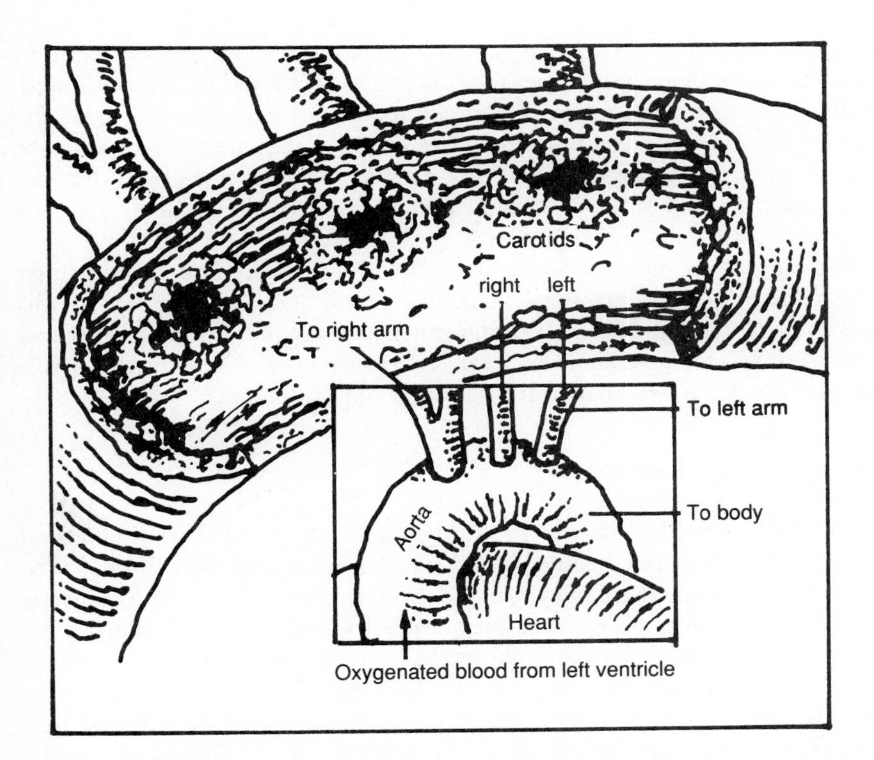

Atherosclerosis. The aorta and carotid arteries are blocked by lipid-containing plaques.

embolus (EM-boh-luss): a thrombus that breaks loose. When it causes sudden closure of a blood vessel, it is an **embolism.**
embol = to insert

heart attack: the event in which an embolus lodges in vessels that feed the heart muscle, causing sudden tissue death; also called **myocardial infarction.**
myo = muscle
cardial = of the heart
infarct = tissue death

stroke: the sudden shutting off of the blood flow to the brain by a thrombus or embolism.

risk factors: factors known to be related to (or correlated with) a disease but not proven to be causal.

diet-heart controversy: the controversy over the questions of whether a high-fat, high-cholesterol diet causes atherosclerosis and whether a low-fat, low-cholesterol diet can prevent it.

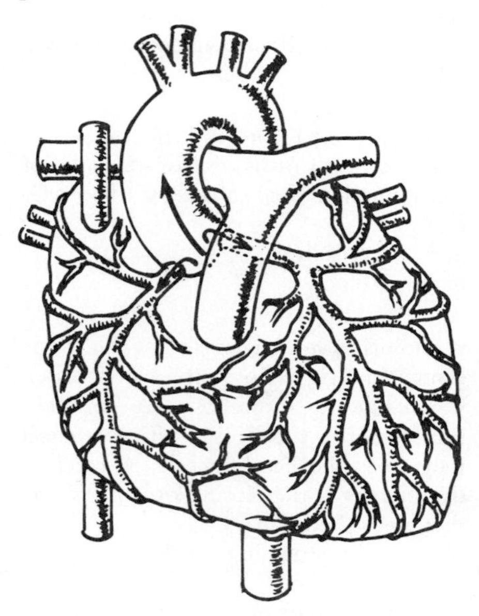

Figure 9–2 The Arteries That Feed the Heart Muscle
Arteries branching from the aorta carry oxygenated blood to every part of the heart muscle. These are the coronary arteries.

A clot can also break loose, becoming an embolus, and travel along the system until it reaches an artery too small to allow its passage. Then the tissues fed by this artery will be robbed of oxygen and nutrients and will die suddenly (embolism). If such a clot lodges in an artery of the heart (Figure 9–2), causing sudden death of part of the heart muscle, we say that the person has had a heart attack. If the clot lodges in an artery of the brain, killing a portion of brain tissue, the event is called a stroke.

On many occasions, it is not clear what has caused a heart attack or stroke. An artery appears to go into spasms, and the blood supply to a portion of the heart muscle or brain is cut off, but examination reveals no visible cause.[11] Much research today is devoted to asking what causes plaques to form, what causes arteries to go into spasms, what governs the activities of platelets and the synthesis of eicosanoids, and why the body allows clot formation unbalanced by clot dissolution. Hypertension makes atherosclerosis worse. A stiffened artery, already strained by each pulse of blood surging through it, is more greatly stressed if the internal pressure is high. Lesions (injured places) develop more frequently, and plaques grow faster.

Atherosclerosis also makes hypertension worse. Hardened arteries are unable to expand with each beat of the heart, so the pressure in them rises. This leads to further hardening of the arteries. Hardened arteries also fail to let blood flow freely through the body's blood pressure-sensing organs, the kidneys; the kidneys respond as if the blood pressure were too low, and raise it further (see the upcoming section, "Hypertension").

Risk Factors for Atherosclerosis

The risk factors for atherosclerosis are listed in Table 9–3. It befits a nutrition book to focus on dietary strategies to reduce them. It should be noted, though, that diet is not the only, and perhaps not even the most important, factor in the causation of CVD. Much controversy has attended the questions of whether diet is of any importance at all in heart disease; whether changes in diet can reduce the risk; and if so, whether such changes should be advocated for everyone, or just for selected high-risk individuals.

The big *diet-related* risk factors for CVD are hypertension, obesity, high blood cholesterol concentrations, and glucose intolerance. The rest of this discussion addresses the areas of disagreement among the experts on diet and atherosclerosis—the so-called diet-heart controversy—and attempts to arrive at suggestions for a personal strategy despite the confusion.

The Diet-Heart Controversy

No one disputes the fact that high blood cholesterol, particularly high blood LDL, predicts CVD, but people do dispute the hypothesis that links *diet* to CVD. The hypothesis is that high blood cholesterol (LDL) is at least partly caused by a diet high in saturated fat and cholesterol, and that reducing the amounts of saturated fat and cholesterol in the diet will lower blood cholesterol and reduce the rate of CVD. Both parts of this hypothesis have some strong support, but pieces are missing. With respect to the first part (whether a high-fat diet elevates blood cholesterol), blood cholesterol can be raised in both animals and people by raising the amounts of saturated fat and cholesterol in their diets—but whether the high blood cholesterol seen among so many people in the real world is *caused* by that aspect of their diets has been impossible to demonstrate. With respect to the second part (whether reducing fat intake will reduce blood cholesterol), it is possible to lower blood cholesterol in both animals and people by reducing the amounts of saturated fat and cholesterol in their diets—but whether this reduces their risks of heart disease has been impossible to demonstrate.

Evidence against the diet-heart hypothesis is as follows. Some major studies have shown no correlation between people's diets and their blood cholesterol levels.[12] The major decline in mortality from CVD seen from 1950 to 1980 did not coincide with a corresponding change in people's saturated fat and cholesterol intakes.[13] Studies in which people have been given diet changes to reduce blood cholesterol to prevent or treat heart disease have not shown a correlation. Finally, studies in which drugs have been used to reduce blood cholesterol have either succeeded in doing so without reducing the heart disease rate or have caused excess deaths from other causes.[14] The high blood cholesterol seen in developed countries may therefore not be related to the saturated fat and cholesterol in the diet, but to other factors in the environment. Some researchers urge investigation into possibilities such as the presence of abnormal oxidized forms of cholesterol that arise in foods when they are processed, *trans*-fatty acids that arise in oils when they are hydrogenated, an excess of vitamin D that is added to foods to "enrich" them, or excess carbon monoxide in the atmosphere that might interfere with the enzymes responsible for handling cholesterol. The prevailing view is the opposite, however; most investigators and many organizations, including the American Heart Association (AHA), see links between the typical high-fat, high-cholesterol diet and heart disease, and they hold out the hope that continued research will confirm and clarify these relationships.

A major trial to test the diet-heart hypothesis was the Multiple Risk Factor Intervention Trial (MRFIT), involving a ten-year-long study of over 12,000 men who were persuaded to quit smoking, control their blood pressure, and make dietary changes to reduce their blood lipid levels. MRFIT was inconclusive: it did not demonstrate that diet changes would reduce the risk of CVD. In fact, it did not even demonstrate that controlling blood pressure would reduce mortality, partly because side effects of the drugs used to control high blood pressure confounded the results.

Another major trial was the Coronary Primary Prevention Trial (CPPT), involving about 4000 men for 7½ years. Initially, 48,000 men, aged 35 to 59, were screened. Those free of CVD, with the highest (top 5 percent) blood cholesterol levels, were chosen for the study—3806 men altogether. All were given diet advice; half were given a cholesterol-lowering drug. The experiment was designed to lower blood cholesterol and measure any resulting changes in heart disease mortality, but it used drugs, not diet, to lower the blood cholesterol.

At the conclusion of the CPPT in 1984, the National Institutes of Health (NIH) researchers stated that the reduction of blood cholesterol (using a drug) in the participants had successfully reduced heart attack risk. For each 1 percent reduction in blood cholesterol, a 2 percent reduction in heart attack risk was seen, they reported. The researchers asserted that for the men given

Table 9–3 Risk Factors for Atherosclerosis

Risk Factors for Atherosclerosis
Smoking
Hypertension
High blood cholesterol levels
Obesity
Glucose intolerance (diabetes)
Lack of exercise
Stress
Heredity
Gender (being male)

the cholesterol-lowering drug in the study, an average 19 percent reduction in risk was seen; for those who had the full degree of cholesterol lowering, the reduction in risk was 49 percent.[15]

The study did not show that *diet* changes had reduced risks of CVD, and did not unequivocally show that cholesterol-lowering drugs had done so, either. Disagreements on the meanings of the numbers of heart attacks and heart attack deaths were particularly troublesome. One critic reviewed the trial design from the start and found several flaws. For example, during the 7½ years of the study, 187 fatal and nonfatal heart attacks occurred in the untreated group and 155 in the treated group. The authors drew the erroneous conclusion that a 19 percent difference had been observed. The difference was actually only 1.6 percent—8.6 percent of the untreated men versus 7 percent of the treated men. The appropriate statistical test showed this difference to be insignificant. Based on this reevaluation of the data, it seemed unreasonable to accept that the results had widespread implications for adults. Only men with the highest blood cholesterol levels—265 milligrams per 100 milliliters and above—had been studied, so any results applied only to them, and perhaps had hardly any significance even for them. No evidence suggested that reducing serum cholesterol in normal persons below 210 milligrams per 100 milliliters would reduce the incidence of CVD in this group.[16] The president of the AHA, however, stated the following view: "The results of the CPPT clearly show the benefit of lowering plasma cholesterol. . . . The AHA has taken the position that a diet recommendation for the healthy U.S. population is warranted- . . . . [People] should consume a diet containing less total fat, saturated fat, and cholesterol."[17]

At the end of 1984, the NIH gathered together a panel of experts to attempt to arrive at a consensus on the question of whether lowering blood cholesterol would help prevent heart disease. The NIH Consensus Conference report concluded that elevated blood cholesterol levels were clearly a major cause of artery disease, and that lowering them would reduce the risk of heart attack in men and probably in women. They agreed that the top 25 percent of blood cholesterol levels should receive top priority in treatment, but they also stated that the whole population had too-high levels, "in large part because of our high dietary intake of kcalories, saturated fat, and cholesterol." They went on to state that "there is no doubt that appropriate changes in our diet will reduce blood cholesterol levels [and] afford significant protection against coronary heart disease."[18] Table 9–4 shows serum cholesterol classifications and recommended follow-up for adults. To assess risk, multiple samples must be taken over time because an individual's serum cholesterol concentration fluctuates; at one cholesterol reading it may appear within the normal range, while at another it may be high.

The NIH Consensus Conference report went on to recommend that:

1. People at high risk (values in the top 10 percent) receive intensive dietary treatment and, if not successful, treatment with drugs.
2. People at moderate risk (the next 15 percent) receive intensive diet treatment (few should need drugs).
3. Everyone from age two up be advised to adopt a preventive diet with 30 percent of kcalories from fat (and close to 10 percent of that from

Table 9–4 Initial Classification and Recommended Follow-up Based on Total Serum Cholesterol Levels for Adults

Cholesterol[a]		Classification
(mg/100 ml)	*(mmol/L)*[b]	
<200	<5.17	Desirable serum cholesterol
200 to 239	5.17 to 6.18	Borderline-high serum cholesterol
240 or higher	6.20 or higher	High serum cholesterol
	Recommended Follow-up	
<200	<5.17	Repeat checkup within 5 yr
200 to 239	5.17 to 6.18	
Without definite coronary heart disease (CHD) or two other CHD risk factors (one of which can be male gender)		Provide dietary information and recheck annually
With definite CHD or two other CHD risk factors (one of which can be male gender)		Perform lipoprotein analysis; take further action based on results
240 or higher	6.20 or higher	Perform lipoprotein analysis; take further action based on results

[a]Based on the average of two cholesterol measurements that have been made within a 7- to 8-wk time frame, and provided that the range between the two tests does not exceed 30 mg/dl (0.77 mmol/L).
[b] To convert cholesterol (mg/100ml) to standard international units (mmol/L), multiply by 0.02586.

Source: Adapted from N. D. Ernst and J. C. LaRosa, Recommendations for treatment of high blood cholesterol: The National Cholesterol Education Program Adult Treatment Panel, *Contemporary Nutrition* 13, no. 1 (1988). (The education program was designed for The National Health, Lung, and Blood Institute of the National Institutes of Health.) Reprinted with permission from General Mills, Minneapolis, Minn., 1988.

polyunsaturated fat and 10 percent from monounsaturated fat), and a maximum of 300 milligrams a day of cholesterol.

4. All people with obesity reduce their weight by means of diet and exercise.
5. People with high blood cholesterol also be counseled to control hypertension and diabetes, abstain from smoking, and exercise regularly.

Additional recommendations addressed education of health professionals and the public, cooperation by the food industry, adoption of informative food labels, universal screening of adults' blood cholesterol levels, further research, and ongoing monitoring and evaluation of these efforts.

The disagreement over diet and heart disease is one of the most prolonged and complicated in 20th century nutrition. The missing piece is still missing. The question of whether dietary fat and cholesterol have any effect on heart disease risk is still unanswered. It is agreed that high blood cholesterol is a major risk factor for atherosclerosis. It is agreed that stringent dietary control can alter blood cholesterol levels somewhat. But a lifelong program of blood cholesterol reduction by diet might increase life expectancy only very slightly, and public health might be better served by reducing smoking and high blood pressure, rather than lowering cholesterol intake, in the general public.[19]

Even among those who agree that dietary fat and cholesterol should be reduced, there are other questions. For example, the NIH conferees noted that the trials had mostly been on middle-aged men, but they believed that women and older men would benefit, too, from a general lowering of blood cholesterol. Two years later, reviews of statistics were leading to doubts about the broad applicability of the recommendations. Using data from the longest available, original diet-and-heart-disease study, the Framingham Study (which followed 5000 people for 30 years), it was clear only that CVD risk for men under 50 was increased at cholesterol levels above 180 milligrams per 100 milliliters. For women, and for men over 50, there was little or no association.[20] Hence the recommendation to reduce cholesterol may be useful only for men under 50, and may be of limited usefulness even to them. (Furthermore, as Chapter 7 already mentioned, it is not considered desirable to reduce children's fat intakes below 30 percent of kcalories except where careful screening indicates unusual hereditary risk.)

Although controversy continues to plague all these areas, many health professionals are abiding by a policy of screening every adult and selected children (the children of high-risk families) for high blood cholesterol. When high blood cholesterol is found, many agree that the first treatment should be diet.[21] Health professionals who accept that assumption then have two more questions to address. The first is about blood cholesterol: how high is too high? The second is about diet: which diet?

How High Is Too High?

Much debate surrounds the question, At what level of blood cholesterol should alarms sound? In this society, all people's blood cholesterol levels rise as they grow older. Should higher blood cholesterol levels in the later years therefore be considered acceptable? Some say yes, but the louder voices say no. Even moderately elevated cholesterol levels increase the odds of dying from CVD, and the danger is present even in people who do not smoke and do not have hypertension. According to one expert, who has worked with heart disease and public health for several decades, the optimal cholesterol level is 180 milligrams per 100 milliliters, and 80 percent of middle-aged men have levels higher than that.[22] Readings from 180 to 240 have been defined as "moderate" by some, but people with readings over 180 have elevated risks that rise as the cholesterol readings rise.

According to the authors of the Whitehall Study, a study of over 15,000 middle-aged men whose mortality was followed over ten years, most deaths related to cholesterol occur among men with cholesterol levels in the *middle* three-fifths of the distribution. That means that if treatment were aimed only at those with the highest levels, it would miss an enormous number of people. The only way to save these lives, according to those who believe lives *can* be saved by lowering cholesterol levels, would be to lower cholesterol levels in the whole population.[23] Other researchers agree that it would not help to treat only the high-risk people, because most heart attacks occur in people at moderate risk.[24] They therefore propound the population approach.

An exactly opposite view is put forth by the authors of a study involving over 10,000 middle-aged Israelis followed for 15 years. The authors found no association between elevated cholesterol and mortality until cholesterol levels rose above 240 milligrams per 100 milliliters, and they therefore viewed it as premature to support a policy of reducing mean cholesterol below 220 in the general population.[25] They therefore advocate the high-risk approach.

The president of the American Heart Association takes the position that cholesterol levels near the low end of the range should be considered risky, even though overt disease does not appear until levels are high. Figure 9–3 shows what may be going on. While a person's blood cholesterol steadily rises from the teen years onward, plaques are silently accumulating in that person's arteries. When 60 percent of the inner walls of the arteries are covered with accumulations of cholesterol, the CVD risk suddenly becomes apparent, but it has been developing for a long time. If this is the case, it can be argued that to minimize CVD risk, the process should be stopped near its beginning, not just short of the danger point. Accordingly, the AHA president recommends that:[26]

population approach: the approach (to prevention of CVD) that argues that the whole population should change its diet.

high-risk approach: the approach (to prevention of CVD) that says that only those with blood cholesterol levels above 240 milligrams per 100 milliliters should change their diets.

Figure 9–3 The Silent Progression of Plaque Formation
As blood cholesterol rises from 100 to 300 mg/100ml, plaques form in the arteries, first covering 25% then 50, then 75, then 100% of their surfaces (straight line and numbers at left). The arteries here are coronary arteries, the ones that feed the heart muscle. However, heart disease symptoms remain at 0 until 60 percent of the artery walls are covered by plaques; then they rise steeply (curved line and numbers at right). Thus it appears that heart disease is nonexistent at cholesterol levels below 200 to 250 mg/100 ml. Actually, it is present but invisible—and progressing.

Most people with a cholesterol level of 200 and no other risk factors reach a critical degree of atherosclerosis at age 70. If their cholesterol level were 250 or 300, they would reach this critical point earlier—at age 60 or 50. Knowing the cholesterol level enables the health professional to predict the age of onset of CVD.

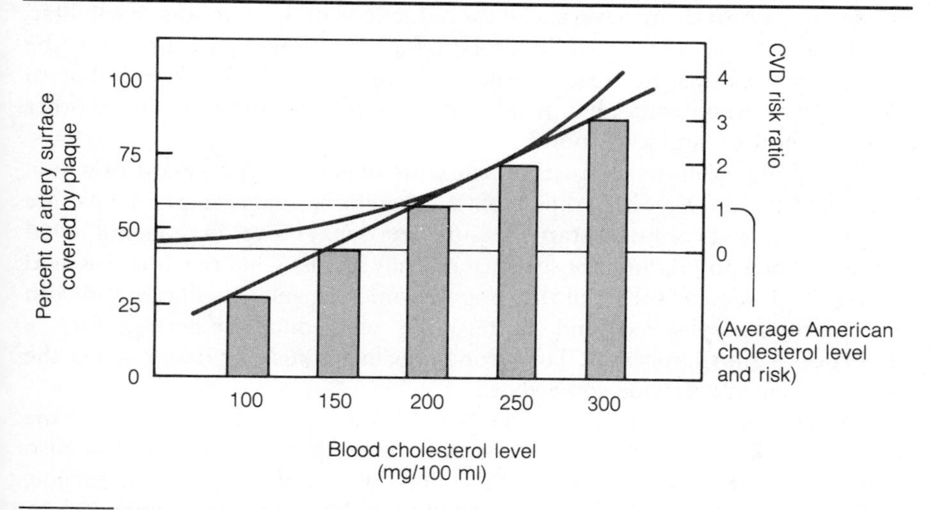

Source: Adapted with permission from S.M. Grundy, Cholesterol and coronary heart disease: A new era, *Journal of the American Medical Association* 256 (1986): 2849–2858, and Figure 6, p. 2851. Copyright 1986, American Medical Association.

- For people over 30, the line should be drawn at 200 milligrams per 100 milliliters. For people under 30, the line should be drawn at 180.
- If total serum cholesterol is above this line, then more-sensitive indicators of risk should be measured: LDL and HDL cholesterol in particular.
- Triglycerides should be measured, too. They may or may not indicate risk for CVD, but they do indicate diabetes, kidney problems, and other conditions that should be treated.

Another constructive suggestion has been made. Rather than arguing over whether to target only those with the highest cholesterol levels or to educate everyone to control their blood cholesterol, why not do both? It has been pointed out that the two are complementary, not contradictory, strategies.[27]

Thus cholesterol screening goes on, in the hope of reducing this risk factor for atherosclerosis. Meanwhile, the reader may wonder what became of the loose ends mentioned earlier. What if the culprit responsible for high blood cholesterol and associated heart disease in the Western world is not fat and cholesterol, but some other factor—oxidized cholesterol, *trans*-fatty acids, vitamin D fortification, a virus, or something else? Investigations pursuing each of these possibilities are under way but have, so far, yielded no findings suggesting that answers will lie there.[28] As mentioned, a piece is still missing from the heart disease puzzle.

Prevention by Diet

This section is about prevention of atherosclerosis, but first a word about diet in its treatment is in order (see Table 9–5), because the public has confused the two. Several disorders of blood lipid regulation, largely or entirely hereditary, known as the hyperlipidemias, can cause atherosclerosis to advance rapidly. Depending on the disorder, different lipoproteins in the blood may be altered in concentration or character, and different dietary measures may be recommended to control them. Several call for reductions in the amounts of total fat, saturated fat, and dietary cholesterol in the diet. These measures may be effective when used to treat people with high blood cholesterol due to particular hyperlipidemias, but people with high blood cholesterol due to other causes do not, of course, respond.

The adult population consists of a mixture of genetic types, many of whom have variants on the hyperlipidemias. Therefore, when experiments are undertaken to test various dietary measures for their efficacy in reducing blood cholesterol in a population, those measures may be just what the doctor would order for a particular subset of that population. That subset will respond with lowered cholesterol levels, and the response will reduce the average for the whole population somewhat. The wrong conclusion may be drawn—that the measures helped everyone somewhat.

The genetic variation among individuals in test populations is one of the factors that has confounded intervention research into CVD. It has also confused the public, who think that "if a diet low in cholesterol is good for Joe, it must be good for me." It is always hard to make dietary recommendations for the public, but it is even harder than usual in the case of blood cholesterol-lowering recommendations for these reasons of genetic variability.

Table 9–5 Diet in Treatment of Hyperlipidemias (examples)

	Type I	Type IIa	Type IIb	Type III	Type IV	Type V
Lipid abnormality	⇑ chylomicrons, ⇑ TG, ⇑ C	⇑ LDL, ⇑ C	⇑ LDL, ⇑ VLDL, ⇑ C, ⇑ TG	⇑ VLDL, ⇑ C, ⇑ TG	⇑ VLDL, normal or ⇑ C, ⇑ TG	⇑ chylomicrons, ⇑ VLDL, ⇑ TG, ⇑ C
			Diet[a]			
Carbohydrate	Not restricted	Not restricted	40% kcal, restrict concentrated sweets	(Same as IIb)	45% kcal, restrict concentrated sweets	50% kcal, restrict concentrated sweets
Fat	25 to 35 g; type not restricted	⇑ poly-unsaturated, ⇓ saturated	40% kcal, ⇑ poly-unsaturated, ⇓ saturated	(Same as IIb)	⇑ poly-unsaturated, ⇓ saturated	30% kcal, ⇑ poly-unsaturated, ⇓ saturated
Cholesterol	Not restricted	Less than 300 mg	Less than 300 mg	(Same as IIb)	300 to 500 mg	300 to 500 mg
Alcohol	Avoid	Physician's discretion	Physician's discretion	(Same as IIb)	Physician's discretion	Avoid

[a]For every type, achieve and maintain ideal body weight.

Key:
⇑—Elevated or increased. **⇓—Lowered or decreased.**
C—Cholesterol. **TG—Triglycerides.**
LDL—Low-density lipoproteins. **VLDL—Very-low-density lipoproteins.**

Suppose that despite the conflicting evidence, a person wanted to adopt a diet consistent with the *possibility* that it might help reduce blood cholesterol and CVD risk. The next question is, What diet to choose? The conferees at the NIH Consensus Conference recommended the AHA diet as the one of choice, but several dissenting opinions were voiced. Alternatives were a vegetarian diet with a high ratio of polyunsaturated to saturated fat, a diet rich in monounsaturated fats, and a diet rich in fish oils.[29] Extremists might propose the Pritikin diet—a diet with only 5 to 10 percent of its kcalories from fat, with negligible cholesterol, and with meat intake strictly limited. (Nathan Pritikin, who advocated this diet, died at 69 of causes unrelated to heart disease, with a blood cholesterol level of less than 100 milligrams per 100 milliliters and no plaques in his arteries at all.)[30]

The person who wanted to take every possible dietary measure to prevent atherosclerosis might choose to do all of the following. Eat a kcalorie-controlled, low-fat, low-cholesterol, high-fiber, high-complex-carbohydrate, adequate, balanced, varied diet. Abstain from too much coffee, alcohol, and sodium. Use a variety of oils, olive oil among them. Use a variety of vegetables, and include raw vegetables daily. Use sprouts and whole grains, including oats. Eat periodic meals of fish. Feel free to use shellfish; they are not as high in cholesterol as has been believed, and they do contain EPA and other fatty acids in the omega-3 series. Use eggs and other animal-protein foods in moderation.

Why do all this? Body weight has proven, in some studies, to be the most important single determinant of blood cholesterol level, but even if weight control does not help ward off heart disease by reducing blood cholesterol, it will help by reducing blood pressure (see the "Hypertension" section).[31] So will eating a low-fat diet. Even if the high-fiber, high-complex-carbohydrate aspect doesn't help by way of cholesterol, it will help by improving glucose tolerance, a major risk factor for CVD. Even if the monounsaturated oils and the omega-3 oils from the fish don't help by way of cholesterol, they may help by favoring the right eicosanoid balance so that clot formation is unlikely. An adequate diet will protect the health of the heart muscle itself; mineral deficiencies precipitate disease of the heart muscle and arteries.[32] Also, don't forget the lifestyle context in which these dietary measures are only a part. Exercise daily, relax, and enjoy yourself.[33]

Prevention by Exercise

The next chapter shows that physical activity promotes longevity. One reason why it does so is doubtless because of its beneficial effect on cardiovascular disease risk. A strong inverse relationship exists between plasma HDL cholesterol concentrations and coronary heart disease.[34] Evidence shows that exercise raises HDL cholesterol concentrations and may therefore help protect against this killer disease.[35] Even without a change in body fatness, regular moderate exercise significantly raised HDL cholesterol concentrations in previously sedentary, overweight men. The combination of exercise and weight loss produced a still greater increase in HDL concentrations. To the previous diet prescription, then, add: exercise regularly.

Hypertension

Chronic elevated blood pressure, or hypertension, is the most prevalent form of cardiovascular disease, believed to affect some 60 million people—more than a third of the entire adult population.[36] It contributes to half a million strokes and over a million heart attacks each year.[37] The higher the blood pressure above normal, the greater the risk of heart disease.

People cannot tell if they have high blood pressure; it presents no symptoms they can feel. Therefore, the most effective single step people can take in self-defense against it is to find out whether they have it. At checkup time, a health care professional can take an accurate blood pressure reading. (Self-test machines in drugstores and other places can mislead people with inaccurate readings.) If a blood pressure reading is above normal, another should be taken at another time before the diagnosis of hypertension is considered definite. Thereafter, additional readings should be taken at regular intervals.

In a blood pressure reading, two numbers are important: the pressure in the arteries during contraction of the ventricles (the "dub" of the heartbeat), which is higher, and the pressure during relaxation of the ventricles (the "lub"), which is lower. The first number is the systolic pressure, and the second is the diastolic pressure.

systolic pressure: the first figure in a blood pressure reading, which represents arterial pressure caused by the contraction of the left ventricle of the heart.

diastolic pressure: the second figure in a blood pressure reading, which represents the arterial pressure when the heart is between beats.

Blood pressure is generally considered normal if it is less than 140 over 90 (140/90).* Above this level, the risks of heart attacks and strokes increase in direct proportion to increasing blood pressure, especially systolic pressure. Mild hypertension, which accounts for 75 percent of all blood pressure problems, involves a systolic pressure in the range of 140 to 160, and a diastolic pressure of 90 to 95. Severe hypertension is anything greater than 160/95.[38]

A word of caution about interpreting blood pressure readings: many factors can affect them. For example, blood pressure rises when a person speaks; is lowest when the person is lying down, higher when sitting up, and higher still when standing; and can appear falsely high if the cuff is too small.

How Hypertension Develops

A certain blood pressure is vital to life in the cells. The pressure of the blood against the walls of the arteries pushes fluids, carrying a cargo of nutrients and oxygen, out of the arteries into the tissues. By the time blood reaches the veins, much of its fluid has exited, and the concentration of cells and dissolved materials in the remaining blood is at a maximum. Fluids carrying wastes from the tissues are attracted by the concentrated blood and seep back into the veins. Thus the cells are nourished and cleansed.

The pressure the blood exerts on the inner walls of the arteries is the result of two forces acting together: the heart's pushing the blood into the arteries, and the smallest arteries and capillaries resisting the blood's flow. The heart's push ensures that the blood circulates through the whole system; the peripheral resistance and resulting pressure ensure that some of the blood's components, including nutrients, are pushed through the capillary walls to feed the tissues. One other factor contributes to blood pressure—the volume of fluid in the vascular system—and that, in turn, is affected by the number of dissolved particles the vascular system contains. Thus the more salt in the blood, the more water there will be. Figure 9–4 shows how exchange of material takes place in the capillaries and the forces that contribute to blood pressure.

peripheral resistance: resistance to the flow of blood caused by a reduced diameter of the vessels at the periphery of the body—the smallest arteries and capillaries.

The kidneys depend on the blood pressure to help them filter waste materials out of the blood. (The pressure has to be high enough to force the blood's fluid out of the capillaries into the kidney's filtering networks.) If the blood pressure is too low, the kidneys, together with the pituitary gland, initiate actions to set things right. A hormone, natriuretic hormone, constricts the peripheral blood vessels; the kidney-initiated renin-angiotensin mechanism leads to sodium and water retention, and the ADH mechanism to water retention directly, increasing the blood volume. With narrower peripheral blood vessels and a greater volume of blood, the blood pressure rises, ensuring continued nourishment and oxygen delivery to the tissues.

natriuretic hormone: a hormone that increases the rate of excretion of sodium in the urine and constricts peripheral blood vessels.
natri = sodium
uresis = urinary excretion

If blood pressure has been low, the blood pressure-raising response of the kidneys is an adaptive and beneficial response. In dehydration, for example, the response ensures that blood pressure is raised until the person can drink

*Blood pressure is read in millimeters of mercury. The silver column that rises on a blood pressure instrument is a column of mercury; the height to which it is pushed is marked off in millimeters.

Figure 9–4 Blood Pressure Regulation
Two major contributors to the pressure inside an artery are the heart's pushing blood into it, and the small-diameter arteries and capillaries at its other end resisting the blood's flow (peripheral resistance). Another determining factor is the volume of fluid in the circulatory system, which depends in turn on the number of dissolved particles in that fluid.

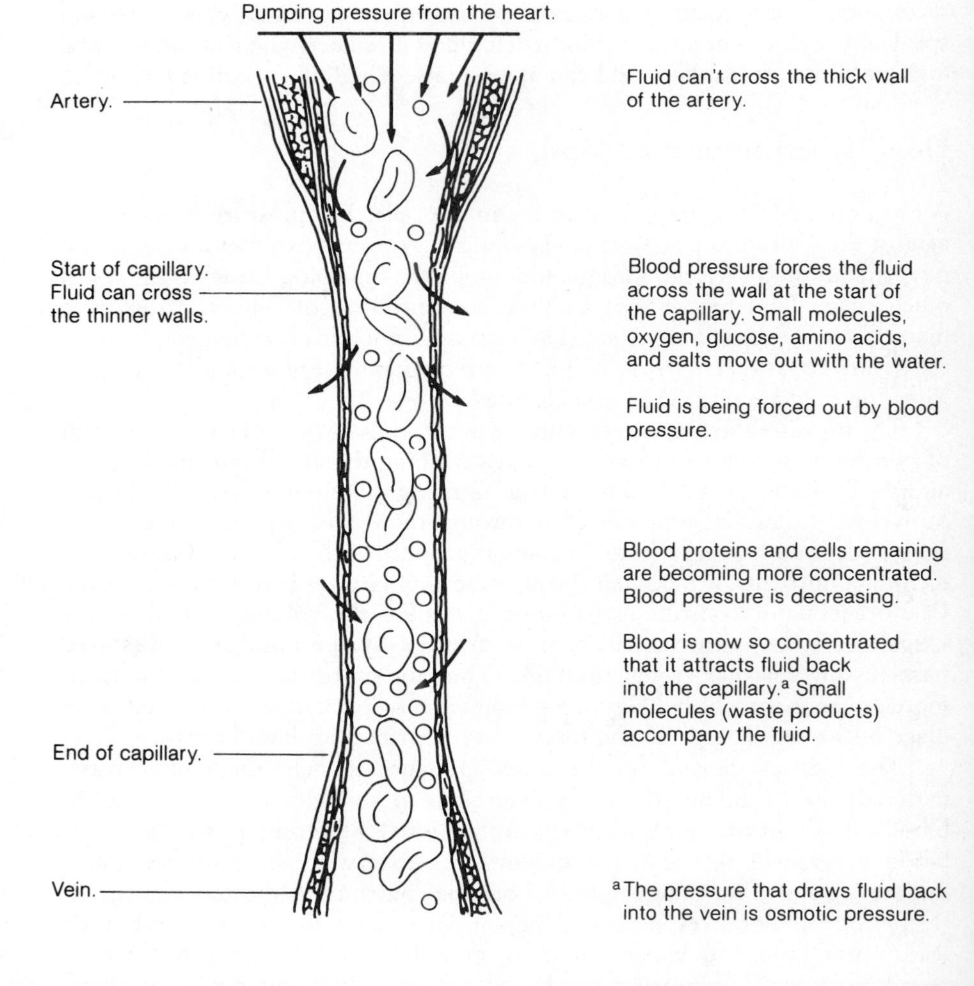

more water. When kidney arteries are hard and inflexible due to disease (such as atherosclerosis), however, the kidneys may be fooled: they respond as in dehydration, but the response is inappropriate. The raising of the blood pressure achieves adequate delivery of nourishment and oxygen to the kidney's tissues, but at the expense of further damage to all the body's arteries, which will lead in time to still-higher blood pressure. This is the cycle by which atherosclerosis and hypertension each make the other worse.

Fewer than 10 percent of all cases of hypertension are *known* to arise from kidney disorders. These cases are classified as secondary hypertension, and treatment focuses on correcting the underlying disorder in the kidneys—infection, injury, or whatever. For the other 90 percent or more of all cases of hypertension, no cause is known; these are classified as cases of essential, or primary, hypertension.

The kidneys may be just as involved in essential hypertension as they are in secondary hypertension, but exactly how they are involved is not known. Essential hypertension may have more than one cause: suspected are genetic, environmental, nutrition, and lifestyle factors. One such factor is atherosclerosis, just described. Another is an imbalance among the eicosanoids that regulate blood pressure by dilating or constricting the narrow arteries.[39] Still another is insulin resistance (discussed later); and another, often linked with it, is obesity. Added weight raises blood pressure; extra adipose tissue means miles of extra capillaries through which the blood must be pumped.

secondary hypertension: hypertension caused by kidney disease, accounting for about 10 percent of all cases of hypertension.

essential (primary) hypertension: hypertension with no known cause, accounting for more than 90% of all cases of hypertension.

The combination of hypertension, atherosclerosis, and obesity puts a severe strain on the heart and arteries, leading to many forms of cardiovascular disease and death. Some of the results were mentioned in the section on atherosclerosis preceding this one. Strain on the heart's pump, the left ventricle, enlarges and weakens it, until finally it fails (heart failure). Pressure in the aorta causes it to balloon out and burst (aneurysm). Pressure in the small arteries of the brain makes them burst and bleed (stroke). The kidneys can be damaged when the heart is unable to adequately pump blood through them (kidney failure).

Risk Factors for Hypertension

Epidemiological studies have identified several risk factors to predict the development of hypertension, including:

- *Age.* Blood pressure levels increase with age; most people who develop essential hypertension do so in their 50s and 60s.
- *Family background.* A family history of hypertension and heart disease raises the risk of developing hypertension two to five times.
- *Obesity.* Obese people are more likely to develop hypertension.
- *Race.* Hypertension is twice as common among blacks as among whites; it tends to develop earlier and become more severe.

While researchers continue looking for the cause or causes, clearly it is urgent to detect and treat hypertension wherever it presents its deadly threat—or better still, to prevent it. A major national effort has been made to identify and treat hypertension. Even mild hypertension can be dangerous; individuals who have it benefit from treatment, showing a reduced incidence of early death and illness.[40] Diet changes alone, even without the drugs used to reduce blood pressure, can bring about these benefits.

Authorities differ on the exact modes of treatment to adopt. Some say drugs should be used rather routinely; others advocate the aggressive use of diet therapy and other lifestyle changes (stress management, weight reduction, regular exercise) in preference to the use of drugs.[41] The discussion that follows focuses on diet and exercise.

Dietary Factors in Hypertension

Many diet-related factors seem to affect blood pressure. Obesity is clearly responsible, at least in some people; so are salt intakes (in some people). Several minerals besides sodium and chloride may be involved—notably potassium, calcium, and magnesium. Dietary fat and alcohol are also suspect as is caffeine. Insulin resistance may also be linked to hypertension.

Obesity Not every obese person is hypertensive, but for those who are, weight reduction significantly lowers blood pressure. This is so even if the person does not go all the way to achieve ideal body weight. Those who are using drugs to control their blood pressure can often reduce or discontinue their use after weight loss.[42] Even a 10-pound weight loss more than doubles the chance that hypertensive people can normalize their blood pressure without drugs, even if they have been maintained on aggressive hypertensive drug therapy for five years.[43]

Researchers have wondered whether the weight loss itself brought the benefits mentioned, or whether, in restricting kcalories, people were actually eating a diet that was lower in sodium, and *that* accounted for the results. Moderation in sodium intake does help normalize blood pressure.[44] However, people who lose weight can lower their blood pressure even without altering their sodium intakes. Thus weight loss alone may be one of the most effective nondrug treatments for hypertension.[45]

Distribution of body fat seems important, too. Although the association between obesity and hypertension is well established, not every obese person is hypertensive, and among those who are both obese and hypertensive, not all respond favorably to weight loss. These differences have prompted researchers to determine other factors that may influence hypertension in obese people. Researchers studied 30 obese women with normal blood pressures, and 30 obese women with hypertension. Upper body (abdominal) obesity was significantly more common among the women with hypertension compared with those without hypertension.[46]

Sodium and sodium chloride Sodium clearly has something to do with blood pressure. In fact, for years, research on populations has seemed to indicate that high sodium intakes were a major factor responsible for people's high blood pressure; but recently that idea has been yielding to the impression that it may be salt (sodium chloride) that is responsible for the effects seen. Overwhelmingly, sodium's accompanying anion in foods and in the body fluids is chloride. Much of the early research that implicated sodium in hypertension's causation may have unwittingly uncovered the effects of sodium's silent partner, chloride. Apparently, sodium in combination with other negative ions causes water retention and certain hormonal responses, but sodium chloride, uniquely, seems to raise blood pressure.[47]

The sodium (and chloride) concentration in the blood and other body tissues is maintained through an elaborate regulatory mechanism involving the kidneys, the adrenal glands, the pituitary gland, and other glands. Most people can therefore safely consume more salt than they need, and rely on these control mechanisms to regulate the excretion and retention of sodium as needed. Genetically salt-sensitive individuals, however, experience high blood pressure from excesses in salt intake. People with chronic renal disease, those whose parents (one or both) have hypertension, blacks, and persons over 50 years of age are most likely to be salt sensitive.[48]

How dietary salt contributes to hypertension in those who are salt sensitive is unknown, but recent research suggests a chain of events that leads to contraction of the small arteries, increasing peripheral resistance. The chain begins with an inherited or acquired defect in the kidneys.

When salt intake exceeds what the kidneys can excrete, sodium, chloride, and other negative ions are retained in the blood, water accompanies them, and the blood volume expands. The extra fluid volume is thought to help bring about the secretion of natriuretic hormone, which is not normally active and which enables the kidneys to excrete some of the excess. The same hormone moves sodium (and chloride and other negative ions) into the smooth muscle cells that line the arteries, and affects their membranes so that they bring in too much calcium, too. The more calcium, the more these muscle cells contract, the more peripheral resistance increases, and the higher the blood pressure rises.[49] In short, the hormone trades a wrong for a right: it reduces the blood volume, but it increases peripheral resistance so that blood pressure stays high.

For salt-sensitive individuals, then, eating a diet low in salt is a wise idea. But for others—the majority of people with hypertension—this may not be an effective diet strategy. Salt restriction does not lower blood pressure in half of the hypertensive people in whom it is tried.[50] It is important to look further to see what other dietary factors might be relevant.

Potassium When sodium is retained in the body, potassium is excreted. Some subjects with normal blood pressure, if fed very large quantities of sodium, ultimately show a rise in blood pressure—but at the same time, their potassium excretion is increasing. Fed potassium simultaneously with the sodium, they do not have a rise in blood pressure.[51] For this reason, many health care practitioners recommend that food sources of potassium be emphasized in the daily diet.

Population studies show sodium being traded for potassium in a different sense. People who eat many foods high in sodium often happen to be eating fewer potassium-containing foods at the same time.[52] Furthermore, as the *same* food goes through several processing steps, it loses potassium as it gains sodium, so its potassium-sodium ratio falls dramatically. For example, a baked potato with skin has about 800 milligrams of potassium and 16 milligrams of sodium. If the potato is processed to frozen hash brown potatoes, it then contains about 700 milligrams of potassium and 50 milligrams of sodium. Finally, in the form of potato chips, the potassium content drops to about 350 milligrams, while the sodium increases to 130 milligrams. Some authorities believe that it is as important to obtain enough potassium as to restrict dietary sodium, both for prevention and treatment of hypertension.[53]

Potassium may be important in heart disease independently of its relationship to sodium. Even in people without high blood pressure, a high potassium intake protects against stroke.[54] People using diuretics to control hypertension should know that some cause potassium excretion and can induce a deficiency.

Those using these drugs must be particularly careful to include rich sources of potassium in their daily diets.*

Calcium Epidemiological research first suggested a relationship between calcium and hypertension by showing that water hardness and cardiovascular death rates were related. Hard water contains high concentrations of several minerals, one of which is calcium, and people who drink it have less hypertension and heart disease than people who drink soft water, high in sodium.

Calcium may play a role in pregnancy-induced hypertension, too. Fewer than 1 in 200 pregnant women suffer from hypertension in areas where average calcium consumption exceeds 1000 milligrams per day. There is a fivefold to tenfold increase in the incidence of pregnancy-induced hypertension where calcium intakes average less than 500 milligrams per day.[55]

Several surveys report that people with hypertension consume less calcium than those with normal blood pressure.[56] Surprisingly, they also consume *less sodium* than those without high blood pressure—but perhaps this is because they are trying to do so. (People on low-salt diets tend to avoid dairy products, since these contain significant amounts of sodium; consequently, they lower their intakes of both sodium and calcium.) In a major survey that undertook to determine the relationship of 17 nutrients and total kcalories to blood pressure levels in over 10,000 individuals 18 to 74 years old, calcium was the nutrient that distinguished those with hypertension from those without it. The survey evaluation revealed that hypertensive people reported consuming about 20 percent less calcium than those without hypertension; there appeared to be an inverse relationship between calcium and blood pressure. Based on these data, researchers estimated that people with calcium intakes below 300 milligrams per day had a twofold to threefold increase in risk of developing hypertension when compared with people consuming 1200 milligrams per day.[57]

Calcium is known to affect blood pressure through its action on the muscle cells of the arteries. An excess of calcium inside these muscle cells causes contraction, which narrows the arteries and elevates blood pressure. The contraction process involves other components as well: hormones, sodium, and potassium. A possible explanation already presented for the calcium-hypertension connection is that when dietary calcium is low, the artery muscle cells "hoard" the calcium and contract, raising blood pressure. If dietary calcium were at least adequate, the cells would limit their intake of calcium, and the arteries would be relaxed, lowering blood pressure.[58]

Calcium may be important in both the prevention and treatment of hypertension. For those at risk of developing it, increasing the amount of calcium in the diet may protect against hypertension. For people already diagnosed with hypertension, obtaining adequate calcium in the diet may lower blood pressure. One study shows that a calcium-rich diet reduced blood pressure in 44 percent of the people with hypertension studied. Even among those with normal blood pressure, 19 percent experienced reduction in their

---

*Another class of drugs used to treat hypertension is the adrenergic blockers. Adrenergic blockers interfere with a neurotransmitter to alter blood pressure.

systolic pressure, the indicator most closely associated with risk of mortality.[59] It is recommended, therefore, that people with hypertension, or at risk of developing it, at least meet the current RDA for calcium—800 milligrams a day for adults. Dairy products are recommended, because they provide not only calcium but also potassium and magnesium, which may also help keep blood pressure normal. Low-fat or nonfat dairy products have the added advantage of being low in fat and helping to curb obesity.

Magnesium Magnesium accompanies calcium in many foods and also in hard water, and hard water has long been known to be geographically associated with a lower blood pressure than that associated with soft water. More direct evidence favoring adequate magnesium intakes as protective against hypertension comes from a study showing that magnesium deficiency causes visible changes in the walls of arteries and capillaries and makes them tend to constrict, a possible mechanism for its hypertensive effect.[60]

Fat Fat is well known to contribute to atherosclerosis, but it has an independent role in relation to blood pressure as well. Diets high in saturated fat are associated with hypertension. Populations that consume small quantities of animal products—vegetarians, for example—have a low incidence of hypertension. When people restrict their total dietary fat and increase the ratio of polyunsaturated to saturated fatty acids in the diet to 1.0 or above, their blood pressure falls, regardless of whether it was their intent to make it do so.[61]

This probably works at least partly by way of eicosanoids that affect both hormonal control of sodium excretion and the relaxation or constriction of the peripheral blood vessels. The essential fatty acid linoleic acid is a precursor of arachidonic acid, which in turn is a precursor of the prostaglandins. Research using animals shows that when dietary linoleic acid is restricted almost to the point of deficiency, synthesis of prostaglandins is suppressed, and blood pressure rises.[62] Linoleic acid may increase the synthesis of the prostaglandins that relax the blood vessels and therefore lower peripheral resistance or slow the long-term vascular changes associated with hypertension.[63] The effects are small and not consistent from one study to the next.[64] Monounsaturated fat may work as well as polyunsaturated fat; consumption of oleic acid from olive oil seems to correlate inversely with blood pressure levels in the Mediterranean.[65]

In one well-controlled study, researchers studied eicosanoid production and blood pressure in mildly hypertensive men who ingested supplements of polyunsaturated fats.[66] Four groups of eight men received either low-dose fish oil supplements (3 grams of omega-3 fatty acids); high-dose fish oil supplements (15 grams of omega-3 fatty acids); safflower oil (39 grams of linoleic acid, an omega-6 fatty acid); or a mixture of oils designed to approximate the fat content of the average diet in the United States. In the men who received the high-dose fish oil supplements, blood pressure fell significantly. The researchers did not attribute the decreased blood pressure to an increased synthesis of the vasodilatory prostaglandins; their synthesis increased initially but was not maintained as blood pressure fell. The researchers assert that the synthesis of vasodilator prostaglandins is not the primary mechanism by which fish oil supplements lower blood pressure. The National Committee on Detection, Evaluation, and Treatment of High Blood Pressure of the NIH recommends increased consumption of fish for hypertension, but states that the health benefits of fish oil supplements are still unsubstantiated.[67]

Many professionals advise hypertensive clients to follow the same fat-controlled diet as is recommended to prevent atherosclerosis—reducing fat to 30 percent of kcalories, obtaining one-third each from polyunsaturated, monounsaturated, and saturated sources; and limiting dietary cholesterol to 300 milligrams a day.

Alcohol Alcohol has several roles in relation to heart disease. In moderate doses, alcohol initially reduces pressure in the peripheral arteries and so reduces blood pressure, but high doses clearly raise blood pressure.[68] In fact, of people with alcoholism, 30 to 60 percent have hypertension.[69] The hypertension is apparently caused directly by the alcohol, and it leads to cardiovascular disease as severe as hypertension caused by any other factor.[70] In a heavy drinker, immediately on withdrawal, the blood pressure soars, but abstinence from drinking restores normal blood pressure in most drinkers after a while.[71] Furthermore, alcohol causes strokes—even *without* hypertension.[72] Advice on alcohol use is quite straightforward, then: drink, if at all, in moderation. *Moderation* means 1 to 2 drinks a day, not more.[73]

Caffeine Although caffeine ingestion produces an acute rise in blood pressure, its association with permanent hypertension is not confirmed.[74] It is possible that people develop a tolerance to caffeine's effects on blood pressure.[75] A randomized, double-blind, crossover study compared the effects of 200 milligrams of caffeine versus placebo on blood pressure in hypertensive men who normally drank at least two cups of coffee per day.[76] The blood pressure of those who ingested the 200 milligrams of caffeine rose significantly. Conversely, a study of over 66,000 people with varying blood pressure and coffee consumption levels found no association between caffeine and high blood pressure.[77]

Insulin resistance Hypertension appears to be an insulin-resistant state. This is so even in the absence of obesity and may reflect the operation of a factor common to both hypertension and diabetes. Perhaps insulin itself is sometimes, in some way, a causal agent of hypertension.[78] Hypertension and elevated blood cholesterol are much more common in people with elevated blood glucose than in others.[79]

A bit of information that suggests the nature of the link is that insulin enhances the kidney's reabsorption of sodium.[80] Another is that insulin stimulates the activity of the hormones associated with the stress response.[81] Since obesity aggravates both hypertension and insulin resistance, and since all three are associated with type II diabetes, it is obvious that preventive dietary efforts effective against any of these will help with all of them. Diabetes is the subject of an upcoming section.

Diet in Prevention of Hypertension

The role of diet in the *treatment* of hypertension is not questioned. The two most effective dietary measures the person with hypertension can take are to reduce weight if overweight, and to reduce salt intake. As for diet in the *prevention* of hypertension, there is less agreement, but many believe that enough evidence is available to warrant a recommendation to the general public to moderately restrict salt intake. They reason that at worst, such a diet cannot be harmful.

The person wishing to avoid hypertension can, however, take many other dietary measures that may be more useful. Start with weight control. Expend energy, so as to earn the right to eat more nutrients—again, exercise. (If that benefit doesn't motivate you, then exercise to improve circulation, reduce weight, improve morale, or make friends—but anyway, exercise.) Eat foods high in potassium (whole foods of all descriptions), high in calcium and magnesium (whole foods and dairy products), low in fat, and high in fiber. Not all the nutrients that affect blood pressure have been studied yet, so vary the diet. Excesses of some dietary components are harmful, so use moderation.[82] Use moderation with respect to alcohol, too. If the recommendations sound familiar, it may be because the same ones are made in relation to almost every other health goal, as well as hypertension control.

Role of Exercise

Exercise helps not only with energy balance and therefore weight control, but also helps directly to reduce hypertension. Although blood pressure rises temporarily at each bout of exercise, the effect in the long run is to lower the blood pressure significantly.[83]

To reduce hypertension, the best kind of exercise is the endurance type, such as jogging, undertaken faithfully as a daily or every-other-day routine, that strengthens the heart and blood vessels and permanently alters body composition in favor of lean over fat tissue. Such exercise training increases the volume of oxygen the heart can deliver to the tissues at each beat, reducing its work load. Such exercise also changes the hormonal climate in which the body does its work (it alters "sympathetic tone"—stress hormone secretion—in such a way as to lower blood pressure). It brings about a redistribution of body water, and it eases transit of the blood through the peripheral arteries.[84] The exercise to seek is that of great enough intensity to elevate the heart rate and speed up breathing, but of low enough intensity to be sustainable for 20 consecutive minutes.

Exercise helps correct raised blood cholesterol levels, too, and if heart and artery disease has already set in, a monitored exercise program may actually help to reverse it.[85] When heart muscle tissue is threatened by a narrowed artery, the heart begins to compensate by finding alternate vessels through which to deliver the blood. These smaller collateral blood vessels in the heart act as a detour around the blockages, and many times they can avert tissue death. Some evidence from studies using animals suggests that the heart forms new collateral vessels in response to exercise, especially in the young.[86] In

collateral blood vessels: small, alternate blood vessels in the heart that form a detour around blocked or narrowed larger arteries, permitting the heart to deliver blood and thus preventing much tissue death. Collateral vessels may also develop in response to exercise.

human beings, the development of such vessels may be a factor in the excellent post-heart attack recovery seen in some heart attack victims who exercise.[87]

A study of about 20 physically active 65-year-old men found they had lower blood pressure than controls, and less heart disease.[88] The men in this study had made the choice to exercise long before they participated in the study. People who exercise regularly generally are more concerned about their health than those who are sedentary. The physically fit men in this study smoked less, consumed fewer drugs, and weighed less than sedentary controls. The comparisons drawn from this study are limited, but do lend support to the assertion that regular physical activity promotes health. Considering that cardiovascular disease is a major killer of adults in the United States each year, the positive influence of regular physical activity on several of the risk factors for this disease cannot be disregarded or taken lightly. Furthermore, physical *inactivity* is positively associated with coronary heart disease. In fact, physical inactivity appears to be as influential a risk factor for coronary heart disease as are cigarette smoking, hypercholesterolemia, and hypertension.[89] More research is needed before it can be concluded that exercise protects against heart disease, but the evidence so far certainly justifies such research.

diabetes: a condition of abnormal carbohydrate metabolism, resulting in too much glucose in the blood and the presence of glucose in the urine. Diabetes is caused by inadequate production of insulin or the failure to use the insulin produced. The common form, **diabetes mellitus**, is a major risk factor in the onset of heart disease and stroke, and it causes many organ diseases.

type I diabetes: a type of diabetes mellitus that begins in childhood and is characterized by a lack of insulin in the blood, thinness, and failure to grow; also called **juvenile-onset diabetes.**

type II diabetes: a type of diabetes mellitus that sets in during adulthood and is characterized by the presence of adequate insulin in the blood, a lack of tissue response to the insulin, and obesity; also called **adult-onset diabetes** or **maturity-onset diabetes.**

potential diabetes: a state in which glucose tolerance is abnormal on testing, while fasting blood glucose remains normal.

latent diabetes: a state in which glucose tolerance test results are normal, but pregnancy, obesity, or serious disease brings on diabetes symptoms.

Diabetes

Diabetes is not one disease, but several. Dietary preventive measures are effective against the major form of diabetes—diabetes mellitus type II, which accounts for over 90 percent of all diabetes in the United States—and that is the type discussed here. When the term *diabetes* is used without specifying the type, it is this type that is meant; this section follows that convention.

Diabetes is a metabolic disorder characterized by hyperglycemia. It may be caused in part by a failure of insulin secretion, but its major characteristic is a failure on the responding end—insulin resistance. Cells fail to respond to insulin by taking up glucose, and the result is excretion of glucose in the urine, an accompanying loss of water, and mobilization of body fat and protein, leading to acidosis, dehydration, and mineral losses. Characteristic cellular changes in the peripheral blood vessels lead to degeneration of the retina and kidney that can lead to blindness and renal disease. Reduced circulation and sensation in the limbs lead to infection and injury, at the worst necessitating amputation. The risk of cardiovascular disease is far higher among people with diabetes than among the general population.

Risk Factors for Diabetes

The major risk factor for diabetes is impaired glucose tolerance. In turn, the major predictors or risk factors for impaired glucose tolerance are obesity, age, and family history.[90] Diabetes is often described as having at least two precursor states, although the distinctions among them are more orderly than the actual states themselves. In potential diabetes, glucose tolerance is abnormal on testing, while fasting blood glucose remains normal. In latent diabetes, test results are normal, but pregnancy, obesity, or serious disease brings on diabetes symptoms. Another state that implies risk of diabetes is gestational diabetes, which also presents a greater-than-normal risk of birth defects.

In any case, impaired glucose tolerance always comes first. About 15 percent of the population between the ages of 20 and 74 has some degree of glucose intolerance, and for people aged 65 to 74 its incidence is over 40 percent.[91] Obesity precedes or accompanies diabetes in 85 percent of cases.

Diabetes is unlike hypertension in that only about half of people who have diabetes are aware that they have it. (In contrast, hypertension awareness is at an all-time high.) People with the precursor condition, impaired glucose tolerance, may never develop the complications of diabetes if they control their blood glucose levels successfully; therefore, efforts at prevention center on this aspect of control.

gestational diabetes: a type of diabetes mellitus that develops during pregnancy and is associated with birth defects in offspring; often a forerunner of type II diabetes.

Prevention of Diabetes

To prevent diabetes, health authorities recommend screening for impaired glucose tolerance. Among those who have it, recommended measures are all of the following:

- Lose weight, if necessary; adjust the diet to balance kcalorie intake against energy output.
- In pregnancy, gain appropriate weight without excess; this helps prevent gestational diabetes.
- Increase activity; regular exercise improves glucose tolerance.
- Replace rapidly absorbed carbohydrates with complex and high-fiber carbohydrates.
- Reduce intakes of saturated fat.
- Control hypertension, if present, by reducing salt, saturated fat, and alcohol intakes and increasing potassium, magnesium, and fiber intakes.

In general, authorities recommend all of the measures useful against CVD and, if necessary, the pharmacological control of blood glucose as well, using the drug tolbutamide.[92]

Cancer

Almost everyone knows someone who has had cancer; according to present statistics, one in every four people in the United States will contract it.[93] Within the last 50 years, the outlook for recovery from most kinds of cancer has improved somewhat, and for a few kinds, it has improved dramatically. But still, after many years spent in the laboratory and untold millions of research dollars invested, no cures exist for many types of cancer, and virtually none for advanced cases. Some experts are beginning to say that resources should be used to support efforts to find causes, and that the hope of the future lies in preventing, not curing, cancer.[94]

Cancer is a complex disease, and its causes are not easily discovered. Prevention is straightforward only when a culprit has been identified. Smoking and other tobacco use, for example, are known to cause cancer; quitting tobacco use helps to prevent it. Some cancers occur because viruses attack the genes. Others involve other environmental factors, such as water and air pollution.[95] The links between nutrition and cancer are not so clear-cut, but research has revealed some connections that can be presented here. Some constituents in foods may be responsible for starting cancer; some may speed its development; and some may protect against cancer.

Development of Cancer

The steps in cancer development are thought to be:

carcinogen (car-SIN-oh-jen): a cancer-causing substance.
carcin = cancer
gen = gives rise to

initiation: an event, probably in the cell's genetic equipment, caused by a carcinogen or by radiation, that can give rise to cancer.

promotion: assistance in the development of cancer. A **promoter** is a substance that does not initiate cancer, but that favors its development once the initiating event has taken place.

1. Exposure to a carcinogen.
2. Entry of the carcinogen into a cell.
3. A change inside the cell—initiation, probably by the carcinogen's altering the cell's genetic equipment.
4. Facilitation of the cancer's growth—promotion, probably involving several more steps before the cell multiplies out of control to form a tumor.
5. Tumor formation and possible spreading to new locations.

The process is long; it may be 20 years or more from the initial event to the appearance of diagnosable cancer.

Most people think that the major diet-related step is the first one, and that the most effective preventive measure is to avoid eating foods that contain carcinogens. In particular, many people have learned to fear food additives, believing that they are responsible for diet-related cancer.

In reality, food additives probably have little to do with the causation of cancer. The law forbids the addition to food of any substances that have ever been shown to cause cancer in any animal. However, foods themselves contain substances that may influence whether or not people get cancer. These substances may promote or inhibit cancer after the initiating event—that is, after the cancer has been started by some other environmental influence.

The overall adequacy of the diet has far-reaching effects in preventing or promoting cancer. At each step, including the initiation process, the immune system plays a role in stopping cancer development. A diet that fails to provide the essential nutrients weakens the body's immune defenses, making cancer more likely. The reverse is also true—an adequate diet supports optimal defenses in every cell and system, to combat and overcome cancer at each stage of development.[96]

epidemiological studies: studies of populations, in which differing disease incidences are correlated with other factors to identify possible causes of the diseases.

case-control studies: studies of individuals with a disease, matched as closely as possible to other individuals without the disease, in the attempt to isolate causes of the disease.

Several kinds of research reveal links between diet and cancer. Epidemiological studies provide one source of information; these help identify environmental factors that correlate with cancer incidence. Another is case-control studies—studies of people who have cancer and of control subjects as closely matched to them as possible in age, occupation, and other key variables; these help identify differences in lifestyle that may account for differing cancer incidences. Another type of research tests animals under controlled laboratory conditions ruling out all variables except the possible cancer cause of interest.

The most powerful research tool, the human intervention trial, is only now beginning to be employed to a limited degree in cancer research.[97] In these trials, human subjects agree to adopt a new behavior and continue it for years so that any effect of the behavior on cancer can be discovered. Each type of study must be interpreted with an awareness of its limitations.

intervention trials: studies in which preventive measures are tried on half of a population to see if they will reduce the incidence of a disease.

Findings from Epidemiological Studies

A thought-provoking finding from studies of populations comes from the comparison of high-risk and low-risk areas. If only 10 people out of 1000 get a certain kind of cancer in location X, while 100 out of 1000 get that same kind of cancer in location Y, researchers are inclined to conclude that 90 percent of the cancers in location Y are caused by some environmental factor and are therefore, in theory, preventable. Comparison of high-risk and low-risk areas suggests that 80 to 90 percent of human cancers may indeed be preventable.[98] Hence the great challenge to nutrition researchers is to discover what dietary factors differ between people who do and do not get cancer, and to what extent such factors may influence the development of cancer.

Japanese immigration to the United States after World War I provided an especially interesting opportunity for cancer research. The Japanese living in Japan develop more stomach cancers and fewer colon cancers than people in the United States and other Western countries. However, when Japanese people have come to the United States, their children have developed both stomach and colon cancers at a rate like that of U.S. citizens. The altered susceptibility of Japanese immigrants is probably not due to pollution, because Japan and the United States are both industrial countries. An obvious candidate is diet.

Although the incidence of colon cancer rises in the immigrants just described, the incidence of breast cancer remains the same as in the homeland, even in second-generation immigrants. No change in breast cancer rates shows up until the third generation.[99] This contrasts with worldwide trends, where breast and colon cancer correlate, rising and falling together in the same population. A possible explanation is that the rate of breast cancer in adulthood reflects the food intakes of childhood, so that it takes more than one generation to bring about a change. Another is that the women who become pregnant adopt the trusted traditional diet of the old country, thereby changing the uterine environment and also their offspring's predisposition to cancer. Maternal diet is known to affect the cancer rates of offspring of mice; the effect could possibly work in people, too.[100]

Other epidemiological studies provide additional clues pointing to diet. For example, Seventh-Day Adventists have rates of most forms of cancer lower than those of the general population. Members of this religion refrain from smoking and using alcohol, but when cancers linked to smoking and alcohol are taken into account, Seventh-Day Adventists still have a mortality rate from cancer about one-half to two-thirds that of the rest of the population, suggesting that diet is responsible. Seventh-Day Adventists' foodways center on a lacto-ovo-vegetarian diet (one that includes milk and eggs). While not all church members are vegetarians, those who eat meat do so sparingly and do not eat biblical "unclean meats," such as pork and shellfish. Most church members also follow health recommendations to avoid caffeine and certain spices. Their low cancer mortality may be due to their low meat intakes, to their high intakes of vegetables and cereal grains, or to some other dietary factor, such as coffee avoidance. Factors other than diet have not been ruled out. Seventh-Day Adventists are of a higher-than-average socioeconomic level, and most are college educated.[101] In an early study of dietary factors and cancers, one pair of investigators studied the diets of people in 37 countries. They documented the food available per person per day, as well as other indicators of lifestyle, such as possession of radios and motor vehicles. They found many correlations, but one of the most interesting showed both breast and colon cancer to be strongly associated with "indicators of affluence, such as a high-fat diet rich in animal protein."[102] Other researchers analyzed almost 100,000 medical records and found that high blood cholesterol (an indicator of a high-fat diet) predicted colon and rectal cancers.[103]

Such an attempt to link dietary components with disease must be approached with caution. An increase in one component of the diet causes increases or decreases in others. If a close correlation is shown between a disease and, say, the consumption of animal protein, it would not prove the critical factor to be the animal protein. It might be increased fat consumption; fat goes with animal protein in foods. Or the disease might occur because of what was crowded out: the vitamins, minerals, or fiber contained in the missing fruits, vegetables, and cereals. Remember, too, that owning radios and motor vehicles was also positively linked to the cancers studied. While it is improbable that the gadgets themselves cause cancer, they do foster a sedentary lifestyle, and evidence points to lack of exercise as a contributor to many diseases, including some forms of cancer.[104]

Another problem inherent in epidemiological studies is that they depend on dietary recall. People tend to have trouble remembering how much of each food they have eaten. Moreover, in the case of cancer studies, the need is not so much to know what the diet is like now as to know what it was like at an earlier time—say, 30 years ago—when the initiating event may have taken place. In this connection, study of the Seventh-Day Adventist vegetarians offers hope of clarifying the relationship between animal protein and cancer. Most can recall exactly when they quit eating meat—they quit when they joined the church.

In general, wherever the diet is high in fat-rich foods, it is simultaneously low in vegetable fiber. The association of fat with cancer is stronger than that of low fiber intake, but fiber may independently help to protect against some cancer—for example, by promoting the excretion of bile from the body, or by speeding up the transit of all materials through the colon so that the colon walls are not exposed for long to cancer-causing substances. That fiber does have an independent protective effect of some kind is supported by evidence from Finland. The Finns eat a high-fat diet, but unlike other such diets, theirs is very high in fiber as well. Their colon cancer rate is low, suggesting that fiber has a protective effect even in the presence of a high-fat diet.[105]

If fat, a meat-rich diet, or both are implicated in the causation of certain cancers, and if fiber, a vegetable-rich diet, or both are associated with prevention, then vegetarians should have a lower incidence of those cancers. They do. The Seventh-Day Adventists have already been mentioned; other vegetarian women also have less breast cancer than do meat eaters. A study of people in Minnesota and Norway found less-frequent use of vegetables in people with colon cancer; a New York study found, specifically, less use of the so-called cruciferous vegetables—cabbage, broccoli, and brussels sprouts—in colon cancer victims. Similarly, careful comparisons of stomach cancer victims' diets with those of case controls show less use of vegetables in the cancer group—in one case, vegetables in general; in another, fresh vegetables; in others, lettuce and other fresh greens, or vegetables containing vitamin C.[106]

cruciferous vegetables: vegetables of the cabbage family.
cruci = cross-shaped (referring to the shape of the flowers)

When environmental causes of another kind of cancer—that of the head and neck—have been sought, the major factor has appeared not to be diet but the combination of alcohol and tobacco consumption. Again, however, some dietary factors have turned up here and there, pointing to a low intake of fruits and raw vegetables in cancer cases. This time, intakes were low specifically of the fruits and vegetables that contribute carotene (the vitamin A precursor) and riboflavin. Carotene and its relatives the retinoids are also important in preventing cancers of epithelial origin, including skin cancer.[107]

Among the known actions of vitamin A are the important roles it plays in maintaining the immune function. A strong immune system may be able to prevent cancers from gaining control, even after they have gotten started in the body. Some studies suggest that this may be one place vitamin A makes its contribution. In Norway, a five-year study showed lung cancer incidence to be 60 to 80 percent lower in men with a high vitamin A intake than in those with a low intake. In Japan, a study of 280,000 people showed lung cancer rates to be 20 to 30 percent lower in smokers who ate yellow or green vegetables daily than in those who did not. In ex-smokers who ingested yellow or green vegetables daily, the reduction was much greater, as if something in the vegetables enhanced the *repair* of damage done by smoking after the initiation of cancer.[108] Other studies of the microscopic events that occur during cancer initiation indicate that vitamin A may also play a role in defending against the earliest cancerous changes.[109] Abundant evidence along these lines makes clear that anyone at risk for the development of cancer should obtain adequate vitamin A and its previtamin carotene. Whether *excess* carotene is extra beneficial against cancer remains for intervention studies to prove or disprove.

In general, studies of populations have suggested that low cancer rates correlate with low meat and high vegetable and grain intakes, but there are exceptions. For example, another religious group, the Mormons, do not limit meat; they consume it in amounts that are typical of most people living in the United States. Still, Mormons experience lower rates of breast and colon cancer than other U.S. citizens do.[110] Case-control studies, in which researchers can control some of the variables, can help to confirm or refute the meat-cancer relationship.

Findings from Case-Control Studies

Case-control studies have generally implicated diet in cancer causation. When 179 Hawaiian Japanese people with colon cancer were carefully matched with 357 Hawaiian Japanese people without cancer, those with cancer were seen to have a strikingly higher consumption of meat, especially beef. An Israeli study showed fiber consumption to be lower in victims of colon cancer than in comparable people who did not have cancer. A study of U.S. blacks with colon cancer showed that they ate less fiber and more saturated fat than others without cancer. These studies and many others like them have led reviewers to the view that a diet "high in total fat, low in fiber, and high in beef [is] associated with an increased incidence of large-bowel cancer in man."[111] Others deny that these findings have established a strong association between fat consumption and colon cancer.[112]

Findings from Studies on Animals: Fat and Cancer

Once population and case-control studies have identified a possible dietary link to cancer, researchers often turn to experiments with laboratory animals. By using animals, researchers can control many variables while manipulating only the diet.

Laboratory studies using animals confirm suspicions that fat, of all dietary components, is most strongly correlated with cancer. For example, it is well known that the number of mammary tumors in rats increases as fat is added to the diet, especially if the fat is unsaturated.[113]

People are often surprised to learn that fat may be far more significant in cancer causation than food additives, which have long been suspected and feared. Fat is thought to act as a promoter that somehow enhances the process by which cancer becomes established in a cell or tissue. A high-fat diet may advance cancer:

- By altering body tissue responsiveness to certain hormones (for example, growth hormone), thus stimulating cell division and the advancement of certain cancers.
- By promoting the secretion of bile into the intestine; bile may then be converted by organisms in the colon into compounds that cause cancer.
- By supplying unsaturated fat, which can split into molecules that can initiate cancer.
- By inhibiting the production of molecules that modulate cell division (prostaglandins).
- By blocking the communications pathways cells use to signal their neighbors to stop dividing.
- By reducing the immune system's effectiveness in destroying cancerous cells.
- By contributing to obesity (a known cancer risk factor).
- By being incorporated into cell membranes and changing them so that they offer less defense against cancer-causing invaders.
- By delivering to tumors the concentrated energy they need to grow rapidly.

Fat in general is associated with increased cancer, but studies seem to point to polyunsaturated fat specifically, whereas diets high in the omega-3 series of polyunsaturated fatty acids seem to *enhance* defenses against cancer. Whether these findings will be substantiated by replication is unclear, but one thing is clear: diets relatively higher in total fat (20 percent or more), and in energy, enhance cancer development.[114] Overnutrition and overweight are directly related to a high cancer risk.[115]

Dietary Prevention of Cancer

Studies are in progress to determine if people who are at risk for certain cancers, or who already have them, can benefit from the same kind of diet that benefits rats—one that is very low in fat.[116] There would appear to be no harm in reducing the fat intake of adults from the widespread 40 or so percent to 20 percent or less of total kcalories, and to bring energy intakes into line with energy needs. (For children, this may be too low. Fat supports growth.) For the average Westerner, accomplishing this would mean choosing low-fat alternative foods; drastically reducing the amount of fat used in food preparation; and refraining from adding fat, such as butter, margarine, or salad dressings, to foods at the table.

Among the suspects for the causation of stomach cancer are nitrosamines, produced in the stomach from nitrites. Vegetables may help keep nitrosamines from forming by contributing vitamin C, which inhibits the conversion of nitrites to nitrosamines. Nitrites are present in the water supply and are present in high concentrations in many vegetables. They are also made in the human body in quantities much larger than those found in food. We can't avoid nitrites, but perhaps we can help to prevent their conversion to nitrosamines (which cause cancer) by eating vitamin C-containing vegetables and fruits along with them. This approach to the prevention of stomach cancer has a strong theoretical basis, but as yet, no actual experiments on human beings have proven it effective. Among unanswered questions: Within what range of stomach acidities is vitamin C effective in preventing nitrosamine formation? How effective is vitamin C in the presence of other agents that promote nitrosamine formation? How much vitamin C has to be present for an effect to occur?[117] It makes sense, for other reasons, to eat foods containing plenty of vitamin C, so health authorities do not hesitate to recommend this as a daily practice.

What other substances might vegetables contribute to help protect the body against cancer? Among other nutrients now known to be important in the functioning of the immune system are vitamin B_6, folate, pantothenic acid, vitamin B_{12}, vitamin E, iron, and zinc. Doubtless, there are others.

In 1980, the Food and Nutrition Board of the National Academy of Sciences stated that not enough evidence was yet available to justify making any recommendations for the dietary prevention of cancer. Two years later, however, under heavy pressure from the public, the National Academy did publish some provisional recommendations.[118] It was as if they were saying, "This is what we are tempted to recommend, but we don't think we can, yet." Among the provisional recommendations were these:

1. Reduce the consumption of both saturated and unsaturated fats.
2. Include fruits (especially citrus fruits), vegetables (particularly carotene-rich and cruciferous vegetables), and whole-grain products in the daily diet.

In addition, they recommended minimizing consumption of smoked and cured foods, on the supposition that the carcinogens they contained might increase the risk of cancer. They also endorsed protection of the food supply against contamination with carcinogens from any source, as well as continued evaluation of food additives for carcinogenic activity. It bears repeating that additives legally permitted in foods do not contribute significantly to people's overall risk of cancer.

Much remains to be learned about the connections between nutrition and cancer. Still, many people working in cancer research believe that enough is known already to warrant taking the tentative preventive steps provisionally recommended by the National Academy. One pair of reviewers says, "The public is looking for answers regarding this diet-cancer link and will look to anyone willing to provide answers, regardless of his/her qualifications. . . . The recommendations offered here constitute no risk and may help lower the incidence of . . . cancers."[119] In other words, it can't hurt, and it may help.

Osteoporosis

osteoporosis (OSS-tee-oh-pore-OH-sis): a disease of older persons in which bone mass is lost and the bone becomes porous; also known as **adult bone loss.**
osteo = bones
poros = porous

The skeletal system changes over time. Bone-building and bone-dismantling cells are constantly remodeling this structure, but with age, bone-building cells are lost, and so bone dismantling outpaces bone building. The result is osteoporosis—a disease that today afflicts close to half of all people over 65. The bones of a person with osteoporosis may become so fragile that simply tossing in bed at night can cause a break. In older people without frank osteoporosis, the reduced number of bone-building cells means that breaks do not heal as well or as rapidly as they once did. The extent to which osteoporosis can be prevented is not yet known. Evidence that a poor calcium intake correlates with its occurrence is conflicting.[120] The association between physical inactivity and accelerated bone loss is well established, and it is known that the rate of osteoporosis development accelerates in women who do not take estrogen replacement therapy at menopause. Osteoporosis is four times more prevalent in women than in men after age 50.

The pelvic bone and spine are especially susceptible to bone loss. Figure 9–5 shows the effect of the loss of spinal bone on a woman's height and posture. It is not inevitable that people "grow shorter" as they age, but it is more likely to happen if they don't take the measures necessary to prevent bone loss.

Calcium Stores and Bone Loss

The skeleton serves as a bank to keep the blood supplied with a constant concentration of calcium. Withdrawals and deposits of calcium are regulated by hormones sensitive to blood levels of calcium. Calcitonin, made in the thyroid gland, is secreted whenever the calcium concentration rises too high; it acts to stop withdrawal from bone and to slow absorption from the intestine. Parathyroid hormone has the opposite effect. Blood calcium, therefore, does not diminish if a person eats too little calcium to balance excretion, but bone calcium does diminish. A person can go without dietary calcium for years and never suffer a noticeable symptom. Only late in life will it suddenly become apparent that bone loss has been occurring throughout the adult years.

It seems logical that a person's childhood calcium intake might affect the timing of osteoporosis development, but conflicting evidence surrounds this

Figure 9–5 Loss of Height in a Woman Caused by Adult Bone Loss
On the left is a woman at menopause and on the right, the same woman 30 years later. Notice that collapse of her vertebrae has shortened her back; the length of her legs has not changed.

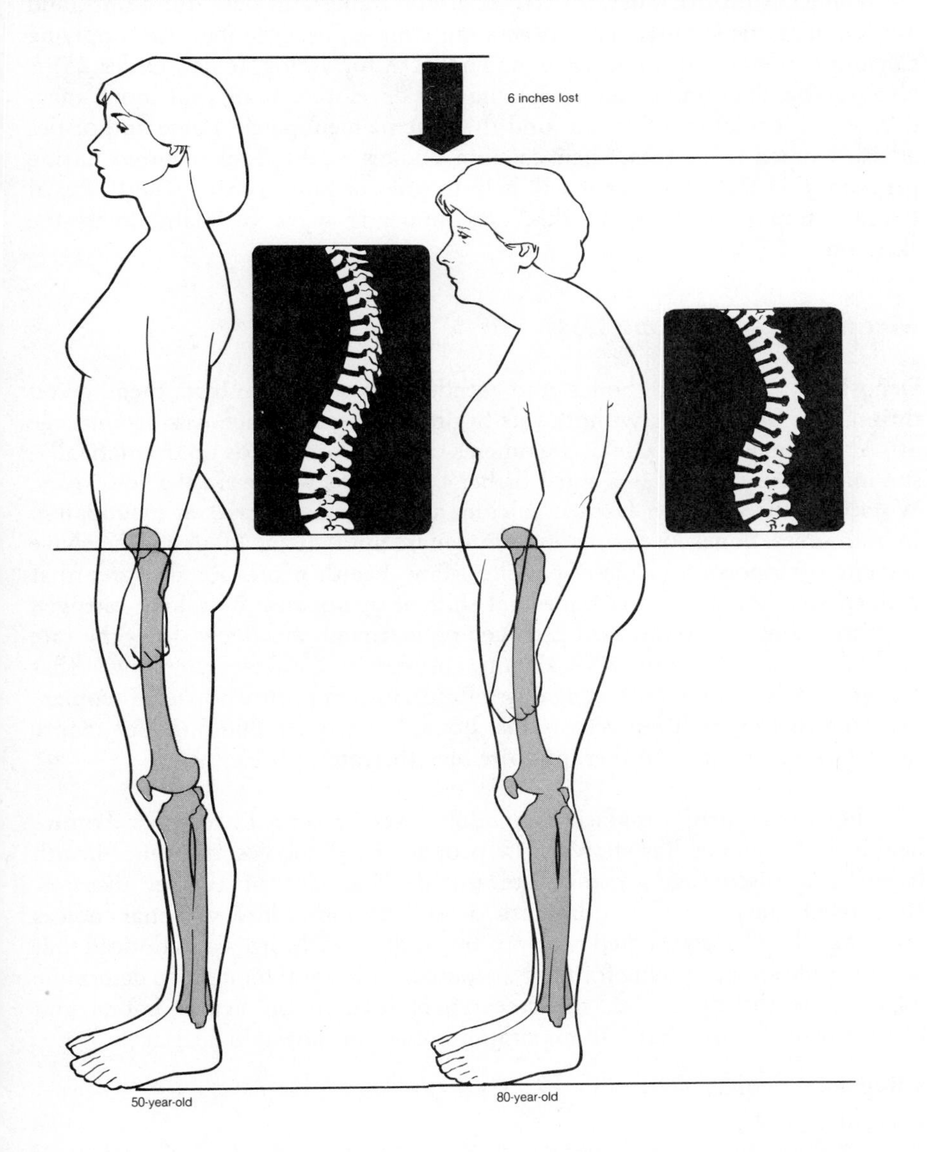

possibility. Although high bone density maintains skeletal integrity and defends against bone loss in later life, and although reduced bone density is the major predictor of osteoporotic fractures, it is not clear that calcium intake controls bone density. Genetically determined hormonal factors are powerful, and adults seem to arrive at their later years with preprogrammed skeletal densities regardless of prior calcium intakes. No one opposes the effort to obtain ample dietary calcium throughout life, but the consensus among bone researchers as of 1984 was that calcium lack was only one factor, and perhaps a minor factor, in osteoporosis development.[121] Other nutrient deficiencies—especially of fluoride and vitamin D—and indeed many other factors contribute to osteoporosis. Besides heredity and hormones, alcohol abuse, lack of exercise, prolonged use of prescription drugs, and other drugs all affect the bones.

The RDA for calcium has been set at 800 milligrams daily for adults, and 1200 milligrams for pregnant or lactating women, because they are supplying calcium for fetal and infant growth. The RDA for young adults (under 25) is also 1200 milligrams to allow for the full development of peak bone mass during the formative years. Around the time of menopause, some authorities advise women to increase their calcium intakes to the level required during pregnancy, in the hope that it will help to prevent bone loss.[122] The Practical Point "Supporting Bone Health," recommends ways of maintaining the skeleton.

Menopause and Bone Loss

Deposits of calcium into bones, and withdrawals of calcium from them, go on throughout life, but the withdrawals begin to exceed the deposits around age 30. At menopause, the rate of a women's calcium loss speeds up dramatically; she may lose as much as a third of her bone mass within only a few years. Worsening the situation further, calcium absorption decreases at menopause. In both sexes, bones lose more and more mass until, at age 60 or so, they have become osteoporotic in many people. Many health professionals agree that women should counter the hormonal shift of menopause by taking estrogen from that time of life onward. Estrogen replacement effectively slows the rate of bone loss and improves calcium absorption.[123] Research shows that estrogen therapy may protect against hip fracture in postmenopausal women. Estrogen therapy will not restore lost bone, but women entering menopause should consider estrogen therapy to reduce the rate of bone loss.

The choices that people make in their adult years affect their present and future health. This chapter has shown how people's food choices and other health habits can discourage or promote the development of certain diseases. Researchers have provided a dramatic demonstration of how personal choices affect health.[124] They studied nearly 7000 adults in California and noticed that some people seemed young for their age; others, old for their age. To determine what made the difference, the researchers focused on health habits and identified six factors that had maximum impact on physiological age:

physiological age: a person's age as estimated from his or her body's health and probable life expectancy.

- Regular, adequate sleep.
- Regularity of meals.
- Regular physical activity.
- Abstinence from smoking.
- Abstinence from, or moderation in, alcohol use.
- Weight control.

The effects of these factors were cumulative. Those who followed all six positive health practices were in better health, even if older by chronological age, than people who failed to do so. The physical health of those who reported all positive health practices was consistently about the same as that of

chronological age: a person's age in years from his or her date of birth.

people *30 years younger* who followed few or none. Some people will become ill or die early no matter what precautions they take; others will be long-lived and healthy without taking any precautions; the large majority, however, will affect their health with the choices they make each and every day. The next chapter focuses on how nutrition, exercise, and other lifestyle practices continue to influence the health and nutrition status of older adults.

▸▸ **PRACTICAL POINT**

Supporting Bone Health

Bodies do much to ensure that their bones will grow and remain strong throughout reproductive life. Given a reasonable dietary supply of calcium, they adjust absorption to meet the needs of different times of life—from more than 60 percent of that ingested during times of rapid growth or high need, to 10 percent or less if supply exceeds demand. Cases are on record of adults' maintaining calcium balance on intakes of 600, 500, 400, and even less than 200 milligrams a day for long periods.[a]

Still, adults can enhance their bodies' work in managing these remarkable adjustments. Because substance abuses of many kinds impair calcium balance, don't smoke, use alcohol or caffeine to excess, or take any medicine unnecessarily. Because the hormone estrogen is a key regulator of calcium balance, and because estrogen secretion becomes abnormal if body fat content falls too low, maintain a reasonable food intake and weight. Because exercise, within reason, places demands on the bones that favor calcium deposition, maintain a schedule of regular exercise.

Beyond these measures, take steps to obtain a calcium intake near the RDA, especially if you have done so since childhood. The body loses calcium when asked to adjust to lower-than-customary intakes, and it takes weeks or months to accomplish this adjustment.

For those who like and can tolerate milk and milk products, they offer the easiest way to obtain needed calcium; two to three servings daily can meet adults' calcium recommendations. People who dislike milk can find substitutes for it. Care is needed, though—*wise* substitutions must be made. Most of milk's many relatives are recommended choices: yogurt, kefir, buttermilk, cheese (especially the low-fat or nonfat varieties) and, for people who can afford the kcalories, ice milk. Even lactose-intolerant individuals can absorb significant quantities of calcium from milk or fermented milk products if they take them in small, divided doses throughout the day. Some highly reputed milk products are less-than-ideal sources, though. Cottage cheese is only fair, 2 cups being equivalent in calcium to 1 cup of milk. Butter, cream, and cream cheese contain negligible calcium, being almost pure fat.

If no calcium-rich milk product is acceptable as is, consider tinkering with it. Make pudding, cream soups, macaroni and cheese. Add chocolate to milk; fruit to yogurt or kefir; or nonfat milk powder to *any* dish—meatloaf, cookies, hamburgers, gravies, soups, casseroles, sauces, milkshakes, even beverages such as coffee or tea, hot or iced. Only five heaping tablespoons are the equivalent of a cup of fresh milk.

▸▸ PRACTICAL POINT *continued*

Equal to milk and milk products in calcium richness are small fish such as sardines or herring prepared with the bones left in the meat; salmon (with the bones); or oysters. Another rich source of calcium is extracts made from bones. (Witches are said to make magic with such brews, but actually, wise food preparers of all cultures favor them, too.) Save the bones from chicken, turkey, pork, or fish dishes, soak them in vinegar, and boil them. The bones release their calcium into the acid medium, and the vinegar boils off, leaving no acid taste behind. Then use the stock in place of water to cook soup, vegetables, rice, or stew. One *tablespoon* of such stock may contain over 100 milligrams of calcium.

Other foods that appear equal to milk in calcium richness are dark green leafy vegetables, and some grains and legumes. Actually, however, many of these provide very little calcium, because they contain binders that hold onto the calcium and prevent its absorption. The presence of calcium binders does not make greens and grains inferior foods; dark greens are a superb source of riboflavin, virtually indispensable for the vegan or anyone else who drinks no milk. Greens also are rich in dozens of other essential nutrients; and grains are excellent sources of complex carbohydrate and fiber.

Next in order of preference among nonmilk sources of calcium are foods that contain large amounts of calcium salts by an accident of processing or by intentional fortification. In the processed category are bean curd (tofu) if a calcium salt is used in its preparation; canned tomatoes (they offer 63 milligrams per cup); stone-ground or self-rising flour; stone-ground whole or self-rising cornmeal; and blackstrap molasses. Among food products specially fortified to add calcium to people's diets, the richest in calcium is high-calcium milk itself (a new product available from major dairies), which provides more calcium per cup than any natural milk—500 milligrams per 8 ounces. Then there is calcium-fortified orange juice, with 300 milligrams per 8 ounces—a good choice, because the bioavailability of its calcium compares favorably with that of milk. Calcium-fortified soy milk can also be prepared so that it contains more calcium than whole cow's milk.[b] Infant formula, based on soy, is fortified with calcium, and no law says adults can't use it in cooking for themselves. The vegan, especially, may find it worthwhile to use such products regularly in ample quantities. Other products that may be fortified if demand warrants are soft drinks, breads, cereals, and others. These are not the equal of "real" foods as calcium sources, but they rank somewhere above supplements in most people's estimates.

Finally, there are supplements—not foods, but packages of nutrient preparations intended to meet the need for calcium without regard to needs for energy or other nutrients. Supplements are a poor source of calcium in comparison with milk, milk products, and other foods, but some authorities consider them acceptable as a last resort.

Calcium comes in combination with a number of different anions in organic and inorganic salts. The organic salts include the lactate, the gluconate, and the citrate; the inorganic salts include the carbonate, the phosphate, and others. The carbonate is 40 percent calcium, the gluconate only 9 percent—so the user has to swallow fewer pills to get the needed calcium from the carbonate. On the other hand, the organic salts are probably better absorbed, especially by older people with reduced gastric acid production. Acid aids in calcium absorption, and calcium carbonate is an antacid, requiring the healthy

▸▸ PRACTICAL POINT *continued*

stomach's secretion of abundant hydrochloric acid to help it get into the system. People with reduced stomach acid secretion may absorb only one-tenth as much calcium from calcium carbonate pills as they would from the citrate. The effect is abolished by taking calcium carbonate with a full breakfast, so this is probably the strategy to adopt.[c]

Regular vitamin-mineral pills contain no significant calcium at all. The label may misleadingly list some number of milligrams of calcium that sounds like a lot, but consumers need to be reminded that the U.S. RDA for calcium is a gram—1000 milligrams.

Calcium pills contain only calcium; they **do not** offer the other nutrients that accompany calcium as fringe benefits in a food such as milk—thiamin, riboflavin, niacin, potassium, phosphorus, vitamin A, and all the rest. Therefore, the person who omits milk and attempts to make up for it by taking calcium supplements still is left with the task of obtaining all the other nutrients, and for some of these, milk is an excellent source. Milk also helps the body absorb calcium, thanks partly to its vitamin D content. Its essential nutrients are present in appropriate amounts; imbalances are not likely, and overdoses are impossible. Milk's long-term safety is assured. The habit of drinking milk is easy to sustain, once established, because it fits with meals, provides kcalories, and can be served in delicious forms. It is low in cost, and fits well into a food budget. Milk drinking squares with the philosophy that using whole foods is preferable to taking supplements; pill taking contradicts that philosophy.

If supplements must be used, here are some guidelines. Be careful to read the calcium contents correctly. A 1-gram pill does not offer 1 gram of calcium, it offers 1 gram of the salt, of which calcium is a part. Take the supplement several times a day in divided doses rather than all at once, to improve absorption, and take it with a meal or snack each time.

At menopause, estrogen levels fall, and bone minerals drain rapidly from the skeleton. Women should consider estrogen therapy at menopause, and possibly starting even before, so as to minimize the early, accelerated loss of calcium from bone that begins as estrogen first begins to decline. No matter how high a woman's calcium intake, it cannot override the influence of the estrogen lack that accelerates bone loss. Importantly, too, women at menopause should continue exercising regularly. No matter how many glasses of milk a person drinks in a day, calcium will not be maximally deposited or retained in bones unless the person *works* them. Bones are like muscles; it takes regular exercise to make them strong and to keep them that way.

---

[a]O. J. Malm, Adaptation to alterations in calcium intake, in *The Transfer of Calcium and Strontium across Biological Membranes*, (New York: Academic Press, 1963), pp. 143–173; D. M. Hegsted, I. Moscoso, and C. Collazos, A study of the minimum calcium requirements of adult men, *Journal of Nutrition* 46 (1952): 181-201.
[b]M. Hirotsuka and coauthors, Calcium fortification of soy milk with calcium-lecithin liposome system, *Journal of Food Science* 49 (1984): 1111–1112, 1127.
[c]R. R. Recker, Calcium absorption and achlorhydria, *New England Journal of Medicine* 313 (1985): 70–73.

Chapter 9 Notes

1. P. O'Sullivan, R. A. Linke, and S. Dalton, Evaluation of body weight and nutritional status among AIDS patients, *Journal of the American Dietetic Association* 85 (1985): 1483–1484.
2. B. M. Dworkin and coauthors, Selenium deficiency in the acquired immunodeficiency syndrome, *Journal of Parenteral and Enteral Nutrition* 10 (1986): 405–407.
3. S. S. Resler, Nutrition care of AIDS patients, *Journal of the American Dietetic Association* 88 (1988): 828–832.
4. U.S. Senate, Select Committee on Nutrition and Human Needs, *Dietary Goals for the United States* (Washington, D.C.: Government Printing Office, February 1977). The *Goals* were revised and reprinted as a second edition in December 1977. We have simplified the wording, but it agrees with both the first and second editions of the *Goals*.
5. U.S. Department of Health and Human Services, Public Health Service, *Promoting Health, Preventing Disease: Objectives for the Nation* (Washington, D.C.: Government Printing Office, 1980), p. 1.
6. A. D. Harries, V. A. Danis, and R. V. Heatley, Influence of nutritional status on immune functions in patients with Crohn's disease, *Gut* 25 (1984): 465–472.
7. J. D. Stinnett, Protein-calorie malnutrition and host defense, in *Nutrition and the Immune Response* (Boca Raton, Fla.: 1983), p. 113.
8. Stinnett, 1983, p. 114.
9. *America's Health: A Century of Progress but a Time of Despair*, a booklet (1983) available from the American Council on Science and Health, 47 Maple St., Summit, NJ 07901.
10. Parts of the discussion on atherosclerosis were adapted from E. M. N. Hamilton, E. N. Whitney, and F. S. Sizer, *Nutrition: Concepts and Controversies*, 4th ed. (St. Paul, Minn.: West, 1988), pp. 131–142.
11. H. Sheldon, *Boyd's Introduction to the Study of Disease*, 9th ed. (Philadelphia: Lea and Febiger, 1984), pp. 347–348.
12. G. V. Mann, Diet-heart: End of an era, *New England Journal of Medicine* 297 (1977): 644–650.
13. Mann, 1977.
14. Mann, 1977.
15. C. Lenfant and B. M. Rifkind, in Diet and heart disease: Responses to the LRC-CPPT findings, *Nutrition Today*, September-October 1984, pp. 22–29.
16. R. E. Olson, in Diet and heart disease: Responses to the LRC-CPPT findings, *Nutrition Today*, November-December 1984, pp. 22–25.
17. A. M. Gotto, Jr., president of the American Heart Association; professor, Department of Medicine, Baylor College of Medicine; and professor, Methodist Hospital, Baylor, Texas.
18. NIH Consensus Conference, Lowering blood cholesterol to prevent heart disease, *Journal of the American Medical Association* 253 (1985): 2080–2086.
19. W. C. Taylor and coauthors, Cholesterol reduction and life expectancy, *Annals of Internal Medicine* 106 (1987): 605–614; A. M. Epstein and G. Oster, Cholesterol reduction and health policy: Taking clinical science to patient care (editorial), *Annals of Internal Medicine* 106 (1987): 621–623; M. H. Becker, The cholesterol saga: Whither health promotion? *Annals of Internal Medicine* 106 (1987): 623–626.
20. K. M. Anderson, W. P. Castelli, and D. Levy, Cholesterol and mortality: 30 years of follow-up from the Framingham Study, *Journal of the American Medical Association* 257 (1987): 2176-2180.
21. A. M. Gotto, Hypercholesterolemia: An assessment of screening and diagnostic techniques, *Modern Medicine*, April 1987, pp. 28–32.
22. J. Stamler, D. Wentworth, and J. D. Neaton, Is the relationship between serum cholesterol and risk of premature death from coronary heart disease continuous and graded? Findings in 356,222 primary screenees of the multiple risk factor intervention trial (MRFIT), *Journal of the American Medical Association* 256 (1986): 2823–2828.
23. G. Rose and M. Shipley, Plasma cholesterol concentration and death from coronary heart disease: 10 year results of the Whitehall Study, *British Medical Journal* 293 (1986): 306–307.
24. D. M. Hegsted, Nutrition: The changing scene (1985 W. O. Atwater Memorial Lecture), *Nutrition Reviews* 43 (1985): 357–367.
25. U. Goldbourt, E. Holtzman, and H. N. Neufeld, Total and high density lipoprotein cholesterol in the serum and risk of mortality: Evidence of a threshold effect, *British Medical Journal* 290 (1985): 1239–1243, as cited in Cholesterol: Mortality risk for CHD linked to a specific threshold, *Modern Medicine*, November 1985, pp. 147, 151.
26. Gotto, 1987.
27. M. F. Oliver, Strategies for preventing coronary heart disease, *Nutrition Reviews* 43 (1985): 257–262.
28. A. M. Pearson and coauthors, Safety implications of oxidized lipids in muscle foods, *Food Technology*, July 1983, pp. 121–129; M. Cleveland, Determination of oxidized cholesterol compounds in commercially processed cow's milk, Ph.D. dissertation, Florida State University, 1986; J. E. Hunter and T. H. Applewhite, Isomeric fatty acids in the U.S. diet: Levels and health perspectives, *American Journal of Clinical Nutrition* 44 (1986): 707–717.
29. E. H. Ahrens, Jr., The diet-heart question in 1985: Has it really been settled? (editorial), *Lancet*, (1985): 1085–1087.
30. J. D. Hubbard, S. Inkeles, and R. J. Barnard, Nathan Pritikin's heart, *New England Journal of Medicine* 313 (1985): 52.
31. Body weight and serum cholesterol, *Nutrition Reviews* 43 (1985): 43–44.
32. Diet, metals, and hidden heart disease, *Science News* 130 (1986): 201.
33. Try a little TLC, *Science 80*, January-February 1980, p. 15.
34. P. W. Wilson and coauthors, Factors associated with lipoprotein cholesterol levels: The Framingham Study, *Arteriosclerosis* 3 (1983): 273–281.
35. G. Sopko and coauthors, The effects of exercise and weight loss on plasma lipids in young obese men, *Metabolism* 34 (1985): 227–236; L. Goldberg and D. L. Elliot, The effect of physical activity on lipid and lipoprotein levels, *Medical Clinics of North America* 69 (1985): 41–55.
36. E. D. Frohlich, Physiological observations in essential hypertension, *Journal of the American Dietetic Association* 80 (1982): 18–20.
37. W. B. Kannel and T. J. Thom, Incidence, prevalence, and mortality of cardiovascular diseases, in *The Heart*, 6th ed., ed. J. W. Hurst (New York: McGraw-Hill, 1986), pp. 557–565.
38. D. A. McCarron and coauthors, Blood pressure and nutrient intake in the United States, *Science* 224 (1984): 1392–1398.
39. H. Sheldon, *Boyd's Introduction to the Study of Disease*, 9th ed. (Philadelphia: Lea and Febiger, 1984), pp. 120–121.

40. Hypertension Detection and Follow-up Program Cooperative Group, The effect of treatment on mortality in "mild" hypertension, *New England Journal of Medicine* 307 (1982): 976-980.
41. N. M. Kaplan, Non-drug treatment of hypertension, *Annals of Internal Medicine* 402 (1985): 359–373.
42. E. Reisin and coauthors, Effect of weight loss without salt restriction on the reduction of blood pressure in overweight hypertensive patients, *New England Journal of Medicine* 298 (1978): 1–6.
43. H. G. Langford and coauthors, Dietary therapy slows the return of hypertension after stopping prolonged medication, *Journal of the American Medical Association* 253 (1985): 657–664.
44. Langford and coauthors, 1985.
45. S. Wassertheil and coauthors, Effective dietary intervention in hypertensives: Sodium restriction and weight reduction, *Journal of the American Dietetic Association* 85 (1985): 423–430.
46. J. Raison and B. Guy-Grand, Body fat distribution in obese hypertensives, in *Metabolic Complications of Human Obesities,* eds. J. Vague and coeditors (Amsterdam: Elsevier Science Publishers, 1985), pp. 67–75.
47. T. W. Kurtz, H. A. Al-Bander, and C. Morris, "Salt-sensitive" essential hypertension in men: Is the sodium ion alone important? *New England Journal of Medicine* 317 (1987): 1043-1048.
48. A. M. Altschul and J. K. Grommet, Sodium intake and sodium sensitivity, *Nutrition Reviews* 38 (1980): 393–402.
49. M. P. Blaustein and J. M. Hamlyn, Sodium transport inhibition, cell calcium, and hypertension: The natriuretic hormone/Na-Ca exchange/hypertension hypothesis, *American Journal of Medicine* 77, no. 4A (1984): 45–59.
50. J. K. Huttunen and coauthors, Dietary factors and hypertension, *Acta Medica Scandinavica* (supplement) 701 (1985): 72–82.
51. G. Kolata, Value of low-sodium diets questioned (Research News), *Science* 216 (1982): 38–39.
52. H. G. Langford, Dietary potassium and hypertension: Epidemiologic data, *Annals of Internal Medicine* 98 (1983): 770-772.
53. Huttunen and coauthors, 1985.
54. K. T. Khaw and E. Barrett-Connor, Dietary potassium and stroke-associated mortality: A 12-year prospective population study, *New England Journal of Medicine* 316 (1987): 235–240.
55. J. M. Belizan and J. Villar, The relationship between calcium intake and edema-, proteinuria, and hypertension-gestosis: An hypothesis, *American Journal of Clinical Nutrition* 33 (1980): 2202–2206.
56. H. Henry and coauthors, Increasing calcium intake lowers blood pressure: The literature reviewed, *Journal of the American Dietetic Association* 85 (1985): 182–185.
57. McCarron and coauthors, 1984.
58. Kolata, 1982.
59. D. A. McCarron and C. D. Morris, Blood pressure response to oral calcium in persons with mild to moderate hypertension: A randomized, double-blind, placebo-controlled, crossover trial, *Annals of Internal Medicine* 103 (1985): 825–831.
60. M. R. Joffres, D. M. Reed, and K. Yano, Relationship of magnesium intake and other dietary factors to blood pressure: The Honolulu heart study, *American Journal of Clinical Nutrition* 45 (1987): 469–475.
61. R. Weinsier, Recent developments in the etiology and treatment of hypertension: Dietary calcium, fat, and magnesium, *American Journal of Clinical Nutrition* 42 (1985): 1331–1338.
62. Weinsier, 1985.
63. Huttunen and coauthors, 1985.
64. P. Bursztyn, Does dietary linolenic acid influence blood pressure? (letter to the editor), *American Journal of Clinical Nutrition* 45 (1987): 1541–1542.
65. P. T. Williams and coauthors, Associations of dietary fat, regional adiposity, and blood pressure in men, *Journal of the American Medical Association* 257 (1987): 3251–3256.
66. H. R. Knapp and G. A. Fitzgerald, The antihypertensive effects of fish oil: A controlled study of polyunsaturated fatty acid supplements in essential hypertension, *New England Journal of Medicine* 320 (1989): 1037–1043.
67. *Report of the National Committee on Detection, Evaluation, and Treatment of High Blood Pressure,* NIH publication no. 88–1088 (Washington, D.C.: Government Printing Office, 1988).
68. J. P. Knochel, Cardiovascular effects of alcohol, *Annals of Internal Medicine* 98 (1983): 849–854.
69. Knochel, 1983.
70. A. L. Klatsky, G. D. Friedman, and M. A. Armstrong, The relationships between alcoholic beverage use and other traits to blood pressure: A new Kaiser Permanente study, *Circulation* 73 (1986): 628–636; G. D. Friedman, A. L. Klatsky, and A. B. Siegelaub, Alcohol intake and hypertension, *Annals of Internal Medicine* 98 (1983): 846–849.
71. Klatsky, Friedman, and Armstrong, 1986.
72. J. S. Gill and coauthors, Stroke and alcohol consumption, *New England Journal of Medicine* 315 (1986): 1041–1046; R. P. Donahue and coauthors, Alcohol and hemorrhagic stroke: The Honolulu heart program, *Journal of the American Medical Association* 255 (1986): 2311–2314, as cited in Alcohol and hemorrhagic stroke, *Lancet,* (1986): 256–257.
73. A. L. Klatsky, M. A. Armstrong, and G. D. Friedman, Relationship of alcoholic beverage use to subsequent coronary artery disease hospitalization, *American Journal of Cardiology* 58 (1986): 710–714.
74. S. Freestone and L. E. Ramsay, Effect of coffee and cigarette smoking on the blood pressure of untreated and diuretic-treated hypertensive patients, *American Journal of Medicine* 73 (1982): 348–353, as cited by N. M. Kaplan, Non-drug treatment of hypertension, *Annals of Internal Medicine* 102 (1985): 359–373.
75. D. Robertson and coauthors, Tolerance to the humoral and hemodynamic effects of caffeine in man, *Journal of Clinical Investigation* 67 (1981): 1111–1117, as cited by Kaplan, 1985.
76. I. B. Goldstein and D. Shapiro, The effects of stress and caffeine on hypertensives, *Psychosomatic Medicine,* May 1987, pp. 226–235.
77. Klatsky, Friedman, and Armstrong, 1986.
78. E. Ferrannini and coauthors, Insulin resistance in essential hypertension, *New England Journal of Medicine* 317 (1987): 378-379.
79. G. L. Burke, L. S. Webber, and S. R. Srinivasan, Fasting plasma glucose and insulin levels and their relationship to cardiovascular risk factors in children: Bogalusa heart study, *Metabolism* 35 (1986): 441–446.
80. Ferrannini and coauthors, 1987; L. Landsberg, Insulin and hypertension: Lessons from obesity (editorial) *New England Journal of Medicine* 317 (1987): 378–379.
81. J. W. Rowe and coauthors, Effect of insulin and glucose infusions on sympathetic nervous system activity in normal man, *Diabetes* 30 (1981): 219–225, as cited by Landsberg, 1987.

82. J. A. Wilber, The role of diet in the treatment of high blood pressure, *Journal of the American Dietetic Association* 80 (1982): 25–29.
83. C. M. Tipton, Exercise, training, and hypertension, *Exercise and Sports Sciences Reviews* 12 (1984): 245–306; R. S. Williams, R. A. McKinnis, and F. R. Cobb, Effects of physical conditioning on left ventricular ejection fraction in patients with coronary artery disease, *Circulation,* July 1984, pp. 69–75.
84. G. Nomura, Physical training in essential hypertension: Alone and in combination with dietary salt restriction, *Journal of Cardiac Rehabilitation* 4 (1984): 469–475.
85. Tipton, 1984.
86. T. B. Jacobs, R. D. Bell, and J. D. Clements, Exercise, age and the development of myocardial vasculature, *Growth* 48 (1984): 148–157.
87. K. Przyklenk and A. C. Groom, Effects of exercise frequency, intensity, and duration on revascularization in the transition zone of infarcted rat hearts, *Canadian Journal of Physiology and Pharmacology* 63 (1985): 273–278.
88. B. Larsson and coauthors, Health and aging characteristics of highly physically active 65-year-old men, *International Journal of Sports Medicine* 5 (1984): 336–340.
89. Protective effect of physical activity on coronary heart disease, *Morbidity and Mortality Weekly Report* 6 (1987): 426–430.
90. L. C. Harlan and coauthors, Factors associated with glucose tolerance in adults in the United States, *American Journal of Epidemiology* 126 (1987): 674–684.
91. M. C. Kovar, M. I. Harris, and W. C. Hadden, The scope of diabetes in the United States population, *American Journal of Public Health* 77 (1987): 1549–1550.
92. J. Tuomilehto and E. Wolf, Primary prevention of diabetes mellitus, *Diabetes Care* 10 (1987): 238–248.
93. Parts of this section are adapted from E. N. Whitney, E. M. N. Hamilton, and S. R. Rolfes, Chapter 16: Nutrition and disease prevention, in *Understanding Nutrition* 5th ed., (St. Paul, Minn.: West, 1990).
94. J. C. Bailar and E. M. Smith, Progress against cancer? *New England Journal of Medicine* 314 (1986): 1226–1232.
95. A. E. Reif, The causes of cancer, *American Scientist* 69 (1981): 437–447.
96. L. A. Poirier, Stages in carcinogenesis: Alteration by diet, *American Journal of Clinical Nutrition* 45 (1987): 185–191.
97. S. Graham, Fats, calories, and calorie expenditure in the epidemiology of cancer, *American Journal of Clinical Nutrition* 45 (1987): 342–346.
98. B. S. Reddy and coauthors, Nutrition and its relationship to cancer, *Advances in Cancer Research* 32 (1980): 238–245.
99. Reddy and coauthors, 1980.
100. G. L. Wolff, Body weight and cancer, *American Journal of Clinical Nutrition* 45 (1987): 168–180.
101. R. L. Phillips and D. A. Snowdon, Dietary relationships with fatal colo-rectal cancer among Seventh-Day Adventists, *Journal of the National Cancer Institute* 74 (1985): 307–317.
102. B. S. Drasar and D. Irving, Environmental factors and cancer of the colon and breast, *British Journal of Cancer* 27 (1973): 167–172.
103. S. A. Tornberg and coauthors, Risks of cancer of the colon and rectum in relation to serum cholesterol and B-lipoprotein, *New England Journal of Medicine* 315 (1986): 1629–1633, as reported in *Modern Medicine,* May 1987, pp. 136, 141.
104. J. E. Vena and coauthors, Occupational exercise and risk of cancer, *American Journal of Clinical Nutrition* 45 (1987): 318-327; R. E. Frisch and coauthors, Lower lifetime occurrence of breast cancer and cancers of the reproductive system among former college athletes, *American Journal of Clinical Nutrition* 45 (1987): 328–335.
105. E. L. Wynder, Dietary habits and cancer epidemiology, *Cancer* 43 (1979): 1955–1961, as cited by S. H. Brammer and R. L. DeFelice, Dietary advice in regard to risk for colon and breast cancer, *Preventive Medicine* 9 (1980): 544–549.
106. Reddy and coauthors, 1980.
107. J. L. Werther, Food and cancer, *New York State Journal of Medicine,* August 1980, pp. 1401–1408.
108. Werther, 1980.
109. L. M. Deluca and E. M. McDowell, Deletion of essential functions and tumorigenesis, *Journal of Nutrition* 116 (1986): 2064–2065.
110. R. Doll and R. Peto, The causes of cancer: Quantitative estimates of avoidable risks of cancer in the United States today, *Journal of the National Cancer Institute* 66 (1981): 1191-1308, as cited by M. W. Pariza, *Diet and Cancer* (Summit, N.J.: American Council on Science and Health, 1985), p. 12.
111. Reddy and coauthors, 1980.
112. L. N. Kolonel and coauthors, Role of diet in cancer incidence in Hawaii, *Cancer Research* (supplement) 43 (1983): 2297s-2402s; W. C. Willet and B. MacMahon, Diet and cancer: An overview, *New England Journal of Medicine* 310 (1984): 697–703.
113. C. W. Welsch, Enhancement of mammary tumorigenesis by dietary fat: Review of potential mechanisms, *American Journal of Clinical Nutrition* 45 (1987): 192–202.
114. I. Clement, Fat and essential fatty acid in mammary carcinogenesis, *American Journal of Clinical Nutrition* 45 (1987): 218–224.
115. M. W. Pariza and R. K. Boutwell, Historical perspective: Calories and energy expenditure in carcinogenesis, *American Journal of Clinical Nutrition* 45 (1987): 151–156.
116. P. Greenwald and coauthors, Feasibility studies of a low-fat diet to prevent or retard breast cancer, *American Journal of Clinical Nutrition* 45 (1987): 347–353.
117. S. A. Kyrtopoulos, Ascorbic acid and the formation of N-nitroso compounds: Possible role of ascorbic acid in cancer prevention, *American Journal of Clinical Nutrition* 45 (1987): 1344–1350.
118. National Research Council, Committee on Diet, Nutrition, and Cancer, *Executive Summary: Diet, Nutrition, and Cancer* (Washington, D.C.: National Academy Press, 1982).
119. National Research Council, 1982.
120. S. R. Cummings and coauthors, Epidemiology of osteoporosis and osteoporotic fractures, *Epidemiological Review* 7 (1985): 178–208; B. L. Riggs and coauthors, Calcium intake and rates of bone loss in women, *Journal of Clinical Investigation* 80 (1987): 979–982.
121. National Institutes of Health, Osteoporosis and calcium, NIH Consensus Conference, *Journal of Nutrition* 116 (1986): 319-322.
122. National Institute of Arthritis, Diabetes, and Digestive and Kidney Diseases, *Osteoporosis: Cause, Treatment, Prevention,* NIH publication no. 83–2226, April 1983.
123. R. P. Heaney and coauthors, Calcium nutrition and bone health in the elderly, *American Journal of Clinical Nutrition* 36 (1982): 986–1013.
124. N. B. Belloc and L. Breslow, Relationship of physical health status and health practices, *Preventive Medicine* 1 (1972): 409–421.

▶ Focal Point 9

Nutrition and the Aging Brain

This focal point discusses a topic of great interest to researchers today—the interactions between nutrition and the aging brain. Because the area of inquiry is new and difficult, this discussion comes to no specific conclusions or practical applications. It asks more questions than it can find answers to—and the occasional answers do not always provide an understanding of the relationships between nutrient deficiencies, cognition, and the aging brain. Many of the research implications are only speculative, but they are fascinating enough to warrant reporting and thinking about.

First, it is necessary to lay groundwork in three areas: the brain, the aging brain, and some of the nutrients essential to brain function. The final section then presents research about nutrient deficiencies, cognition, and the aging brain.

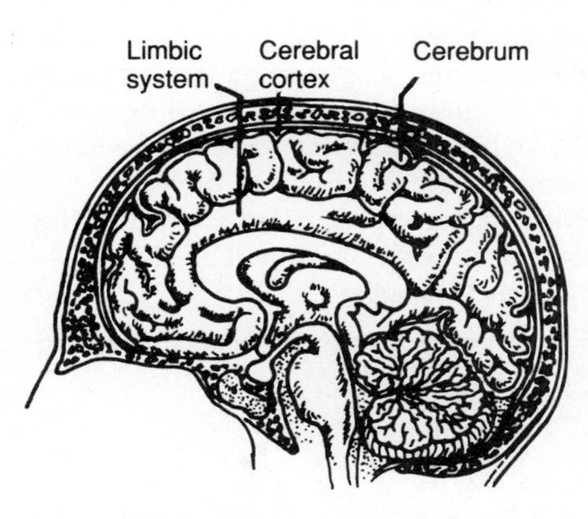

Figure FP9–1 A Section of the Human Brain
The cerebral cortex is the area in which thinking takes place; the cerebrum coordinates motor activity; and the limbic system is the seat of the emotions and learning.

The Brain

The old adage "You can't tell a book by its cover" seems especially apropos of the brain. To look at it, the brain seems an unimpressive lump of convoluted matter, rather like a large sea sponge. Microscopically, the brain takes on intrigue as a fantastic communications network more intricate than any man-made computer, possessing billions of connections between its cells. Functionally, it is the body's awesome master, coordinating muscular movement, life-support functions, sensory integrations, thought and memory, and human emotions. Though it accounts for only about 2 percent of an adult's body weight, and though it expends no energy in physical movement, the brain receives 20 percent of the body's blood flow and uses one-quarter of the available oxygen.

Figure FP9–1 shows a midsection of a human brain. The outermost layer, the cerebral cortex, confers the ability to think, remember, and reason. Underlying the cortex is the site of motor coordination (the cerebrum). The centermost structures of the brain make up the limbic system, the seat of the emotions and learning. Although structurally distinct, the various brain parts communicate with one another through physical connections much like telephone wires. Each part acts on information it receives from a variety of sources: from the other brain parts, from the outside world, from stored memory. The brain, of course, dictates the actions of the body at large, everything from orchestrating the many subtle muscular movements needed to play a flute to instructing the blood vessels to expand or contract.

The cells of the brain are of two types: the neurons, which specialize in transmitting information; and the glial cells or glia, which play a supporting

cerebral cortex: the outer surface of the cerebrum.

cerebrum: the largest part of the brain; the cerebrum controls voluntary muscle function and higher mental functions.

limbic system: a group of brain structures whose primary responsiblity is the regulation of emotional behavior.

neuron: a nerve cell; the structural and functional unit of the nervous system. Neurons initiate and conduct nerve transmissions.

glia (GLY-ah): cells that surround, support, and nourish the neurons.

Figure FP9–2 The Structure of a Neuron
The mature human brain is composed of about 100 billion nerve cells, or neurons, which communicate with one another using a combination of electrical and chemical signals. Each neuron is a long, slim cell with receiving structures (dendrites) at one end; a transmitting structure (the axon) at the other; and a bulge, the cell body, between. Electrical impulses arise in the dendrites, pass through the cell body, and continue down the axon to its terminal. When an impulse arrives at the axon terminal, neurotransmitter molecules are released, thus transferring the signal to the next neuron through the fluid that separates their membranes.

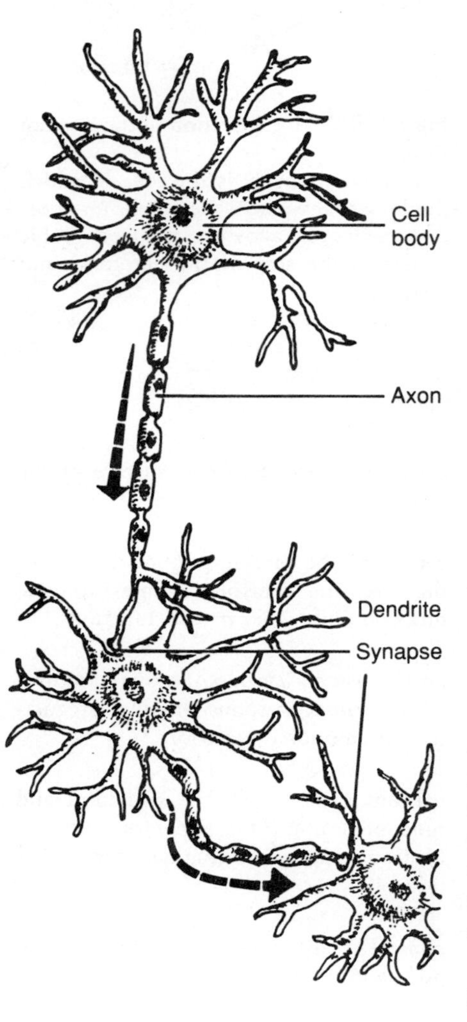

role. Like benevolent caretakers, the glial cells embrace each neuron and sustain its life by providing nutrients and removing wastes via their own internal metabolism.

The structure of a neuron reveals its function (see Figure FP9–2). Each neuron possesses thousands of short, spidery projections, the dendrites, through which it receives messages from other neurons. Each neuron also has a long, fiberlike structure, an axon, through which it sends messages to other nerve cells or to target organs. The axons of some neurons in the brain are so long that they run all the way down the spinal cord, where they terminate in thousands more projections, disseminating information to as many receiving cells. The brain contains 100 *billion* neurons, each with thousands of sending sites and thousands of receiving sites.

A neuron functions somewhat as a gun does—that is, it either fires a transmission or remains silent. Neurons do not partially fire. Subtle shadings of human experience—for example, how hard a pianist strikes a key—are determined by how many neurons fire and how rapidly. A strong reaction is the result of many neurons' firing in rapid succession. For a lesser reaction, fewer fire in less rapid succession. A neuron fires whenever it receives instructions to do so. When the signals received at the dendrites add up to a threshold level that can make the neuron fire, a chemical chain reaction involving the transfer of electrically charged ions (sodium and potassium) across the axon's membrane generates an electrochemical signal. That signal travels down the cell membrane along the length of the axon, somewhat as an electrical current travels down a wire. When a nerve impulse reaches the end of an axon, it transfers the signal to the next neuron through the fluid that separates their membranes. To do this, the neuron releases chemicals, called neurotransmitters, into the fluid-filled space, the synapse. Neurotransmitters send and receive messages from cell to cell. Figure FP9–3 demonstrates how a nerve impulse crosses from one neuron to the next.

Neurons make their own neurotransmitters from the nutrient supplies in the blood that reach the brain. For a nutrient, or any substance, to reach the brain, it must have received special clearance, for the body protects the brain from harmful influences more completely than any other organ.

Three barriers offer the brain protection. First, the GI tract cells selectively allow absorption of specific substances. Second, the substances absorbed circulate through the liver, which selectively removes toxins, drugs, and excess quantities of nutrients before allowing the blood to reach other parts of the body. Third, the brain presents its own barrier: a molecular sieve. Called the blood-brain barrier, this protective device normally lets in only those substances the brain cells particularly need: glucose (or ketones), oxygen, amino acids, and other nutrients.

The brain cells make complex molecules for themselves out of the simple building blocks they accept from the passing blood supply. If there is a *deficiency* of an essential nutrient in the blood supply, the brain's supply falls short; if there is an *excess,* or if substances are circulating in the body that the brain does not need, the contents of brain cells do not usually reflect these fluctuations. Thus, in general, substances in the brain do not necessarily reflect blood concentrations, but the neurotransmitters are an exception.

At least some of the neurotransmitters are unusual in being subject to precursor control; the neurons respond to a larger or smaller supply of

Figure FP9–3 Nerve Impulse Transmission

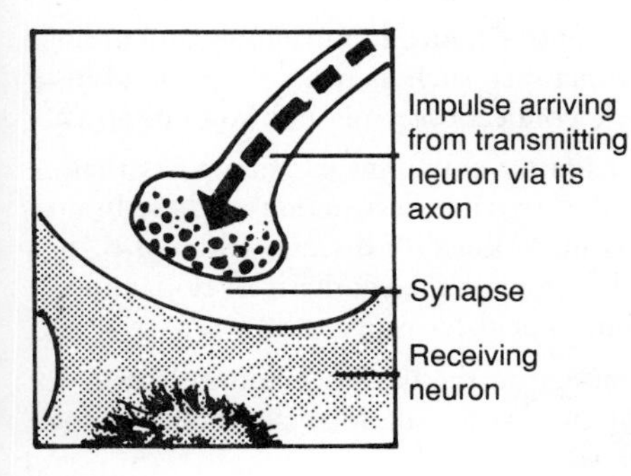

A. The impulse arrives at the end of the first neuron. Clustered just inside the neuron's ending are a multitude of little sacs (vesicles) filled with neurotransmitters.

B. The vesicles fuse with the neuron's membrane, releasing the neurotransmitters into the gap between cells (synapse).

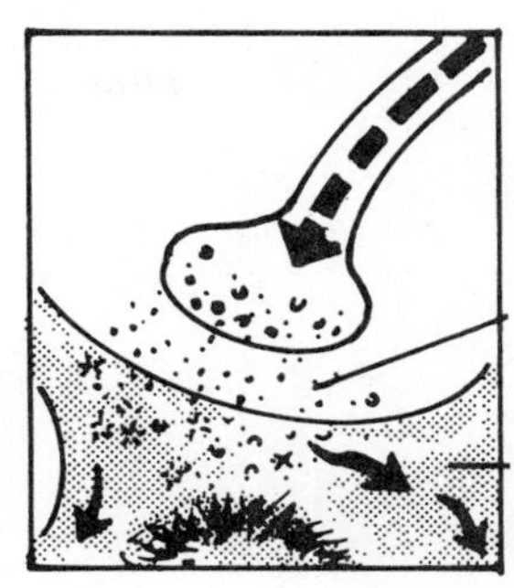

C. The neurotransmitters arrive at the receiver cell and (in this instance) stimulate it to generate an impulse that will travel along its length. Simultaneously, enzymes in the synapse destroy the molecules of neurotransmitters, or the transmitter cell takes up the neurotransmitters to reuse them. Total elapsed time: a fraction of a second.

building blocks (precursors) by making larger or smaller amounts of neurotransmitters. Furthermore, these building blocks (nutrients derived from food) are able to penetrate the blood-brain barrier. Thus the food you eat can influence your brain chemistry, to the extent that the food produces high concentrations of the precursor nutrients in an available form. These facts link nutrition to brain activity in some intriguing ways.

The neurons store their neurotransmitters in saclike vesicles at the terminal ends of their axons. When the time comes to transmit a message, a neuron opens its vesicles and releases the contents into the space (the synapse) between the two cell membranes. The neurotransmitters traverse the synapse and come to rest on specialized membranes of the receiving cell's dendrites (see Figure FP9–3). The receiving cell, recognizing the chemical message, becomes either excited or inhibited, and as mentioned, will fire a signal of its own when excitatory messages have built up to the needed threshold.

The neurotransmitters' life in the synapse is short. After transmission of an impulse, the neuron that released the neurotransmitters whisks them back in, or enzymes in the synaptic space destroy them. This wipes clean the synaptic slate for the next rapid-fire message. All of this synaptic activity occurs at lightning speed.

The brain is thought to contain dozens of different types of neurotransmitters, each conveying a unique message. Which message the neighboring cell receives depends upon which type of neurotransmitter is released, and whether that cell's membrane is equipped to recognize it. Some neurotransmitters excite the receiving neurons to fire, and some inhibit such excitation. The sum of many such messages simultaneously arriving at a single receiving cell determines whether that cell will fire or remain silent. Some examples of neurotransmitters and their effects are listed in Table FP9–1.

dendrites: fibrous branches emanating from the cell body of a neuron; dendrites receive incoming signals.

axon: a major fiber emanating from the cell body of a neuron; the axon transports signals away from the cell body.

neurotransmitter: a chemical agent released by one neuron that acts upon a second neuron or upon a muscle or gland cell and alters its electrical state or activity.

synapse: an anatomically specialized junction between two neurons where the activity in one neuron influences the excitability of the second.

Table FP9–1 Neurotransmitters and Their Functions

Neurotransmitter	Function
Acetylcholine	A neurotransmitter that inhibits (slows down) involuntary functions, such as heart rate; found in nerve-muscle connections; role in brain unknown.
Dopamine	A neurotransmitter abundant in brain parts that regulate motor activity; destruction of these brain parts results in Parkinson's disease. Vitamin B_6 is an essential factor for the synthesis of dopamine from the amino acid tyrosine.
Endorphins	Peptides that act on the brain's opiate receptors to inhibit pain and cause sedation; also called *endogenous opiates.*
GABA (gamma-aminobutyric acid)	An inhibitory neurotransmitter of the nervous system, it is abundant in the brain's cortex (reasoning portion), affecting more brain neurons than any other single neurotransmitter. Vitamin B_6 is an essential factor for GABA synthesis in the brain.
Glycine	A small amino acid that acts as the major inhibitory neurotransmitter in the spinal cord and brain stem.
Histamine	A protein that causes inflammation of tissues as part of the immune response, it also acts as a neurotransmitter in the brain's limbic system (regulates emotions).
Norepinephrine	A neurotransmitter that excites neurons responsible for emergency reactions (accelerated heartbeat, elevated blood pressure, increased muscle tension).
Serotonin	A neurotransmitter that has inhibitory action on the brain stem; involved in sleep and relaxation. Vitamin B_6 is an essential factor for the synthesis of serotonin from the amino acid tryptophan. Iron is an essential factor for the degradation of serotonin.
Substance P	A peptide that excites nerves that transmit pain.

The human nervous system is a communications system of almost unlimited capacity and has many ways to compensate for injury and disease. Yet it is also vulnerable to injury, disease, and aging, just as the rest of the body is—and in some ways, even more so. The rest of this discussion focuses on the aging brain.

The Aging Brain

The choices people make throughout their lives about diet, alcohol and drug use, smoking, and exercise either improve or harm their health, including the health of their brains. Over time, the health effects of these choices accumulate, affecting each organ and in turn, the whole body. The brain, like all of the body's organs, is influenced by both genetic and environmental factors that can enhance or diminish its amazing capacities. One of the challenges researchers face when studying the aging process in human beings is to distinguish among disease processes, normal age-related physiological changes, and changes that are the result of cumulative, extrinsic factors such as diet.

The brain ages in some characteristic ways. The number of neurons decreases as people age, and so does blood flow to the brain. When nerve cells in one part of the cerebral cortex diminish in number, hearing and speech are affected. Losses of neurons in other parts of the cortex can impair memory and cognitive function. When the number of neurons in the hindbrain (which houses the cerebellum) diminishes, balance and posture are affected. Losses of neurons in other parts of the brain affect still other functions. Lately, much attention has focused on the *abnormal* aging of the brain, called senile dementia of the Alzheimer's type (SDAT), also known as primary degenerative dementia of senile onset or chronic brain syndrome.[1] SDAT is a devastating degenerative disease for which a cause and cure are not yet known. SDAT may be the most common acquired progressive brain syndrome, afflicting 5 percent of the population by the age of 65 and 20 percent of those over 80.[2]

cerebellum: large lower posterior part of the brain; center of unconscious control of skeletal muscles.

senile dementia of the Alzheimer's type (SDAT): a degenerative disease of the brain involving memory loss and major structural changes in neuron networks; also known as **primary degenerative dementia of senile onset** or **chronic brain syndrome,** and often simply called **Alzheimer's disease.**

senile dementia (SEE-nile dee-MEN-she-ah): loss of brain function beyond the normal loss of physical adeptness and memory that occurs with aging.

Physicians conduct preliminary tests to rule out other possible disorders, but they can only diagnose SDAT with certainty upon autopsy. Its characteristic symptoms make its presence known: gradual loss of memory, loss of the ability to communicate, loss of physical capabilities, and eventually death.

The cause of SDAT continues to elude researchers, although it appears that there are genetic factors involved. Consequently, researchers have yet to find a cure for this devastating degenerative disease. Treatment involves providing relief and support to both the clients and their families.

Some SDAT characteristics may be relevant to nutrition. For example, normally, as blood flow to the brain diminishes with age, the brain compensates by absorbing more glucose and oxygen. In SDAT, no such compensation occurs, and glucose and oxygen concentrations decline. Whether the brain's diminished capacity to get glucose and oxygen causes or results from SDAT remains unclear. Another abnormality of interest involves the extremely low concentrations of the enzyme that makes the neurotransmitter acetylcholine from choline and acetyl CoA.

The role of aluminum, if any, in the development of SDAT has not yet been defined. Brain concentrations of aluminum in SDAT people do exceed normal brain concentrations by some 10 to 30 times—but blood and hair levels remain normal, indicating that the accumulation is caused by something in the brain itself, not by something in the environment. However, an epidemiological survey found that the risk of SDAT was 1½ times greater in areas where water aluminum concentrations were high compared with areas where concentrations were low, indicating that environmental aluminum may have a role.[3] Researchers are investigating the relationship between dietary aluminum and SDAT in individuals; aluminum cookware can increase the aluminum content of foods slightly, but whether this significantly affects the progress of SDAT is unknown.

Clinicians now recognize that much of the cognitive loss and forgetfulness generally attributed to aging is due in part to extrinsic, and therefore controllable, factors. In some instances, the degree of cognitive loss is extensive and attributable to a specific disorder than may respond to treatment; many dementias are treatable.

The rest of this discussion explores the concept that moderate, long-term nutrient deficiencies may contribute to the loss of memory and cognition that some older adults may experience. The exploration begins with a brief description of the roles of certain nutrients in brain metabolism. The last section presents research that examines the relationship between moderate nutrient deficiencies and cognition.

Nutrition and the Aging Brain

The normal development of the brain depends on an adequate supply of nutrients. For example, severe malnutrition in early childhood curtails the normal increase in brain cell number.[4] The importance of adequate nutrition to the brain continues long after its development, however. As mentioned earlier, the ability of neurons to synthesize specific neurotransmitters depends in part on the availability of precursor proteins and amino acids that are obtained from the diet. For example, the neurotransmitter serotonin derives from the amino acid tryptophan (see Figure FP9–4).

The enzymes involved in neurotransmitter synthesis require vitamins and minerals for activity.[5] For example, the vitamin B_6 coenzyme, pyridoxal phosphate, is needed for the synthesis of the compound gamma-aminobutyric acid (GABA), which acts as the neurotransmitter at the synapses of inhibitory

Figure FP9–4 The Making of a Neurotransmitter
In the brain, tryptophan is converted to the neurotransmitter serotonin in two steps. Serotonin is then converted to the inactive product 5-HIAA, and 5-HIAA leaves the brain to be excreted. The chemical drawings shown here are simplified by omission of the Cs and Hs in the ring structures.

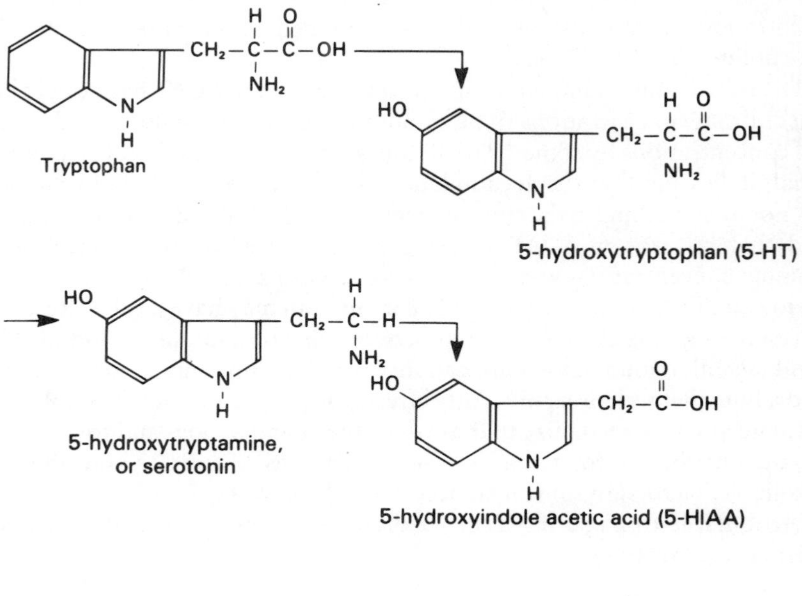

Tryptophan

5-hydroxytryptophan (5-HT)

5-hydroxytryptamine, or serotonin

5-hydroxyindole acetic acid (5-HIAA)

cells in the nervous system. Vitamin B_6 deficiency interferes with GABA synthesis in the brain. Low concentrations of GABA are associated with convulsions, a symptom common to vitamin B_6 deficiency.

Research on animals and human beings clearly shows that severe dietary deficiencies of thiamin, niacin, vitamin B_6, vitamin B_{12}, folate, and vitamin C impair mental ability, including memory.[6] Trace elements such as iodine, iron, copper, and zinc also support normal brain function.[7] For example, many of the enzymes that participate in the synthesis and catabolism of neurotransmitters depend on iron. Iron deficiency may thus interfere with the ability of the neurons to transmit signals to other neurons by way of impaired neurotransmitter synthesis or degradation. Zinc is also essential to many enzymatic systems, including those involved in DNA synthesis, repair, replication, and transcription. One researcher speculates that zinc deficiency impairs enzyme synthesis, causing a cascade of effects in the central nervous system.[8]

Researchers have begun exploring the possibility that the memory impairments observed in people and animals with severe nutrient deficiencies could develop in older adults who have experienced moderate (subclinical) deficiencies for prolonged periods of time. Diet is one of the more controllable components of people's lives. If research reveals significant links between moderate nutrient deficiencies and loss of cognitive function with aging, then through diet the loss may be preventable, or at least diminished or delayed.

Nutrition and Cognition

Research on both animals and human beings has contributed to present knowledge about nutrition and cognitive function. Caution is necessary when applying the results of animal research to human beings, but animal research enables scientists to conduct studies on the brain that would otherwise be impossible, and to determine directions for future research. Further research is needed to determine whether mild, long-term nutrient deficiencies result in brain cell changes similar to those observed in, for example, rats with severe deficiency, but the studies reported here represent a start in that direction. In human beings, multiple nutrient deficiencies are more likely than individual deficiencies, because nutrients occur in foods in combination. A lifetime of multiple nutrient deficiencies could conceivably contribute to mental impairment in later life by way of brain cell degeneration.

The studies that follow each take a different approach. Together, they show that nutrient deficits do hasten the onset of impaired cognitive functioning, and they give the impression that adequate nutrition would forestall, if not reverse, the observed degeneration.

Water-soluble vitamins One group of researchers studied the relationship between nutrition status and cognitive functioning in more than 200 healthy, independent-living men and women over 60 years of age.[9] The researchers evaluated nutrient intakes by way of diet records, and nutrient status by way of biochemical tests. Participants in the study were given two tests of cognitive function. One was a test of short-term memory (Wechsler Memory Test) and the other, a test of problem-solving ability (Halstead-Reitan Categories Test). Participants with low blood concentrations of vitamin C or vitamin B_{12} scored worse than better-nourished participants on both tests. Those with low blood concentrations of riboflavin or folate scored worse in the problem-solving test.

It is important to note that even the lowest scores in this study were still within the normal range for men and women of the same age. The researchers point out that the relationship between poor cognition and poor nutrition might be compared with the question of whether the chicken or the egg came first. It is possible that poor cognition is itself a risk factor for poor nutrition. People with impaired cognition might be less adept at providing optimal nutrition for themselves. However, the participants in this study had no history of dementia or impaired mental status, and so the researchers concluded that poor nutrition status might contribute to poor cognitive functioning in healthy elderly people. Studies of older populations at greater risk of malnutrition than this population might provide more insight into the relationship between nutrition and the aging brain. (Another possibility: both apparent poor cognition and poor nutrition may result from the same underlying cause—namely, depression.)

Vitamin B_{12} Memory impairment due to vitamin B_{12} deficiency can precede the blood symptoms of deficiency by years.[10] Dietary deficiency of vitamin B_{12} is rare, but inadequate absorption is common, accounting for more than 95 percent of deficiencies seen.[11] As many as 50 percent of people over the age of 60 are affected by atrophic gastritis, a condition that can impair vitamin B_{12} absorption.[12]

Evidence that vitamin B_{12} deficiency accounts for some cognitive deficits in older people comes from a study that revealed abnormal short-term memory in more than two-thirds of clients with pernicious anemia.[13] Treatment with vitamin B_{12} restored memory within one month in three-fourths of the clients. The researchers recommend that a diagnosis of senile dementia should not be made, even in the absence of anemia, until vitamin B_{12} status is determined biochemically.

Thiamin It is estimated that as many as 3 million people over the age of 60 abuse alcohol.[14] As Focal Point 8 describes, alcohol displaces food from the diet and alters normal nutrient metabolism, causing multiple nutrient deficiencies. Thiamin deficiency is common in alcoholics and is at least partially responsible for the mental symptoms that can accompany alcoholism.[15]

Surveys indicate that 5 percent of people over the age of 60 have impaired thiamin status.[16] Small amounts of thiamin are present in nearly all whole foods, but people with low food energy intakes or those who do not eat foods from all four food groups risk thiamin deficiency. As mentioned earlier, energy intake declines with age, and older people frequently fail to eat foods from all four groups.[17]

It is clear that mental impairment occurs with severe thiamin deficiency, but it is unclear whether mental impairment accompanies moderate deficiency. Limited research suggests that older people may be more susceptible to the effects of subclinical thiamin deficiency than younger people.[18] When thiamin was restricted in the diets of ten women between the ages of 52 and 72, irritability, fatigue, and headaches occurred after 12 days. Women between the ages of 18 and 21 did not experience these symptoms with the same level of thiamin restriction.

Copper and vitamin B_6 One study showed that changes in the brains of young rats subjected to dietary deficiencies of copper and vitamin B_6 resembled the degenerative changes that occur in the brains of human beings as they age.[19] The rats in this study ate diets extremely deficient in vitamin B_6, an unlikely situation for human beings; but vitamin B_6 intakes of many older adults vary, and many people are ingesting amounts well below recommendations.[20]

Iron Iron deficiency is prevalent worldwide, especially among infants, children, and young women. Research on children, as mentioned, shows a relationship between iron deficiency and cognitive function.[21] Researchers examining iron status and cognition in college students found a relationship between body iron stores and cognition.[22] The exact relationship is unclear, but iron status appeared to influence tasks dominated by the left side of the brain differently from tasks dominated by the right side. For example, higher iron status was associated with better word fluency performance (a left brain–dominated task), but with poorer performance of right brain–dominated tasks. (All of the students in this study were right-handed.) As Chapter 6 pointed out, animal research supports a relationship between iron deficiency and cognition.[23] Recall that the offspring of mice fed an iron-deficient diet were less responsive (they reared on their hind legs and stood immobile less frequently) to adverse, novel stimuli than were the offspring of iron-sufficient mice.

Iron deficiency occurs among older adults. Perhaps most significant with respect to iron and cognitive function in later life are the effects of the long-term moderate deficiencies so common in women prior to menopause—especially women with either repeated pregnancies or limited food energy intakes. In view of the widespread occurrence of iron deficiency, its role in mental function deserves further research.

Zinc Surveys indicate that zinc intakes in the United States are below recommended levels.[24] As many as one-fourth of older women have zinc intakes less than one-half of the recommended allowance.[25] A study in England showed that people with senile dementia had much lower levels of zinc than those without dementia.[26]

As the number of people over age 65 continues to grow, the urgent need for solutions to the problems that this major portion of the population faces becomes apparent. Senile dementia and other losses of brain function afflict millions of older adults. The search for ways to prevent or delay brain dysfunction should be a high priority for researchers investigating the relationships between nutrition and aging. The findings reported here on possible relationships between dementia and nutrition offer promise as a reasonable approach to a devastating problem.

It is clear that severe nutrient deficiencies impair cognition, but the effects of moderate deficiencies are less clear. It is also clear that a person's nutrition status affects the health and functioning of the whole body. Eating a nutritious, balanced diet throughout life seems a minimal effort toward promoting continued health and the enjoyment of later life.

In addition, there is much people can do, besides obtaining adequate nutrition, to support a high quality of life into old age. By practicing stress-management skills, maintaining physical fitness, participating in activities of interest, and cultivating spiritual health, a person can grow old gracefully.

Focal Point 9 Notes

1. A. Cherking, Effects of nutritional factors on memory function, in *Nutritional Intervention in the Aging Process,* eds. H. J. Armbrecht, J. M. Prendergast, and R. M. Coe (New York: Springer-Verlag, 1984), pp. 229–249.
2. M. S. Claggett, Nutritional factors relevant to Alzheimer's disease, *Journal of the American Dietetic Association* 89 (1989): 392–396.
3. C. N. Martyn and coauthors, Geographical relation between Alzheimer's disease and aluminum drinking water, *Lancet* 1 (1989): 59–62.
4. M. Winick and P. Rosso, The effect of severe early malnutrition on cellular growth of human brain, *Pediatric Research* 10 (1976): 57–61.
5. W. M. Lovenberg, Biochemical regulation of brain function, *Nutrition Reviews* (supplement), 44 (1986): 6–11.
6. K. Yoshimura and coauthors, Animal experiments on thiamine avitaminosis and cerebral function, *Journal of Nutritional Science and Vitaminology* 22 (1976): 429–437, as cited by Cherkin, 1984; M. K. Horwitt, Niacin, in *Modern Nutrition in Health and Disease,* 6th ed. eds. R. S. Goodhart and M. S. Shils (Philadelphia: Lea and Febiger, 1980), pp. 204–208; C. S. Russ and coauthors, Vitamin B_6 status of depressed and obsessive-compulsive patients, *Nutrition Reports International* 27 (1983): 867–873; J. S. Goodwin, J. M. Goodwin, and P. J. Garry, Association between nutritional status and cognitive functioning in a healthy elderly population, *Journal of the American Medical Association* 249 (1983): 2917–2921.
7. H. Sandstead, A brief history of the influence of trace elements on brain function, *American Journal of Clinical Nutrition* 43 (1986): 293–298.
8. F. M. Burnet, A possible role of zinc in the pathology of dementia, *Lancet* 1 (1981): 186–188.
9. Goodwin, Goodwin, and Garry, 1983.
10. Cherkin, 1984.
11. V. Herbert, Recommended dietary intakes (RDI) of vitamin B_{12} in humans, *American Journal of Clinical Nutrition* 45 (1987): 671–678.
12. M. Siurala and coauthors, Prevalence of gastritis in a rural population, *Scandinavian Journal of Gastroenterology* 3 (1968): 211–223, as cited by P. Suter and R. M. Russell, Vitamin requirements of the elderly, *American Journal of Clinical Nutrition* 45 (1987): 501–512.
13. R. W. Strachan and J. G. Henderson, Psychiatric syndromes due to avitaminosis B_{12} with normal blood and marrow, *Journal of Medicine* 34 (1965): 303–317, as cited by Cherkin, 1984.
14. P. J. Bloom, Alcoholism after sixty, *American Family Physician* 28 (1983): 111–113.
15. M. Victor, Alcohol and nutritional diseases of the nervous system, *Journal of the American Medical Association* 167 (1958): 65–71.
16. F. L. Iber and coauthors, Thiamin in the elderly—Relation to alcoholism and to neurological degenerative disease, *American Journal of Clinical Nutrition* 36 (1982): 1067–1082.
17. V. Holt, J. Nordstrom, and M. B. Kohrs, Food preferences of older adults (abstract), *Journal of the American Dietetic Association* 87 (1987): 1597.
18. H. G. Oldham, Thiamin requirements of women, *Annals of the New York Academy of Sciences* 378 (1982): 542–549, as cited by R. H. Haas, Thiamin and the brain, in *Annual Review of Nutrition,* eds. R. E. Olson, E. Beutler, and H. P. Broquist (Palo Alto, Calif.: Annual Reviews, 1988), pp. 483–515.
19. E. J. Root and J. B. Longenecker, Brain cell alterations suggesting premature aging induced by dietary deficiency of vitamin B_6 and/or copper, *American Journal of Clinical Nutrition* 37 (1983): 540–552.
20. P. J. Garry and coauthors, Nutritional status in a healthy elderly population: Dietary and supplemental intakes, *American Journal of Clinical Nutrition* 36 (1982): 319–331; J. R. Turnlund, Copper nutriture, bioavailability, and the influence of dietary factors, *Journal of the American Dietetic Association* 88 (1988): 303–308.
21. E. Pollitt and coauthors, Iron deficiency and behavioral development in infants and preschool children, *American Journal of Clinical Nutrition* 43 (1986): 555–565.
22. D. M. Tucker and coauthors, Iron status and brain function: Serum ferritin levels associated with asymmetries of cortical electrophysiology and cognitive performance, *American Journal of Clinical Nutrition* 39 (1984): 105–113.
23. J. Weinberg, Behavioral and physiological effects of early iron deficiency in the rat, in *Iron Deficiency: Brain Biochemistry and Behavior,* eds. E. Pollitt and R. L. Leibel (New York: Raven Press, 1982), pp. 93–123.
24. K. Patterson and coauthors, Zinc, copper, and manganese intake and balance for adults consuming self-selected diets, *American Journal of Clinical Nutrition* 40 (1984): 1397–1403.
25. Garry and coauthors, 1982. At the time of this research the zinc RDA for women was 15 milligrams; even with the lower current standards, many of the women's intakes would still fall short of recommendations.
26. R. Hullin, Zinc deficiency: Can it cause dementia? *Therapaecia,* September 1983, pp. 26, 27, 30.

Nutrition and the Aging Process

10

Married Love by Oscar Nemon.

The number of people in the United States reaching 65 years of age is greater each year than the year before, as Figure 10–1 shows. As of 1990, 12.7 percent of the population is 65 years old.[1] As the number of older adults grows, so does the science of aging research, led by the National Institute on Aging, one of the National Institutes of Health. Research is helping to dispel myths about aging; to answer questions about how nutrition, exercise, and other lifestyle habits affect aging; and to enable health care workers to provide better care for older adults, thereby enhancing the quality of life. The aging process is, and will continue to be, a subject of intense study.

The Aging Process

Just as all parts of the body do not grow at the same rate, neither do they stop growing at the same rate. Some continually grow throughout adult life—skin, hair, nails, and the lining of the digestive tract, for example. In contrast, growth in height stops at about age 18 in females and age 20 in males, and sexual maturity often arrives before that time.[2]

What happens once an individual is fully grown and developed? Does the body attain a certain physiological state and remain there until it receives a signal to begin changing again? Although the physical appearances of some people seem to suggest this is so, it is not. With the exception of the central nervous system, the cells of all other body systems continuously undergo cycles of growth, division, degeneration, death, and replacement. Depending on the body system, time of life, and extent of damage, the growth and replacement of damaged cells occur at different rates and to greater or lesser degrees. Studies of different organ systems indicate that many of them attain maximum

Figure 10–1 The Changing Profile of America's Age
People over 65 make up a larger percentage of the population than formerly, and their numbers are growing faster than those of other age groups.

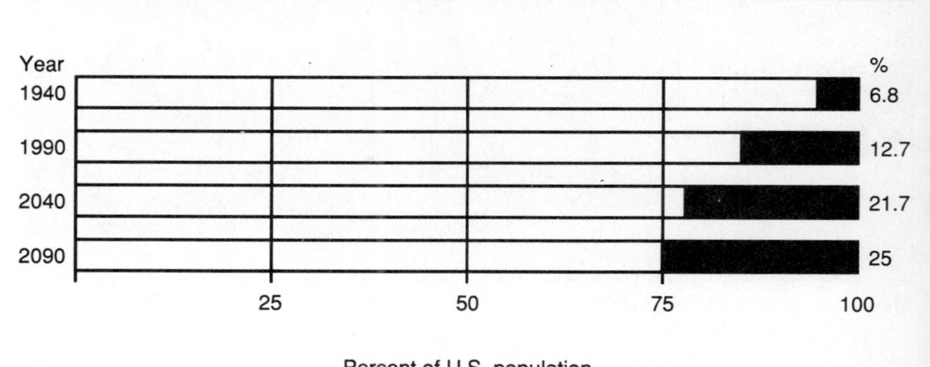

Source: From data in K. Flieger, Why do we age? *FDA Consumer*, October 1988, pp. 20–25.

functional efficiency between 20 and 30 years of age.[3] Thereafter, as the human body ages, physiological functions and capacities diminish. The physiological changes that occur after maturity, and in some cases even before, are those associated with the aging process. Many of the changes in physiological functions are inevitable, due simply to the passage of time and genetic factors. Aging is a natural, inevitable process—a biological clock that winds down at a genetically determined rate and stops ticking at a genetically determined point in time. The probability of dying increases exponentially after age 30.[4]

Scientists studying the aging process are interested in determining which physiological changes are inevitable and which ones can be prevented or slowed down by changes in environment and lifestyle habits. Progress has been made in this respect; within the limits set by heredity, the adoption of healthful lifestyle habits such as eating nutritious food, exercising regularly, and minimizing daily stress can slow the aging process.

The first part of this chapter discusses the influence of nutrition, genetics, exercise, and stress on longevity and the aging process, with emphasis on the role of nutrition. Later sections focus on the nutrient needs and eating habits of older adults.

longevity: long duration of life.

Nutrition and Longevity

Juan Ponce de León's search for the fountain of youth was by no means an original quest. As far back as recorded history, human beings have sought to live longer and remain young while doing so. The search is as relentless today as it has ever been, spurred on by quacks who claim to possess anti-aging remedies based on the latest scientific findings. In reality, such claims are usually based on exaggerations or oversimplifications of scientific data. Some are based on nothing more than the perpetrator's greed. Great strides have been made by scientists who study the aging process, especially with regard to the effects of nutrition on aging, but so far, the discovery of a dietary regimen or a single nutrient supplement that will prolong the life span of human beings has eluded scientists.[5] The scientific study of the aging process in human beings is among the youngest of the scientific disciplines, for it is only in this century that human beings have achieved a life expectancy that permits scientific research.[6] What has been learned so far about effects of nutrition and environment on longevity provides incentive for researchers to keep asking questions about how and why human beings age.

life span: the age attained by a single member of a species. In human beings the maximum documented life span is about 115 years.

life expectancy: the average number of years expected in a population at a specific age, usually at birth. In 1900 the life expectancy at birth was 45 years; by 1983 the life expectancy for baby boys had increased to 71 years and for baby girls, to 78 years. These changes in life expectancy have no effect on maximum life span.

Animal studies The first evidence that diet might increase longevity surfaced more than half a century ago.[7] Researchers studied the effects of early food restriction on the long-term survival of rats. Upon weaning, groups of rats were fed diets adequate in all the nutrients essential for growth but short in kcalories. Durations of restriction were 300, 500, 700, or 1000 days, and growth ceased for those periods. Then kcalories were increased, and growth resumed. Controls were allowed to eat and grow normally. All of the rats from experimental groups experienced a delay in disease development and outlived controls. An often unmentioned postscript to this finding is that many of the food-restricted animals died early on, and that the survivors were permanently growth retarded, even upon resumption of normal food intake. This drastic side effect of early starvation limits the potential usefulness of these studies to human beings.

Subsequent research using various experimental designs has produced similar results. For example, one researcher restricted experimental rats' food intakes to 30 percent of those of controls fed ad lib.[8] This level of restriction was not as great as in the previous study. The food-restricted rats had a lower incidence of disease and lived nearly twice as long as controls.

One group of researchers looked specifically at the onset of disease in food-restricted animals as compared with controls.[9] They examined the effects of two levels of food restriction—33 and 46 percent. Compared with controls, nearly twice as many rats in the restricted groups lived to 800 days. More severe food restriction did not increase the number of survivors. Disease incidence was lower in the food-restricted groups. The restricted animals contracted the same diseases as the controls, but did so at later times of life.

In the studies just described, food restriction began as soon as the animals were weaned—that is, at three weeks. Researchers have also investigated the effects of food restriction initiated later in life on the longevity of rats. When rats were full-fed initially for a period of three months, followed by a three-month food restriction period, their survival times coincided with the survival times of rats restricted from weaning throughout life.[10] In this study, early food restriction was no more beneficial than later restriction. However, when compared with no restriction at all, early, short-term food restriction appears to promote longevity. In a different study, animals restricted upon weaning for less than three months survived longer than those unrestricted throughout life.[11] Furthermore, in the same study, animals who were allowed to eat freely for ten weeks following weaning, and then restricted for the rest of their lives, had lower mortality rates than both those restricted immediately after weaning and controls.

Thus full feeding early in life to ensure growth and development, followed by later food restriction, results in greater longevity than early food restriction. In view of knowledge about the importance of nutrition during critical periods, this is not surprising. Life expectancy responded not only to timing of restriction but also to both body weight reduction and diet composition. Among the restricted diets, only those with the highest protein-to-energy ratio (1:5) increased longevity.

mortality rate: death rate; ratio of number of deaths to a given population.

Even when food restriction was begun as late as mid-life in rats, mortality rates were lower in restricted rats than in unrestricted rats. The level of restriction was important: severe food restriction increased mortality, and slight increases in food intake later in life also shortened life expectancy.[12] When rats ate self-selected diets following weaning, little difference in food intake was noted during the first four weeks between animals who lived about 500 days and those who lived about 1000 days. Beyond four weeks, however, a 10 percent increase in food intake brought on increased body weight and resulted in an 8 percent decrease in survival time.

In summary, results of studies of food restriction, body weight, and longevity in rats lend support to the premise that nutrition can significantly influence life expectancy. The results of the studies discussed so far look like this:

- Food energy restriction (but with all nutrients essential to growth provided) immediately following weaning increased longevity of rats, but the animals were immature and growth retarded.

- ▸ Less severe food energy restriction immediately following weaning lowered disease incidence and prolonged life.
- ▸ Food energy-restricted rats contract the same diseases as rats fed ad libitum, but do so later in life.
- ▸ Food energy restriction initiated later than just after weaning was equally as beneficial as, or even more beneficial than, early restriction in terms of prolonging survival.
- ▸ Finally, moderate restriction beginning as late as mid-life increases life expectancy of rats.

As discussed throughout this text, extrapolation of the results of animal studies to human beings is often unrealistic, or at best, difficult. Some studies are not directly applicable to human beings—for example, those suggesting that lean animals that consumed low-energy diets early in life survived longer than well-fed, heavier controls. Restricting the diets of infants or young children impairs growth and development. On the other hand, animal studies suggesting that moderate energy restriction later in life increases life expectancy apply positively to human beings.

Epidemiological studies The relationships between diet, body weight, and longevity are less clear-cut for human beings, but not completely obscure. A study of the relationship between physical health and a number of independent variables supports the nutrition-longevity connection. Recall from Chapter 9 the study conducted in southern California in which different health practices were analyzed to determine their influence on the physical health of about 7000 adults.[13] Of the six factors that had the greatest influence on physiological age, three were related to nutrition: alcohol consumption, regularity of meals, and weight control. (The other three were sleep, smoking, and exercise habits.) For both men and women in this study, overweight, especially high degrees of overweight, correlated with poor health. Irregular eating habits also correlated with poor health. The physical health status of those who followed all six practices beneficial to health was comparable to that of people 30 years younger who followed none of them.

Insurance companies have compiled data on weights, heights, and mortality rates of policyholders for years. Their findings and those from the Framingham Heart Study indicate that the more overweight an individual is, the greater the risk of mortality.[14] The same is true for severely underweight individuals. Unlike controlled studies of animals, however, these studies involve many other factors—genetics, disease state, and medical care—that influence longevity in human beings.

Overweight individuals, especially obese individuals, have an increased mortality risk; despite this, many live long lives. Thirty to 60 percent of U.S. adults (depending on race, sex, and socioeconomic status) between the ages of 45 and 54 years are overweight or obese, according to national health surveys.[15] The incidence of obesity decreases with advancing age. The differences between obese individuals who remain healthy and live long lives and those who succumb to disease or death prematurely remain to be determined. That many overweight individuals enjoy long, healthy lives contradicts the assumption that body weight alone is responsible for the increased mortality of obese people. Nevertheless, anthropometric data confirm that people who live to advanced ages are leaner than younger persons.[16]

Evidence supporting a correlation between food energy restriction and longevity in human beings is limited, but not totally absent. For example, Japan maintains excellent food intake data on its population by conducting annual nutrition surveys.[17] Survey data include disease and mortality incidences, in addition to nutrition information. Data from 1972 showed that the average food energy intake of people living on the island of Okinawa was 17 to 40 percent lower than that of people living on mainland Japan. The foods consumed, mainly fish and vegetables, were highly nutritious. The Okinawan people have only about a 60 percent incidence of cerebral disease, heart disease, and cancer compared with people in the rest of Japan.[18] Okinawa also has up to 40 times the number of centenarians as the rest of Japan. Thus the Okinawan people who consume a low-energy, high-nutrient diet throughout their lives live long, healthy lives. Other factors such as climate and genetics must be considered as well, but at least for this population of Japanese people, a low-energy, nutrient-dense diet appears to be associated with enhanced longevity.

The children and adults of Okinawa are shorter and lighter compared with the people of mainland Japan, consistent with consuming a low-kcalorie diet throughout life. Because so many factors must be considered, and because research in the area of longevity and the aging process in human beings is in its youth, it would be premature to reach any conclusions about how energy restriction, lower disease incidence, and body weight interact and contribute to the long, healthy lives of these individuals. Still, some pieces of this puzzle appear to be in place.

HDL cholesterol is often used as a positive indicator of health, because evidence for an inverse relationship between HDL cholesterol and body weight and between HDL cholesterol and coronary heart disease risk is strong.[19] Researchers studied the effects of a hypocaloric (1200 kcalories), nutrient-dense diet on changes in blood lipids of 40 obese women.[20] The women consumed the diet for nine months, losing an average of 30 pounds during that time. All of the women had plasma HDL cholesterol concentrations below those of normal-weight women at the start of the study. With weight reduction, a significant rise in plasma HDL cholesterol concentrations occurred, independent of changes in other blood lipid concentrations. HDL cholesterol concentrations are also strongly and inversely related to coronary heart disease risk for men and women between 50 and 79 years of age.[21] Thus the maintenance of a reasonable weight for height appears to promote HDL cholesterol concentrations most consistent with coronary heart disease prevention.

Another study of human beings investigated the effects of energy restriction on mortality. A group of 60 healthy people over the age of 65 who resided in a religious institution for the elderly were fed a diet containing 2300 kcalories every odd day of each month for three years.[22] On the even days, the diet consisted of a liter of milk and 500 grams of fresh fruit, amounting to about 1,000 kcalories. A control group received the 2300-kcalorie diet every day. Those on the alternating diet had a lower incidence of illness (as indicated by admissions to the infirmary) and half as many deaths as the control group during the study period. This more energy-restricted diet, which was nutrient dense, appeared to support disease resistance.

Although evidence from animal studies suggests that energy restriction and lower body weight promote longevity, similar carefully controlled studies of human beings are extremely difficult to carry out. Clearly, more research is needed to understand the metabolic means by which energy restriction and body weight contribute to longevity. The following sections discuss how genetics, physical activity, and stress influence longevity.

Lifelong Genetic and Environmental Influences

Many people mistakenly assume that the aging process expresses itself similarly in everyone—that all old people are physiologically much the same. Nothing could be further from the truth. As people get older, each one becomes less and less like anyone else. Even when researchers group aging people together based on whatever diseases they may have contracted, many remaining elements clutter the picture of the natural aging process. The older that people are, the more time has elapsed for such factors as diet, genetics, exercise, and everyday stress to influence the physical and psychological progress of aging. Each of these factors can interact with one or more of the others, and each can affect the aging process positively or negatively. Nutrition exemplifies this concept. Severe malnutrition promotes illness and death, while sound nutrition is an important component in the prevention of some diseases. Genetics, physical activity, and stress are but a few of the other life factors that influence the aging process; nutrition's effects occur within this context.

Genetics The chances of an individual's succumbing to infections, cardiovascular disease, cancer, or a number of other diseases, and the times at which these diseases afflict the individual, actually have little to do with chance and much to do with inheritance and environment. The relationship between genetics and environmental factors is often synergistic; for instance, cigarette smoking augments the risk of coronary heart disease in genetically predisposed people.[23]

To assess the role that genetics and environment play in disease and longevity, researchers study families with twins and adopted children. As they do so, they keep uncovering evidence in support of the concept that both genetics and environment determine an individual's susceptibility to diseases.[24]

A revealing study of almost 1000 adult adoptees illustrates this point.[25] Researchers evaluated the risks of dying from all causes or from specific groups of causes between the ages of 16 and 58 years for adoptees with an adoptive or biologic parent who had died of the same cause before the age of 50 or 70. These risks were compared with the adoptees' risks of dying from the same causes when their parents were still alive between 50 and 70. The results were startling. Death of a biologic parent before age 50 from natural causes doubled the mortality rate for adoptees. Death of an adopted parent from natural causes before age 50 had no effect on the mortality rate for adoptees. The mortality rate was five times greater for adoptees with a biologic parent who died from infection before age 50 or 70 compared with adoptees whose adoptive parents died from infection. A similar large increase in mortality rate was observed for adoptees whose biologic parents died from vascular causes, while deaths of adoptive parents from vascular causes only slightly increased the risk of death for the adoptees from the same causes. Thus susceptibility to infective and vascular diseases seems to be largely a matter of heredity. In contrast, deaths of adoptive parents from cancer before the age of 50 increased the adoptees' mortality from cancer fivefold, whereas deaths of biologic parents from cancer had no effect on adoptees' death rates from cancer. Thus cancer appears to be environmentally caused.

The authors offer some important cautions in generalizing the results of their study. The population studied was homogeneous—all white and of similar cultural background. It is also important to note that adopted individuals are usually not able to change their environments to counteract genetic tendencies toward diseases, since they are usually unaware of such tendencies. In contrast, biologic children are aware and can take preventative measures against diseases they might inherit. This difference may heighten the contrast in studies of adoptive versus biologic children and overstate the impact of genetics on disease. Despite this, these studies have provided valuable insight into the roles of genetics and environment in disease and longevity. As researchers learn more about the genes and environmental factors that influence disease susceptibility, so do they learn more about how to help susceptible families avoid diseases. A person cannot change an inherited genetic map, but can control nutrition, exercise, and other lifestyle habits to enhance the quality of life and make more likely the attainment of maximal life expectancy.

Fitness One of the most promising areas of research today centers on the relationships between exercise, health, and longevity. Research continues to confirm what athletes, trainers, and most people who exercise on a regular and frequent basis probably already know: the mental and physical benefits of regular physical activity are many and remarkable. These include increased self-esteem, reduced body fatness and increased lean body tissue, reduced blood pressure, slowed cardiovascular aging, and a reduced risk of cardiovascular disease.[26] Much current research focuses on the relationships between exercise, cardiovascular disease, and longevity.

The exact effect of exercise on longevity is still to be determined, but an extensive study of more than 16,000 Harvard alumni indicates that regular exercise can prolong life.[27] The men were between 35 and 74 years of age and were studied for 12 to 16 years. Those who expended 2000 or more kcalories in exercise per week (equal to walking or running about 20 miles per week) had death rates 25 to 33 percent lower than those who were less active. Death rates declined as energy expended on physical activity increased from less than 500 kcalories to 3500 kcalories per week. Furthermore, lack of exercise seemed to influence risk of death more than heredity. In this study, the death of one or both parents before age 65 increased the risk of death among alumni, but the risk was lower for those who were physically active. Thus exercise may offset the influence of premature parental death. The mortality rates of physically active men were lower with or without regard to smoking, hypertension, or extremes in body weight.

Physical activity slows cardiovascular aging, positively affects risk factors for heart disease, and may even add years to life. It also promotes lean body tissue and reduces body fatness. The ever more numerous benefits derived from regular exercise emphasize the importance of making physical activity a priority in everyone's life.

Stress The body has remarkable ability to maintain a steady internal condition—homeostasis—by adapting to stressors that would otherwise cause life-threatening changes. Both physical stressors (pain, heat, illness) and psychological stressors (divorce, moving, death of a loved one) elicit the body's stress response. Figure 10–2 shows the elaborate physiological steps by which the body responds to stressors, using the nervous and hormonal systems to bring about a state of readiness in every body part. The effects all favor

stressor: a demand placed on the body to adapt.

stress: any threat to a person's well-being. The threat may be physical or psychological, desired or feared, but the reaction is always the same.

Figure 10–2 The Stress Reaction
The stress reaction begins when the brain perceives a threat to the body's equilibrium. There follows a chain of events that acts through both nerves and hormones to bring about a state of readiness in every body part.

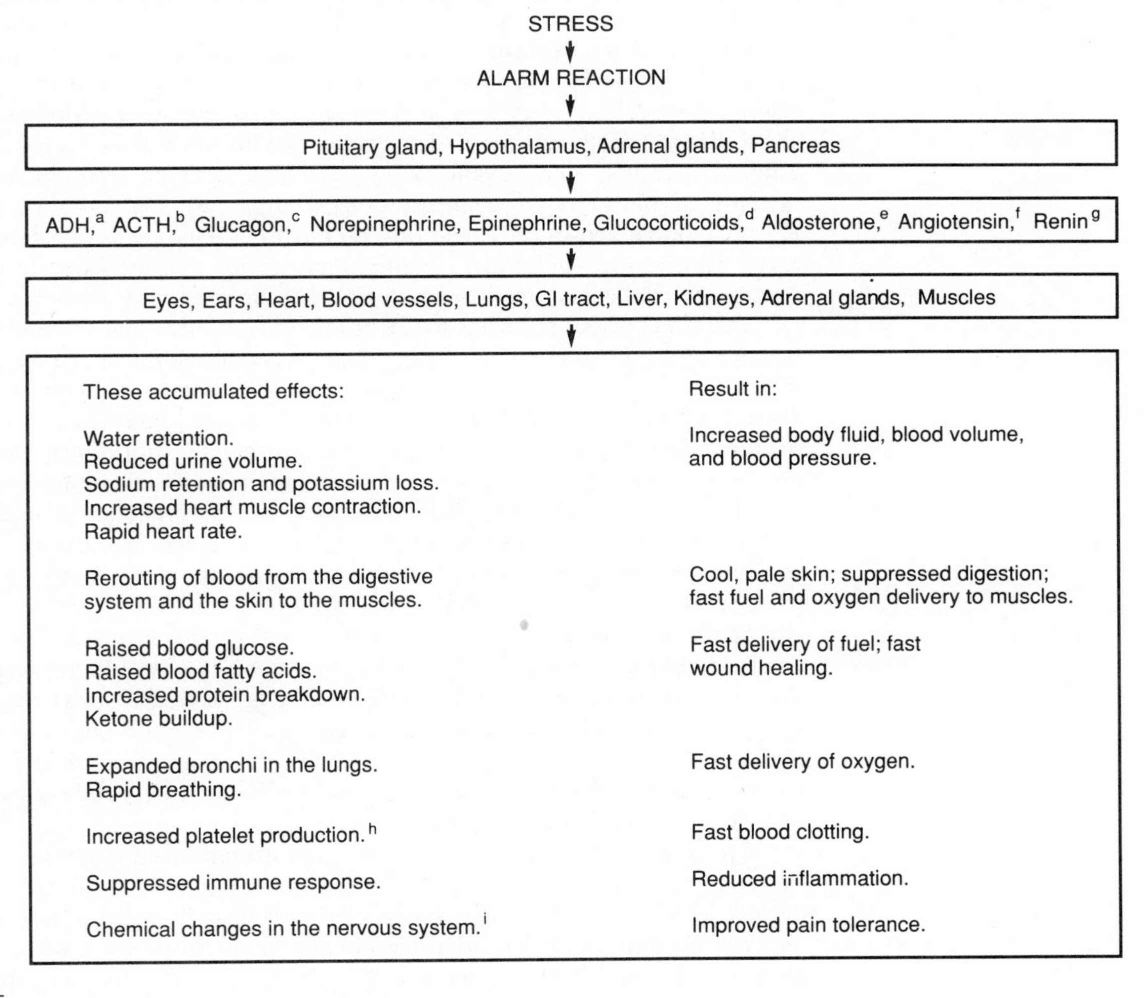

[a]ADH (antidiuretic hormone) prevents water loss in urine.
[b]ACTH (adrenocorticotropic hormone) stimulates the adrenal glands.
[c]In tandem with the release of glucagon is the reduction of insulin secretion.
[d]Glucocorticoids are hormones from the adrenal glands, affecting the body's management of glucose.
[e]The hormone aldosterone, from the adrenal glands, is involved in blood pressure regulation.
[f]The hormone angiotensin is involved in blood pressure regulation.
[g]The enzyme renin functions to raise blood pressure by activating angiotensin.
[h]Platelets are small, cell-fragment-like bodies in the blood that help with blood clotting if injury occurs.
[i]The brain produces opiatelike substances called endorphins that dull the sensation of pain.

stress response: the body's response to stress, mediated by both nerves and hormones initially; begins with an *alarm reaction*, proceeds through a stage of *resistance*, and then leads to *recovery* or, if prolonged, to *exhaustion*. This three-stage response has also been termed the **general adaptation syndrome.**

physical action (fight or flight). If stress is prolonged or severe, and especially if physical action is not a permitted response to the stressor, then it can drain the body of its reserves and leave it weakened, aged, and vulnerable to illness.

Over 30 years ago Dr. Hans Selye described the limited capacity of the body to withstand stress as *adaptation energy*. According to his description, the length of the human life span appears to be primarily determined by the amount of available adaptation energy. Constant exposure to any stressor will use it up.[28] Death occurs when an individual runs out of adaptation energy.

Some kinds of stress are challenging, while others are draining, depending on an individual's ability to cope with them. It seems that aging itself affects how the body responds to stressors. The older a person is, the more stress the person has been exposed to, and the more adaptation energy is required to respond to a given stressor.

Older individuals have less ability to adapt to both external and internal disturbances. For an example of external stressors, animals show a reduced tolerance to cold as they age, and epidemiological studies of human beings show that the death rate from heat stroke rises abruptly after age 60.[29] As for internal stressors, with advancing age, glucose tolerance declines. A marked increase in blood sugar concentration occurs after a glucose challenge.[30] Fasting blood glucose concentrations, on the other hand, rise only slightly with age.[31] Due to the age-related decline in adaptation energy, when the body is stressed or challenged, the ability to restore homeostasis is diminished.

The stress response begins in the brain, and it is here that researchers have begun to look in trying to understand exactly what adaptation energy is and why, with aging, adaptive capacity declines. The hormone-secreting adrenal gland is intimately involved in the stress reaction (see Figure 10–2). One part of the adrenal gland, the adrenal cortex, secretes hormones known as glucocorticoids, which increase fuel concentrations in the blood (providing needed energy). When researchers examine the sites of regulation of glucocorticoid release in the brains of animals, they find that repeated exposure of the regulating brain cells to the glucocorticoids reduces the ability of the brain cells to respond to stress.[32] This diminished response is reflected in three measurable characteristics.[33] First, higher concentrations of the stress hormones are needed to elicit a response from the brain; second, the adaptation response is slower; and third, recovery is delayed—that is, once the stress response is turned on, it takes longer to turn it off. The researchers also point out that the metabolic changes induced by glucocorticoids in response to stress, such as increases in blood glucose concentrations and heart rate, while essential for dealing with stress, are harmful if prolonged.

As aging progresses, inevitable changes in structure and function of each of the body's organs contribute to the decline in adapting ability of each of these organs. The decreased capacity for homeostasis of each organ in turn imposes an even greater decline in adaptive ability on the body as a whole. When disease strikes and further impairs function, the reduced ability to adapt makes the aging individual more vulnerable to death than the younger, more adaptive person.

In summary, the longevity of an individual depends on the influences and interactions of many life factors, including nutrition, genetics, physical activity, and stressors. Each of these elements, along with others, exerts independent effects on health and longevity; and each, in turn, influences the others. As discussed earlier, energy-restricted animals live longer than others. Human populations that eat low-kcalorie, nutrient-dense diets live longer than others. Obese individuals who eat low-kcalorie, nutrient-dense diets achieve higher blood HDL concentrations and lower heart disease risks. Early nutrition and exercise intervention can sometimes overcome the genetic predisposition to disease. Finally, an individual who is strong, healthy, and well-nourished can withstand stress better than one who is not.

Assessment of Nutrition Status

Because people's physiology changes as they age, standards used to assess their nutrition status change, too. A whole new set of considerations applies in evaluation of the nutritional health of older adults. Unfortunately, many of the measurements used to evaluate nutrition status derive from young or middle-aged adults, and are unreliable when applied to elderly people.[34] Specific standards for healthy elderly people need to be established to permit more accurate nutrition status assessment.

Historical Data

The assessor who takes care and patience in obtaining thorough diet, medical, socioeconomic, and drug history information on older adults will save time and frustration when evaluating the later components of nutrition assessment. Many life circumstances of older adults influence their nutrition status, and many of these can be ascertained from complete history information. Researchers have divided factors contributing to malnutrition in older adults into primary and secondary causes.[35] Primary causes of malnutrition include:

- Ignorance of the need for a nutritionally sound diet.
- Financial restrictions that limit the range of foods purchased.
- Physical disabilities that cause the older people to be housebound.
- Social isolation (loneliness) that weakens the incentive to eat.
- Mental disorders that limit people's ability to provide balanced meals for themselves.

Secondary causes of malnutrition include:

- Malabsorption from intestinal diseases.
- Alcoholism, which both interferes with nutrient intake and impairs nutrient metabolism.
- Therapeutic drugs, which interfere with the absorption and metabolism of some nutrients.

Diet history information will indicate diet adequacy or inadequacy, while medical, socioeconomic, and drug histories will alert the assessor to possible causes of inadequacies. Because loneliness and social isolation are major causes of malnutrition in the elderly, Practical Point: Mealtimes for One offers suggestions to enhance nutrition for people who eat alone. The earlier in life people learn to excel at providing outstanding nutrition for themselves when necessary, the less vulnerable they will be to the negative nutrition impact of times of solitude, whenever they occur in life.

▸▸ PRACTICAL POINT

Mealtimes for One

Singles of all ages face problems concerning food purchasing, storing, and preparing. Large packages of meat and vegetables are often suitable for a family of four or more, and even a head of lettuce can spoil before one person can use it all. Many singles live in small dwellings, some without kitchens and freezers—for them, purchasing and storage problems are compounded. Following is a collection of ideas gathered from single people who have devised answers to these problems.

Buy only what you will use: the small cans of vegetables may be expensive, but it is also expensive to let the unused portion of a large can spoil before using. Buy only three pieces of each kind of fresh fruit: a ripe one, a medium one, and a green one. Eat the first right away and the second soon, and let the last one ripen to eat days later. Don't be timid about asking the grocer to break open a family-sized package of wrapped meat or fresh vegetables.

Think up a variety of ways to use a vegetable when you must buy it in large quantity. For example, you can divide a head of cauliflower into thirds. Cook one third and eat it as a hot vegetable. Put another third into a salad dressing marinade for use as an appetizer, and save the rest to use raw in salad. Make mixtures, using what you have on hand. A thick stew prepared from any leftover vegetables and bits of meat, with some added onion, pepper, celery, and potatoes, makes a complete and balanced meal—except for milk. Note the uses of powdered milk that follow: you could add some to your stew.

Buy fresh milk in the sizes best suited for you. If your grocery store doesn't carry pints or quarts of milk, ask if it can stock them for you, or try a nearby service station or convenience store.

Set aside a place for rows of glass jars containing shelf staple items that you can't buy in single-serving quantities—rice, tapioca, lentils and other dried beans, flour, cornmeal, nonfat dry milk, and cereal, to name only a few possibilities. (Remember, light destroys riboflavin, so use opaque jars for enriched pasta and dry milk.) Place each jar, tightly sealed, in the freezer for one night to kill any eggs or organisms before storing it on the shelf. Then the jars will keep bugs out of the foods indefinitely. The jars make an attractive display and will remind you of possibilities for variety in your menus. Cut the directions-for-use labels from the packages and store them in the jars.

Learn to use nonfat dry milk. It is the greatest convenience food there is. Not only does it offer much more calcium than any other food; it also is fortified with vitamins A and D. Dry milk can be stored on the shelf for several months at room temperature. It can be mixed with water to make fluid milk in as small a quantity as you like. One person says he keeps a jar of nonfat dry milk next to his stove and "dumps it into everything": hamburgers, gravies, soups, casseroles, sauces, even beverages such as iced coffee. The taste is negligible, but five "dumpings" of a heaping tablespoon each are the equivalent of a cup of fresh milk. Ask a friend who is a member of Weight Watchers to give you some recipes for delicious milk shakes and ice cream using nonfat dry milk. Their recipes are for single servings.

Make soup stock from leftover pork and chicken bones soaked in vinegar and then boiled. The bones release their calcium into the acid medium, and the vinegar boils off. One *tablespoon* of such stock may contain over 100 milligrams of calcium.[a] Then cook something in this stock every day: vegetables, rice, stew—and of course, make soups with it.

Cook for several meals at a time. For example, boil three potatoes with skins. Eat one hot, mashed with chives. When the others have cooled, use one to make a potato-cheese casserole ready to be put into the oven for the next evening's meal. Slice the third one into a covered bowl, and pour over it the juice from pickles. The pickled potato will keep several days in the refrigerator and can be used in a salad.

Experiment with stir-fried foods. Use a frying pan if you don't have a wok. A variety of vegetables and meats can be enjoyed this way; inexpensive vegetables such as cabbage and celery are delicious when crisp-cooked in a little oil with soy sauce or lemon added. Cooked, leftover vegetables can be dropped in at the last minute. There are frozen mixtures of Chinese or Polynesian vegetables available in the larger grocery stores. Bonus: only one pan to wash.

Depending on your freezer space, make twice or even six times as much as you need of a dish that takes time to prepare: a casserole, vegetable pie, or meat loaf. Freeze individual portions in containers that can be microwaved or oven heated for serving later. Be sure to date these so you will use the oldest first.

Buy a loaf of bread and immediately store half, well wrapped, in the freezer (not the refrigerator, which will make it stale). If you have space in your freezing compartment, buy frozen vegetables in the large bags rather than in the small cartons. You can take out the exact amount you need and close the bag tightly with a rubber band or spring clothespin. You can season the bag of vegetables by adding some fresh herbs. For low-cost elegance, buy bunches of fresh parsley, dill, oregano, or basil; keep just a few sprigs out for use during the next few days; and chop up the rest to add to the frozen vegetable bags. Wrap individual portions of meat (purchased in bulk) in thick aluminum foil, not freezer paper: the foil can become the liner for the pan in which you bake or broil the meat, thus saving cleanup work. Don't label them individually, but put them all in a brown bag marked "hamburger" or "chicken thighs," along with the date. The bag is easy to locate in the freezer, and you'll know when your supply is running low.

Although the suggestions here will help the single person with the mechanics of food chores, they meet only a part of the need. Dr. Jack Weinberg, professor of psychiatry at the University of Illinois, wrote perceptively:

> It is not *what* the older person eats but *with whom* that will be the deciding factor in proper care for him. The oft-repeated complaint of the older patient that he has little incentive to prepare food for only himself is not merely a statement of fact but also a rebuke to the questioner for failing to perceive his isolation and aloneness

▸▸ PRACTICAL POINT *continued*

and to realize that food . . . for one's self lacks the condiment of another's presence which can transform the simplest fare to the ceremonial act with all its shared meaning.[b]

Even for nutrition's sake, it is important to attend to loneliness at mealtimes; the person who is living alone must learn to connect food with socializing. Cook for yourself with the idea that you are also preparing for guests you might want to invite. Or turn this suggestion around: invite guests, and make enough food so that you will have some left for yourself at a later meal. With a wide variety of leftovers on hand, you can invite one or another single friend on the spur of the moment to "Come over and share my frozen dinners with me tonight." Most elderly people—and, in fact, most everyone—respond best not when they are made to feel dependent, such as when they are cared for in nursing homes, but when they share close proximity with others and interact with them.[c] So, go ahead and invite an older person in for a meal, but ask that person to bring the bread or otherwise include him or her in the planning.

Of course, there are times when being alone at mealtime is inevitable, and it is appropriate to make them special, because preparing a meal for yourself is a way of taking good care of yourself. For a special meal alone, try this: set the table with tablecloth; napkin; a full complement of utensils; and a wildflower, if you can. Set a pot of stew with vegetables and fresh herbs on low heat to cook, and make a salad. Go settle in a stuffed chair and enjoy a book, some soothing music, and a crackling cold glass of lemoned sparkling water, until the rich aroma of stew calls you to dinner. After serving your plate, light a candle, dim the lights, savor the food, and relish some of the best company you will ever have—your own.

---

[a]A. Rosanoff and D. H. Calloway, Calcium source in Indochinese immigrants (correspondence), *New England Journal of Medicine* 306 (1982): 239–240.
[b]J. Weinberg, Psychologic implications of the nutritional needs of the elderly, *Journal of the American Dietetic Association* 60 (1972): 293–296.
[c]W. A. McIntosh and P. A. Shifflett, Influence of social support systems on dietary intake of the elderly, *Journal of Nutrition for the Elderly* 4 (1984): 5–18.

Anthropometric Measurements

Useful anthropometric measures for older adults include height, weight, frame size, and fatfold thickness. Universally accepted weight-for-height standards for people over 65 years of age are not available. The 1983 Metropolitan Life Insurance Company tables are widely used (see Table A-2 in Appendix A), but for older adults, usual weight and a history of recent weight changes may be more useful for assessing nutrition status. The assessor must consider the changes in body composition (lean body mass declines, body fat increases) that accompany aging to evaluate anthropometric data accurately. Men steadily gain weight until about age 45, and women gain until about age 60, when average body weights begin to decline.[36] Based on fatfold measures, patterns of body fat gains and losses parallel weight changes in both men and women. Triceps fatfold measures are easy to obtain and are the most widely used.

Figure A-3 in Appendix A shows how to measure the triceps fatfold, and Table A-5 gives triceps fatfold percentiles for males and females. For greater accuracy in assessing body fat percentages of elderly people, the assessor should take fatfold measures at three different body sites. This will pick up fatness averages that might be missed if body fat has become unequally distributed. Appendix A describes the procedure and provides percent fat estimates based on the sum of three measures. Because average body weight and body fat measures change considerably as people age, it is important that the appropriate age-specific standards, when available, be used to ensure accuracy. Frame size can be determined from wrist circumference and elbow breadth, as shown in Appendix A.

Physical Examinations

A careful examination of older adults for physical signs of malnutrition can offer the assessor important clues to nutrition status. Ill-fitting dentures or missing teeth indicate that people may have trouble chewing foods and therefore may have poor nutrition status. Muscle wasting and edema may be the physical signs of protein-energy malnutrition, not just signs of aging. The same may be true of the symptoms of iron deficiency: pallor, fatigue, and weakness. The interpretation of physical data for older people requires awareness of the differences between signs of age and signs of malnutrition.

Biochemical Analyses

Normal ranges for blood test results are usually derived from findings in young adults and may not be applicable to older people. About 5 percent of blood tests performed on healthy adults fall outside the normal range. In contrast, a study in which over 50 different blood tests were performed on about 70 people over age 63 found that 15 percent of the test results were outside the normal range, despite the absence of other symptoms.[37] These findings support the assertion that specific biochemical standards for older adults are needed.

Biochemical assessment of older adults should include tests of iron status (hemoglobin and hematocrit), protein (serum albumin), and glucose status (blood glucose). If specific vitamin deficiencies are suspected, the appropriate biochemical tests should be performed to help confirm diagnosis. Tables in Appendix A list biochemical tests useful for assessing nutrition status.

Serum albumin concentrations do not reflect short-term changes in protein and energy intake, but are more indicative of long-term changes. One study found, however, that serum albumin concentrations are the simplest and best single predictor of mortality in very elderly people (85 years of age and up).[38] The authors advised that serum albumin can provide early identification of those older adults at increased risk of death. Serum albumin concentrations less than 30 grams per liter were more closely associated with death.

When interpreting biochemical tests for elderly people, the assessor must consider additional factors, such as disease conditions that can influence the results. For example, the anemia of chronic disease has features identical to anemias of protein-energy malnutrition.[39] The interpretation of biochemical findings in older adults is extremely complex.

Nutrition Status of Older Adults

Overall, it appears that the nutrition status of older adults as a group, in the United States, is better than previously thought. Specific nutrient deficiencies among individuals, however, point to the need for programs and interventions designed for subgroups of this population.

Nutrient Intakes

Studies and surveys of nutrient intakes show that people voluntarily reduce their energy intakes as they age.[40] In some instances these intakes decline below recommendations.[41] As energy intakes fall, the likelihood of deficiencies of nutrients increases, especially among the poor, whose diets are less nutrient dense than the diets of more privileged people.[42]

Probably the single group among the elderly most prone to protein-energy malnutrition is those who are ill, and especially those ill enough to require hospitalization. A typical finding is that PEM affects from a fourth to a half of all people admitted to the hospital for medical or surgical treatment.[43] The incidence of PEM among older people in the hospital for two weeks or more is higher than this.[44]

Specific vitamin deficiencies among the elderly are not all well characterized. They may not be recognized, or they may be uncommon. If there are vitamin deficiencies among the elderly, they seem not to be overt or obvious, but rather to be subclinical deficiencies that can adversely influence or promote chronic diseases of aging.

In the case of vitamin A, limited research suggests an age-related *increase* in vitamin A absorption. One study showed that older people absorb more vitamin A than younger ones do.[45] Another study indicated no age-related change in vitamin A absorption.[46] When the vitamin A intake and plasma retinol concentrations of healthy older adults and young adults were compared, the older adults had higher mean plasma vitamin A concentrations despite little difference in dietary intake between the two groups.[47] Only one of the 300 older people in this study had a low plasma vitamin A concentration. The authors of the study concluded that age-specific criteria should be used when interpreting plasma vitamin A concentrations, and the vitamin A status of older adults remains to be defined.

Folate absorption and metabolism do not change as people age, unless specific diseases cause such changes.[48] Swedish researchers determined the folate intakes of 35 people between the ages of 60 and 67 and monitored folate status for 16 years by measuring blood folate.[49] They found that folate intakes between 100 and 200 micrograms per day supported normal folate status in these people.

Dietary deficiencies of vitamin B_{12} are rare—inadequate absorption accounts for more than 95 percent of deficiencies seen.[50] The vitamin B_{12} intakes of older adults vary; one study found that one-fourth of older men and more than one-third of older women had intakes below the RDA.[51]

Despite low intakes of vitamin B_{12}, most older adults maintain normal serum concentrations.[52] Vitamin B_{12} serum concentrations decline slightly with advancing age, but stay within the normal range.[53] Absorption of the vitamin does not change with age in healthy adults free of conditions that impair absorption.[54] The lower vitamin B_{12} serum concentrations of older adults may be due to a greater prevalence of atrophic gastritis and gastric atrophy with advancing age. Up to 50 percent of those over age 60 are affected.[55] These conditions impair production of intrinsic factor and may also affect vitamin B_{12} absorption by other mechanisms.[56]

atrophic gastritis: inflammation of the stomach accompanied by a decrease in the number of gastric glands and reduced production of hydrochloric acid.

gastric atrophy: complete atrophy of the stomach mucosa and total lack of hydrochloric acid; results in pernicious anemia.

Vitamin D illustrates the generalization that vitamin deficiencies among the elderly tend to be hidden but can accelerate disease processes. Vitamin D is the most important regulator of the calcium content of bone and is therefore critical to the maintenance of a healthy skeletal structure. As many as 10 million older adults in the United States have reduced bone mass, making them vulnerable to wrist, hip, and spine fractures.[57]

Vitamin D deficiency is a greater threat to aged individuals than it is to younger ones for several reasons. First, because older adults consume less food due to lower energy needs, food sources of vitamin D are limited. Intakes of the vitamin are higher in the United States, where milk is vitamin D fortified, than elsewhere, but many older adults drink little or no milk.[58] As many as one-third of older adults have vitamin D intakes of less than half of the RDA.[59] Further compromising the vitamin D status of many older adults is their limited exposure to sunlight, especially among those in nursing homes and institutions.[60] Finally, the potential for vitamin D deficiency in older people is favored by age-related changes in vitamin D metabolism. The most significant of these changes to bone health is a deficiency, or reduced response, of the enzyme that stimulates production of the active form of vitamin D, 1,25-$(OH)_2D_3$, in the kidney.[61]

The active form of vitamin D is known as vitamin D hormone; it is the only hormone known that stimulates calcium absorption from the intestine. Interference with the metabolism of vitamin D to its active form, then, inhibits calcium absorption. Low blood calcium concentration promotes mobilization of calcium from bone by way of parathyroid hormone. Thus bone calcium is sacrificed to meet the calcium needs of other body tissues. Some researchers suggest that older adults suffering from loss of bone mass should be treated with exogenous active vitamin D to improve calcium absorption.[62] The extent of crippling bone disorders among older adults speaks to the importance of adequate vitamin D and calcium nutrition throughout life to minimize the impact of age-related changes in the metabolism of these nutrients.

Milk and other dairy foods are the best sources of dietary calcium, as well as of vitamin D, for those who can tolerate them. Some older people purposely avoid these foods, however, because they experience stomach discomfort from them. They may find the guidelines in Chapter 9's Practical Point helpful in finding alternative sources of calcium; sunlight exposure, in moderation, can help with vitamin D. Milk and milk products may also be too costly for some people, but these problems can be relieved by food programs for older adults (see Practical Point: Assistance Programs).[63] Availability of foods, companionship, and education contribute to changed attitudes of participants, as well as to improved nutrient intakes and status (including those of calcium and vitamin D).

Another nutrient of concern for older people is magnesium. Marginal intakes of magnesium increase the risk of cardiovascular disease.[64] In some areas, the magnesium content of the water inversely correlates with death rates from heart disease.[65]

▸▸ PRACTICAL POINT

Assistance Programs

The responsibility for support in old age cannot be left entirely to the individual. Three major federal programs can help older persons with money problems, at least a little. Under Social Security, employees and employers pay into a fund from which the employee collects benefits at retirement. The Food Stamp Program enables people who qualify to obtain stamps with which to buy food. The Supplemental Security Income Program is aimed at directly improving the financial plight of the very poor, by increasing a person's or family's income to the defined poverty level. This sometimes helps older people retain their independence.

Food banks have been established in some areas to help older people stretch their food dollars. A food bank project buys industry's "irregulars"—products that have been mislabeled, underweighted, redesigned, or mispackaged and would ordinarily therefore be thrown away. Nothing is wrong with this food; the industry can credit it as a donation. As government money dwindles, the nutrition status of low-income people of all ages depends more and more on private efforts such as food banking.

Another major program to benefit the elderly is Title IIIC of the Older Americans Act of 1965, known as the Nutrition Program for the Elderly. The major goals of this program are to provide:

- Low-cost, nutritious meals.
- Opportunities for social interaction.
- Homemaker education and shopping assistance.
- Counseling and referral to other social services.
- Transportation services.

The program is intended to improve older people's nutrition status and enable them to avoid medical problems, continue living in communities of their own choice, and stay out of institutions.

Sites chosen for congregate meals under this program must be accessible to most of the target population. Volunteers may also deliver meals to those who are homebound either permanently or temporarily; these efforts are known as Meals on Wheels. The program ensures nutrition, but its recipients miss out on the social benefit of the congregate meal sites; every effort is made to persuade them to come to the shared meals, if they can. Despite these programs, many eligible people are still missing meals and are malnourished simply because they don't know of the programs available.[a] Identification of such people should become a higher priority.

Preparation for the later years should include financial planning and developing of social skills to avert loneliness. It helps, too, to give of one's talents in volunteer or paid work; some think that being needed and contributing to humanity is the key to happiness in the later years. Intellectual pursuits can give meaning to the days. Each person must practice adjusting to change, especially when it comes without consent, to allow continued control over life. The goal is to enjoy life fully, with optimal health of mind and body.

When the care needs of an elderly person—say, someone with Alzheimer's disease—make staying at home impossible, there are facilities that can give medical and other care. A variety of options exist; for example, some facilities provide assistance and limited medical care during the day, while other family members attend to their own daytime needs. These facilities can provide the needed daily care, along with activities and a peer group for friendships. All these factors work together to give the elderly stimulating experiences; in such a setting, mealtimes are often a highlight. Another familiar alternative is the nursing home. A nursing home serves people who need constant medical care. The next Practical Point offers guidance on nutrition in nursing homes.

▸▸ PRACTICAL POINT *continued*

---

[a]Many frail elderly get inadequate food, according to Cornell study, *Journal of the American Dietetic Association* 86 (1986): 647.

Despite the variety of magnesium-rich foods available, magnesium intakes of adults in the United States are low, especially when energy intakes are low.[66] One survey of over 37,000 people found that only one-fourth had magnesium intakes that equaled or exceeded the RDA.[67] Older adults whose energy intakes are lower than those of younger people are likely to have marginal magnesium intakes.

Iron deficiency among older adults of normal health is not as prevalent as it is during other stages of life, but it is still cause for concern among certain groups of older persons.[68] Older adults with low food energy intakes risk iron deficiency, as do people of low socioeconomic status.[69] Chronic blood loss due to ulcers and poor iron absorption due to reduced stomach acid secretion also increase the likelihood of iron deficiency. Still, the prevalence of iron deficiency is slightly lower among older individuals, partly because women's iron needs decrease after menopause.

In contrast to iron intakes, zinc intakes of older adults are well below the RDA. As many as 95 percent of older adults may not get the zinc they need, and many, especially women, miss the mark by more than half.[70] Zinc intakes decrease with advancing age, in line with decreasing energy intakes.[71] It is possible that improved absorption compensates for low zinc intakes, but this has not been confirmed; indeed, some research suggests that older adults absorb zinc less efficiently than younger people do.[72] The bright side of the zinc story is that intakes below the RDA may be adequate for some healthy older adults.[73] For those who are not healthy, questions about how chronic illness, marginal dietary zinc intakes, and inefficient absorption affect health and nutrition status need to be answered.

Zinc deficiency impairs immune function—a finding that may have significant implications for older adults, who are susceptible to infectious diseases.[74] Researchers studying 100 adults between 60 and 89 years of age found that plasma zinc concentrations correlated with lymphocyte responses to antigens.[75] Those with low zinc concentrations had depressed lymphocyte responses. Further research is needed to ascertain the relationship between zinc and immune function and whether improved zinc nutriture enhances cellular immunity. For older adults who have a high incidence of infectious diseases and cancer, the potential benefits justify continued study.

Adverse Influences on Adults' Nutrition

Among the negative impacts on nutrition that people experience during their lives are those that are self-imposed by their choices to ingest substances other than foods. Chief among these are drugs (including medical drugs), supplements, nicotine, and alcohol.

Drug-nutrient interactions Older adults are avid users of over-the-counter (OTC) and prescription medications. People over the age of 65 take about 25 percent of all the OTC and prescription drugs that are sold. Almost half of older adults surveyed in Boston were using medications, often several at a time.[76] Older adults often go to more than one doctor, each one prescribing a different medication to treat various conditions.

The huge quantity of prescription and OTC drugs available today enables many people to enjoy long, healthy lives. These same drugs exert a significant influence on the nutrient status of older adults. Several hundred interactions among drugs, nutrients, and foods are known. The chances that adverse interactions will result in nutrient deficiencies increase if drugs are taken over a long time period, if several drugs are taken simultaneously, or if the person is in poor nutrition status to begin with. Older adults whose nutrition status is marginal and who take several drugs for long times are subject to problems. Age-related physiological changes also contribute to the likelihood of adverse reactions.[77] For example, body composition changes with age; lean body mass decreases proportionately to body fat. Drugs that usually disperse to lean body tissue may end up in unexpectedly high concentrations in the bloodstream. Elimination of drugs from the body may be impaired due to changes in the kidney.[78]

Most drugs interact with one or more nutrients in several ways. Table A-1 in Appendix A identifies commonly used drugs and their effects on nutrition. Drugs may:

- Alter food intake by depressing or stimulating the appetite.
- Reduce the absorption of nutrients from foods.
- Alter the metabolism and excretion of nutrients.

Table 1–7 in Chapter 1 summarizes the mechanisms by which these interactions occur. Foods and nutrients can interfere with the action of drugs, too. Prescription instructions take this into account and should be strictly followed, but while some older people are conscientious, others are notorious for not always following prescription instructions. They adjust their own doses based on their own hunches; they mix and match drugs; and they share drugs with one another. The hazards of these practices are considerable. The abuse of physician-prescribed medications can impair nutrition status.

Over-the-counter drugs are no less harmful to nutrition than prescription drugs are. For example, many older adults use laxatives to relieve constipation. Daily laxative use speeds up food transit time through the intestine so that many vitamins do not have time to be absorbed. Oil laxatives impair absorption of calcium and vitamin D, further compromising the already marginal calcium status of many older adults.

Many cancer drugs, such as methotrexate, are vitamin antagonists. They have a chemical structure similar to that of a vitamin and can displace the vitamin or metabolite from an enzyme, blocking a normal metabolic pathway.

Arthritis sufferers are often heavy users of aspirin, an agent that prevents passage of vitamin C from the plasma to the tissues. When vitamin C concentrations are measured in the plasma of these people, no deficit is seen, and yet the tissues may be experiencing a deficit. Aspirin may also induce iron deficiency by causing stomach erosions and bleeding.

Older people are heavy users of both diuretics and antacids. Diuretics may be used to promote intentional excretion of sodium, but other minerals, such as potassium, magnesium, and zinc, may be lost as well. Regular use of antacids causes the body to excrete some nutrients as wastes rather than absorb them.

Older people also buy many drug-containing products in response to popular fads. Frauds abound—from wrinkle creams to sexual potency enhancers—and while many are harmless, some are not. All put a strain on the budgets of people with fixed incomes.

Knowledge about the medications older adults use is essential to the management of their nutrition status and health. Older adults who must take medication on a regular basis because of various illnesses are understandably concerned about alleviating their conditions or improving their health, while those who are not ill want to remain as healthy as possible. For many reasons, then, some older adults rely on nutrient supplements to improve their health.

Supplement use Advertisers target older people for supplements and "health foods" by claiming that their products prevent disease and promote longevity, claims that attract older people. Despite this, and much to their credit, older adults are, for the most part, reasonable in their approach—most avoid health-food stores, or they buy less there than others.[79]

Not all older adults resist supplement use. A study in a California retirement community showed that 72 percent of those surveyed were taking supplements—mostly vitamins C and E—but that the choices of supplements were not related to the users' dietary intakes.[80] The vitamins they were taking were not the ones they may have needed. Furthermore, some of the older people in this study consumed toxic levels of vitamin A (more than 25,000 IU per day) and more than ten times the RDA of vitamins C and E.

The wrong nutrient supplements are often taken for the wrong reasons. Another study found that supplement use correlated with medical problems and living alone.[81] Certain diseases or health problems may require supplements, but in this study the vitamins taken were often inappropriate and not prescribed by a health care professional. For example, vitamin E was taken to prevent constipation and vitamin C, to prevent arthritis.

Smoking and alcohol use Among the most common drugs that affect appetite is nicotine in tobacco; it is an appetite suppressant. Many people complain of weight gain upon quitting smoking. The problem of regulating food intake, however, is a small price to pay for quitting compared with the health risks incurred by not quitting. Smoking also alters the metabolism of vitamin C; excretion of the vitamin is hastened in smokers, who may need more than nonsmokers to normalize blood concentrations.[82] Blood lipids are changed by smoking to favor cholesterol buildup, increasing the likelihood of heart disease.[83]

The most common drug that affects nutrition in older people is alcohol. Alcohol, like nicotine, is an appetite suppressant. It also provides kcalories, thereby displacing nutrient-containing foods from the diet. Alcohol reduces the absorption of thiamin, folate, and vitamin B_{12}. Focal Point 8 discusses the effects of alcohol abuse.

Energy and Nutrient Needs of Older Adults

As mentioned, the older people get, the more dissimilar they become. This truth is not reflected in the current RDA for older adults: they still use only one age grouping (51 years and up) for all nutrients. The RDA are therefore the same for a 55-year-old man and a 90-year-old man. Furthermore, most of the recommendations are extrapolated from data obtained from studies of younger people. Recommendations for younger people do not take into account age-related changes, such as the reduction of absorption of some nutrients.

It is difficult to estimate the nutrient needs of a population as diverse, both physiologically and otherwise, as the elderly. Compared with younger people, more of the elderly are afflicted with diseases or disorders, or affected by drugs, that influence nutrition status. Diverse living and economic conditions also influence nutrient needs. All of these factors complicate the process of defining appropriate nutrient recommendations for older adults. One group of researchers has developed a list of factors to be considered in developing RDA for older adults.[84] The RDA should:

- Be aimed at the maintenance of *optimal* functioning of body systems, and set at levels that may prevent age-related diseases such as osteoporosis.
- Possibly be individualized according to health and history, rather than generalized to an age range.
- Take into account how nutrients are affected by other food components, such as fiber, and how nutrients are affected by drugs that older people are likely to take.
- Take into account toxic effects of nutrient supplements that may build up with long-term use and become apparent in older people.

Until more information is available on which to develop improved RDA for nutrients for older adults, the present RDA must suffice. The following sections note nutrients of concern.

Water

Water constitutes more than one-half of an adult's body weight. It is the most abundant nutrient in the human body and the most essential to life. The body has several homeostatic mechanisms for regulating water balance, and one is under conscious control—that is, thirst.

The thirst mechanism lags behind water lack. A water deficiency that develops slowly can switch on drinking behavior in time to prevent serious dehydration, but one that develops rapidly may not. Thirst itself does not remedy a water deficiency; a person has to notice the thirst, pay attention, and take the time to drink. Dehydration is a risk for elderly people, who may not notice or pay attention to their thirst, or who are unable to obtain water because of immobility. One study indicates that elderly people may not feel thirsty when their bodies need fluid, and may therefore be at risk of dehydration.[85] Many older adults limit their fluid intakes to avoid frequent trips to the bathroom.[86]

You can tell from the color of your urine if you're getting enough water. Bright or dark yellow urine is too concentrated; you need more water. Pale yellow, almost colorless urine is dilute enough; your water intake is ample.

Water needs vary greatly, depending on age (infants need more per unit of body weight than adults), climate, activity level, and health status. Under normal conditions in a moderate climate, 6 to 8 cups per day is a reasonable allowance for an older adult.

Water recommendation for adults:
1 to 1½ oz/kg actual body weight.

Energy-Yielding Nutrients

Energy needs decrease with advancing age. This is attributed to two main characteristics of aging. First, lean body mass declines as people age, resulting in a lower basal metabolic rate.[87] Second, aging is usually accompanied by reduced physical activity (though it need not be). The lower energy expenditure of the elderly means that less energy is required to maintain weight. This is reflected in the energy RDA for men and women, which decrease slightly starting at age 51.[88] It is important to note that changes in metabolic rate and physical activity vary among individuals, making the energy RDA for some older adults, such as physically active ones, inappropriate.[89]

Energy RDA for adults:
2900 kcal (25 to 50 yr, males).
2200 kcal (25 to 50 yr, females).
2300 kcal (51 yr and over, males).
1900 kcal (51 yr and over, females).

The limited energy allowance for older adults demands that the foods selected be of high nutrient density. Foods high in sugar, fat, or alcohol use up the energy allowance at the expense of essential nutrients. Table 10–1 shows the basis for a food pattern that supplies energy slightly below the RDA figures for people over 50. A person following this pattern would have an additional 250 to 300 kcalories to select other foods and beverages.

Table 10–1 Eating Patterns for Older People

	Number of Portions	
Food	*Woman (1600 kcal)*	*Man (2050 kcal)*
Milk (nonfat)	3	4
Vegetable	4	6
Fruit	4	6
Bread	6	8
Meat (lean)	4	6
Fat	6	7

Note: Food selections should be the same as those described in Table 8–7 in Chapter 8. These eating patterns supply about 55% of food energy as carbohydrate, 20% as protein, and 25% as fat.

Protein RDA for adults:
63 g/day (25 yr and over, males).
50 g/day (25 yr and over, females).

Protein The protein RDA at present is the same for older adults as it is for younger ones.[90] Results of some studies of protein requirements of older adults imply, however, that they should be higher. Researchers studying nitrogen balance in people between 70 and 90 years of age found that when they consumed high-quality protein at the RDA level of 0.8 grams per kilogram of body weight, many did not maintain nitrogen balance.[91] Other researchers have concluded that older men have the same protein requirements as younger ones, 0.8 grams per kilogram of body weight.[92]

In general, most adults consume 14 to 18 percent of their energy as protein.[93] This represents an intake slightly more than the RDA, assuming food energy intakes equal to the RDA. Protein needs increase with illness, so some older people may need to derive more of their food energy from protein. However, too much protein can be as damaging as too little, increasing calcium excretion and stressing the kidneys.

The protein needs of older adults appear to be about the same as, or even greater than, those of younger persons. Because energy needs decrease, the protein-containing foods chosen must be of high quality but not high in kcalories. Foods such as lean meats, poultry, and fish; nonfat milk; and low-fat cottage cheese are examples of protein-rich, low-kcalorie foods.

Fat Foods that are high in fat must be limited in older people's diets, since fat carries more than twice the kcalories per gram that protein and carbohydrate do. Fat restriction is recommended to protect against the development of atherosclerosis, cancer, and other diseases as well. An intake of about 25 percent of kcalories from fat is appropriate for older adults. Most of this should be in the form of polyunsaturated and monounsaturated fat to provide essential fatty acids while limiting saturated fat intake.

Carbohydrates Abundant carbohydrate is needed to protect protein from being used as energy. Complex carbohydrates such as vegetables, whole grains, and fruits are rich in fiber and essential vitamins and minerals.

Dietary fiber can alleviate constipation—a condition prevalent among older adults, and especially among nursing home residents.[94] One group of researchers found that fiber intakes of older adults are not as low as might be expected.[95] They compared the dietary fiber intakes of independent-living older adults and nursing home residents. The researchers discovered that the fiber intakes of the two groups were similar, and were comparable to the fiber content of typical diets. Constipation was more prevalent among the nursing home residents, whereas the bowel habits of the independent-living adults in this study were normal—suggesting that the dietary fiber intake (18 grams per day) of this group was adequate. A slight increase in dietary fiber for nursing home residents who are physically inactive may improve bowel function.

Fiber is beneficial in weight control efforts, too. Diets that are high in fiber-rich foods present more bulk and satiety than energy-dense diets, and make it easier to stop eating before excess kcalories have been consumed. A positive recommendation any adult can adopt in order to control kcaloric intake is to eat more foods high in fiber.

Vitamins

Researchers need more information about the vitamin requirements of older adults. As they learn more about the roles of specific vitamins in disease prevention and development and about how age-related physiological changes affect vitamin metabolism, they will come closer to consensus on this topic. Meanwhile, the RDA for some vitamins are likely to keep on changing in the future as the result of new discoveries.[96] The following sections discuss vitamins of particular interest for older adults.

Some nutrition authorities propose lowering the vitamin A RDA.[97] Research into the role of carotenes (vitamin A precursors) in preventing cancer should perhaps also be considered before final decisions about vitamin A allowances for older adults are made.

Vitamin A RDA for adults:
1000 RE/day (males).
800 RE/day (females).

Despite the possibility that some older adults malabsorb vitamin B_{12}, the malabsorption does not necessitate higher allowances for the vitamin.[98] Previously healthy, well-fed adults normally have large liver stores of the vitamin that can help compensate for the decreased absorption in those with atrophic gastritis. Those who can no longer absorb the vitamin from food, as indicated by serum concentrations, require injections.

Vitamin B_{12} RDA for adults:
2 μg/day.

Changes in lifestyle and eating habits, as well as age-related metabolic changes, influence vitamin D status in older adults. Increases in physical activity, outdoor sun exposure, and food intake of vitamin D by older persons can improve vitamin D status and bone health.

Vitamin D RDA for adults:
5 μg/day.

The determination of vitamin requirements for older adults is confounded by physiological diversity, disease conditions, medical and diet experiences, and lack of research. The older people become, the more their histories influence their vitamin requirements, and the more difficult it becomes to generalize about these requirements. The diversity of this large group of individuals seems to demand attention to specific subgroups within it—those with chronic diseases, those who are institutionalized, and those who are physically active and in good health.[99]

Minerals

Establishing mineral requirements of older adults, or of anyone for that matter, poses even more problems than establishing vitamin requirements does. Among the problems are the inadequacy of food composition tables regarding some of the minerals. The accuracy of dietary intake studies is hindered when it is not known how much of a certain nutrient is present in foods. This, in turn, precludes determining the appropriate intake for maintaining health. Mineral bioavailability is affected by interactions with many other nutrients and drugs, thus presenting another problem for researchers trying to determine mineral requirements. Relevant to older individuals' requirements are questions about which aging processes are normal and which might be prevented or alleviated by nutrition intervention. In addition, physiological changes that accompany aging influence nutrient needs. For example, the iron requirements of postmenopausal women fall abruptly as menstruation ceases.

Iron RDA for adults:
10 mg/day (25 yr and over, males).
15 mg/day (25 to 50 yr, females).
10 mg/day (51 yr and over, females).

Among the minerals, calcium is of particular interest. Controversy surrounds the question of what the appropriate calcium intake for older adults is.

Calcium RDA for adults:
800 mg/day (25 yr and over).

Chapter 9 discusses calcium and osteoporosis.

Some researchers argue that the current RDA is too low for postmenopausal women, especially in light of the prevalence of osteoporosis and the age-related changes in calcium absorption.[100] Others contend that people adapt to lower calcium intakes by way of increased absorption, so higher intakes are not needed.[101] Calcium absorption increases in response to low intakes less readily in older individuals than in younger ones, however.[102]

As mentioned previously, mineral bioavailability is affected by other dietary components, such as fiber and the minerals themselves. Large doses of zinc, as can be obtained from supplements, interfere with calcium absorption, while calcium may impair magnesium absorption.[103] Such interactions present an added challenge to researchers trying to determine mineral requirements of healthy individuals. When age-related metabolic changes and disease conditions are superimposed, the task becomes overwhelmingly difficult, but not impossible.

The ever-increasing number of older people in the world creates an urgency to know more about how their nutrient needs differ from those of younger people, and how such knowledge can enhance their health. In the meantime, people judge for themselves how to manage their nutrition, and some turn to supplements.

Supplements for Older Adults

Do older people need supplements? If so, which ones? A look at the needs of the older population provides some clues. As described, energy needs decrease with age, so older adults need more nutrients per kcalorie from their food than they did when they were younger; they need to consume a more nutrient-dense diet. Unfortunately, many older people do just the opposite: they consume nutrient-poor diets. Recall the discussions about vitamin D, calcium, and zinc. To obtain the recommended amounts of these nutrients in a small number of kcalories demands that the nutrient density of the foods chosen must be extraordinarily high. Medication use and disease conditions of many older people push their nutrient needs up further.

Can supplements meet the needs of older people? The answer is sometimes, depending on the nutrient being supplemented. For instance, iron supplements for anemia, when recommended by a health care professional, are beneficial. However, in most cases, the money people spend on supplements would be better spent on nutritious foods.

The National Institute on Aging (NIA) states that a well-balanced diet will provide most healthy older people with the nutrients they need.[104] According to the NIA, a well-balanced diet is the core of the eating patterns given earlier (in Table 10–1): this diet consists of at least 2 servings of milk or milk products; 2 servings of protein-rich foods such as lean meat, fish, and legumes; 4 servings of fruits and vegetables; and 4 servings of whole-grain breads and cereals. The chosen foods should be low in fat and high in fiber.

Older adults with certain diseases or health problems, or those facing surgery, may require supplements. A health care professional should be consulted beforehand and given all medical information available to ensure the appropriate supplement and the right dosage. Appendix E compares the vitamin and mineral contents of supplements commonly available in the United States.

Older adults with food energy intakes less than about 1500 kcalories should probably take vitamin-mineral supplements—not megavitamins, but just once-daily types of supplements. Many older adults fall into this category and should consider this precaution, but food is the best source of nutrients for everybody. Supplements are just that—supplements to foods, not substitutions for them. For anyone who is motivated to obtain the best possible health, it is never too late to learn to exercise regularly, eat well, and adopt other lifestyle changes to achieve that goal. Inactive adults, rather than taking supplements, would do well to become active and earn the right to eat more food.

As the end of this chapter approaches, it is appropriate to include a final section about the foods that older people do and do not eat, and to look at the reasons why. All the research and discussion about nutrient requirements and nutrition status are of little use if the knowledge gained cannot be practically applied to benefit the people being studied. Foods deliver nutrients to people, and most people eat them not for the nutrients they contain but rather because they taste good, because they are convenient, because they are equated with comfort and security, or for any number of other reasons.

Food Choices and Eating Habits

Strategies and feeding programs to improve people's nutrition status must be based on knowledge of their food preferences and eating patterns, if they are to afford any benefit. Menus and feeding programs for older adults must take into consideration not only the food likes and dislikes of this diverse group of people but their home environments, economic status, and medical conditions as well. For example, a person living without a refrigerator or stove has little use for foods that require refrigeration or cooking.

Information about specific subgroups of older people is lacking, making it difficult to interpret existing research. For instance, nutrition surveys seldom differentiate among older people living alone, those living with others, and those in institutions.[105] Evidence shows that these factors play a significant role in food practices of older adults.[106] One study found higher nutrition scores for diets of older people who lived independently and could maintain traditional eating habits.[107] In a different study, work experience, education, housing (federally funded versus privately owned), and gender influenced food intakes.[108] The results indicated that older adults most likely to be at risk of malnutrition were women, those with the least education, those living alone in federally funded housing, and those who had recently experienced changes in lifestyle.

This section looks at the food choices of people living independently; nutrition for people in nursing homes is the subject of the accompanying Practical Point. Financial status plays a key role in food intake. A study of poor older people in New York found that one-fourth of them ate fewer than seven hot meals a week, and 16 percent went without food for one or more days a week.[109] In contrast, a study in Boston found that middle-income older adults ate three meals a day, did not snack, and had protein intakes similar to those of younger people.[110]

Despite the diversity in living conditions and economic status among older adults, some consistency in food choices emerges. For one thing, older people eat limited amounts of fresh plant foods—fruits and, especially, vegetables. When almost 500 participants in a meal program were surveyed

▸▸ **PRACTICAL POINT**

Nutrition in Nursing Homes

When inquiring into nursing homes or day care centers for older adults, a person should find out some things about the food service:

- Can the people choose their own food?
- How often are the menus repeated (is the cycle monotonous)?
- How often are fresh fruits and vegetables served? Is the food kept appropriately hot or cold until serving?
- Is a plate check conducted regularly—at least once a week—to discover what the elderly person is consuming?
- Does the staff keep track of each person's weight?
- Is there good communication between the nursing staff and the dietitian so that the dietitian will know if someone is not eating?
- Is the elderly person encouraged and helped to go to the dining room to eat in order to enjoy other people's company?
- Is the dining room attractive?
- Does someone help those who can't manage feeding themselves?
- Are minced meats offered to those who have problems with their dentures?
- Are religious and ethnic dietary requests honored?
- How high a proportion of the foods is prepackaged? (No guide can be given for what proportion is desirable, but it should be remembered that processed foods are low in vitamin content and high in salt.)

Other questions that the investigator will want to ask have to do with the general atmosphere of the facility, in recognition of the effect of social climate on a person's appetite. A nursing home or day care center that views participants as persons, not as patients, gets a mark in its favor.

Opinions differ on the philosophy to adopt for institutional menus. Managing a multitude of different special diets is difficult and expensive, and one authority recommends a "liberalized geriatric diet" for most cases, rather than modified diets. Based on the assumption that older people "should have the right to choose the food they eat," this general, liberal approach provides in one package the key characteristics of several special diets:[a]

- 1500 to 2000 kcalories per day, mostly from nutrient-dense foods, with simple desserts.
- Minimal salt used in preparation.
- 65 to 70 grams protein per day from 2 cups milk and 4 to 6 ounces meat or alternate.
- At least 6 milligrams iron per day (the RDA for older people is 10 milligrams per day).
- Generous amounts of natural fiber.

▸ Fluid intake of 64 ounces per day.

Further modifications are essential for people with severe disease conditions.

[a]E. Luros, A rational approach to geriatric nutrition, *Dietetic Currents, Ross Timesaver* 8 (November-December 1981).

---

about their food likes and dislikes, nine of the top ten most-disliked foods were vegetables.[111] In a study designed to determine eating habits of older adults, a list of core foods was developed based on how frequently specific foods were eaten.[112] Few fruits and vegetables were among the common core foods. Women eat more fruits and vegetables than men do. In both of these studies and in another one, potatoes were well liked and eaten frequently.[113] Orange juice and bananas were the favorite fruits. Nonfat milk is not well liked by older adults, probably because they are not accustomed to its taste.

Breakfast appears to be older people's favorite meal, and many of them eat a nutritious breakfast daily.[114] In the study mentioned previously, a large percentage of the core foods were breakfast foods such as cereals, eggs, orange juice, and whole-wheat bread.[115] Cheeses and meats are also favorite foods.

Taste and health beliefs exert greater influence on older people's food selections than convenience or price do, although these are influential as well.[116] The importance of diet and health beliefs in food selections is evidenced by surveys indicating that older people are heavy users of bran cereals, egg substitutes, and decaffeinated coffee.[117] Older people are less likely to diet to lose weight than younger people are, and more likely to diet for medical reasons such as blood glucose control, cholesterol control, and sodium control.

Knowledge about the kinds of foods older people prefer and the reasons why they select or reject foods should be used for developing nutrition intervention programs and acceptable food products. Most older people are independent, productive, health-conscious consumers who know what they want from the foods they purchase. The myth that they subsist on tea and toast is without foundation. They spend more money per person on food they eat at home than other age groups and less money on food away from home.[118] Manufacturers would be wise to cater to their preferences by providing good-tasting, nutritious foods in easy-to-open, single-serving packages with labels that are easy to read.

Researchers studying nutrition and aging are challenged by the physiological and psychosocial diversity of older adults. Life factors such as nutrition, genetics, physical activity, and stress contribute to the diversity and make the study of the aging process complex. Many of the health problems older people experience are presently attributed to normal, age-related processes, perhaps to a greater extent than is valid.[119] Research that focuses on how life factors affect aging and disease processes is vital to ensuring that more and more people can look forward to long, healthy lives.

Chapter 10 Notes

1. K. Flieger, Why do we age? *FDA Consumer,* October 1988, pp. 20–25.
2. D. Sinclair, *Human Growth after Birth* (New York: Oxford University Press, 1985), pp. 29–31, 102–122.
3. A. E. Harper, Nutrition, aging, and longevity, *American Journal of Clinical Nutrition* 36 (1982): 737–749.
4. Harper, 1982.
5. J. F. Fries, Aging, natural death, and the compression of morbidity, *New England Journal of Medicine* 303 (1980): 130–135, as cited by E. L. Schneider and J. D. Reed, Life extension, *New England Journal of Medicine* 312 (1985): 1159–1168.
6. Schneider and Reed, 1985.
7. C. M. McCay, M. F. Crowell, and L. A. Maynard, The effect of retarded growth upon the length of life span and upon the ultimate body size, *Journal of Nutrition* 10 (1935): 63–79.
8. M. H. Ross, Nutrition and longevity in experimental animals, in *Nutrition and Aging,* ed. M. Winick (New York: Wiley, 1976), pp. 43–55.
9. B. N. Berg and H. S. Simms, Nutrition and longevity in the rat: II. Longevity and onset of disease with different levels of food intake, *Journal of Nutrition* 71 (1960): 255–263.
10. G. A. Nolen, Effect of various restricted dietary regimens on growth, health, and longevity of albino rats, *Journal of Nutrition* 102 (1972): 1477–1494.
11. M. H. Ross, Length of life and caloric intake, *American Journal of Clinical Nutrition* 25 (1972): 834–838.
12. Ross, 1972; M. H. Ross and G. Bras, Food preference and length of life, *Science* 190 (1975): 165–167.
13. N. B. Belloc and L. Breslow, Relationship of physical health status and health practices, *Preventive Medicine* 1 (1972): 409-421.
14. T. Harris and coauthors, Body mass index and mortality among nonsmoking older persons, *Journal of the American Medical Association* 259 (1988): 1520–1524.
15. R. W. Jeffery and coauthors, Prevalence of overweight and weight loss behavior in a metropolitan adult population: The Minnesota Heart Survey experience, *American Journal of Public Health* 74 (1984): 349–352; M. R. C. Greenwood and V. A. Pittman-Waller, Weight control: A complex, various, and controversial problem, in *Obesity and Weight Control: The Health Professional's Guide to Understanding and Treatment,* eds. R. T. Frankle and M. Yang (Rockville, Md.: Aspen, 1988), pp. 3–15.
16. E. D. Schlenker, Obesity and the lifespan, in *Nutrition, Physiology, and Obesity,* ed. R. Schemmel (Boca Raton, Fla.: CRC Press, 1980), pp. 151–166; Y. Kagawa, Impact of westernization on the nutrition of Japanese: Changes in physique, cancer, longevity and centenarians, *Preventive Medicine* 7 (1978): 205–217.
17. Kagawa, 1978.
18. Kagawa, 1978.
19. R. Carmena and coauthors, Changes in plasma high-density lipoproteins after body weight reduction in obese women, *International Journal of Obesity* 8 (1984): 135–140.
20. R. Carmena and coauthors, Changes in plasma high-density lipoproteins after body weight reduction in obese women, *International Journal of Obesity* 8 (1984): 135–140.
21. P. W. F. Wilson and coauthors, Factors associated with lipoprotein cholesterol levels: The Framingham Study, *Arteriosclerosis* 3 (1983): 273–281.
22. E. A. Vallejo, Restricted diet on alternate days in the nutrition of the aged, *Revista Clinica Espanola* 63 (1956): 25–27.
23. K. T. Khaw and E. Barrett-Connor, Family history of heart attack: A modifiable risk factor, *Circulation* 74 (1986): 239–244.
24. R. R. Williams, Nature, nurture, and family predisposition, *New England Journal of Medicine* 318 (1988): 769–770.
25. T. I. A. Sorensen and coauthors, Genetic and environmental influences on premature death in adult adoptees, *New England Journal of Medicine* 318 (1988): 727–732.
26. B. Larsson and coauthors, Health and aging characteristics of highly physically active 65-year-old men, *International Journal of Sports Medicine* 5 (1984): 336–340; T. W. Wright and coauthors, Cardiac output in male middle-aged runners, *Journal of Sports Medicine* 22 (1982): 17–22; G. Sopko and coauthors, The effects of exercise and weight loss on plasma lipids in young obese men, *Metabolism* 34 (1985): 227–236.
27. R. S. Paffenbarger and coauthors, Physical activity, all-cause mortality, and longevity of college alumni, *New England Journal of Medicine* 314 (1986): 605–611.
28. H. Selye, *The Stress of Life* (New York: McGraw-Hill, 1956; rev. ed. 1976), as cited by F. S. Sizer and E. N. Whitney, *Life Choices: Health Concepts and Strategies* (St. Paul, Minn.: West, 1987), pp. 33–52.
29. Harper, 1982; D. M. Driscoll, The relationship between weather and mortality in the major metropolitan areas in the United States, 1962–1965, *International Journal of Biometeorology* 15 (1971): 23–39, as cited by D. M. Watkin, The physiology of aging, *American Journal of Clinical Nutrition* 36 (1982): 750–758.
30. J. W. Rowe, Physiologic interface of aging and nutrition, in *Nutrition and Aging,* eds. M. L. Hutchinson and H. N. Munro (Orlando, Fla.: Academic Press, 1986), pp. 11–21.
31. Rowe, 1986.
32. P. W. Landfield, R. K. Baskin, and T. A. Pitler, Brain aging correlates: Retardation by hormonal-pharmacological treatments, *Science* 214 (1981): 581–584.
33. R. M. Sapolsky, L. C. Krey, and B. S. McEwen, The adrenocortical stress-response in the aged male rat: Impairment of recovery from stress, *Experimental Gerontology* 18 (1983): 55-64, as cited by M. L. Zoler, Hormones and aging: Turning off "the aging switch," *Geriatics* 38 (1983): 107–112.
34. K. Kinsell and G. G. Harrison, Concepts of sensitivity and specificity: Their relevance to nutritional assessment of the elderly, Report of the Third Ross Roundtable on Medical Issues, in *Assessing the Nutritional Status of the Elderly: State of the Art,* ed. D. E. Redfern (Columbus, Ohio: Ross Laboratories, 1982), pp. 2–6.
35. A. N. Exton-Smith, Nutritional status: Diagnosis and prevention of malnutrition, in *Metabolic and Nutritional Disorders in the Elderly,* eds. A. N. Exton-Smith and F. I. Caird (Bristol, Tenn.: John Wright, 1980), pp. 66–76, as cited by H. Munro, P. M. Suter, and R. M. Russell, Nutritional requirements of the elderly, *Annual Review of Nutrition* 7 (1987): 23–49.
36. Schlenker, 1980.
37. J. A. Jernigan and coauthors, Reference values for blood findings in relatively fit elderly persons, *Journal of the American Geriatric Society* 28 (1980): 308–314, as

cited in Abnormal blood tests in elderly patients, *Ross Timesaver: Geriatric Medicine Currents* 2 (1981): 1–2.

38. N. Agarwal and coauthors, Predictive ability of various nutritional variables for mortality in elderly people, *American Journal of Clinical Nutrition* 48 (1988): 1173–1178.
39. D. A. Lipschitz and C. O. Mitchell, Hematologic measurements in nutritional assessment of the elderly, Report of the Third Ross Roundtable on Medical Issues, in *Assessing the Nutritional Status of the Elderly: State of the Art,* ed. D. E. Redfern (Columbus, Ohio: Ross Laboratories, 1982); pp. 38–41.
40. H. N. Munro, Major gaps in nutrient allowances, *Journal of the American Dietetic Association* 76 (1980): 137–141.
41. M. Stiedemann, C. Jansen, and I. Harrill, Nutritional status of elderly men and women, *Journal of the American Dietetic Association* 73 (1978): 132–138.
42. S. Abraham and coauthors, Dietary intake of persons 1–74 years of age in the United States, *Advance Data from Vital and Health Statistics* 6 (1977): 1–15.
43. S. R. Jambert and A. R. Juansing, Protein-calorie malnutrition in the elderly, *Journal of the American Geriatrics Society* 28 (1980): 272–275.
44. Jambert and Juansing, 1980.
45. S. D. Krasinski and coauthors, Aging changes vitamin A absorption characteristics (abstract), *Gastroenterology* 88 (1985): 171, as cited by P. M. Suter, and R. M. Russell, Vitamin requirements of the elderly, *American Journal of Clinical Nutrition* 45 (1987): 501–512.
46. M. J. Yiengst and N. W. Shock, Effect of oral administration of vitamin A on plasma levels of vitamin A and carotene in aged males, *Journal of Gerontology* 4 (1984): 205–211, as cited by Suter and Russell, 1987.
47. P. J. Garry and coauthors, Vitamin A intake and plasma retinol levels in healthy elderly men and women, *American Journal of Clinical Nutrition* 46 (1987): 989–994.
48. Suter and Russell, 1987.
49. M. Jagerstad and A. K. Westesson, Folate, *Scandinavian Journal of Gastroenterology* 14 (1979): 196–202.
50. V. Herbert, Recommended dietary intakes (RDI) of vitamin B_{12} in humans, *American Journal of Clinical Nutrition* 45 (1987): 671-678.
51. P. J. Garry and coauthors, Nutritional status in a healthy elderly population: Dietary and supplemental intakes, *American Journal of Clinical Nutrition* 36 (1982): 319–331.
52. Garry and coauthors, 1982.
53. P. J. Garry, J. S. Goodwin, and W. C. Hunt, Folate and vitamin B_{12} status in a healthy elderly population, *Journal of the American Geriatrics Society* 32 (1984): 719–726.
54. R. M. Russell, Implications of gastric atrophy for vitamin and mineral nutriture, in *Nutrition and Aging,* eds. M. L. Hutchinson and H. N. Munro (Orlando, Fla.: Academic Press, 1986), pp. 59–69.
55. M. Siurala and coauthors, Prevalence of gastritis in a rural population, *Scandinavian Journal of Gastroenterology* 3 (1968): 211–223, as cited by Suter and Russell, 1987.
56. Russell, 1986.
57. M. F. Holick, Vitamin synthesis by the aging skin, in *Nutrition and Aging,* eds. M. L. Hutchinson and H. N. Munro, (Orlando, Fla.: Academic Press, 1986), pp. 45–58.
58. A. M. Parfitt and coauthors, Vitamin D and bone health in the elderly, *American Journal of Clinical Nutrition* 36 (1982): 1014–1031.
59. Garry and coauthors, 1982.
60. C. Lamberg-Allardt, Vitamin D intake, sunlight exposure and 25-hydroxy vitamin D levels in the elderly during one year, *Annals of Nutrition and Metabolism* 28 (1984): 144–150, as cited in Vitamin D status of the elderly: Contributions of sunlight exposure and diet, *Nutrition Reviews* 43 (1985): 78–80; P. Lips and coauthors, Determinants of vitamin D status in patients with hip fracture and in elderly control subjects, *American Journal of Clinical Nutrition* 46 (1987): 1005–1010.
61. D. M. Slovik and coauthors, Deficient production of 1,25-dihydroxyvitamin D in elderly osteoporotic patients, *New England Journal of Medicine* 305 (1981): 372–374; Y. Tanaka and H. F. DeLuca, Rat renal 25-hydroxyvitamin D_3 1- and 24-hydroxylases: Their in vivo regulation, *American Journal of Physiology* 246 (1984): E168-E173.
62. H. F. DeLuca, Significance of vitamin D in age-related bone disease, in *Nutrition and Aging,* eds. M. L. Hutchinson and H. N. Munro (Orlando, Fla.: Academic Press, 1986), pp. 217–234.
63. M. B. Kohrs, Effectiveness of nutrition intervention programs for the elderly, in *Nutrition and Aging,* eds. M. L. Hutchinson and H. N. Munro (Orlando, Fla.: Academic Press, 1986), pp. 139–167.
64. M. S. Seelig, *Magnesium Deficiency in the Pathogenesis of Disease* (New York: Plenum, 1980), as cited by P. O. Wester, Magnesium, *American Journal of Clinical Nutrition* (supplement) 45 (1987): 1305–1312.
65. H. Karppanen, Epidemiological aspects of magnesium deficiency in cardiovascular diseases, *Magnesium Bulletin* 8 (1986): 199–203.
66. F. Lazicki and coauthors, Magnesium intakes, balances, and blood levels of adults consuming self-selected diets, *American Journal of Clinical Nutrition* 40 (1984): 1380–1389.
67. E. M. Pao and S. J. Mickle, Problem nutrients in the United States, *Food Technology* 35 (1981): 58–69, as cited by Wester, 1987. Note that the RDA refers to the 1980 edition; however, the 1989 magnesium RDA for men did not change, and for women, it declined only slightly.
68. W. Mertz, Trace elements and the needs of the elderly, in *Nutrition and Aging,* eds. M. L. Hutchinson and H. N. Munro (Orlando, Fla.: Academic Press, 1986), pp. 71–83.
69. S. R. Lynch and coauthors, Iron status of elderly Americans, *American Journal of Clinical Nutrition* 36 (1982): 1032–1045.
70. Garry and coauthors, 1982. Note that at the time of this research the zinc RDA for women was 15 milligrams; even with the lower current standards, many of the women's intakes would still fall short of recommendations.
71. H. H. Sandstead and coauthors, Zinc nutriture in the elderly in relation to taste acuity, immune response, and wound healing, *American Journal of Clinical Nutrition* 36 (1982): 1046–1059.
72. J. R. Turnlund and coauthors, Stable isotope studies of zinc absorption and retention in young and elderly men (abstract), *Journal of the American Dietetic Association* 86 (1986): 1762.
73. Turnlund and coauthors, 1986.
74. Sandstead and coauthors, 1982.
75. J. D. Bogden and coauthors, Zinc and immunocompetence in the elderly: Baseline data on zinc nutriture and immunity in unsupplemented subjects, *American Journal of Clinical Nutrition* 46 (1987): 101–109.
76. R. B. McGandy and coauthors, Nutritional status survey of healthy noninstitutionalized elderly: Nutrient intakes from three-day diet records and nutrient

supplements, *Nutrition Research* 6 (1986): 785–798, as cited by H. N. Munro, P. M. Suter, and R. M. Russell, Nutritional requirements of the elderly, in *Annual Review of Nutrition* 7 (1987): 23–49.

77. A. Hecht, Medicine and the elderly, *FDA Consumer,* September 1983, pp. 20–22.

78. Hecht, 1983.

79. L. Yung, I. Contento, and J. D. Gussow, Use of health foods by the elderly, *Journal of Nutrition Education* 3 (1984): 127–131.

80. G. E. Gray and coauthors, Vitamin supplement use in a southern California retirement community, *Journal of the American Dietetic Association* 86 (1986): 800–802.

81. B. S. Ranno, G. M. Wardlaw, and C. J. Geiger, What characterizes elderly women who overuse vitamin and mineral supplements? *Journal of the American Dietetic Association* 88 (1988): 347–348.

82. A. B. Kallner, D. Hartmann, and D. H. Hornig, On the requirements of ascorbic acid in man: Steady-state turnover and body pool in smokers, *American Journal of Clinical Nutrition* 34 (1981): 1347–1355.

83. E. Koop, *The Health Consequences of Smoking,* a report (1983) of the surgeon general available from the Superintendent of Documents, U.S. Government Printing Office, Washington, D.C. 20402.

84. E. L. Schneider and coauthors, Recommeneded Dietary allowances and the health of the elderly, *New England Journal of Medicine* 314 (1986): 157–160.

85. P. A. Phillips and coauthors, Reduced thirst after water deprivation in healthy elderly men, *New England Journal of Medicine* 311 (1984): 753–759.

86. A. Leaf, Dehydration in the elderly, *New England Journal of Medicine* 311 (1984): 791–792.

87. J. J. Cunningham, A reanalysis of the factors influencing basal metabolic rate in normal adults, *American Journal of Clinical Nutrition* 33 (1980): 2372–2374.

88. Food and Nutrition Board, *Recommended Dietary Allowances,* 10th ed. (Washington, D.C.: National Academy of Sciences, 1989), pp. 32–33.

89. D. H. Calloway and E. Zanni, Energy requirements and energy expenditure of elderly men, *American Journal of Clinical Nutrition* 33 (1980): 2088–2092.

90. Food and Nutrition Board, 1989, pp. 58–60.

91. V. R. Young, M. Gersovitz, and H. N. Munro, Human aging: Protein and amino acid metabolism and implications for protein and amino acid requirements, in *Nutritional Approaches to Aging Research,* ed. G. B. Moment (Boca Raton, Fla.: CRC Press, 1982), pp. 47–81.

92. A. H. R. Cheng and coauthors, Comparative nitrogen balance study between young and aged adults using three levels of protein intake from a combination wheat-soy-milk mixture, *American Journal of Clinical Nutrition* 31 (1978): 12–22.

93. Food and Nutrition Board, 1989, pp. 68–69.

94. H. Kallman, Constipation in the elderly, *American Family Physician* 27 (1983): 179–184, as cited by E. J. Johnson and coauthors, Dietary fiber intakes of nursing home residents and independent-living older adults, *American Journal of Clinical Nutrition* 48 (1988): 159–164.

95. Johnson and coauthors, 1988.

96. P. M. Suter and R. M. Russell, Vitamin requirements of the elderly, *American Journal of Clinical Nutrition* 45 (1987): 501-512.

97. J. A. Olson, Recommended dietary intakes (RDI) of vitamin A in humans, *American Journal of Clinical Nutrition* 45 (1987): 704-716.

98. Herbert, 1987.

99. Schneider and coauthors, 1986.

100. R. P. Heaney, R. R. Recker, and P. D. Saville, Calcium balance and calcium requirements in middle-aged women, *American Journal of Clinical Nutrition* 30 (1977): 1603–1611H; H. Spencer, L. Kramer, and D. Osis, Factors contributing to calcium loss in aging, *American Journal of Clinical Nutrition* 36 (1982): 776–787; H. Spencer, Minerals and mineral interactions in human nutrition, *Journal of the American Dietetic Association* 86 (1986): 864–867.

101. A. R. P. Walker, The human requirement of calcium: Should low intakes be supplemented? *American Journal of Clinical Nutrition* 25 (1972): 518–530, as cited by L. H. Allen, The role of nutrition in the onset and treatment of metabolic bone disease, *Nutrition Update* 1 (1983): 263–282.

102. R. P. Heaney and coauthors, Calcium nutrition and bone health in the elderly, *American Journal of Clinical Nutrition* 36 (1982): 986–1013.

103. H. Spencer, Minerals and mineral interactions in human beings, *Journal of the American Dietetic Association* 86 (1986): 864–867.

104. C. Lecos, Diet and the elderly, *FDA Consumer,* November 1984, pp. 7.

105. Food and nutrient intakes of individuals in one day in the United States: Spring 1977, USDA/SEA Nationwide Food Consumption Survey, 1977–78, Preliminary Report No. 2, September 1980, as cited by M. Krondl and coauthors, Food use and perceived food meanings of the elderly, *Journal of the American Dietetic Association* 80 (1982): 523–529.

106. J. S. Atkins and coauthors, Cluster analysis of food consumption patterns of older Americans, *Journal of the American Dietetic Association* 86 (1986): 616–624.

107. M. Clark and L. M. Wakefield, Food choices of institutionalized vs. independent-living elderly, *Journal of the American Dietetic Association* 66 (1975): 600–604.

108. P. O'Hanlon and coauthors, Socioeconomic factors and dietary intake of elderly Missourians, *Journal of the American Dietetic Association* 82 (1983): 646–653.

109. D. A. Roe, D. F. Williamson, and E. A. Frongillo, Supplemental Nutrition Assistance Program, final report, 1984–85 survey of elderly recipients of SNAP home-delivered meals in New York State (Ithaca, N.Y.: Division of Nutritional Sciences, Cornell University, 1985).

110. J. Wurtman, Aging and eating: Good news for some, *Science News* 130 (1986): 148.

111. V. Holt, J. Nordstrom, and M. B. Kohrs, Food preferences of older adults (abstract), *Journal of the American Dietetic Association* 87 (1987): 1597.

112. M. Krondl and coauthors, Food use and perceived food meanings of the elderly, *Journal of the American Dietetic Association* 80 (1982): 523–529.

113. M. Fanelli and K. J. Stevenhagen, Characterizing consumption patterns by food frequency methods: Core foods and variety of foods in diets of older Americans, *Journal of the American Dietetic Association* 85 (1985): 1570–1575.

114. M. Chou, Selling to older Americans (abstract), *Journal of the American Dietetic Association* 80 (1982): 277.

115. Krondl and coauthors, 1982.

116. Krondl and coauthors, 1982; Fanelli and Stevenhagen, 1985.

117. S. B. Sellery, New product opportunities: Diet food for older Americans (abstract) *Journal of the American Dietetic Association* 85 (1985): 128.

118. Chou, 1982.

119. J. W. Rowe and R. L. Kahn, Human aging: Usual and successful, *Science* 237 (1987): 143–149.

Appendix A

Nutrition Assessment Procedures, Standards, and Forms

This appendix provides a sample of the procedures, standards, and forms commonly used in nutrition assessment. The appropriate uses of each of these, except some measures of protein-energy malnutrition (PEM) and of iron status, are discussed in the opening chapter of this text. For PEM, the discussion and standards for adults are here, while the standards for children are in Chapter 7. For iron, the text in Chapter 1 provides an overview, and the detailed discussion is here.

General Assessment

Chapter 1 described the interviews used in collecting *historical data.* To go along with that discussion, Forms A–1 through A–5 ascertain pertinent information and Table A–1 presents examples of drug-nutrient interactions.

Chapter 1 also described the standard *anthropometric measurements* used in nutrition assessment. To go with that discussion, Tables A–2 through A–7 and Figures A–1 through A–3 present standards and procedures.

Chapter 1 described the *physical examinations* used in nutrition assessment. To go with that discussion, Table A–8 lists symptoms of vitamin and mineral imbalances.

Chapter 1 also described *biochemical analyses* as part of nutrition assessment. Table A–9 lists biochemical tests useful in nutrition assessment of vitamin and mineral status. For PEM and iron assessment, the discussions follow.

Assessment of Protein-Energy Malnutrition (PEM)

The most common tests used in hospitals today for nutrition assessment help to uncover PEM. These tests include serum albumin, serum transferrin, total lymphocyte count, and urinary creatinine excretion.

Serum albumin Albumin accounts for over 50 percent of the total serum proteins. It helps to maintain fluid and electrolyte balance and to transport many nutrients, hormones, drugs, and other compounds. Albumin synthesis depends on the existence of functioning liver cells and on an appropriate supply of amino acids. Because there is so much albumin in the body and because it is not broken down quickly, albumin concentrations change slowly. Therefore, albumin is a useful indicator of prolonged depression of the protein

(continued on p. 461)

Form A–1 History

Name ______________________ Today's date ______________
Address ______________________ Age ______________
______________________ Sex ______________
______________________ Phone ______________
Date of last medical checkup ______________ Height ______________
Reason for coming in ______________ Weight ______________
______________ Usual Weight ______________

PERSONAL DATA

1. Last grade of school completed ______ Still in school? ______________
2. Are you employed? ______ Occupation ______________
3. Does someone else live at your home? ______ Who? ______________
4. Do you smoke in any way? ______ How much? ______________
5. Have you recently lost or gained more than 10 lb? ______ If yes, please explain how ______________
6. Are you pregnant? ______ How many months? ______________
7. How many pregnancies have you carried to term? ______________
8. Are your menstrual periods normal? ______ If not, please explain ______________
____
9. Have you been told that you have (check any that apply):
Diabetes ______ High blood pressure ______ Hardening of the arteries ______
Lung disease ______ Kidney disease ______ Liver disease ______ Ulcers ____
Cancer ______ Other ______________
10. Do you eat at regular times each day? ______ How many times per day? ____
11. Do you usually eat snacks? ______ When? ______________
12. Where do you usually eat your meal?
Morning __________ Noon __________ Night __________
With whom?
Morning __________ Noon __________ Night __________
13. Would you say your appetite is good? ______ Fair? ______ Poor? ______
If poor, please explain ______________
14. What foods do you particularly dislike? ______________
15. Are there foods you don't eat for other reasons? ______________
16. Do you have any difficulty eating? ______________
17. How would you describe your feelings about food? ______________
____
____
18. Who prepares your meals? ______________
19. Are you, or is any member of your family, on a special diet? ______________
If yes, who and what kind? ______________
20. Do you drink alcohol? ______ How many drinks per day? ______________
Do you ever drink alcohol excessively? ______ How often? ______________
21. Do you take any kind of medication, either prescribed by a doctor or over-the-counter, for any condition? ______________
22. How would you describe your exercise habits?
Kind of exercise? ______________ How intense? ______________
How long at a time? ______________ How often? ______________
23. Are there any other facts about your lifestyle that you think might be related to your nutritional health? ______ Explain ______________
____

Form A–2 Food Intake for a 24-Hour Recall

Name and address ________________________ Date ____________

____

____

Did you take a vitamin/mineral supplement? ________________________

If yes, what kind? ________________________ Dose? ____________

Please record the amount and type of foods and beverages consumed today.

Food	Amount (c, tbsp, or piece)	Description (how cooked, how served)

Form A–3 Food Frequency Checklist

The following information will help us to understand your regular eating habits so that we may offer you the best service possible. If you have any doubt about some items, be sure to underestimate the "goodness" of your habits rather than to overestimate.

1. How many times *per week* do you eat the following foods? Circle the appropriate number:

	PER WEEK	
Poultry	0 <1 1 2 3 4 5 6 7 8 9 >9	____
Fish	0 <1 1 2 3 4 5 6 7 8 9 >9	____
Hot dogs	0 <1 1 2 3 4 5 6 7 8 9 >9	____
Bacon	0 <1 1 2 3 4 5 6 7 8 9 >9	____
Lunch meat	0 <1 1 2 3 4 5 6 7 8 9 >9	____
Sausage	0 <1 1 2 3 4 5 6 7 8 9 >9	____
Pork or ham	0 <1 1 2 3 4 5 6 7 8 9 >9	____
Salt pork	0 <1 1 2 3 4 5 6 7 8 9 >9	____
Liver	0 <1 1 2 3 4 5 6 7 8 9 >9	____
Beef or veal	0 <1 1 2 3 4 5 6 7 8 9 >9	____
Other meats (which?) ________	0 <1 1 2 3 4 5 6 7 8 9 >9	____
Eggs	0 <1 1 2 3 4 5 6 7 8 9 >9	____
Fast foods	0 <1 1 2 3 4 5 6 7 8 9 >9	____

2. How many times *per day* do you eat the following foods? Circle the appropriate number:

	PER DAY	
Bread, toast, rolls, muffins	0 <1 1 2 3 4 5 6 7 8 9 >9	____
Milk (including on cereal)	0 <1 1 2 3 4 5 6 7 8 9 >9	____
Yogurt or tofu	0 <1 1 2 3 4 5 6 7 8 9 >9	____
Cheese or cheese dishes	0 <1 1 2 3 4 5 6 7 8 9 >9	____
Sugar, jam, jelly, syrup, honey	0 <1 1 2 3 4 5 6 7 8 9 >9	____
Butter or margarine	0 <1 1 2 3 4 5 6 7 8 9 >9	____

3. How many times *per week* do you eat the following foods? Circle the appropriate number:

	PER WEEK	
Fruit or fruit juice	0 <1 1 2 3 4 5 6 7 8 9 >9	____
Vegetables other than potatoes	0 <1 1 2 3 4 5 6 7 8 9 >9	____
Potatoes and other starchy vegetables	0 <1 1 2 3 4 5 6 7 8 9 >9	____
Salads or raw vegetables	0 <1 1 2 3 4 5 6 7 8 9 >9	____
Cereal (which kind?) ________	0 <1 1 2 3 4 5 6 7 8 9 >9	____
Pancakes or waffles	0 <1 1 2 3 4 5 6 7 8 9 >9	____
Rice or other cooked grains	0 <1 1 2 3 4 5 6 7 8 9 >9	____
Noodles (macaroni, spaghetti)	0 <1 1 2 3 4 5 6 7 8 9 >9	____
Crackers or pretzels	0 <1 1 2 3 4 5 6 7 8 9 >9	____
Sweet rolls or doughnuts	0 <1 1 2 3 4 5 6 7 8 9 >9	____
Cooked dried beans or peas	0 <1 1 2 3 4 5 6 7 8 9 >9	____
Peanut butter or nuts	0 <1 1 2 3 4 5 6 7 8 9 >9	____
Milk or milk products	0 <1 1 2 3 4 5 6 7 8 9 >9	____
TV dinners, pot pies, other prepared meals	0 <1 1 2 3 4 5 6 7 8 9 >9	____
Sweet bakery goods (cakes, cookies)	0 <1 1 2 3 4 5 6 7 8 9 >9	____
Snack foods (potato or corn chips)	0 <1 1 2 3 4 5 6 7 8 9 >9	____
Candy	0 <1 1 2 3 4 5 6 7 8 9 >9	____
Soft drinks (which?) ________	0 <1 1 2 3 4 5 6 7 8 9 >9	____
Coffee or tea	0 <1 1 2 3 4 5 6 7 8 9 >9	____
Frozen sweets (which?) ________	0 <1 1 2 3 4 5 6 7 8 9 >9	____
Instant meals such as breakfast bars or diet meal beverages (which?) ________	0 <1 1 2 3 4 5 6 7 8 9 >9	____
Wine	0 <1 1 2 3 4 5 6 7 8 9 >9	____
Beer	0 <1 1 2 3 4 5 6 7 8 9 >9	____
Whiskey, vodka, rum, etc.	0 <1 1 2 3 4 5 6 7 8 9 >9	____

4. What specific kinds of the following foods do you eat most often? Include the name of the food; whether it is fresh, canned, or frozen; and how it is prepared.

   Fruits and fruit juices ________________
   Vegetables ________________
   Milk and milk products ________________
   Meats ________________
   Breads and cereals ________________
   Desserts ________________
   Snack foods ________________

5. Please list the names of any liquid, powder, or pill form of vitamin or mineral product you take, and state how often you take it. Please list also any diet supplement you use (such as protein milk shakes or brewer's yeast), how much you use, and how often you use it. ________________

6. Is there anything else we should know about your food/nutrient intake? ________
____

Form A–4 Food Diary

Name ____________________
Date ____________________

Time	Place	With Whom	Emotional State	Hungry or Not Hungry	Food Eaten (amount)

(etc.)

status of the blood and internal organs. Standards for determining the severity of serum albumin depletion are given in Table A–10.

Many other conditions besides malnutrition can depress albumin concentration, including eclampsia, liver disease, advanced kidney disease (nephrotic syndrome), infection, cancer, and burns. Therefore, as is true for all nutrition assessment measurements, albumin alone cannot determine protein status, but rather serves as one indicator among many.

Transferrin Transferrin is a protein that transports iron between the intestine and sites of hemoglobin synthesis and degradation. Researchers consider it a more sensitive indicator of protein malnutrition than albumin, because it responds more promptly to changes in protein intake and has a smaller body pool.

Transferrin concentration is inversely related to iron stores; its concentration is high in iron deficiency and low when iron storage is excessive. Therefore, the transferrin level is useful as an indicator of protein status only when iron nutrition is normal. Liver disease, nephrotic syndrome, and burns cause decreases in transferrin levels; pregnancy and blood loss elevate values. Standards for determining the severity of transferrin depletion are given in Table A–10.

Lymphocyte count Various forms of PEM and individual nutrient deficiencies depress the immune system. The total number of lymphocytes appears to decrease as protein depletion occurs, so the total lymphocyte count is one useful index in nutrition assessment. The standard is 2500 mm^3; values below 1500 mm^3 are considered depleted. White blood cell (WBC) volume, and red and white blood cell counts are routinely measured in hospital tests, so the total lymphocyte count can be derived from these as follows:

$$\text{Total lymphocyte count (mm}^3) = \text{WBC (mm}^3) \times \%\ \text{lymphocytes.}$$

(continued on p. 463)

Form A–5 Drug History

1a. Do you have any health problems for which you are taking prescription medications at the present time? Yes ___ No ___
If yes:

Health problem	*Proprietary name of drug*	*Generic name of drug*	*Dose frequency*	*Duration of intake*

1b. Are you taking any other medication that a doctor has prescribed (name of drug unknown, reason for taking unknown)?
Yes ___ No ___
If yes:

Description of drug	*Dose*	*Frequency*	*Duration of intake*

2a. Have you taken prescription medication for any of the health problems listed below within the past three months?
Yes ___ No ___
If yes:

Health problem	*Drug name*	*Duration of intake*	*When discontinued*	*Reason for stopping*	*Still taking**
Asthma					
Arthritis					
High blood pressure					
Fluid retention					
Infection (specify)					
Tuberculosis					
Malaria					
Psoriasis					
Colitis					
High cholesterol					
Parkinson's disease					
Liver disease					
Kidney disease					
Blood disease					
Bone disease					
Gout					
Blood clots					
Diabetes					
Other (specify)					

*Check (√) if still taking.

2b. Have you taken any other medication within the past three months that a doctor has prescribed (name of drug unknown, reason for taking unknown)? Yes ___ No ___
If yes:

Description of drug	*Dose*	*Frequency*	*Duration of intake*

3a. Do you take medications, self-prescribed, for any reason? Yes ___ No ___
If yes:

Complaint	*Constantly*	*Frequently*	*Occasionally*
Constipation			
Indigestion			
Headaches			
Nervousness			
Insomnia			
Pain			
Menstrual cramps			
Colds and sinus trouble			
Other (state)			

Form A–5—*Continued*

3b. If your response to 3a is positive in one or more categories, what medication do you take to relieve these complaints, and how much do you need to gain relief?

Complaint	*Drug*	*Dose*	*Frequency*	*Duration*
Constipation				
Indigestion				
Headaches				
Nervousness				
Insomnia				
Pain				
Menstrual cramps				
Colds and sinus trouble				
Other (state)				

4. Are you taking birth control pills now ? Yes ___[a] No ___
If yes:
Name:
Duration of intake:
Have you taken birth control pills within the past six months? Yes ___ No ___
If yes:
Name:
Duration of intake:
Date discontinued:
Reason for stopping:

Source: Adapted from D. A. Roe, *Drug-Induced Nutritional Deficiencies* (Westport, Conn.: AVI Publishing, 1985).

[a]A yes answer to this question would indicate reduced menstrual blood loss, possible consequent iron conservation, and reduced risk of pregnancy. Chapter 2's discussion on oral contraceptives shows that their overall effect on nutrition status is minimal.

Urinary creatinine excretion Creatinine is a breakdown product of creatine phosphate (phosphocreatine, or PC), an energy fuel that is present specifically in skeletal muscle. Its excretion occurs at a constant rate determined by the amount of skeletal muscle, and the amount excreted therefore reflects skeletal muscle mass. As skeletal muscle atrophies during malnutrition, creatinine excretion decreases.

Standards for creatinine excretion, based on sex and height, are given in Tables A–11 and A–12. Assessors use these standards and measured urinary creatinine to derive the creatinine-height index (CHI):

$$\text{CHI} = \frac{\text{measured urinary creatinine (24-hr sample)}}{\text{standard creatinine for height and sex}} \times 100.$$

The CHI is a percentage of the standard; generally, acceptable values are 90 to 100 percent. Children suffering from PEM have a low CHI. Standards for children are based on expected creatinine excretion of healthy children of normal height (see Chapter 7).

(continued on p. 469)

Table A–1 Examples of Possible Drug-Nutrient Interactions for Selected Commonly Used Drugs

Drug	Possible Effect on Nutrition Status				
	Decreases Absorption	*Raises Blood Concentrations*	*Lowers Blood Concentrations*	*Increases Excretion*	*Other*
Antacids	Phosphorus				Thiamin[a]
Antibiotics	Fats Amino acids Carbohydrates Folate Vitamin B_{12} Fat-soluble vitamins Calcium Iron Potassium Magnesium Zinc			Potassium Niacin Riboflavin Folate Vitamin C	Vitamin K[b]
Aspirin			Folate	Vitamin C Thiamin Vitamin K	Iron[c]
Caffeine				Calcium Magnesium	Cholesterol[d]
Diuretics		Zinc Calcium	Potassium Chloride Magnesium Phosphorus Folate Vitamin B_{12}	Calcium Sodium Thiamin Potassium Chloride Magnesium	Zinc[e]
Laxatives	Fat Glucose Vitamin D Calcium Potassium Fat-soluble vitamins Carotene				
Oral contraceptives	Folate	Vitamin A Copper Iron	Vitamin B_6 Riboflavin Folate Vitamin B_{12} Vitamin C		Riboflavin[f] Vitamin B_6[f] Calcium[g]

[a]Antacids may increase the destruction of thiamin.
[b]Some antibiotics may decrease intestinal synthesis of vitamin K.
[c]Aspirin use may cause blood loss, thus compromising iron status.
[d]Large doses of caffeine may increase blood cholesterol concentrations.
[e]Some diuretics may decrease zinc storage in the liver.
[f]Some oral contraceptives may increase the requirements for riboflavin and vitamin B_6.
[g]Some oral contraceptives may increase the absorption of calcium.

Source: Adapted from R. E. Hodges, *Nutrition in Medical Practice* (Philadelphia: Saunders, 1980), pp. 323–331; R. C. Theuer and J. J. Vitale, Drug and nutrient interactions, in *Nutritional Support of Medical Practice,* eds. H. A. Schneider, C. F. Anderson, and D. B. Coursin (Hagerstown, Md.: Harper & Row, 1977), pp. 297–305; D. A. Roe, *Drug-Induced Nutritional Deficiencies* (Westport, Conn. AVI Publishing, 1985).

Table A–2 Standard Height–Weight Tables

Weights at ages 25 to 29 based on lowest mortality. Weights in pounds according to frame (in indoor clothing weighing 5 lb for men or 3 lb for women; shoes with 1-inch heels). For frame size standards, see Table A–4.

Men					Women				
Height		*Small*	*Medium*	*Large*	*Height*		*Small*	*Medium*	*Large*
Feet	*Inches*	*Frame*	*Frame*	*Frame*	*Feet*	*Inches*	*Frame*	*Frame*	*Frame*
5	2	128–134	131–141	138–150	4	10	102–111	109–121	118–131
5	3	130–136	133–143	140–153	4	11	103–113	111–123	120–134
5	4	132–138	135–145	142–156	5	0	104–115	113–126	122–137
5	5	134–140	137–148	144–160	5	1	106–118	115–129	125–140
5	6	136–142	139–151	146–164	5	2	108–121	118–132	128–143
5	7	138–145	142–154	149–168	5	3	111–124	121–135	131–147
5	8	140–148	145–157	152–172	5	4	114–127	124–138	134–151
5	9	142–151	148–160	155–176	5	5	117–130	127–141	137–155
5	10	144–154	151–163	158–180	5	6	120–133	130–144	140–159
5	11	146–157	154–166	161–184	5	7	123–136	133–147	143–163
6	0	149–160	157–170	164–188	5	8	126–139	136–150	146–167
6	1	152–164	160–174	168–192	5	9	129–142	139–153	149–170
6	2	155–168	164–178	172–197	5	10	132–145	142–156	152–173
6	3	158–172	167–182	176–202	5	11	135–148	145–159	155–176
6	4	162–176	171–187	181–207	6	0	138–151	148–162	158–179

Source: Reproduced with permission of Metropolitan Life Insurance Company. Source of basic data: *1979 Build Study,* Society of Actuaries and Association of Life Insurance Medical Directors of America, 1980.

Table A–3 How to Determine Body Frame by Elbow Breadth

To make a simple approximation of frame size, do the following. Extend the arm, and bend the forearm upward at a 90° angle. Keep the fingers straight, and turn the inside of the wrist away from the body. Place the thumb and index finger on the two prominent bones on *either side* of the elbow. Measure the space between the fingers against a ruler or a tape measure.[a] Compare the measurements with the following standards.

These standards represent the elbow measurements for medium-framed men and women of various heights. Measurements smaller than those listed indicate a small frame, and larger measurements indicate a large frame.

Men		Women	
Height in 1-Inch Heels	*Elbow Breadth*	*Height in 1-Inch Heels*	*Elbow Breadth*
5 ft 2 inches to 5 ft 3 inches	2½ to 2⅞ inches	4 ft 10 inches to 4 ft 11 inches	2¼ to 2½ inches
5 ft 4 inches to 5 ft 7 inches	2⅝ to 2⅞ inches	5 ft 0 inches to 5 ft 3 inches	2¼ to 2½ inches
5 ft 8 inches to 5 ft 11 inches	2¾ to 3 inches	5 ft 4 inches to 5 ft 7 inches	2⅜ to 2⅝ inches
6 ft 0 inches to 6 ft 3 inches	2¾ to 3⅛ inches	5 ft 8 inches to 5 ft 11 inches	2⅜ to 2⅝ inches
6 ft 4 inches and over	2⅞ to 3¼ inches	6 ft 0 inches and over	2½ to 2¾ inches

[a]For the most accurate measurement, have your physician measure your elbow breadth with calipers.

Source: Courtesy of Metropolitan Life Insurance Company.

Figure A–1 How to Measure Wrist Circumference
The wrist circumference is measured as shown below.

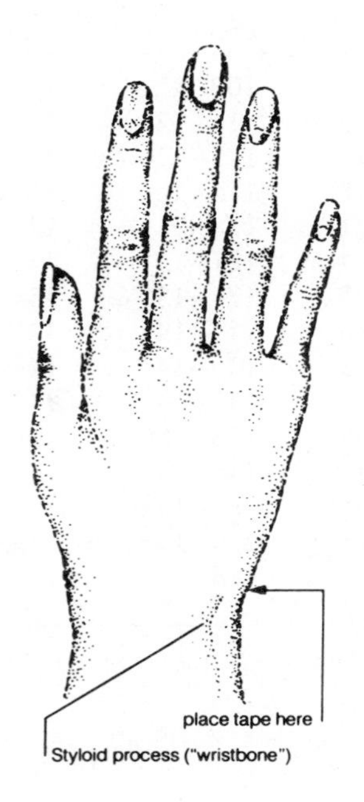

Table A–4 Standards for Frame Size from Height-Wrist Circumference Ratios (r)

	Male r Values[a]	Female r Values[a]
Small	>10.4	>11.0
Medium	9.6 to 10.4	10.1 to 11.0
Large	<9.6	<10.1

[a] $r = \dfrac{\text{height (cm)}}{\text{wrist circumference (cm)}^{b}}$.

[b] The wrist is measured where it bends (distal to the styloid process), on the right arm.

Source: Adapted from J. P. Grant, Patient selection, *Handbook of Total Parenteral Nutrition* (Philadelphia: Saunders, 1980), p. 15.

Table A–5 Percentile Classifications of Triceps Fatfolds for Males and Females (millimeters)

	Males					Females				
Age	*5th*	*25th*	*50th*	*75th*	*95th*	*5th*	*25th*	*50th*	*75th*	*95th*
1–1.9	6	8	10	12	16	6	8	10	12	16
2–2.9	6	8	10	12	15	6	9	10	12	16
3–3.9	6	8	10	11	15	7	9	11	12	15
4–4.9	6	8	9	11	14	7	8	10	12	16
5–5.9	6	8	9	11	15	6	8	10	12	18
6–6.9	5	7	8	10	16	6	8	10	12	16
7–7.9	5	7	9	12	17	6	9	11	13	18
8–8.9	5	7	8	10	16	6	9	12	15	24
9–9.9	6	7	10	13	18	8	10	13	16	22
10–10.9	6	8	10	14	21	7	10	12	17	27
11–11.9	6	8	11	16	24	7	10	13	18	28
12–12.9	6	8	11	14	28	8	11	14	18	27
13–13.9	5	7	10	14	26	8	12	15	21	30
14–14.9	4	7	9	14	24	9	13	16	21	28
15–15.9	4	6	8	11	24	8	12	17	21	32
16–16.9	4	6	8	12	22	10	15	18	22	31
17–17.9	5	6	8	12	19	10	13	19	24	37
18–18.9	4	6	9	13	24	10	15	18	22	30
19–24.9	4	7	10	15	22	10	14	18	24	34
25–34.9	5	8	12	16	24	10	16	21	27	37
35–44.9	5	8	12	16	23	12	18	23	29	38
45–54.9	6	8	12	15	25	12	20	25	30	40
55–64.9	5	8	11	14	22	12	20	25	31	38
65–74.9	4	8	11	15	22	12	18	24	29	36

Source: Adapted from A. R. Frisancho, New norms of upper limb fat and muscle areas for assessment of nutritional status, *American Journal of Clinical Nutrition* 34 (1981): 2540–2545.

Table A–6 Percent Fat Estimates for Women from Sum of Triceps, Iliac Crest, and Thigh Fatfolds[a]

	Age to the Last Year								
Sum of Fatfolds (mm)	*Under 22*	*23 to 27*	*28 to 32*	*33 to 37*	*38 to 42*	*43 to 47*	*48 to 52*	*53 to 57*	*Over 58*
23–25	9.7	9.9	10.2	10.4	10.7	10.9	11.2	11.4	11.7
26–28	11.0	11.2	11.5	11.7	12.0	12.3	12.5	12.7	13.0
29–31	12.3	12.5	12.8	13.0	13.3	13.5	13.8	14.0	14.3
32–34	13.6	13.8	14.0	14.3	14.5	14.8	15.0	15.3	15.5
35–37	14.8	15.0	15.3	15.5	15.8	16.0	16.3	16.5	16.8
38–40	16.0	16.3	16.5	16.7	17.0	17.2	17.5	17.7	18.0
41–43	17.2	17.4	17.7	17.9	18.2	18.4	18.7	18.9	19.2
44–46	18.3	18.6	18.8	19.1	19.3	19.6	19.8	20.1	20.3
47–49	19.5	19.7	20.0	20.2	20.5	20.7	21.0	21.2	21.5
50–52	20.6	20.8	21.1	21.3	21.6	21.8	22.1	22.3	22.6
53–55	21.7	21.9	22.1	22.4	22.6	22.9	23.1	23.4	23.6
56–58	22.7	23.0	23.2	23.4	23.7	23.9	24.2	24.4	24.7
59–61	23.7	24.0	24.2	24.5	24.7	25.0	25.2	25.5	25.7
62–64	24.7	25.0	25.2	25.5	35.7	26.0	26.7	26.4	26.7
65–67	25.7	25.9	26.2	26.4	26.7	26.9	27.2	27.4	27.7
68–70	26.6	26.9	27.1	27.4	27.6	27.9	28.1	28.4	28.6
71–73	27.5	27.8	28.0	28.3	28.5	28.8	28.0	29.3	29.5
74–76	28.4	28.7	28.9	29.2	29.4	29.7	29.9	30.2	30.4
77–79	29.3	29.5	29.8	30.0	30.3	30.5	30.8	31.0	31.3
80–82	30.1	30.4	30.6	30.9	31.1	31.4	31.6	31.9	32.1
83–85	30.9	31.2	31.4	31.7	31.9	32.2	32.4	32.7	32.9
86–88	31.7	32.0	32.2	32.5	32.7	32.9	33.2	33.4	33.7
89–91	32.5	32.7	33.0	33.2	33.5	33.7	33.9	34.2	34.4
92–94	33.2	33.4	33.7	33.9	34.2	34.4	34.7	34.9	35.2
95–97	33.9	34.1	34.4	34.6	34.9	35.1	35.4	35.6	35.9
98–100	34.6	34.8	35.1	35.3	35.5	35.8	36.0	36.3	36.5
101–103	35.3	35.4	35.7	35.9	36.2	36.4	36.7	36.9	37.2
104–106	35.8	36.1	36.3	36.6	36.8	37.1	37.3	37.5	37.8
107–109	36.4	36.7	36.9	37.1	37.4	37.6	37.9	38.1	38.4
110–112	37.0	37.2	37.5	37.7	38.0	38.2	38.5	38.7	38.9
113–115	37.5	37.8	38.0	38.2	38.5	38.7	39.0	39.2	39.5
116–118	38.0	38.3	38.5	38.8	39.0	39.3	39.5	39.7	40.0
119–121	38.5	38.7	39.0	39.2	39.5	39.7	40.0	40.2	40.5
122–124	39.0	39.2	39.4	39.7	39.9	40.2	40.4	40.7	40.9
125–127	39.4	39.6	39.9	40.1	40.4	40.6	40.9	41.1	41.4
128–130	39.8	40.0	40.3	40.5	40.8	41.0	41.3	41.5	41.8

[a]Percent fat calculated by the formula of Siri: % fat = [(4.95/BD) − 4.5] × 100, where BD = body density; W. E. Siri, Body composition from fluid spaces and density, in *Techniques for Measuring Body Composition*, eds. J. Brozek and A. Hanschel (Washington, D.C.: National Academy of Sciences, 1961), pp. 223–224.

Source: Adapted from M. L. Pollock, D. H. Schmidt, and A. S. Jackson, Measurement of cardiorespiration fitness and body composition in the clinical setting, *Comprehensive Therapy* 6 (September 1980): 12–27.

Table A–7 Percent Fat Estimates for Men from Sum of Chest, Abdominal, and Thigh Fatfolds[a]

	Age to the Last Year								
Sum of Fatfolds (mm)	*Under 22*	*23 to 27*	*28 to 32*	*33 to 37*	*38 to 42*	*43 to 47*	*48 to 52*	*53 to 57*	*Over 58*
8–10	1.3	1.8	2.3	2.9	3.4	3.9	4.5	5.0	5.5
11–13	2.2	2.8	3.3	3.9	4.4	4.9	5.5	6.0	6.5
14–16	3.2	3.8	4.3	4.8	5.4	5.9	6.4	7.0	7.5
17–19	4.2	4.7	5.3	5.8	6.3	6.9	7.4	8.0	8.5
20–22	5.1	5.7	6.2	6.8	7.3	7.9	8.4	8.9	9.5
23–25	6.1	6.6	7.2	7.7	8.3	8.8	9.4	9.9	10.5
26–28	7.0	7.6	8.1	8.7	9.2	9.8	10.3	10.9	11.4
29–31	8.0	8.5	9.1	9.6	10.2	10.7	11.3	11.8	12.4
32–34	8.9	9.4	10.0	10.5	11.1	11.6	12.2	12.8	13.3
35–37	9.8	10.4	10.9	11.5	12.0	12.6	13.1	13.7	14.3
38–40	10.7	11.3	11.8	12.4	12.9	13.5	14.1	14.6	15.2
41–43	11.6	12.2	12.7	13.3	13.8	14.4	15.0	15.5	16.1
44–46	12.5	13.1	13.6	14.2	14.7	15.3	15.9	16.4	17.0
47–49	13.4	13.9	14.5	15.1	15.6	16.2	16.8	17.3	17.9
50–52	14.3	14.8	15.4	15.9	16.5	17.1	17.6	18.2	18.8
53–55	15.1	15.7	16.2	16.8	17.4	17.9	18.5	18.1	19.7
56–58	16.0	16.5	17.1	17.7	18.2	18.8	19.4	20.0	20.5
59–61	16.9	17.4	17.9	18.5	19.1	19.7	20.2	20.8	21.4
62–64	17.6	18.2	18.8	19.4	19.9	20.5	21.1	21.7	22.2
65–67	18.5	19.0	19.6	20.2	20.8	21.3	21.9	22.5	23.1
68–70	19.3	19.9	20.4	21.0	21.6	22.2	22.7	23.3	23.9
71–73	20.1	20.7	21.2	21.8	22.4	23.0	23.6	24.1	24.7
74–76	20.9	21.5	22.0	22.6	23.2	23.8	24.4	25.0	25.5
77–79	21.7	22.2	22.8	23.4	24.0	24.6	25.2	25.8	26.3
80–82	22.4	23.0	23.6	24.2	24.8	25.4	25.9	26.5	27.1
83–85	23.2	23.8	24.4	25.0	25.5	26.1	26.7	27.3	27.9
86–88	24.0	24.5	25.1	25.7	26.3	26.9	27.5	28.1	28.7
89–91	24.7	25.3	25.9	25.5	27.1	27.6	28.2	28.8	29.4
92–94	25.4	26.0	26.6	27.2	27.8	28.4	29.0	29.6	30.2
95–97	26.1	16.7	27.3	27.9	28.5	29.1	29.7	30.3	30.9
98–100	26.9	27.4	28.0	28.6	29.2	29.8	30.4	31.0	31.6
101–103	27.5	28.1	28.7	29.3	29.9	30.5	31.1	31.7	32.3
104–106	28.2	28.8	29.4	30.0	30.6	31.2	31.8	32.4	33.0
107–109	28.9	29.5	30.1	30.7	31.3	31.9	32.5	33.1	33.7
110–112	29.6	30.2	30.8	31.4	32.0	32.6	33.2	33.8	34.4
113–115	30.2	30.8	31.4	32.0	32.6	33.2	33.8	34.5	35.1
116–118	30.9	31.5	32.1	32.7	33.3	33.9	34.5	35.1	35.7
119–121	31.5	32.1	32.7	33.3	33.9	34.5	35.1	35.7	36.4
122–124	32.1	32.7	33.3	33.9	34.5	35.1	35.8	36.4	37.0
125–127	32.7	33.3	33.9	34.5	35.1	35.8	36.4	37.0	37.6

[a]Percent fat calculated by the formula by Siri: % fat = [4.95/BD] − 4.5] × 100, where BD = body density; W. E. Siri, Body composition from fluid spaces and density, in *Techniques for Measuring Body Composition,* eds. J. Brozek and A. Hanschel (Washington, D.C.: National Academy of Sciences, 1961), pp. 223–224.

Source: Adapted from M. L. Pollock, D. H. Schmidt, and A. S. Jackson, Measurement of cardiorespiration fitness and body composition in the clinical setting, *Comprehensive Therapy* 6 (September 1980): 12–27.

Creatinine excretion is also used to determine whether other urinary lab test results are appropriate to the size of the individual's skeletal muscle mass. The measurement of urinary creatinine requires a 24-hour urine collection, which may be difficult to obtain. The test is invalid if the subject shows signs of kidney disease, since the disease might reduce the body's ability to excrete creatinine.

Iron Assessment

For an overview of iron assessment, see Chapter 1. The specific tasks are discussed here.

Serum ferritin In the first stage of iron deficiency, iron stores diminish. Serum ferritin measures provide a noninvasive estimate of iron stores. Such information is most valuable to iron assessment. Table A–13 shows serum ferritin cutoff values that indicate iron store depletion in children and adults. In infants, the reliability of serum ferritin for diagnosing iron deficiency is uncertain; normal serum ferritin values are often present in conjunction with iron-responsive anemia.[1]

A decrease in transport iron characterizes the second stage of iron deficiency. This is detected by an increase in the iron-binding capacity of the protein transferrin and a decrease in serum iron. These changes are reflected by the transferrin saturation, which is calculated from the ratio of the other two values, as described below.

Total iron-binding capacity (TIBC) Iron travels through the blood bound to the protein transferrin. TIBC is a measure of the total amount of iron that transferrin can carry. Lab technicians measure iron-binding capacity directly. TIBC values greater than 400 micrograms per 100 milliliters indicate iron deficiency.*

Serum iron Lab technicians can also measure serum iron directly. Elevated values indicate iron overload; reduced values indicate iron deficiency. Table A–14 shows acceptable and deficient values for serum iron.

Transferrin saturation The percentage of transferrin that is saturated with iron is an indirect measure; the mathematical equation derives it from the serum iron and total iron-binding capacity measures, as follows:

$$\%\ \text{transferrin} = \frac{\text{serum iron} \times 100}{\text{total iron-binding capacity}}.$$

Table A–15 shows deficient and acceptable transferrin saturation values for various age groups.

(continued on p. 483)

*To convert iron-binding capacity (μg/100 ml) to standard international units (μmol/L), multiply by 0.1791; 400 μg/100 ml = 71 μmol/L.

Figure A–2A Body Mass Index for Specified Heights and Weights—Women
Find your height in the top line and follow the column down to your weight, as listed in the left column. Use the key below to determine if your weight is within the acceptable range.

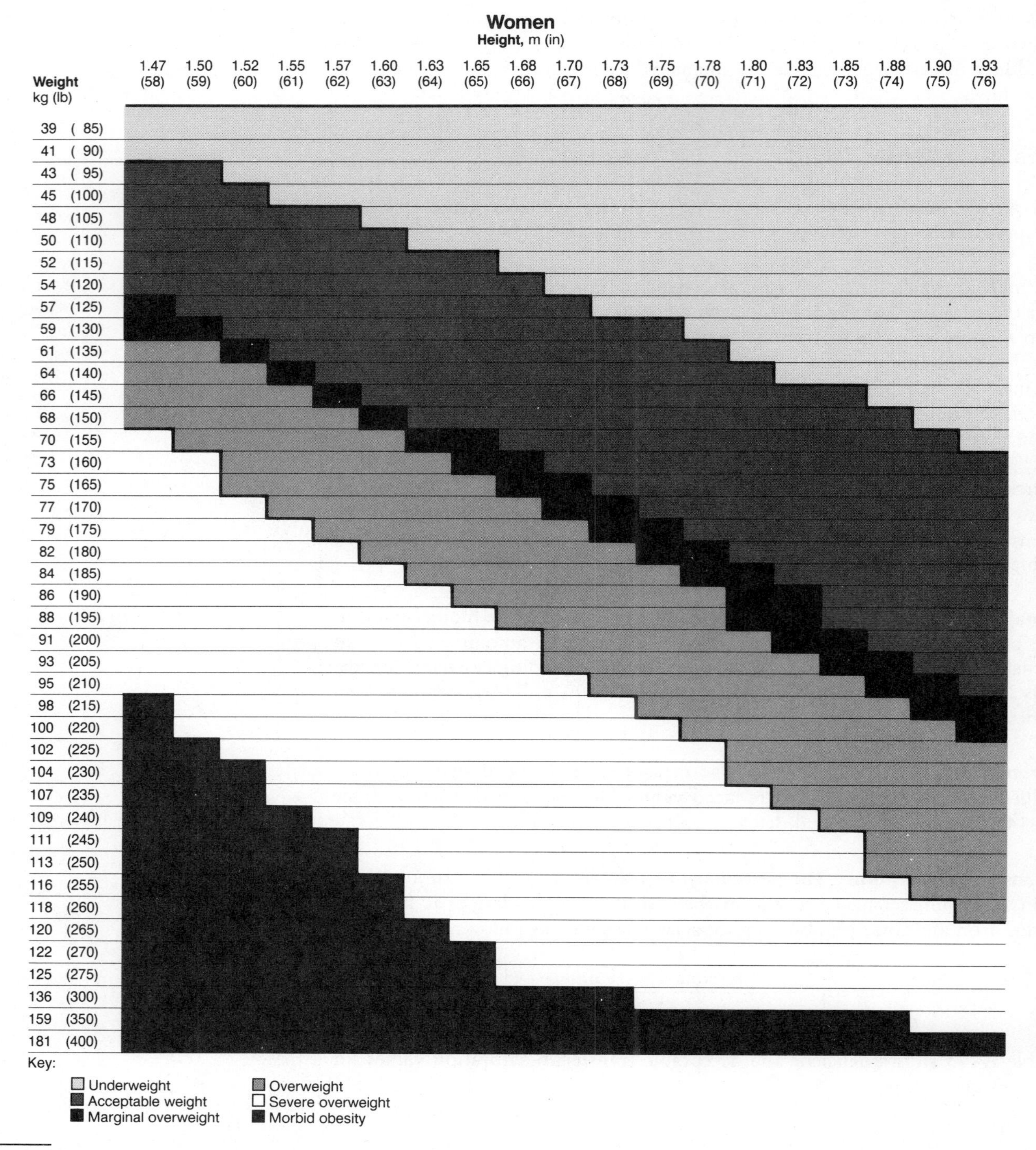

Figure A–2B Body Mass Index for Specified Heights and Weights—Men
Find your height in the top line and follow the column down to your weight, as listed in the left column. Use the key below to determine if your weight is within the acceptable range.

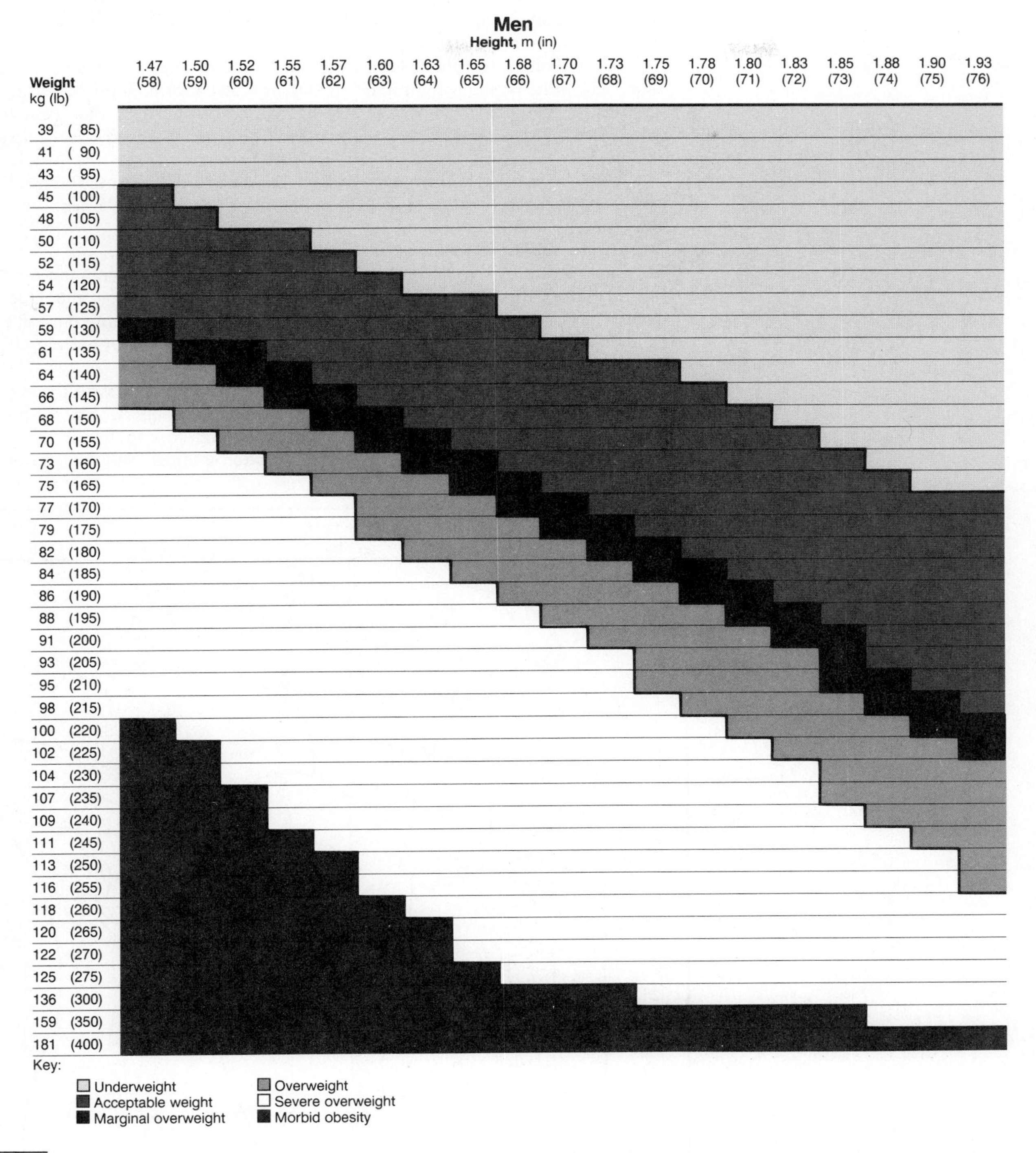

Source: Used with permission of Ross Laboratories, Columbus, OH 43216, from *Dietetic Currents,* Vol. 16, no. 2, pp. 8–9, © 1989 Ross Laboratories.

Figure A–3 How to Measure the Triceps Fatfold

The most commonly used site for measuring fatfold thickness is the triceps area, because the upper midarm is easily accessible. The average of three fatfold measurements should be recorded. To measure fatfold, a trained technician follows a standard procedure using reliable calipers, as illustrated here, and then refers to a table of standards to estimate total body fat. The standard procedure is as follows:

A. Find the midpoint of the arm:

1. Ask the subject to bend his or her arm at the elbow and lay the hand across the stomach. (If he or she is right handed, measure the left arm, and vice versa.)
2. Feel the shoulder to locate the acromial process. It helps to slide your fingers along the clavicle to find the acromial process. The olecranon process is the tip of the elbow.
3. Place a measuring tape from the acromial process to the tip of the elbow. Divide this measurement by 2, and mark the midpoint of the arm with a pen.

B. Measure the fatfold:

1. Ask the subject to let his or her arm hang loosely to the side.
2. Grasp a fold of skin and subcutaneous fat between your thumb and forefinger slightly above the midpoint mark. Gently pull the skin away from the underlying muscle. (This step takes a lot of practice. If you want to be sure you don't have muscle as well as fat, ask the subject to contract and relax his or her muscle. You should be able to feel if you are pinching muscle.)
3. Place the calipers over the fatfold at the midpoint mark, and read the measurement to the nearest 1.0 mm in 2 to 3 sec. (If using plastic calipers, align the pressure lines, and read the measurement to the nearest 1.0 mm in 2 to 3 sec.)
4. Repeat steps 2 and 3 twice more. Add the three readings, and then divide by 3 to find the average.

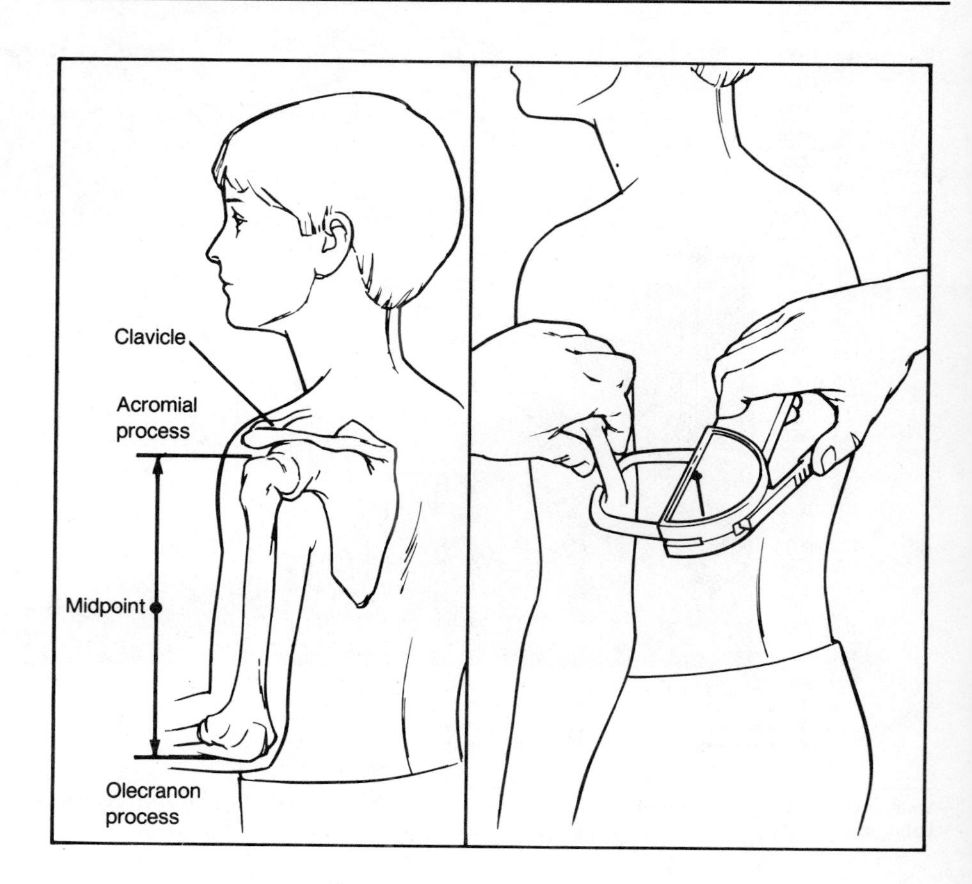

Table A–8 Symptoms of Vitamin and Mineral Imbalances

Vitamin Name	Deficiency Disease Name	Deficiency Symptoms	Toxicity Symptoms
		Bones/Teeth[a]	
Vitamin A	Hypovitaminosis A	Cessation of bone growth, change in shapes of bones, painful joints; malfunctioning of enamel-forming cells, development of cracks in teeth and tendency to decay, atrophy of dentin-forming cells	Increased activity of osteoclasts[b] causing decalcification, joint pain, fragility, stunted growth, and thickening of long bones; increase of pressure inside skull, mimicking brain tumor; headaches
		Blood	
		Microcytic anemia, often masked by dehydration	Loss of hemoglobin and potassium by red blood cells, cessation of menstruation, slowed clotting time, easily induced bleeding
		Eyes[c]	
		Night blindness, change in epithelial tissue caused by failure to secrete mucopolysaccharide (hyperkeratinization), drying (xerosis), triangular gray spots on eye (Bitot's spots), irreversible drying (keratomalacia), and degeneration of the cornea causing blindness (most severe)	
		Skin[d]	
		Plugging of hair follicles with keratin, forming white lumps (hyperkeratosis)	Dryness; itching; peeling; rashes; dry, scaling lips; cracking and bleeding of lips; nosebleeds; loss of hair; brittle nails
		Digestive System	
		Changes in lining, diarrhea	Nausea, vomiting, abdominal pain, diarrhea, weight loss
		Immune System	
		Depression of immune reactions	Stimulation of immune reactions
		Nervous/Muscular System	
		Brain and spinal cord growth too fast for stunted skull and spine, paralysis caused by injury to brain and nerves	Loss of appetite, irritability, fatigue, insomnia, restlessness, headache, blurred vision, nausea, vomiting, muscle weakness, interference with thyroxin
		Respiratory Tract	
		Changes in lining, infections	
		Urogenital Tract	
		Changes in lining that favor calcium deposition, resulting in kidney stones and bladder disorders; infections of bladder and kidney; infections of vagina	
		Reproductive System	
			Amenorrhea[e]
		Liver[f]	
			Jaundice,[g] enlargement, massive accumulation of fat and vitamin A
		Spleen	
			Enlargement

Table A–8—*Continued*

Vitamin Name	Deficiency Disease Name	Deficiency Symptoms		Toxicity Symptoms
			Bones/Teeth	
Vitamin D	Rickets, osteomalacia	*Rickets*	*Osteomalacia*	
		Faulty calcification, resulting in misshapen bones (bowing of legs) and retarded growth; enlargement of ends of long bones (knees, wrists); deformities of ribs (bowed, with beads or knobs);[h] delayed closing of fontanel, resulting in rapid enlargement of head (see accompanying figure); slow eruption of teeth; teeth not well formed, with a tendency to decay	Softening effect; deformities of limbs, spine, thorax, and pelvis; demineralization; pain in pelvis, lower back, and legs; bone fractures	Increased calcium withdrawal
			Blood	
		Decreased calcium and/or phosphorus	Decreased calcium and/or phosphorus, increased alkaline phosphatase[i]	Increased calcium and phosphorus concentration
			Nervous/Muscular Systems	
		Lax muscles resulting in protrusion of abdomen; muscle spasms	Involuntary twitching, muscle spasms	Loss of appetite, headache, weakness, fatigue, excessive thirst, irritability, apathy
			Excretory System	
		Increased calcium in stools, decreased calcium in urine		Increased excretion of calcium in urine, kidney stones, irreversible renal damage
			Other	
		Abnormally high secretion of parathormone		Calcification of soft tissues (blood vessels, kidneys, heart, lungs, tissues around joints), death
Vitamin E			*Blood/Circulatory System*	
		Red blood cell breakage,[j] anemia		Augments effects of anticlotting medication
			Digestive System	
				General discomfort

Fontanel
A fontanel is the open space in the top of a baby's skull before the bones have grown together. In rickets, closing of the fontanel is delayed.

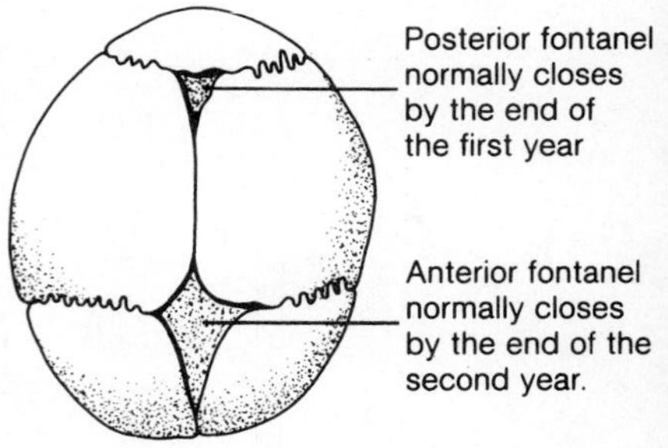

Posterior fontanel normally closes by the end of the first year

Anterior fontanel normally closes by the end of the second year.

Table A–8—*Continued*

Vitamin Name	Deficiency Disease Name	Deficiency Symptoms	Toxicity Symptoms
		Nervous/Muscular Systems	
Vitamin E (*continued*)		Degeneration, weakness, difficulty walking, severe pain in calf muscles	Headache, weakness, dizziness, fatigue, visual abnormalties
		Other	
		Fibrocystic breast disease	
Vitamin K		*Blood/Circulatory System*	
		Hemorrhaging	Interference with anticlotting medication, possible jaundice caused by vitamin K analogues
		Blood/Circulatory System	
Thiamin	Beriberi	Edema, enlarged heart, abnormal heart rhythms, heart failure	Rapid pulse
		Nervous/Muscular Systems	
		Degeneration, wasting, weakness, painful calf muscles, low morale, difficulty walking, loss of ankle and knee jerk reflexes, mental confusion, paralysis	Weakness, headaches, insomnia, irritability
		Mouth, Gums, Tongue	
Riboflavin	Ariboflavinosis	Cracks at corners of mouth,[k] magenta tongue	
		Nervous System and Eyes	
		Hypersensitivity to light,[l] reddening of cornea	
		Other	
		Skin rash	Interference with anticancer medication
		Digestive System	
Niacin	Pellagra	Diarrhea	Diarrhea, heartburn, nausea, ulcer irritation, vomiting
		Mouth, Gums, Tongue	
			Inflammed, swollen, smooth tongue[m]
		Nervous System	
		Irritability, loss of appetite, weakness, dizziness, mental confusion progressing to psychosis or delirium	
		Skin	
		Bilateral symmetrical dermatitis, especially on areas exposed to sun	Painful flush and rash, itching, burning, excessive sweating
		Other	
			Abnormal liver function, low blood pressure

Table A–8—*Continued*

Vitamin Name	Deficiency Disease Name	Deficiency Symptoms	Toxicity Symptoms
		Blood/Circulatory System	
Vitamin B_6		Microcytic anemia	Bloating
		Mouth, Gums, Tongue	
		Smooth tongue,[m] cracked corners of the mouth[k]	
		Nervous/Muscular Systems	
		Abnormal brain wave pattern, irritability, muscle twitching, convulsions	Depression, fatigue, irritability, headaches, numbness, damage to nerves leading to loss of reflexes and sensation, difficulty walking
		Skin	
		Irritation of sweat glands, dermatitis	
		Other	
		Kidney stones	
		Blood/Circulatory System	
Folate		Macrocytic or megaloblastic anemia	
		Digestive System	
		Heartburn, diarrhea (loss of villi and their enzymes), constipation	Diarrhea
		Immune System	
		Suppression, frequent infections	
		Mouth, Gums, Tongue	
		Smooth, red tongue[m]	
		Nervous System	
		Depression, mental confusion, fainting, fatigue	Insomnia, irritability
		Other	
			Masking of vitamin B_{12}—deficiency symptoms
		Blood/Circulatory System	
Vitamin B_{12}	Pernicious anemia[n]	Macrocytic or megaloblastic anemia	
		Mouth, Gums, Tongue	
		Smooth tongue[m]	
		Nervous System	
		Fatigue, degeneration of peripheral nerves progressing to paralysis	
		Skin	
		Hypersensitivity	
		Digestive System	
Pantothenic acid		Vomiting, intestinal distress	Occasional diarrhea

Table A–8—*Continued*

Vitamin Name	Deficiency Disease Name	Deficiency Symptoms	Toxicity Symptoms
		Nervous System	
Pantothenic acid (*continued*)		Insomnia, fatigue	
		Other	
			Water retention (infrequent)
		Blood/Circulatory System	
Biotin		Abnormal heart action	
		Digestive System	
		Loss of appetite, nausea	
		Nervous/Muscular Systems	
		Depression, muscle pain, weakness, fatigue	
		Skin	
		Drying, scaly dermatitis, loss of hair	
		Blood/Circulatory System	
Vitamin C	Scurvy	Microcytic anemia, atherosclerotic plaques, pinpoint hemorrhages	Blood cell breakage in certain racial groups[o]
		Digestive System	
			Nausea, abdominal cramps, diarrhea
		Immune System	
		Depression, frequent infections	
		Mouth, Gums, Tongue	
		Bleeding gums, loosened teeth	
		Nervous/Muscular Systems	
		Muscle degeneration and pain, hysteria, depression	Headache, fatigue, insomnia
		Skeletal System	
		Bone fragility, joint pain	
		Skin	
		Rough skin, blotchy bruises	Hot flashes, rashes
		Other	
		Failure of wounds to heal	Interference with medical tests, aggravation of gout symptoms, excessive urination, kidney stones,[p] (deficiency symptoms may appear at first on withdrawal of high doses)
Major Minerals			
Calcium		Stunted growth in children, implicated in bone loss (osteoporosis) in adults	Excess calcium is excreted except in hormonal imbalance states (not caused by nutritional deficiency)
Phosphorus		Deficiency unknown	

Table A–8—*Continued*

Mineral Name	Deficiency Symptoms	Toxicity Symptoms
Magnesium	Weakness; confusion; depressed pancreatic hormone secretion; if extreme, convulsions, bizarre muscle movements (especially of the eye and facial muscles), hallucinations, and difficulty in swallowing;[q] in children, growth failure	Not known; large doses have been taken in the form of the laxative Epsom salts without ill effects except diarrhea
Sodium	Muscle cramps, mental apathy, loss of appetite	Hypertension[r]
Chloride	Growth failure in children; muscle cramps, mental apathy, loss of appetite	Vomiting
Potassium	Deficiency accompanies dehydration; causes muscular weakness, paralysis, and confusion	Muscular weakness; vomiting; if given into a vein, can stop the heart
Sulfur	None known; protein deficiency would occur first	Would occur only if sulfur amino acids were eaten in excess; this (in animals) depresses growth
Trace Minerals		
	Eyes	
Iron	Blue sclerae	
	GI Tract	
	Lactose intolerance, and possibly intolerance to other sugars; increased risk of lead and cadmium poisoning	
	Immune System	
	Reduced resistance to infection (lowered immunity)	Infections
	Nervous/Muscular Systems	
	Reduced work productivity, tolerance to work, and voluntary work; reduced physical fitness; weakness; fatigue; impaired cognitive function (children); reduced learning ability; increased distractibility (inability to pay attention); impaired visual discrimination; impaired reactivity and coordination (infants)	
	Skin	
	Itching; pale nailbeds, eye membranes, and palm creases; concave nails; impaired wound healing	

Table A–8—*Continued*

Mineral Name	Deficiency Symptoms	Toxicity Symptoms
	General	
Iron (*continued*)	Reduced resistance to cold, inability to regulate body temperature, pica (clay eating, ice eating)	
	Blood	
Zinc	Tendency to atherosclerosis, elevated ammonia levels, decreased alkaline phosphatase, decreased insulin concentration	Anemia: reduced hemoglobin production
	Bones	
	Growth retardation, abnormal collagen synthesis	Growth in length, but without normal zinc content
	Cells/Metabolism	
	Decreased DNA synthesis, impaired cell division and protein synthesis	Raised LDL, lowered HDL
	Digestive System	
	Diarrhea, vomiting, decreased calcium and copper absorption	Reduced sense of smell, reduced sensitivity to the taste of salt, weight loss, delayed glucose absorption, diarrhea, nausea, impaired folate absorption
	Eyes	
	Night blindness	
	Glandular System	
	Delayed onset of puberty, small gonads in males, decreased synthesis and release of testosterone, abnormal glucose tolerance, reduced synthesis of adrenocortical hormones, altered thyroid function	
	Immune System	
	Altered skin test responses, reduced numbers of white blood cells and antibody-forming cells, thymus atrophy, increased susceptibility to infection	Fever, elevated white blood cell count
	Kidney	
		Renal failure
	Liver/Spleen	
	Enlargement	
	Nervous/Muscular Systems	
	Anorexia (poor appetite), mental lethargy, irritability	Muscular pain and incoordination, heart muscle degeneration, exhaustion, dizziness, drowsiness

Table A–8—*Continued*

Mineral Name	Deficiency Symptoms	Toxicity Symptoms
	Reproductive System	
Zinc (*continued*)	Impaired reproductive function (rats), low sperm counts	Reproductive failure
	Skin	
	Generalized hair loss; lesions; rough, dry appearance; slow healing of wounds and burns	
Iodine	Enlargement of the thyroid gland, weight gain, mental and physical retardation of an infant	Enlargement of the thyroid gland, depressed thyroid activity
Copper	Anemia, bone changes (rare in human beings)	Vomiting, diarrhea
Fluoride	Susceptibility to tooth decay	Fluorosis (discoloration of teeth), nausea, diarrhea, chest pain, itching, vomiting
Selenium	Anemia (rare); heart disease	Digestive system disorders, loss of hair and nails, skin lesions, nervous system disorders, tooth damage
Chromium	Diabetes-like condition marked by an inability to use glucose normally; associated with coronary artery disease	Unknown as a nutrition disorder; occupational exposures damage skin and kidneys
Cobalt	Unknown in human beings except in vitamin B_{12} deficiency	Unknown as a nutrition disorder; occupational exposures damage skin and red blood cells
Molybdenum	Unknown	Enzyme inhibition
Manganese	(In animals): poor growth, nervous system disorders, reproductive abnormalities	Nervous system disorders

[a]Focal Point 5 describes vitamin A's role in tooth formation in more detail.
[b]Osteoclasts are the cells that destroy bone during its growth. Those that build bone are osteoblasts.
[c]The eyes' symptoms of vitamin A deficiency are collectively known as *xerophthalmia*.
[d]A related toxicity condition, hypercarotenemia, is caused by the accumulation of too much of the vitamin A precursor beta-carotene in the blood, which turns the skin noticeably yellow. Hypercarotenemia is not, strickly speaking, a toxicity symptom.
[e]Elevated serum carotene concentrations are associated with amenorrhea.
[f]If liver impairment is severe, the "classic" signs seen in skin and hair may be masked.
[g]A symptom of liver disease, in which bile and related pigments spill into the bloodstream and the skin yellows, is *jaundice* (JAWN-diss).
[h]Bowing of the ribs causes the symptom known as *pigeon breast*. The beads that form on the ribs resemble rosary beads; thus this symptom is known as *rachitic* (ra-KIT-ik) *rosary* ("the rosary of rickets").
[i]Alkaline phosphatase is an enzyme in the blood that rises during bone resorption.
[j]The breaking of red blood cells is called *erythrocyte hemolysis*.
[k]Cracks at the corners of the mouth are termed *cheilosis* (kee-LOH-sis).
[l]Hypersensitivity to light is *photophobia*.
[m]Smoothness of the tongue is caused by loss of its surface structures and is termed *glossitis* (gloss-EYE-tis).
[n]The name *pernicious anemia* refers to the vitamin B_{12} deficiency caused by lack of intrinsic factor, but not to that caused by inadequate dietary intake.
[o]Groups susceptible to vitamin C toxicity are Sephardic Jews, black Americans and Africans, and Asians.
[p]People who have a tendency toward gout and those who have a genetic abnormality that alters the break down of vitamin C are prone to forming kidney stones. Vitamin C is inactivated and degraded by several routes, and sometimes a product along the way is oxalate, which can form stones in the kidneys. People can also have oxalate crystals in their kidneys that are not due to vitamin C overdoses.
[q]A still more severe deficiency causes tetany, an extreme, prolonged contraction of the muscles similar to that caused by low blood calcium.
[r]Chapter 9 describes the role of sodium in the development of hypertension.

Table A–9 Biochemical Tests Useful for Assessing Vitamin and Mineral Status

Nutrient	Assessment Tests
Vitamins	
Vitamin A	Retinol binding protein, serum vitamin A
Thiamin	Erythrocyte (red blood cell) transketolase activity, urinary thiamin
Riboflavin	Erythrocyte glutathione reductase activity, urinary riboflavin
Vitamin B_6	Urinary xanthurenic acid excretion after tryptophan loan test, urinary vitamin B_6, erythrocyte transaminase activity
Niacin	Urinary metabolites NMN (N-methyl nicotinamide) or 2–pyridone, or preferably both expressed as a ratio
Folate	Free folate in the blood, erythrocyte folate (reflects liver stores), urinary formiminoglutamic acid (FIGLU), vitamin B_{12} status (because folate assessment tests alone do not distinguish between the two deficiencies)
Vitamin B_{12}	Serum vitamin B_{12}, erythrocyte vitamin B_{12}, urinary methylmalonic acid, synthesis test—DUMP test (from the abbreviation for the chemical name of DNA's raw material, deoxyuridine monophosphate)
Biotin	Serum biotin, urinary biotin
Vitamin C	Serum or plasma vitamin C,[a] leukocyte vitamin C, urinary vitamin C
Vitamin D	Serum alkaline phosphatase
Vitamin E	Serum tocopherol, erythrocyte hemolysis
Vitamin K	Blood clotting time (prothrombin time)
Minerals	
Potassium	Serum potassium
Magnesium	Serum magnesium
Iron	Hemoglobin, hematocrit, serum ferritin, total iron-binding capacity (TIBC), transferrin saturation, erythrocyte protoporphyrin, serum ferritin, mean corpuscular volume (MCV), mean corpuscular hemoglobin concentration (MCHC), serum iron
Iodine	Serum protein-bound iodine, radioiodine uptake
Zinc	Plasma zinc, hair zinc

[a]Vitamin C shifts unpredictably between the plasma and the white blood cells known as leukocytes; thus a plasma or serum determination may not accurately reflect the body's pool. The appropriate clinical test may be a measurement of leukocyte vitamin C. A combination of both tests may be more reliable than either one alone.

Source: Adapted from A. Grant and S. DeHoog, *Nutritional Assessment and Support,* 3rd ed., 1985 (available from Anne Grant and Susan DeHoog, P.O. Box 25057, Northgate Station, Seattle, WA 98125).

Table A–10 Standards for Serum Protein Test Results

	Indicator			
Degree of	*Albumin*		*Transferrin*	
Depletion	*(g/100 ml)*	*(nmol/L)*[a]	*(mg/100 ml)*	*(g/L)*[b]
Mild	2.8 to 3.4	104 to 126	150 to 200	1.50 to 2.00
Moderate	2.1 to 2.7	79 to 100	100 to 149	1.00 to 1.49
Severe	<2.1	79	<100	<1.00

[a]To convert albumin (g/100 ml) to standard international units (nmol/L), multiply by 37.06.
[b]To convert transferrin (mg/100 ml) to standard international units (g/L), multiply by 0.01.

Table A–11 Standards for Creatinine-Height Index—Men

Height		Small Frame			Medium Frame			Large Frame		
		Ideal Weight	*Creatinine*		*Ideal Weight*	*Creatinine*		*Ideal Weight*	*Creatinine*	
Inches	*Cm*	*(kg)*	*(g/24 hr)*	*(mmol/d)*[a]	*(kg)*	*(g/24 hr)*	*(mmol/d)*[a]	*(kg)*	*(g/24 hr)*	*(mmol/d)*[a]
61	154.9	52.7	1.21	10.7	56.1	1.29	11.4	60.7	1.40	12.4
62	157.5	54.1	1.24	11.0	57.7	1.33	11.8	62.0	1.43	12.6
63	160.0	55.4	1.27	11.2	59.1	1.36	12.0	63.6	1.46	12.9
64	162.5	56.8	1.31	11.6	60.4	1.39	12.3	65.2	1.50	13.3
65	165.1	58.4	1.34	11.8	62.0	1.43	12.6	66.8	1.54	13.6
66	167.6	60.2	1.39	12.3	63.9	1.47	13.0	68.9	1.59	14.1
67	170.2	62.0	1.43	12.6	65.9	1.52	13.4	71.1	1.64	14.5
68	172.7	63.9	1.47	13.0	67.7	1.56	13.8	72.9	1.68	14.9
69	175.3	65.9	1.52	13.4	69.5	1.60	14.1	74.8	1.72	15.2
70	177.8	67.7	1.56	13.8	71.6	1.65	14.6	76.8	1.77	15.6
71	180.3	69.5	1.60	14.1	73.6	1.69	14.9	79.1	1.82	16.1
72	182.9	71.4	1.64	14.5	75.7	1.74	15.4	81.1	1.87	16.5
73	185.4	73.4	1.69	14.9	77.7	1.79	15.8	83.4	1.92	17.0
74	187.9	75.2	1.73	15.3	80.0	1.85	16.4	85.7	1.97	17.4
75	190.5	77.0	1.77	15.6	82.3	1.89	16.7	87.7	2.02	17.9

[a]To convert urinary creatinine measures (g/24 hr) to standard international units (mmol/d), multiply by 8.840.

Source: A. Grant and S. DeHoog, *Nutritional Assessment and Support,* 3rd ed., 1985 (available from Anne Grant and Susan DeHoog, P.O. Box 25057, Northgate Station, Seattle, WA 98125).

Table A–12 Standards for Creatinine-Height Index—Women

Height		Small Frame			Medium Frame			Large Frame		
		Ideal Weight	*Creatinine*		*Ideal Weight*	*Creatinine*		*Ideal Weight*	*Creatinine*	
Inches	*Cm*	*(kg)*	*(g/24 hr)*	*(mmol/d)*[a]	*(kg)*	*(g/24 hr)*	*(mmol/d)*[a]	*(kg)*	*(g/24 hr)*	*(mmol/d)*[a]
56	142.2	43.2	0.79	7.0	46.1	0.83	7.3	50.7	0.91	8.0
57	144.8	44.3	0.80	7.1	47.3	0.85	7.5	51.8	0.93	8.2
58	147.3	45.4	0.82	7.2	48.6	0.88	7.8	53.2	0.96	8.5
59	149.8	46.8	0.84	7.4	50.0	0.90	8.0	54.5	0.98	8.7
60	152.4	48.2	0.87	7.7	51.4	0.93	8.2	55.9	1.01	8.9
61	154.9	49.5	0.89	7.9	52.7	0.95	8.4	57.3	1.03	9.1
62	157.5	50.9	0.92	8.1	54.3	0.98	8.7	58.9	1.06	9.4
63	160.0	52.3	0.94	8.3	55.9	1.01	8.9	60.6	1.09	9.6
64	162.5	53.9	0.97	8.6	57.9	1.04	9.2	62.5	1.13	10.0
65	165.1	55.7	1.00	8.8	59.8	1.08	9.5	64.3	1.16	10.3
66	167.6	57.5	1.04	9.2	61.6	1.11	9.8	66.1	1.19	10.5
67	170.2	59.3	1.07	9.5	63.4	1.14	10.1	67.9	1.22	10.8
68	172.7	61.4	1.11	9.8	65.2	1.17	10.3	70.0	1.26	11.1
69	175.2	63.2	1.14	10.1	67.0	1.21	10.7	72.0	1.30	11.5
70	177.8	65.0	1.17	10.3	68.9	1.24	11.0	74.1	1.33	11.8

[a]To convert urinary creatinine measures (g/24 hr) to standard international units (mmol/d), multiply by 8.840.

Source: A. Grant and S. DeHoog, *Nutritional Assessment and Support,* 3rd ed., 1985 (available from Anne Grant and Susan DeHoog, P.O. Box 25057, Northgate Station, Seattle, WA 98125).

Table A–13 Standards for Serum Ferritin Test Results

Group	Serum Ferritin Deficient Values (ng/ml)
Children (3 to 14 yr)	<10
Adolescents and adults	<12
Pregnant women	<10

Table A–14 Standards for Serum Iron Test Results—Infants, Children, and Adults

		Deficient		Acceptable	
Age (yr)	Sex	*(μg/100 ml)*	*(μmol/L)*[a]	*(μg/100 ml)*	*(μmol/L)*
<2	M-F	<30	<5.3	≥30	≥5.3
2 to 5	M-F	<40	<7.1	≥40	≥7.1
6 to 12	M-F	<50	<8.9	≥50	≥8.9
>12	M	<60	<10.7	≥60	≥10.7
	F	<40	<7.1	≥40	≥ 7.1

[a]To convert μg/100 ml to standard international units (μmol/L), multiply by 0.1791.

Table A–15 Standards for Percent Transferrin Saturation—Infants, Children, and Adults

		% Transferrin Saturation	
Age (yr)	Sex	*Deficient*	*Acceptable*
<2	M-F	<15	≥15
2 to 12	M-F	<20	≥20
>13	M	<20	≥20
	F	<15	≥15

The third stage of iron deficiency occurs when the supply of transport iron diminishes to the point that it limits hemoglobin production. It is characterized by increases in erythrocyte protoporphyrin, a decrease in mean corpuscular volume (MCV), and deceased hemoglobin concentration and hematocrit.

Erythrocyte protoporphyrin The iron-containing portion of the hemoglobin molecule is heme. Heme is a combination of iron and protoporphyrin. Protoporphyrin accumulates in the blood when iron supplies are inadequate for the formation of heme. Lab technicians can measure erythrocyte protoporphyrin directly in a blood sample. The cutoffs for abnormal values of erythrocyte protoporphyrin are shown in Table A–16.

Table A–16 Standards for Erythrocyte Protoporphyrin and Mean Corpuscular Volume (MCV)—Infants, Children, and Adults

Age (yr)	Erythrocyte Protoporphyrin *(μg/dl RBC)*	MCV *(fl)*[a]
1 to 2	>80	<73
3 to 4	>75	<75
5 to 10	>70	<76
11 to 14	>70	<78
15 to 74	>70	<80

[a]MCV is measured in femtoliters (fl).

Source: Expert Scientific Working Group, Summary of a report on assessment of the iron nutritional status of the United States population, *American Journal of Clinical Nutrition* 42 (1985): 1318–1330.

Mean corpuscular volume (MCV) A direct or calculated measure of the MCV determines the average size of a red blood cell (RBC). Such a measure helps to classify the type of nutrient anemia. The equation to calculate mean corpuscular volume is:

$$\text{MCV} = \frac{\text{hematocrit}}{\text{RBC count}} \times 10.$$

The cutoffs for abnormal values of MCV that indicate iron deficiency are shown in Table A–16.

Hemoglobin Hemoglobin is a more direct measure of iron deficiency than hematocrit, but its usefulness in assessing iron status is limited, because blood hemoglobin concentrations fall with other nutrient anemias as well. Table A–17 provides hemoglobin values used in nutrition assessment.

Hematocrit Hematocrit is commonly used to diagnose iron deficiency, even though it is an inconclusive measure of iron status. To measure the hematocrit, a clinician spins a volume of blood in a centrifuge to separate the red blood cells from the plasma. The packed red cell volume is the hematocrit and is expressed as a percentage of the total blood volume. Table A–18 provides values used to assess hematocrit status. Low values indicate incomplete hemoglobin formation, which is manifested by microcytic (abnormally small celled), hypochromic (abnormally lacking in color) red blood cells.

Table A–17 Standards for Hemoglobin Test Results—Infants, Children and Adults (grams per 100 milliliters)[a]

Category	Sex	Deficient	Acceptable
<2 yr	M-F	<9.0	≥10.0
2 to 5 yr	M-F	<10.0	≥11.0
6 to 12 yr	M-F	<10.0	≥11.5
13 to 16 yr	M	<12.0	≥13.0
	F	<10.0	≥11.5
>16 yr	M	<12.0	≥14.0
	F	<10.0	≥12.0
Pregnancy			
Trimester 2		<9.5	≥11.0
Trimester 3		<9.0	≥10.5

[a]To convert g/100 ml to standard international units, multiply by 10.

Table A–18 Standards for Hematocrit Test Results—Infants, Children, and Adults

Category	Sex	Deficient[a]	Acceptable[a]
< 2 yr	M-F	<0.28	≥0.31
2 to 5 yr	M-F	<0.30	≥0.34
6 to 12 yr	M-F	<0.30	≥0.36
13 to 16 yr	M	<0.37	≥0.40
	F	<0.31	≥0.36
>16 yr	M	<0.37	≥0.44
	F	<0.31	≥0.38
Pregnancy			
Trimester 2		<0.30	≥0.35
Trimester 3		<0.30	≥0.33

[a]To convert hematocrit values (%) to standard international units, multiply by 0.01.

Appendix A Note

1. P. R. Dallman, Diagnostic criteria for iron deficiency in *Iron Nutrition Revisited—Infancy, Childhood, Adolescence: Report of the 82nd Ross Conference on Pediatric Research* (Columbus, Ohio: Ross Laboratories, 1981), pp. 3–12.

Appendix B

Recommended Nutrient Intakes (RDA, U.S. RDA, RNI)

Many countries have developed nutrient standards. Those of the United States (the Recommended Dietary Allowances, or RDA) and Canada (the Recommended Nutrient Intakes, or RNI) are examples. The main RDA table is presented on the inside front cover of this book. The energy RDA are presented here in Table B–1. The estimated safe and adequate daily dietary intakes of selected vitamins and mineral appear in Table B–2 and Table B–3 presents the estimated minimum requirements of electrolytes.

Table B–1 Median Heights and Weights and Recommended Energy Intakes (United States)

Age	Weight		Height			Average Energy Allowance		
(years)	*(kg)*	*(lb)*	*(cm)*	*(inches)*	*REE*[a] (kcal/day)	*Multiples of REE*[b]	*kcal per kg*	*kcal per day*[c]
Infants								
0.0–0.5	6	13	60	24	320		108	650
0.5–1.0	9	20	71	28	500		98	850
Children								
1–3	13	29	90	35	740		102	1,300
4–6	20	44	112	44	950		90	1,800
7–10	28	62	132	52	1,130		70	2,000
Males								
11–14	45	99	157	62	1,440	1.70	55	2,500
15–18	66	145	176	69	1,760	1.67	45	3,000
19–24	72	160	177	70	1,780	1.67	40	2,900
25–50	79	174	176	70	1,800	1.60	37	2,900
51 +	77	170	173	68	1,530	1.50	30	2,300
Females								
11–14	46	101	157	62	1,310	1.67	47	2,200
15–18	55	120	163	64	1,370	1.60	40	2,200
19–24	58	128	164	65	1,350	1.60	38	2,200
25–50	63	138	163	64	1,380	1.55	36	2,200
51 +	65	143	160	63	1,280	1.50	30	1,900
Pregnant (2nd and 3rd trimesters)								+300
Lactating								+500

[a]REE (resting energy expenditure) represents the energy expended by a person at rest under normal conditions.
[b]Recommended energy allowances assume light to moderate activity and were calculated by multiplying the REE by an activity factor.
[c]Average energy allowances have been rounded.

Source: *Recommended Dietary Allowances,* 10th ed., © 1989 by the National Academy of Sciences, National Academy Press, Washington, D.C.

Table B–2 Estimated Safe and Adequate Daily Dietary Intakes of Selected Vitamins and Minerals[a]

Age (years)	Vitamins: *Biotin* (*μg*)	Vitamins: *Pantothenic Acid* (*mg*)
Infants		
0–0.5	10	2
0.5–1	15	3
Children		
1–3	20	3
4–6	25	3–4
7–10	30	4–5
11 +	30–100	4–7
Adults	30–100	4–7

Age (years)	Trace Elements[b] *Chromium* (*μg*)	*Molybdenum* (*μg*)	*Copper* (*mg*)	*Manganese* (*mg*)	*Fluoride* (*mg*)
Infants					
0–0.5	10–40	15–30	0.4–0.6	0.3–0.6	0.1–0.5
0.5–1	20–60	20–40	0.6–0.7	0.6–1.0	0.2–1.0
Children					
1–3	20–80	25–50	0.7–1.0	1.0–1.5	0.5–1.5
4–6	30–120	30–75	1.0–1.5	1.5–2.0	1.0–2.5
7–10	50–200	50–150	1.0–2.0	2.0–3.0	1.5–2.5
11 +	50–200	75–250	1.5–2.5	2.0–5.0	1.5–2.5
Adults	50–200	75–250	1.5–3.0	2.0–5.0	1.5–4.0

[a]Because there is less information on which to base allowances, these figures are not given in the main table of the RDA and are provided here in the form of ranges of recommended intakes.
[b]Because the toxic levels for many trace elements may be only several times usual intakes, the upper levels for the trace elements given in this table should not be habitually exceeded.

Source: *Recommended Dietary Allowances,* 10th ed., © 1989 by the National Academy of Sciences, National Academy Press, Washington, D.C.

Table B–3 Estimated Sodium, Chloride, and Potassium Minimum Requirements of Healthy Persons

Age (years)	Sodium[a] (mg)	Chloride (mg)	Potassium[b] (mg)
Infants			
0.0–0.5	120	180	500
0.5–1.0	200	300	700
Children			
1	225	350	1,000
2–5	300	500	1,400
6–9	400	600	1,600
Adolescents	500	750	2,000
Adults	500	750	2,000

[a]Sodium requirements are based on estimates of needs for growth and for replacement of obligatory losses. They cover a wide variation of physical activity patterns and climatic exposure but do not provide for large, prolonged losses from the skin through sweat.
[b]Dietary potassium may benefit the prevention and treatment of hypertension and recommendations to include many servings of fruits and vegetables would raise potassium intakes to about 3,500 mg/day.

Source: *Recommended Dietary Allowances,* 10th ed., © 1989 by the National Academy of Sciences, National Academy Press, Washington, D.C.

Table B–4 presents the U.S. RDA for infants, children, adults, and pregnant or lactating women. The Food and Drug Administration derived the U.S. RDA from the 1968 RDA for use on food labels.

The Canadian recommendations are in Tables B–5 and B–6. The Canadian recommendations differ from the RDA in some respects, partly because of differences in interpretation of the data they were derived from and partly because conditions in Canada differ somewhat from those in the United States.

Table B–4 U.S. Recommended Daily Allowances (U.S. RDA)

Nutrient	Infants	Adults and Children over 4 yr	Children Under 4 yr	Pregnant or Lactating Women
Protein (g)	18[a]	45[a]	20[a]	
Vitamin A (RE)	300	1000	500	1600
Vitamin D[b] (IU)	400	400	400	400
Vitamin E[b] (IU)	5.0	30	10	30
Vitamin C (mg)	35	60	40	60
Folate (mg)	0.1	0.4	0.2	0.8
Thiamin (mg)	0.5	1.5	0.7	1.7
Riboflavin (mg)	0.6	1.7	0.8	2.0
Niacin (mg)	8	20	9	20
Vitamin B_6[b] (mg)	0.4	2.0	0.7	2.5
Vitamin B_{12}[b] (μg)	2.0	6.0	3.0	8.0
Biotin[b] (mg)	0.5	0.3	0.15	0.3
Pantothenic acid[b] (mg)	3	10	5	10
Calcium (g)	0.6	1.0	0.8	1.3
Phosphorus[b] (g)	0.5	1.0	0.8	1.3
Iodine[b] (μg)	45	150	70	150
Iron (mg)	15	18	10	18
Magnesium[b] (mg)	70	400	200	450
Copper[b] (mg)	0.6	2.0	1.0	2.0
Zinc[b] (mg)	5	15	8	15

[a]If protein efficiency ratio of protein is equal to or better than that of casein.
[b]Optional for adults and children 4 yr or over in vitamin and mineral supplements.

Source: U.S. Department of Health and Human Services, Public Health Service, Food and Drug Administration, Office of Public Affairs, 5600 Fishers Lane, Rockville, Maryland 20857, HHS publication no. (FDA) 81–2146, revised March 1981.

Table B–5 Recommended Nutrient Intakes for Canadians, 1983 (formerly Canadian *Dietary Standard, 1975*)

Age	Sex	*Weight (kg)*	*Protein (g/day)*[a]	Fat-Soluble Vitamins: *Vitamin A (RE/day)*[b]	Fat-Soluble Vitamins: *Vitamin D (μg/day)*[c]	Fat-Soluble Vitamins: *Vitamin E (mg/day)*[d]
Months						
0–2	M-F	4.5	11[e]	400	10	3
3–5	M-F	7.0	14[e]	400	10	3
6–8	M-F	8.5	16[e]	400	10	3
9–11	M-F	9.5	18	400	10	3
Years						
1	M-F	11	18	400	10	3
2–3	M-F	14	20	400	5	4
4–6	M-F	18	25	500	5	5
7–9	M	25	31	700	2.5	7
	F	25	29	700	2.5	6
10–12	M	34	38	800	2.5	8
	F	36	39	800	2.5	7
13–15	M	50	49	900	2.5	9
	F	48	43	800	2.5	7
16–18	M	62	54	1,000	2.5	10
	F	53	47	800	2.5	7
19–24	M	71	58	1,000	2.5	10
	F	58	41	800	2.5	7
25–49	M	74	57	1,000	2.5	9
	F	59	41	800	2.5	6
50–74	M	73	57	1,000	2.5	7
	F	63	41	800	2.5	6
75+	M	69	57	1,000	2.5	6
	F	64	41	800	2.5	5
Pregnancy (additional)						
1st trimester			15	100	2.5	2
2nd trimester			20	100	2.5	2
3rd trimester			25	100	2.5	2
Lactation (additional)			20	400	2.5	3

Note: Recommended intakes of certain nutrients are not listed in this table because of the nature of the variables upon which they are based. For nutrients not shown, the following amounts are recommended: thiamin, 0.4 mg/1000 kcal (0.48/5000 kJ); riboflavin, 0.5 mg/1000 kcal (0.6 mg/5000 kJ); niacin, 7.2 NE/1000 kcal (8.6 NE/5000 kJ); vitamin B_6, 15 μg, as pyridoxine, per gram of protein; phosphorus, same as calcium. Recommended intakes during periods of growth are taken as appropriate for individuals representative of the midpoint in each age group. All recommended intakes are designed to cover individual variations in essentially all of a healthy population subsisting upon a variety of common foods available in Canada.

[a]The primary units are expressed per kilogram of body weight. The figures shown here are examples.

[b]One retinol equivalent (RE) corresponds to the biological activity of 1 μg of retinol, 6 μg of beta-carotene, or 12 μg of other carotenes.

[c]Expressed as cholecalciferol or ergocalciferol.

[d]Expressed as d-α-tocopherol equivalents, relative to which β- and γ-tocopherol and α-tocotrienol have activities of 0.5, 0.1, and 0.3, respectively.

[e]Assumption that the protein is from breast milk or is of the same biological value as that of breast milk and that between 3 and 9 mo adjustment for the quality of the protein is made.

Table B–5—*Continued*

Age	Water-Soluble Vitamins			Minerals				
	Vitamin C (mg/day)	*Folate (μg/day)*	*Vitamin B_{12} (μg/day)*	*Calcium (mg/day)*	*Magnesium (mg/day)*	*Iron (mg/day)*	*Iodine (μg/day)*	*Zinc (mg/day)*
Months								
0–2	20	50	0.3	350	30	0.4[f]	25	2[g]
3–5	20	50	0.3	350	40	5	35	3
6–8	20	50	0.3	400	50	7	40	3
9–11	20	55	0.3	400	50	7	45	3
Years								
1	20	65	0.3	500	55	6	55	4
2–3	20	80	0.4	500	70	6	65	4
4–6	25	90	0.5	600	90	6	85	5
7–9	35	125	0.8	700	110	7	110	6
	30	125	0.8	700	110	7	95	6
10–12	40	170	1.0	900	150	10	125	7
	40	180	1.0	1,000	160	10	110	7
13–15	50	150	1.5	1,100	210	12	160	9
	45	145	1.5	800	200	13	160	8
16–18	55	185	1.9	900	250	10	160	9
	45	160	1.9	700	215	14	160	8
19–24	60	210	2.0	800	240	8	160	9
	45	175	2.0	700	200	14	160	8
25–49	60	220	2.0	800	250	8	160	9
	45	175	2.0	700	200	14[h]	160	8
50–74	60	220	2.0	800	250	8	160	9
	45	190	2.0	800	210	7	160	8
75+	60	205	2.0	800	230	8	160	9
	45	190	2.0	800	220	7	160	8
Pregnancy (additional)								
1st trimester	0	305	1.0	500	15	6	25	0
2nd trimester	20	305	1.0	500	20	6	25	1
3rd trimester	20	305	1.0	500	25	6	25	2
Lactation (additional)	30	120	0.5	500	80	0	50	6

[f]It is assumed that breast milk is the source of iron up to 2 mo of age.
[g]Based on the assumption that breast milk is the source of zinc for the first 2 mo.
[h]After the menopause the recommended intake is 7 mg/day.

Source: Health and Welfare Canada, *Recommended Nutrient Intakes for Canadians* (Ottawa: Canadian Government Publishing Centre, 1984), Table X.1, pp. 179–180.

Table B–6 Average Energy Requirements (Canada)

Age	Sex	Average Height (cm)	Average Weight (kg)	Requirements[a] (kcal/kg)[b]	(MJ/kg)[b]	(kcal/day)	(MJ/day)	(kcal/cm)	MJ/cm
Months									
0–2	M-F	55	4.5	120–100	0.50–0.42	500	2.0	9	0.04
3–5	M-F	63	7.0	100–95	0.42–0.40	700	2.8	11	0.05
6–8	M-F	69	8.5	95–97	0.40–0.41	800	3.4	11.5	0.05
9–11	M-F	73	9.5	97–99	0.41	950	3.8	12.5	0.05
Years									
1	M-F	82	11	101	0.42	1100	4.8	13.5	0.06
2–3	M-F	95	14	94	0.39	1300	5.6	13.5	0.06
4–6	M-F	107	18	100	0.42	1800	7.6	17	0.07
7–9	M	126	25	88	0.37	2200	9.2	17.5	0.07
	F	125	25	76	0.32	1900	8.0	15	0.06
10–12	M	141	34	73	0.30	2500	10.4	17.5	0.07
	F	143	36	61	0.25	2200	9.2	15.5	0.06
13–15	M	159	50	57	0.24	2800	12.0	17.5	0.07
	F	157	48	46	0.19	2200	9.2	14	0.06
16–18	M	172	62	51	0.21	3200	13.2	18.5	0.08
	F	160	53	40	0.17	2100	8.8	13	0.05
19–24	M	175	71	42	0.18	3000	12.4		
	F	160	58	36	0.15	2100	8.8		
25–49	M	172	74	36	0.15	2700	11.2		
	F	160	59	32	0.13	1900	8.0		
50–74	M	170	73	31	0.13	2300	9.6		
	F	158	63	29	0.12	1800	7.6		
75+	M	168	69	29	0.12	2000	8.4		
	F	155	64	23	0.10	1500	6.0		

[a]Requirements can be expected to vary within a range of ±30%; based on expected patterns of activity.
[b]First and last figures are averages at the beginning and at the end of the 3-mo period.

Source: Health and Welfare Canada, *Recommended Nutrient Intakes for Canadians* (Ottawa: Canadian Government Publishing Centre, 1984), Table II.1, pp. 22–23.

Appendix C

Food Group Plans

Nutrient recommendations such as those presented in Appendix B are one of the tools used in planning diets, but other tools are needed to translate nutrients into food. Food group plans tell people what kinds of foods to eat and how much to eat. They help to make the diet adequate and balanced. The most familiar food group plan fits all foods into four groups and a miscellaneous category, as shown in Table C–1. Each food group contains foods that are similar in origin and that supply a characteristic array of nutrients. These foods are whole foods that form the foundation of a healthy diet: milk and milk products, fruits and vegetables, starches and grains, and meats and meat alternates. The plan suggests for adults a two, four, four, two pattern of servings from each group, although other patterns are possible and desirable for individuals.

The miscellaneous category contains foods that do not fit into the four food groups. Among them are butter, margarine, cream, salad dressing, catsup, jam and jelly, coffee, tea, herbs, soft drinks, alcoholic beverages, and others. These items may contain a few nutrients, but they are so greatly diluted with fat, sugar, or water, or used in such small quantities, that they make little contribution to nutrition.

Other food group plans are available. Canada has one of its own—Canada's Food Guide, shown in Table C–2. Table C–3 shows a food group plan specifically for vegetarians. The modified food group plan shown in Table C–4 may be appropriate for those who spend many kcalories in vigorous physical activity.

Table C–1 The Four Food Group Plan

Milk and Milk Products
Calcium, riboflavin, protein, vitamin B_{12}, (vitamin D and vitamin A, when fortified).

2 servings per day for adults.
3 servings per day for children.
4 servings per day for teenagers, pregnant/lactating women, women past menopause.
5 servings per day for pregnant/lactating teenagers.

Serving = 1 c milk or yogurt; 1/4 c Parmesan cheese or process cheese spread; 2 c cottage cheese; 1 1/2 c ice cream or ice milk; 2 oz process cheese food; 1 1/3 oz cheese.
□ Nonfat milk, buttermilk, low-fat milk, plain yogurt.
▨ Whole milk, cheese, fruit-flavored yogurt, cottage cheese.
■ Custard, milk shakes, pudding, ice cream.

Breads and Cereals
Riboflavin, thiamin, niacin, iron, protein, magnesium, folate, fiber.

4 servings per day.

Serving = 1 slice bread; 1/2 to 3/4 c cooked cereals, rice, or pastas; 1 oz ready-to-eat cereals.

□ Whole grains (wheat, oats, barley, millet, rye, bulgur), whole-grain and enriched breads, rolls, tortillas.
▨ Rice, cereals, pastas (macaroni, spaghetti), bagels.
■ Pancakes, muffins, cornbread, biscuits, presweetened cereals.

Vegetables and Fruits
Vitamin A, vitamin C, riboflavin, folate, iron, magnesium, low in fat, no cholesterol.

4 servings per day.

Serving = 1/2 c or typical portion (such as 1 medium apple, 1/2 grapefruit, or 1 wedge lettuce).

□ Apricots, bean sprouts, broccoli, brussels sprouts, cabbage, cantaloupe, carrots, cauliflower, cucumbers, grapefruit, green beans, green peas, leafy greens (spinach, mustard, and collard greens), lettuce, mushrooms, oranges, orange juice, peaches, strawberries, tomatoes, winter squash.
▨ Apples, bananas, canned fruit, corn, pears, potatoes.
■ Avocados, dried fruit, sweet potatoes.

Meat and Meat Alternates
Protein, phosphorus, vitamin B_6, vitamin B_{12}, zinc, magnesium, iron, niacin, thiamin.
2 servings per day for adults, children, teenagers.
3 servings per day for pregnant/lactating women/teenagers.

Serving = 2 to 3 oz lean, cooked meat, poultry, or fish; 1 oz meat, poultry, or fish = 1 egg, 1/2 to 3/4 c legumes, 2 tbsp peanut butter, 1/4 to 1/2 c nuts or seeds.
□ Poultry, fish, lean meat (beef, lamb, pork), dried peas and beans, eggs.
▨ Beef, lamb, pork, refried beans.
■ Hotdogs, luncheon meats, peanut butter, nuts.

Miscellaneous Group
Sugar, fat (vitamin E), salt, alcohol, kcalories.

No serving sizes are provided because servings of these foods are not recommended. Concentrate on the four food groups that provide nutrients; the foods in the miscellaneous group will find their way into your diet as ingredients in prepared foods, or added at the table, or just as "extras." Note that some of the following items could be placed in more than one group or in a combination group. For example, potato chips are high in both salt and fat; doughnuts are high in both sugar and fat.

□ Miscellaneous foods, not high in kcalories, include spices, herbs, coffee, tea, and diet soft drinks.
■ Foods high in fat include margarine, salad dressing, oils, mayonnaise, cream, cream cheese, butter, gravy, and sauces.
■ Foods high in salt include potato chips, corn chips, pretzels, pickles, olives, bouillon, prepared mustard, soy sauce, steak sauce, salt, and seasoned salt.
■ Foods high in sugar include cake, pie, cookies, doughnuts, sweet rolls, candy, soft drinks, fruit drinks, jelly, syrup, gelatin desserts, sugar, and honey.
■ Alcoholic beverages include wine, beer, and liquor.

Key:
□ Foods generally lowest in kcalories.
▨ Foods moderate in kcalories.
■ Foods highest in calories.

Table C–2 Canada's Food Guide

Food Group	Servings/Day (adult)
Milk and milk products	2[a]
Meat, fish, poultry, and alternates	2
Fruits and vegetables	4 to 5[b]
Breads and cereals	3 to 5

[a]A serving is 250 ml, or about 1 c. Milk group servings differ; for children up to age 11, 2 to 3 servings; adolescents, 3 to 4 servings; pregnant and nursing women, 3 to 4 servings.
[b]Include at least two vegetables.

Source: Health and Welfare Canada, *Canada's Food Guide Handbook* (Ottawa: Canadian Government Publishing Centre, 1985).

Table C–3 Four Food Group Plan for Vegetarians

Food Group	Servings/Day (adult)
Milk and milk products	2[a]
Protein-rich foods	2[b]
Legumes	2[c]
Fruits and vegetables	4[d]
Breads and cereals (whole-grain only)	4

[a]If not using milk or milk products, use soy milk fortified with calcium and vitamin B_{12}.
[b]Examples of protein-rich foods: cheeses and tofu.
[c]Legumes (2 c daily) should be eaten in addition to protein-rich foods, to help women meet iron requirements.
[d]Include 1 c dark greens daily to help women meet iron requirements.

Source: Adapted from *Vegetarian Food Choices* (Gainesville, Fla.: Shands Teaching Hospital and Clinics, Food and Nutrition Service, University of Florida, 1976).

Table C–4 Modified Four Food Group Plan

Food Group	Servings/Day (adult)
Milk and milk products	2
Meat, fish, or poultry	2[a]
Legumes	2[b]
Fruits and vegetables	4
Breads and cereals (whole-grain only)	4

Note: Designed to increase intakes of needed nutrients, especially iron, vitamin B_6, zinc, magnesium, and vitamin E, within about 2200 kcal/day.
[a]Servings size is 3 oz, not 2 to 3 oz, as in the Four Food Group Plan.
[b]Servings size is ¾ c.

Source: J. C. King and coauthors, Evaluation and modification of the basic four food guide, *Journal of Nutrition Education* 10 (1978): 27–29.

Appendix D

Table of Infant Formula Composition

Table D–1 compares the nutrient composition of three milk-based infant formulas and three soy-based infant formulas. All infant formulas are designed to resemble breast milk and must meet an American Academy of Pediatrics standard for nutrient composition. Milk-based formulas are intended for full-term, healthy infants. Soy-based formulas are designed for infants with milk sensitivity or lactose intolerance. Special formulas are available for premature infants or infants with medical conditions requiring special nutrition treatment.

Table D–1 Comparison of Nutrients in Infant Formulas

	Cow's Milk–Based Infant Formulas			Soy Protein Formulas		
Nutrient	*Enfamil*[a]	*Similac*[b]	*SMA*[c]	*Isomil*[d]	*Nursoy*[e]	*Prosobee*[f]
Energy (kcal/100 ml)	68	68	68	68	68	68
Carbohydrate (g/100 ml)	6.9 (lactose)	7.2 (lactose)	7.2 (lactose)	6.8 (corn syrup solids, sucrose)	6.9 (sucrose)	6.8 (corn syrup solids)
Protein (g/100 ml)	1.5	1.5	1.5	1.8	2.1	2.0
Casein	40	82	40	—	—	—
Whey protein	60	18	60	—	—	—
Soy protein	—	—	—	100	100	100
Fat (g/100 ml)	3.8	3.6	3.6	3.7	3.6	3.6
Minerals (mg/L)						
Calcium	460	510	420	710	600	630
Chloride	420	510	375	440	375	550
Iron	13	12	12	12	12	13
Phosphorus	320	390	280	510	420	500
Potassium	720	810	560	950	700	780
Sodium	180	220	150	320	200	290
Vitamins (per 100 ml)						
Vitamin A (IU)	200	250	200	250	200	200
Thiamin (μg)	50	65	67	40	67	50
Riboflavin (μg)	100	100	100	60	100	60
Niacin (mg)	0.8	0.7	0.5	0.9	0.5	0.8
Vitamin B_6 (μg)	40	40	42	40	42	40
Vitamin B_{12} (μg)	0.15	0.21	0.13	0.30	0.20	0.20
Folate (μg)	10	10	5	10	5	10
Vitamin C (mg)	5.2	5.5	5.5	5.5	5.5	5.2
Vitamin D (IU)	40	40	40	40	40	40
Vitamin E (IU)	2.0	1.5	1.0	1.5	0.95	2.0
Vitamin K (μg)	*	*	5.5	*	10	10
Pantothenic acid (mg)	0.30	0.30	0.21	0.50	0.30	0.30

*Not available
[a]Mead Johnson, *Pediatric Products Handbook* (Evansville, Ind.: Mead Johnson and Company, 1986), pp. 7–8.
[b]Ross Laboratories, *Milk-Based and Soy-Based Formulations* (Columbus, Ohio: Ross Laboratories, 1979).
[c]Wyeth Laboratories, *Wyeth Hospital Infant Feeding System* (Philadelphia, Penn.: Wyeth Laboratories, 1986), p. 20.
[d]Ross Laboratories, 1979.
[e]Wyeth Laboratories, 1986.
[f]Mead Johnson, 1986.

Appendix E

Vitamin/Mineral Supplements Compared

The following tables are useful for comparing the essential vitamin and mineral contents of prenatal, infant, and child supplements commonly available in the United States (Tables E–1 and E–2). Notice that a blank column has been provided for the addition of locally available products you may wish to compare with those shown here.*

Not all ingredients in vitamin/mineral preparations are of proven benefit. To facilitate meaningful comparison, the tables list only the nutrients known to be essential in human nutrition. Other nutrients and compounds found on the labels of these supplements are listed in the table notes.

When a supplement that supplies certain nutrients is needed, these tables will ease the task of selecting an appropriate one. Notice, for example, that the iron and calcium contents of the supplements listed here for children (Table E–2) vary considerably. Some contain no iron at all, while others provide more than 100 percent of a child's RDA. Many of the supplements contain no calcium, either.

*These tables are reprinted with permission from L. K. DeBruyne and S. R. Rolfes, *Selection of Supplements,* a 1986 monograph in the *Nutrition Clinics* series available from J. B. Lippincott, Route 3, Box 20B, Hagerstown, MD 21740.

Table E–1 Supplements for Pregnant Women

Vitamin/Mineral	Company/Product				
	Bronson[a] *Prenatal*	**Lederle** *Filibon F.A.*	**Lederle** *Filibon Forte*	**Lederle** *Filibon*	**Lederle** *Materna*[b]
Vitamins					
Vitamin A (IU)	8000	8000	8000	5000	8000
Vitamin D (IU)	400	400	400	400	400
Vitamin E (IU)	50	30	45	30	30
Vitamin C (mg)	150	60	90	60	100
Thiamin (mg)	5	1.7	2	1.5	3
Riboflavin (mg)	5	2.0	2.5	1.7	3.4
Vitamin B_6 (mg)	15	4	3	2	10
Vitamin B_{12} (μg)	15	8	12	6	12
Niacin (mg)	30	20	30	20	20
Folate (mg)	0.8	1	1	0.4	1
Minerals					
Calcium (mg)	250[cd]	250[c]	300[c]	125[c]	250[c]
Iron (mg)	60	45	45	18	60
Magnesium (mg)	75	100	100	100	25
Zinc (mg)	20[d]	0	0	0	25[e]
Copper (mg)	1[d]	0	0	0	2[e]
Iodine (μg)	150	150	200	150	150
Manganese (mg)	1	0	0	0	5
Cost per day[f]					

[a]Bronson Prenatals also contains 250 μg biotin and 20 mg pantothenic acid.
[b]Lederle Materna also contains 30 μg biotin, 10 mg pantothenic acid, 0.25 mg chromium, and 0.25 mg molybdenum.
[c]Carbonate salt.
[d]Sulfate salt.
[e]Oxide salt.
[f]Divide the total retail price for the container by the number of doses per container. For example, XYZ Vitamins are sold in bottles of 100 tablets, and the recommended dose is 2 tablets per day; there are 50 doses in the bottle. At $5.00 per bottle, XYZ Vitamins cost $.10 per day.

Table E–1—*Continued*

Vitamin/Mineral	Company/Product					
	Mead-Johnson *Natalins*	**Mead-Johnson** *Natalins Rx*[g]	**Mission** *Prenatal*[h]	**Mission** *Prenatal F.A.*[h]	**Mission** *Prenatal H.P.*[h]	**Mission** *Prenatal Rx*[h]
Vitamins						
Vitamin A (IU)	8000	8000	4000	4000	4000	8000
Vitamin D (IU)	400	400	400	400	400	400
Vitamin E (IU)	30	30	0	0	0	0
Vitamin C (mg)	90	90	100	100	100	240
Thiamin (mg)	1.7	2.6	5	5	5	4
Riboflavin (mg)	2	3	2	2	2	2
Vitamin B_6 (mg)	4	10	3	10	25	20
Vitamin B_{12} (μg)	8	8	2	2	2	8
Niacin (mg)	20	20	10	10	10	20
Folate (mg)	0.8	1	0.4	0.8	0.8	1
Minerals						
Calcium (mg)	200[c]	200[c]	50[cij]	50[cij]	50[cij]	175[ck]
Iron (mg)	45	60	30	30	30	60
Magnesium (mg)	100	100	0	0	0	0
Zinc (mg)	0	15[e]	0	15[d]	0	15[d]
Copper (mg)	0	2[e]	0	0	0	2[e]
Iodine (μg)	150	150	0	0	0	300
Manganese (mg)	0	0	0	0	0	0
Cost per day[f]						

[g]Mead-Johnson Natalins Rx also contains 50 μg biotin and 15 mg pantothenic acid.
[h]Mission Prenatals (regular, F.A., and H.P.) also contain 1 mg pantothenic acid; Mission Prenatal Rx contains 10 mg pantothenic acid.
[i]Gluconate salt.
[j]Lactate salt.
[k]Ascorbate salt.

Table E–1—*Continued*

Vitamin/Mineral	Company/Product						
	Parke-Davis *Natabec F.A. Kapseals*	Parke-Davis *Natafort Filmseal*	Parke-Davis *Natabec Kapseals*	Parke-Davis *Natabec Rx Kapseals*	Stuart *Prenatal*	Stuart *Stuartnatal 1 + 1*	Other
Vitamins							
Vitamin A (IU)	4000	6000	4000	4000	4000	4000	
Vitamin D (IU)	400	400	400	400	400	400	
Vitamin E (IU)	0	30	0	0	12	12	
Vitamin C (mg)	50	120	50	50	100	120	
Thiamin (mg)	3	3	3	3	1.5	1.5	
Riboflavin (mg)	2	2	2	2	1.7	3	
Vitamin B_6 (mg)	3	15	3	3	2.6	10	
Vitamin B_{12} (μg)	5	6	5	5	4	12	
Niacin (mg)	10	20	10	10	18	20	
Folate (mg)	0.1	1	0	1.0	0.8	1.0	
Minerals							
Calcium (mg)	600[c]	350[c]	600[c]	600[c]	200[d]	200[d]	
Iron (mg)	150	65	30	30	60	65	
Magnesium (mg)	0	100	0	0	0	0	
Zinc (mg)	0	25[e]	0	0	25[e]	25[e]	
Copper (mg)	0	0	0	0	0	2	
Iodine (μg)	0	150	0	0	0	0	
Manganese (mg)	0	0	0	0	0	0	
Cost per day[f]							

Source: Adapted from V. Newman, R. B. Lyon, and P. O. Anderson, Evaluation of prenatal vitamin-mineral supplements, *Clinical Pharmacy* 6 (1987): 770–777.

Table E–2 Supplements for Infants and Children

	Company/Product					
	Lederle[a]	**Miles Laboratories**	**Mead-Johnson**	**Mead-Johnson**	**Radiance**[b]	**Chocks**
	Centrum Jr.	*Flinstones with Iron*	*Poly-Vi-Sol*	*Poly-Vi-Sol Iron and Zinc*	*Chewable for Children*	*Bugs Bunny plus Iron*
Age of Intended Users	*Children over 4*	*Children over 2*	*Infants*	*Children and Adults*	*Children 2 to 12*	*Children over 2*
Recommended Daily Dose	*1 Chewable*	*1 Chewable*	*1-ml Dropper*	*1 Chewable*	*1 Chewable*	*1 Chewable*
Vitamins						
Vitamin A (IU)	5000	2500	1500	2500	4000	2500
Vitamin D (IU)	400	400	400	400	400	400
Vitamin E (IU)	15	15	5	15	3.4	15
Vitamin C (mg)	60	60	35	60	60	60
Thiamin (B_1) (mg)	1.5	1.05	0.5	1.05	2	1.05
Riboflavin (B_2) (mg)	1.7	1.2	0.6	1.2	2.4	1.2
Vitamin B_6 (mg)	2	1.05	0.4	1.05	2	1.05
Vitamin B_{12} (μg)	6	4.5	2	4.5	10	4.5
Niacin (mg)	20	13.5	8	13.5	10	13.5
Folate (mg)	0.4	0.3	—	0.3	—	0.3
Minerals						
Calcium (mg)	—	—	—	—	19	—
Phosphorus (mg)	—	—	—	—	—	—
Iron (mg)	18	15	—	12	12	15
Potassium (mg)	1.6	—	—	—	4	—
Magnesium (mg)	25	—	—	—	22	—
Zinc (mg)	10	—	—	8	—	—
Copper (mg)	2	—	—	—	0.2	—
Iodine (μg)	—	—	—	—	—	—
Manganese (mg)	1	—	—	—	—	—
Cost per day[c]						

[a]Lederle chewables also contain 1.4 mg chlorine (recommended daily dose for children 2 to 4 is ½ tablet).
[b]Radiance chewables also contain 2 mg pantothenic acid, 10 mg biotin, 2 mg inositol, and 2 mg of a choline compound.
[c]Divide the total retail price for the container by the number of doses per container. For example, XYZ Vitamins are sold in bottles of 100 tablets, and the recommended dose is 2 tablets per day; there are 50 doses in the bottle. At $5.00 per bottle, XYZ Vitamins cost $.10 per day.

Table E–2—*Continued*

Neolife[d]	J. B. Williams[e]	Upjohn	Amway[f]	Shaklee[g]	Richardson Vicks[h]	Ross Labs	Miles Labs	Other
Vita Squares	*Popeye with Mins/Iron*	*Unicap*	*Nutrilite*	*Vita–Lea Children*	*Life–Stage Children*	*Vi–Daylin with Iron*	*Flintstones Complete*	
Children over 2	*Children*	*Children*	*Children*	*Children over 4*	*Children 4 to 12*	*Children under 4*	*Children over 4*	
3 Chewables	*1 Chewable*	*1 Chewable*	*1 Chewable*	*2 Chewables*	*2 Chewables*	*1 Dropper*	*1 Chewable*	
3000	2500	5000	2500	2000	5000	1500	5000	
400	400	400	400	400	400	400	400	
9	15	15	10	10	30	5	30	
60	60	60	40	60	60	35	60	
1.5	1.05	1.5	0.7	1.1	1.5	0.5	1.5	
1.5	1.2	1.7	0.8	1.2	1.7	0.6	1.7	
1.2	1.05	2	0.7	1.5	2	0.4	2	
3	4.5	6	3	3	6	1.5	6	
10	13.5	20	9	14	20	8	20	
0.2	0.3	0.4	0.2	0.3	0.4	—	0.4	
—	—	—	—	130	—	—	100	
—	—	—	—	100	—	—	100	
3	15	—	5	10	9	10	18	
—	—	—	—	—	—	—	—	
—	40	—	—	60	—	—	20	
—	12	—	—	1.5	—	—	15	
0.002	1.5	—	—	0.2	—	—	—	
75	105	—	—	15	—	—	150	
1	—	—	—	—	—	—	2.5	

[d]Neolife chewables also contain 10 mg pantothenic acid, 0.075 mg inositol, 0.05 mg of a choline compound, 15 mg biotin, and 15 mg para-aminobenzoic acid (PABA).
[e]Williams chewables also contain 2.5 mg pantothenic acid and 37.5 mg biotin.
[f]Amway chewables also contain 5 mg pantothenic acid.
[g]Shaklee chewables also contain 4 mg pantothenic acid and 0.1 mg biotin.
[h]Richardson Vicks chewables also contain 10 mg pantothenic acid and 0.3 mg biotin.

Appendix F

Recommended Resources

Reliable nutrition information and support is not always easy to come by, especially if you do not know where to look. The agencies, groups, and organizations listed here can provide reliable information or offer support for those with problems related to specific topics.

General

To obtain a copy of the *Surgeon General's Report on Nutrition and Health,* write to:

Superintendent of Documents
Government Printing Office
Washington, DC 20402

To obtain a copy of *Exchange Lists for Meal Planning,* write to:

American Dietetic Association
216 West Jackson Boulevard, Suite 800
Chicago, IL 60606

To obtain a copy of *Diet and Health: Implications for Reducing Chronic Disease Risk,* write to:

National Academy Press
2101 Constitution Avenue
Washington, DC 20418

The USDA's Food and Nutrition Service (FNS) administers the Food Stamp Program; the national school lunch and school breakfast programs; the special Supplemental Food Program for Women, Infants, and Children (WIC); and the food distribution, child care food, summer food service, and special milk programs. Write to:

FNS, USDA
500 12th Street SW
Washington, DC 20250

The World Health Organization (WHO) is an independent agency of the United Nations concerned with the health of the world. WHO and the Food and Agriculture Organization (FAO) of the United Nations have issued international dietary standards. FAO has many publications about food safety and nutrition. Write to:

World Health Organization
1211 Geneva 27
Switzerland

Food and Agriculture Organization
North American Regional Office
1325 C Street SW
Washington, DC 20025

For information about Canada's food group system, write to:

Canadian Dietetic Association
385 Yonge Street
Toronto, Ontario M4T 1Z5 Canada

Alcohol

The following organizations offer information, programs, and support for anyone concerned with alcoholism:

Alcoholics Anonymous World Services
P.O. Box 459
Grand Central Station
New York, NY 10017

Al-Anon Family Group Headquarters
P.O. Box 182
Madison Square Station
New York, NY 10010

National Council on Alcoholism
733 Third Avenue
New York, NY 10017

The National Clearinghouse for Alcohol Information maintains a state-by-state list of most private and public treatment facilities:

National Clearinghouse for Alcohol Information
Box 2345
Rockville, MD 20850

Anorexia Nervosa/Bulimia

The following organizations provide information and help to people with anorexia nervosa or bulimia and their families to find qualified therapists and support groups:

American Anorexia/Bulimia Association, Inc.
133 Cedar Lane
Teaneck, NJ 07666

Anorexia Nervosa and Associated Disorders, Inc.
P.O. Box 7
Highland Park, IL 60035

Anorexia Nervosa and Related Eating Disorders, Inc.
P.O. Box 5102
Eugene, OR 97404

Bulimia/Anorexia Self-help
6125 Clayton Avenue, Suite 215
St. Louis, MO 63139

National Anorexic Aid Society, Inc.
5796 Karl Road
Columbus, OH 43229

Breastfeeding

La Leche League was founded by a group of women who had successfully breastfed their infants and realized many mothers need someone to talk to about breastfeeding. There are more than 40,000 groups internationally. Write to:

La Leche League International, Inc.
9616 Minneapolis Avenue
Franklin Park, IL 60131

Bulimia (see Anorexia Nervosa/Bulimia)

Children

The American Academy of Pediatrics establishes standards for infant formula composition and recommendations for feeding infants and children. Its address is:

American Academy of Pediatrics
P.O. Box 1034
Evanston, IL

To find out how you can influence children's television, write to:

Action for Children's Television (ACT)
46 Austin Street
Newtonville, MA 02160

Children's Foundation
1420 New York Avenue NW, Suite 800
Washington, DC 20005

Dental

To obtain government publications on dental health, write to:

National Institute of Dental Health
9000 Rockville Pike
Building 31, Room 2C36
Bethesda, MD 20892

To obtain information from the dental professional organization, write to:

American Dental Association
211 East Chicago Avenue
Chicago, IL 60611

To find out how to initiate a fluoride tablet program in a school, write to:

National Caries Program
National Institute of Dental Research
Westwood Building, Room 549
5333 Westbard Avenue
Bethesda, MD 20205

Infant Formulas and Foods

Infant formula and food companies will provide nutrient composition information on all their products, as well as how-to information for feeding infants. Write to:

Gerber Products Company
445 State Street
Fremont, MI 49412

H. J. Heinz
Consumer Relations
P. O. Box 57
Pittsburgh, PA 15250

Mead-Johnson Nutritional Division
2404 Pennsylvania Avenue
Evansville, IN 47721

Ross Laboratories
Director of Professional Services
625 Cleveland Avenue
Columbus, OH 43216

Wyeth Laboratories
Philadelphia, PA 19101

Obesity

Overeaters Anonymous (OA) and Weight Watchers International offer nutritionally sound weight loss programs, as well as information and support for anyone interested in losing weight and enhancing health. Write to:

Overeaters Anonymous
2190 190th Street
Torrence, CA 90504

Weight Watchers International
800 Community Drive
Manhasset, NY 11030

Pregnancy

The March of Dimes Birth Defects Foundation has offices in many cities around the country. Its goal is to fight birth defects through preventive care and education. It offers consumer and professional materials about the most common birth defects, as well as materials about prenatal care and breastfeeding. It also serves as a referral service for those needing prenatal care or genetic counseling. Write to:

March of Dimes Birth Defects Foundation
National Headquarters
1275 Mamaroneck Avenue
White Plains, NY 10605

The Supplemental Food Program for Women, Infants, and Children (WIC) is a government program that provides nutritious supplemental foods and nutrition education to low-income pregnant and lactating women, as well as their infants and children. WIC is one of the most effective and successful government health and nutrition programs available. State and county health departments can provide information about the nearest WIC location. For more information, write to:

FNS, USDA
500 12th Street SW
Washington, DC 20250

Prevention

The American Institute for Cancer Research publishes dietary guidelines for lowering cancer risk and is dedicated to funding research and education programs on the relationship of diet and cancer. For more information, write to:

American Institute for Cancer Research
803 W. Broad Street
Falls Church, VA 20046

The American Heart Association publishes pamphlets and recipe booklets about controlling dietary fat and preventing heart disease. Its address is:

American Heart Association
National Center
7320 Greenville Avenue
Dallas, TX 75231

Index

This index lists primarily topics that received significant mention in the text. Pages inclusive (for example, 334–336) indicate major discussions; pages followed by an "*n*" (for example, 394*n*) refer to footnotes.

A

B

C

D

E

F

G

N

O

R

S

T

U

V

W

About the Authors

Sharon Rady Rolfes, M.S., R.D., received her B.S. in psychology and criminology in 1974 and her M.S. in nutrition and food science in 1982 at the Florida State University. She is a founding member of Nutrition and Health Associates and serves on the board of directors. Her publications include the fifth edition of *Understanding Nutrition, Life Cycle Nutrition: Conception through Adolescence,* the second edition of *Understanding Normal and Clinical Nutrition,* and several *Nutrition Clinics* on topics such as cancer, heart disease, vegetarian diet planning, hypoglycemia, and dental health. Her current projects include the production of the third edition of *Understanding Normal and Clinical Nutrition,* and the review of a National Science Foundation education project.

Linda Kelly DeBruyne, M.S., R.D., received her B.S. in 1980 and her M.S. in 1982 in nutrition and food science at the Florida State University. She serves on the board of directors of Nutrition and Health Associates, an information resource center in Tallahassee, Florida. Her publications include the 1989 textbook, *Life Cycle Nutrition: Conception through Adolescence* and several *Nutrition Clinics* on topics such as vegetarian diet planning, nutrition and behavior, nutrition and fitness, nutrition and the aging brain, and vitamin supplements. As a contributing editor of *Nutrition Forum* and *Healthline,* two national newsletters, she has written articles on fasting, variety and balance in diets, fish oils and heart disease, and vegetarian diets. Her current project is the textbook *The Fitness Triad: Conditioning, Diet, and Stress Management*. As a consultant for a group of Tallahassee pediatricians, she teaches infant nutrition classes to parents.

Eleanor Noss Whitney, PhD., R.D., the editor of this text, received her B.A. in biology from Radcliffe College in 1960 and her Ph.D. in biology with an emphasis on genetics from Washington University, St. Louis, in 1970. Formerly an associate professor at the Florida State University, she now devotes full time to research, writing, and consulting in nutrition and health. Her publications include articles in *Science,* the *Journal of Nutrition, Genetics,* and other journals, and the textbooks, *Understanding Nutrition, Nutrition: Concepts and Controversies, Life Choices: Health Concepts and Strategies,* and others. She is president of Nutrition and Health Associates.